W9-CRB-515

ACS SYMPOSIUM SERIES **612**

Geochemical Transformations of Sedimentary Sulfur

Murthy A. Vairavamurthy, EDITOR
Brookhaven National Laboratory

Martin A. A. Schoonen, EDITOR
State University of New York—Stonybrook

ASSOCIATE EDITORS

Timothy I. Eglinton, *Woods Hole Oceanographic Institution*

George W. Luther III, *University of Delaware*

Bernard Manowitz, *Brookhaven National Laboratory*

Developed from a symposium sponsored
by the Division of Geochemistry, Inc.,
at the 208th National Meeting
of the American Chemical Society,
Washington, DC,
August 21–25, 1994

American Chemical Society, Washington, DC 1995

Library of Congress Cataloging-in-Publication Data

Geochemical transformations of sedimentary sulfur / Murthy A. Vairavamurthy, editor, Martin A. A. Schoonen, editor.

   p.   cm.—(ACS symposium series, ISSN 0097–6156; 612)

"Developed from a symposium sponsored by the Division of Geochemistry, Inc. at the 208th National Meeting of the American Chemical Society, Washington, D.C., August 21–25, 1994."

Includes bibliographical references and indexes.

ISBN 0–8412–3328–4

1. Sulphur—Congresses. 2. Sediments (Geology)—Congresses. 3. Diagenesis—Congresses.

I. Vairavamurthy, Murthy A., 1951– . II. Schoonen, Martin A. A., 1960– . III. American Chemical Society. Division of Geochemistry. IV. American Chemical Society. Meeting (208th: 1994: Washington, D.C.) V. Series.

QE516.S1G46   1995
553.6′68—dc20   95–41151
                                                    CIP

This book is printed on acid-free, recycled paper.

Foreword

THE ACS SYMPOSIUM SERIES was first published in 1974 to provide a mechanism for publishing symposia quickly in book form. The purpose of this series is to publish comprehensive books developed from symposia, which are usually "snapshots in time" of the current research being done on a topic, plus some review material on the topic. For this reason, it is necessary that the papers be published as quickly as possible.

Before a symposium-based book is put under contract, the proposed table of contents is reviewed for appropriateness to the topic and for comprehensiveness of the collection. Some papers are excluded at this point, and others are added to round out the scope of the volume. In addition, a draft of each paper is peer-reviewed prior to final acceptance or rejection. This anonymous review process is supervised by the organizer(s) of the symposium, who become the editor(s) of the book. The authors then revise their papers according to the recommendations of both the reviewers and the editors, prepare camera-ready copy, and submit the final papers to the editors, who check that all necessary revisions have been made.

As a rule, only original research papers and original review papers are included in the volumes. Verbatim reproductions of previously published papers are not accepted.

Contents

STUDIES OF SULFUR SPECIATION
IN SEDIMENTARY SYSTEMS

BIOGEOCHEMICAL TRANSFORMATIONS

ISOTOPIC EFFECTS DURING SULFUR
TRANSFORMATIONS

THERMOCHEMICAL SULFATE REDUCTION

Preface

THE STUDY OF SULFUR in environmental and geochemical systems is blossoming because improved techniques and modeling approaches promise rapid advances to many longstanding questions and problems. Current research has both academic and practical value and is active in university, government, and industrial laboratories.

An important and major quantitative part of the global sulfur cycles is related to the transformations of sulfur in sedimentary systems. From a geochemical viewpoint, the study of the sedimentary sulfur cycle is of much significance because it is intimately intertwined with that of the diagenesis of organic matter. Microbial sulfate reduction generates hydrogen sulfide and derived elemental sulfur and polysulfides that become incorporated into sediments as (1) metal sulfides leading largely to pyrite and (2) organically bound sulfur in molecular structures that survive in kerogens and coals after continued alterations throughout diagenesis. How depositional environments affect these processes is of major interest. The presence of sulfur in the form of metal sulfides, organic sulfur compounds, elemental sulfur, and hydrogen sulfide causes problems when exploring for, exploiting, refining, and burning fossil fuels. Therefore, the oil, gas, and coal industries and national governments have a vital interest in understanding the many complex processes and relationships involved in the transformations of sulfur in marine and nonmarine sediments.

Because the research is diverse and conducted in many different laboratories, interactions among researchers are limited and insufficient. Hence, the American Chemical Society symposium "Geochemical Transformations of Sedimentary Sulfur" was organized last year to bring together organic geochemists, inorganic geochemists, and biogeochemists from all over the world with a common interest in sedimentary sulfur transformations to discuss recent research and to explore their imbricating interests. This book, based on that symposium, presents an interdisciplinary overview of recent research on the complex geochemical processes related to sedimentary sulfur transformations. We hope that the book will encourage researchers to look beyond that of their own specializations and interact and collaborate by way of a multidisciplinary approach to solve manifold problems in sulfur geochemistry.

Following the introduction, twenty-four chapters are divided into seven core sections, each representing a separate theme related to sedimentary sulfur transformations, from low-temperature transformations in early diagenesis to the thermal reactions during later diagenesis and catagenesis. The introductory chapter gives a historical background to the geochemistry of sedimentary sulfur and an overview of the subject. The following three sections deal with the three major pathways of H_2S transformations during early diagenesis: (1) formation of organic sulfur, (2) formation of iron sulfides, and (3) oxidation. Sulfur–organic matter interactions receive major attention, and there are major discussions of the incorporation of sulfur into organic matter and the role of sulfur in preserving and transforming sedimentary organic matter. The section on the geochemistry of iron sulfides in sedimentary systems highlights recent efforts, both laboratory and field-oriented, directed toward obtaining a fuller understanding of the geochemical mechanisms involved in their formation. Despite numerous past studies, the kinetics and mechanism of sulfide oxidation are only partially understood, and they continue to be the focus of emphasis, as illustrated by the studies in this section.

The fourth section deals with studies of organic and inorganic sulfur speciation in sedimentary systems. Biogeochemical transformations constitute the fifth section, and the sixth section describes the use of stable isotopes of sulfur to delineate geochemical pathways. The last section deals with recent advances in thermochemical sulfate reduction, a process for forming H_2S during catagenesis that is thought to be the main cause for forming sour gas fields. Chapters in this section highlight recent emphasis on understanding the geochemical variables controlling thermochemical sulfate reduction and efforts to model these processes.

The book includes contributions from internationally acclaimed experts in their areas of specialization. In assembling this volume, a major objective was to produce a comprehensive reference book and guide for researchers and graduate students of different specializations working in the broad area of sedimentary sulfur geochemistry. Thus, both overviews and original research papers are included; the review chapters lead the sections on organic sulfur geochemistry, geochemistry of iron sulfides, H_2S oxidation, biogeochemical transformations, and isotopic effects.

Acknowledgments

This book never would have materialized if not for the enthusiasm and cooperation of the authors and the encouragement of the former chairperson of the Geochemistry Division, Inc., Mary L. Sohn. I thank my coeditor, Martin Schoonen, for editing some of the manuscripts and for his cooperation in organizing the symposium. I am grateful to my associate editors, Bernard Manowitz, Timothy Eglinton, and George Luther, for their advice during the organization of the symposium, the preparation of this volume, and help in reviewing and editing some manuscripts. Many geochemists enthusiastically reviewed manuscripts and provided many helpful suggestions and criticisms that greatly improved the quality of the book; on behalf of all the editors, I thank them. In addition, Sinninghe Damsté, Henrik Fossing, Edward Leadbetter, Tim Lyons, and Wilson L. Orr devoted considerable time as chairpersons for the different sessions of the symposium. Wilson Orr graciously accepted my invitation to deliver the keynote address at the symposium and was most helpful during the preparation of this volume. My grateful appreciation is due to Avril Woodhead whose skillful editing improved the quality of some manuscripts.

Acknowledgement is due to the Petroleum Research Fund, administered by the ACS, for partial financial support that enabled some foreign speakers to participate in the symposium, and to Martin Schoonen for his efforts in obtaining it. I am indebted to William C. Luth, program manager, Division of Geosciences and Engineering, Office of Basic Energy Sciences, U.S. Department of Energy, for his understanding and encouragement during the preparation of this volume. Finally, I thank Rhonda Bitterli for her guidance throughout the editorial process and the rest of the ACS Books Department staff for their assistance during the production of this volume.

MURTHY A. VAIRAVAMURTHY
Department of Applied Science
Geochemistry Program
Applied Physical Sciences Division
Brookhaven National Laboratory
Upton, NY 11973

August 11, 1995

Chapter 1

Geochemical Transformations of Sedimentary Sulfur: An Introduction

Murthy A. Vairavamurthy[1], Wilson L. Orr[2], and Bernard Manowitz[1]

[1]Department of Applied Science, Geochemistry Program, Applied Physical Sciences Division, Brookhaven National Laboratory, Upton, NY 11973
[2]Earth and Energy Science Advisors, P.O. Box 3729, Dallas, TX 75208

The study of chemical transformations of sulfur in sedimentary systems is a frontier research area in geochemistry, and a fuller knowledge of this field is crucial to many geochemical disciplines. Current research on sulfur geochemistry is diverse, spanning topics in low temperature transformations in early diagenesis to the thermal reactions during later diagenesis to catagenesis. It embraces organic, inorganic and biochemical aspects. The present volume provides an interdisciplinary view of recent research on the complex geochemical processes related to sedimentary sulfur transformations based on the 1994 American Chemical Society Symposium. This leading chapter gives an overview of the major geochemical pathways of sedimentary sulfur transformations, emphasizing marine systems.

Studies of sulfur from geochemical, biogeochemical, and environmental perspectives have accelerated in recent years because of the increased recognition of the role this element plays in these ecologically important systems. Not only is sulfur essential to living organisms, being a component of amino acids, coenzymes, and vitamins, but it is involved in many complex economic, ecological, and environmental aspects of life on earth. The geochemistry of sulfur in sedimentary systems is of importance from both academic and practical viewpoints, and is a very active area of research in university, government and industrial laboratories. The American Chemical Society Symposium "Geochemical Transformations of Sedimentary Sulfur" held in August 1994 in Washington, DC was organized to bring together organic geochemists, inorganic geochemists, and biogeochemists from all over the world with a common interest in sedimentary sulfur transformations to discuss the recent researches and to explore their imbricating interests. This volume, based on that symposium, presents an interdisciplinary view of recent research on the complex geochemical processes related to sedimentary sulfur transformations.

The atmospheric sulfur cycle has drawn considerable attention recently because of increasing anthropogenic emissions of sulfur gases (largely sulfur dioxide) from burning fossil fuels. The sulfur gases, and the aerosols formed from them, cause acidic deposition, and affect the radiation balance of the atmosphere, climatic change, and human health. These concerns prompted studies of natural biogenic emissions of sulfur compounds to assess the background atmospheric levels upon which anthropogenic emissions are superimposed (1,2). New data and models have improved our understanding of the sources and fates of sulfur compounds in the atmosphere. Dimethyl sulfide (DMS) from marine phytoplankton is the major biogenic compound emitted from the oceans (3-5), and is proposed as a major source of cloud condensation nuclei in the remote atmosphere, and as an important control on global climate (6-8). The 1989 ACS Symposium volume (Series 393), "Biogenic sulfur in the environment" deals with this area of research (9).

The impact of sulfur in more localized pollution problems also has been widely studied; in particular, research has focused on the toxicity associated with the release of sulfides and their oxidation products from mine wastes. Past and present mining disposal practices have resulted in widespread degradation of the quality of surface water and ground water largely due to the oxidation of sulfides. This area of research was the topic of the 1994 ACS Symposium volume (Series 550) "Environmental geochemistry of sulfide oxidation" (10).

All fossil fuels contain sulfur but amounts vary from traces to more than 10%, and it is bound in diverse molecular structures. In general, the quantity and molecular composition of sulfur in crude oils reflect largely those properties of the source rock from which they were generated (11). The sulfur content of source-rock kerogen (the macro-molecular component of consolidated sedimentary organic matter that is insoluble in common organic solvents) may reach or slightly exceed 14% by weight (S_{org}/C = ca. 0.08; about 1 sulfur atom for every 12 carbon atoms) (12). Orr proposed the designation "Type II-S kerogen" to distinguish sulfur-rich kerogens (S_{org}/C > 0.04; > 8 wt % S) from "classical" Type II kerogens (S_{org}/C 0.02-0.04; ca. 4-8 wt % S) (13). Recent attempts to accommodate large amounts of sulfur (Type II-S) in molecular models for kerogen invoke abundant sulfur cross-linkages (including polysulfide linkages) throughout the molecular network (14). Polysulfide linkages in kerogen may influence the rate and timing of petroleum generation from source rocks, inasmuch as thermal stress can more easily cleave -S-S- bonds than carbon-associated bonds (see the review chapter 2 by Aizenshtat et al.; chapter 7 by Krein and Aizenshtat; chapter 8 by Nelson et al., this volume). Consequently, sulfur-rich kerogens are likely to generate petroleum at a lower temperature than sulfur-lean kerogens. Thus, the abundance of sulfur in sedimentary organic matter may be an important factor in changing the kinetics of maturation reactions and, therefore, in controlling the depth at which petroleum is generated.

Sulfur in the form of metal sulfides, organic sulfur compounds, elemental sulfur, or hydrogen sulfide causes problems in exploring, exploiting, refining, and burning fossil fuels. Removing sulfur from high-sulfur crude oils, coals, and natural gas to make them environmentally acceptable poses special engineering problems and economic costs. The sulfur content in crude oils varies from less than 0.05 to more than 14 weight percent, but few commercial crude oils exceed 4% (15). Oils with less than 1% sulfur usually are classified as low-sulfur, and those above 1% as high-sulfur (*16*).   Because of technical and economic concerns, the industry prefers to refine low-sulfur crude oils; the sulfur level processed depends on a refinery's design and operation.   The world's potential reserves of high-sulfur oils greatly exceed the known reserves of conventionally produced low-to-moderate-sulfur crudes.   Use of these potential resources will be required more and more to meet future energy demands; consequently, processing methods must be modified.   Because of the practical impacts of the presence of sulfur in fossil fuels, there has been much recent interest in obtaining a fuller understanding of the origin and evolution of sulfur in fossil fuels.   The 1990 ACS Symposium Volume (Series 429) "Geochemistry of sulfur in fossil fuels" covers major progress in this area of research (*17*).

Geochemical Focus

The sedimentary sulfur cycle has received much attention because it is intimately intertwined with that of carbon (i.e. the diagenesis of organic matter).   Also, both the carbon and sulfur cycles exert major controls on the level of atmospheric O_2, the oxidation state of minerals in surficial rocks, and life processes in general (*18-20*).   Most earlier studies on sedimentary sulfur emphasized inorganic aspects, such as pyrite formation (*21-31*), and   biogeochemical transformations (*32-36*).   However, there is a paucity of information on the geochemistry of sedimentary organic sulfur.   Zobell (1963) reviewed the organic geochemistry of sulfur, emphasizing microbiological aspects (*37*).   However, until recently, the incorporation of sedimentary sulfur into organic matter often was ignored or considered insignificant.   This is surprising because the sulfurization of organic matter is the major concern about sulfur in fossil fuels.   Now it is thought that organic matter, like iron, is an important sink for sedimentary sulfur because of the common occurrence of organic sulfur in sedimentary systems and the overwhelming evidence suggesting that most of it is of geochemical origin (see the review by Sinninghe Damsté and de Leeuw (*38*) and references *39-42*).   Various aspects of the   recent research on sedimentary organic sulfur geochemistry also is covered in the book "Geochemistry of sulfur in fossil fuels" (*17*).

Studies of the global biogeochemical sulfur cycle also were among the important projects initiated by the Scientific Committee on Problems of the Environment (SCOPE), International Council of Scientific Unions (ICSU) in its mission to achieve a fuller understanding of the major biogeochemical cycles, including those of carbon, nitrogen, and phosphorus, which determine the composition of the atmosphere, and the fertility of land and waters.   SCOPE's efforts resulted in the publication of a series of four books dealing with the global cycling of sulfur (*43-46*).   These are: (1) "The global biogeochemical sulfur cycle", SCOPE 19, 1983, (2) "Evolution of the global biogeochemical sulfur cycle", SCOPE 39, 1989, (3) "Stable isotopes: natural and anthropogenic sulfur in the environment", SCOPE 43, 1991, and (4) "Sulfur cycling on the continents: wetlands, terrestrial ecosystems and associated water bodies", SCOPE 48, 1992 .

This present book, concerned with sedimentary sulfur geochemistry, has a broad scope in that it deals with different fundamental themes on the transformations of sulfur in sedimentary systems, from low temperature transformations in early diagenesis to the thermal reactions during later diagenesis to catagenesis. (*Diagenesis* refers to biological, physical and chemical alteration of organic debris before it is subject to the pronounced effect of temperature, and before the main stage of oil and gas generation. *Catagenesis* follows diagenesis, and refers to the changes, including oil and gas generation, occurring with deeper burial at temperatures from about 50 ^{0}C to a range of about 150-200 ^{0}C) (*16,47*). The topics cover organic, inorganic and biogeochemical aspects of sulfur transformations, including sulfur - organic matter interactions, the formation of iron sulfides in sedimentary systems, oxidative transformation of hydrogen sulfide, biogeochemical conversions, the use of stable isotopes of sulfur to delineate geochemical pathways, and thermochemical sulfate reduction. We present here an overview of the major geochemical pathways of sedimentary sulfur transformations, particularly in marine systems.

The Geochemical Sedimentary Sulfur Cycle

The sedimentary sulfur cycle is the major component of the global sulfur cycle. The oceans, covering about 70 % of the earth's surface and containing dissolved sulfate at a concentration of about 28.7 mM, represent one of the largest pools of sulfur amounting to about 1.3×10^9 teragrams (Tg $= 10^{12}$ g) (*48*). The main sulfur input to the oceans comes from river water, carrying products from the chemical and mechanical weathering of continental rocks. Atmospheric input is mainly recycled oceanic sulfur from seaspray, with the products of volcanic degassing, and anthropogenic emissions of sulfur dioxide transported from land largely accounting for the influx from the lithosphere. The main sink for this oceanic sulfur is sediment, suggesting a reason why sedimentary-sulfur transformations are a major quantitative part of the sulfur cycle (*49,50*).

Sedimentary Sinks for Sulfur: Major Pathways. Major pathways for removing sulfate from the oceans to the sediments are (1) the precipitation of sulfate minerals in evaporites, (2) the formation of sulfide minerals, particularly pyrite, and (3) the formation of organic sulfur. Although most earlier studies recognized only the first two processes as important (*49,50*), sulfur geochemists increasingly concur that the third also constitutes a major sink. The relative importance of each pathway depends highly on the factors affecting the local depositional environment, including (1) water depth and circulation pattern; (2) the distribution of oxygen throughout the water column, and at the sediment-water interface; (3) total sedimentation rate; (4) input of organic matter; and (5) input of clastic minerals and their reactive iron content.

Evaporites are formed under special conditions restricting circulation (such as basins), precipitating sulfate minerals, such as gypsum and anhydrite, by the rapid evaporation of seawater. Such conditions of semi-isolation from seawater input, and high net evaporation occur only sporadically in geological settings, but when they do occur, dissolved salts are precipitated from seawater at a very rapid rate (*51-53*). Estimates of the earth's current crustal reservoirs of sulfur minerals indicate that about $200\text{-}250 \times 10^{18}$ moles of sulfur occur as gypsum sulfate in evaporite deposits; in comparison, the oceans contain only about $40\text{-}42 \times 10^{18}$ moles of sulfur as dissolved sulfate (*54*). Although evidence from sedimentary rocks suggests that marine evaporites were important sinks for sulfate from the Recent to the late Precambrian, their rate of formation in today's oceans, in the last few million years, probably is quantitatively insignificant (*49,50,55*). Thus, the burial of pyrite and organic sulfur constitutes the major sinks for oceanic sulfur in the modern oceans. (We discuss the geochemical controls on the formation of pyrite and organic sulfur later in the Section, "Major Geochemical Pathways of Sulfur Transformations During Early Diagenesis"). To complete the global redox cycle of sulfur, pyrite and organic sulfur eventually are uplifted into the zone of weathering, and there react with atmospheric O_2 to form sulfate, which ultimately is returned to the ocean. It is debated whether there is a steady-state balance of the input and output fluxes of sulfur in the present-day oceans, although concentrations have fluctuated in the geologic past (for example: Middle Precambrian: 18 mM; Cambrian: 40 mM; Eocene: 30 mM) (*50,55*).

The Removal of Sulfate in Carbonate Minerals: a Minor Sink. Some seawater sulfate is precipitated with carbonate minerals (for example, calcite and aragonite), with concentrations ranging from approximately 200 to 24,000 ppm (*56,57*; chapter 18 by Staudt and Schoonen, this volume). Studies by Pingitore et al. suggest that sulfate probably substitutes for carbonate in the structural lattice, and it is not present as gypsum or anhydrite inclusions (*57*). According to Staudt and Schoonen, the removal of sulfate in marine calcium carbonate components accounts for about 5% of weathered sulfate input to the oceans (chapter 18, this volume). Upon burial and recrystallization of carbonates, most of the included sulfate is released into the sediments, where it can be reduced and incorporated into organic matter or metal sulfides.   Thus, although precipitation with carbonates is only a minor sink for seawater sulfate, it has other geochemical implications.

The Removal and Transformation of Sulfur in Hydrothermal Systems. In marine hydrothermal systems associated with volcanism or tectonic activity (for example, the mid-ocean ridge system), geothermal energy is used to forming hydrogen sulfide, leading to the deposition of copious amounts of iron sulfide near the vents (*58-60*). Leaching from crustal basalts was suggested as the main source of this H_2S, but reduction of seawater sulfate by iron (II) in crustal rocks also is important (*61,62*). When seawater percolates down through the cracks into the crust, most of the sulfate precipitates as anhydrite; the sulfate remaining in solution is then reduced to H_2S. The anhydrite so deposited subsequently redissolves out of vents and re-enters the oceans in cooler waters when the crust moves off-axis (*63*); thus, $CaSO_4$ deposition in the vents may not be an important mechanism for removing sulfate from the oceans. Since most of the hydrogen-sulfide sulfur comes from deeper in the crust, it is not clear whether hydrothermal systems are net sinks or sources for sulfur in the oceans. A unique feature of these systems is the presence of chemolithotrophic bacteria which oxidize the sulfide with oxygen from seawater to generate energy for reducing carbon dioxide to organic carbon (*64*). These bacteria are the primary food source supporting very large populations of specifically adapted invertebrates (for example, the tube worm *Riftia pachyptila*) in the immediate vicinity of the vents. The global significance of sulfur transformations in the vent systems is unclear.

The Formation H_2S from Bacterial Reduction of Sulfate: a Pivotal Process for Geochemical Sulfur Transformations

Biological processes play a key role in controlling the transformations of sedimentary sulfur, particularly during early diagenesis, which are greatly facilitated by its ability to exist in several oxidation states between -2 and +6. The ability of sulfur to catenate (i.e. to form multiple sulfur-sulfur linkages), and to form mixed-valent species further augment the formation of a large variety of both organic and inorganic compounds. Seawater sulfate is the starting sulfur compound for all the major pathways of sulfur transformations in the marine environment. The initial step in these transformations is the reduction of sulfate, which is often mediated by living organisms. There are two major biogeochemical pathways for this reduction: (1) assimilatory sulfate reduction by autotrophic organisms (for example, algae in the photic zone of the water column, chemosynthetic bacteria in hydrothermal systems) for biosynthesizing organic-sulfur compounds, and (2) dissimilatory sulfate reduction mediated by anaerobic bacteria for gaining energy, but not for biosynthesis. The former involves a complex, enzyme-mediated, multi-step process, and cysteine usually is the first organic-sulfur metabolite produced (65).

Energy for all life processes, other than primary production by autotrophs, is obtained by the oxidation of organic matter with available electron acceptors. Thermodynamically, the most favorable one is molecular oxygen; however, bacteria may use other electron acceptors, depending on the supply of oxygen (66). Thus, in organic-rich sediments, SO_4^{2-} becomes the dominant terminal electron acceptor used by bacteria to oxidize organic matter, producing hydrogen sulfide (H_2S), when the available oxygen near the sediment-water interface is rapidly consumed by a large downward flux of organic matter (see Widdel and Hansen (67) for a detailed review on sulfate reducing bacteria). Although MnO_2, NO_3^-, and Fe_2O_3 are more preferable energetically than SO_4^{2-} as electron acceptors, they usually are less important because of their limited supply in sediments (66). The rate of bacterial sulfate reduction can be large, reaching hundreds of millimoles per square meter per day.

In the bacterial reduction of sulfate, the sulfate molecule containing the lighter isotope of sulfur (i.e. $^{32}SO_4^{2-}$) is favored over that containing the heavier atom (i.e. $^{34}SO_4^{2-}$) because of kinetic isotopic effects (68,69). This discrimination causes fractionation of the sulfur isotopes in precursor sulfate during reduction, producing isotopically lighter hydrogen sulfide, and simultaneously enriching the residual sulfate with isotopically heavier sulfur. The extent of isotope fractionation correlates inversely with the rate of sulfate reduction per cell; thus, large fractionations occur during slow rates of bacterial reduction (in chapter 22, this volume, Fry et al. discuss factors affecting the degree of fractionation using results from lake sediments). The distinct isotopic composition of this reduced sulfur allows geochemists to unravel the associated transformation pathways (see chapter 21 by Anderson and Pratt, this volume).

Major Geochemical Pathways of Sulfur Transformations During Early Diagenesis

The amount of H_2S generated in bacterial sulfate reduction is the primary factor that determines the major pathways of transformations of sedimentary sulfur, and those of organic matter and iron. Figure 1 is a scheme of transformations resulting from the intrinsic reactivity of reduced sulfur, summarizing the progression of changes from early diagenesis to the late stages of catagenesis. Three major pathways define the fate of H_2S during early diagenesis: (1) oxidation, (2) formation of iron sulfides, and (3) formation of organic sulfur. Oxidation occurs through chemical and biological pathways, and is prominent at oxic-anoxic interfaces where sulfide encounters oxidants, such as oxygen and oxidized metal ion species (for example, MnO_2) (70,71). Several partial oxidation products of hydrogen sulfide are formed, such as elemental sulfur, polysulfides, sulfite, thiosulfate, and polythionates, and eventually, sulfate (see chapter 14 by Yao and Millero; chapter 15 by Vairavamurthy and Zhou, this volume). Since most intermediates, particularly polysulfides, sulfite, and thiosulfate, are strong sulfur nucleophiles, together with hydrogen sulfide they form the pool of reactive sulfur species in sediments. The sulfur nucleophiles (H_2S and derived oxidation intermediates, including polysulfides) react with either active iron or organic molecules to become incorporated into sediments as (1) metal sulfides leading eventually to pyrite, and (2) organically bound sulfur in molecular structures that survive. (with alterations) in kerogens and coals.

Hydrogen sulfide, generated by sulfate-reducing bacteria, is the main source of sulfur for pyrite because the isotopic composition of the pyritic sulfur and the co-existing pore-water hydrogen sulfide usually are similar (72). Despite numerous studies, the geochemical mechanisms involved in the formation of pyrite are still unclear (see chapter 9, this volume, by Rickard , Schoonen and Luther for a review). Recent laboratory simulations and field studies directed towards understanding the sedimentary formation of iron sulfides are given in subsequent chapters by Furukawa and Barnes (chapter 10); Wang and Morse (chapter 11); and Cornwell and Sampou (chapter 12). The major factor that limits the rate of H_2S production by sulfate-reducing bacteria in marine sediments is the amount of organic matter that they can metabolize (22,73). Therefore, the primary environmental variables controlling the extent of pyrite formation are (1) the amount and reactivity of the organic matter deposited into sediments, (2) the concentration and the rate of deposition of the reactive iron, and (3) the rate of replenishment of the sulfate ion from the overlying water (49). Hence, pyrite formation is more important in shallow and shelf sites than in the deep sea. Also, pyrite content is much lower in carbonate sediments than in siliciclastic sediments mainly because the latter contain abundant reactive iron, principally occurring as ferric oxide and oxyhydroxide coatings on clay particles (73).

Since sulfate reduction occurs mainly in near-surface sediments, it is believed that reactive sulfur becomes locked up in pyrite and organic sulfur during the very early stages of diagenesis. For organic molecules, such reactions with reduced sulfur are particularly important because sulfur helps to form macromolecular geopolymers (humic polymers → kerogen) through di- and polysulfide cross linking that provide a mechanism for preserving sedimentary organic matter (see chapter 5 by Schouten et al., and chapter 7 by Krein and Aizenshtat, this volume). In chapter 3 in this

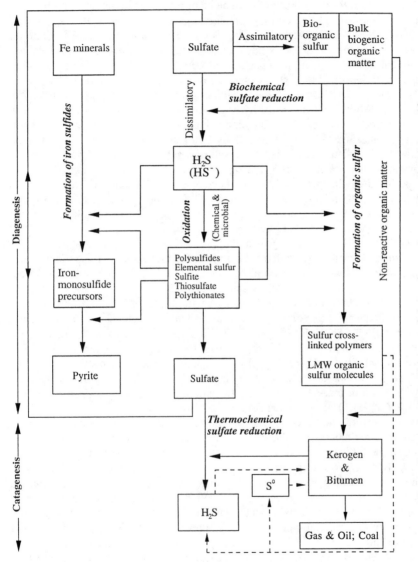

Figure 1. A schematic diagram of the major pathways of sedimentary sulfur trans-formations (solid lines represent known pathways, and dotted lines represent potential pathways not yet established).

volume, Vairavamurthy et al. present evidence, from measuring sulfur speciation in the organic-rich sediments off the coast of Chile, that organic polysulfides and sulfonates (products formed from the reactions of sulfite and thiosulfate with organic molecules) occur maximally at or near the sediment-water interface in the vicinity of the oxic-anoxic interface, in agreement with previous studies. Because this interface occurs usually above the hydrogen sulfide maximum in the sediment column, the oxidation intermediates have the first chance to quench the reactive molecules in the deposited organic matter rather than hydrogen sulfide. This geochemical mechanism may be the reason for the dominance of organic polysulfides in geochemical systems.

The contribution of biosynthesized organic sulfur, incorporated in assimilatory sulfate reduction, to the sedimentary organic sulfur is not clear. Because most of the assimilatory sulfur is contained in proteinaceous substances, generally it is believed that the amount that can survive diagenesis is small. Thus, almost all of the sedimentary organic sulfur may be of geochemical origin. However, organically bound sulfur always is enriched in ^{34}S (often by more than 10 per mil) relative to co-existing pyrite, suggesting that it is not entirely derived from the same diagenetic source of sulfur species that forms pyrite (72; in chapter 21, Anderson and Pratt discuss the use of isotopic information for clarifying the origin of organic sulfur in marine sediments). Thus, a part of the organically bound sulfur, at least in near-surface sediments, may be of biosynthetic origin derived from detrital organic matter.

Reactive iron is believed to react faster than organic matter with available reduced sulfur. Hence, H_2S and derived oxidation intermediates should be available largely for reactions with sedimentary organic matter in organic-rich sediments, where reactive iron is limited. Thus, the geochemical incorporation of sulfur into organic matter usually is enhanced in iron-poor sediments, such as those of carbonate platforms, and in clastic-starved basins which accumulate pelagic materials, particularly cocolith, diatom, and radiolarian oozes (11).

The quantitative importance of organic sulfur as an important global sink is not easily evaluated because of a lack of information on the organic sulfur content of sedimentary organic matter worldwide. Organic sulfur was neglected in many studies of the global sulfur cycle, assuming that total reduced sulfur in sediments is largely pyrite. However, we now know that, locally, organic sulfur is a major sink where organic matter is abundant, microbial sulfate-reduction is active, and the amount of reactive iron is limited; such cases are common in carbonates and diatomites where the sediment is organic-rich. Furthermore, the major and most prolific petroleum basins worldwide are found in carbonate or evaporitic sequences (74,75). These systems generally have the highest sulfur content in their oils and associated gases. Thus, the organic sulfur sink may be much more important quantitatively than was generally assumed.

The Formation of H_2S from Chemical Reduction of Sulfate During Catagenesis.

During catagenesis, in the relatively high temperature regimes of sediment burial between 100 and 200 °C, organic matter is oxidized abiotically by sulfate, producing H_2S and CO_2 (76,77) (Figure 1). This high-temperature reduction is known as "thermochemical sulfate reduction" (TSR). The temperature span during catagenesis is probably too warm for and, thus, not conducive to bacterial reduction of sulfate, although recent studies suggest that it can extend to temperatures as high as 110 °C (78). Usually, optimal rates for bacterial sulfate reduction occur between 0 - 50 °C. Although the net result of TSR is very similar to that to that of dissimilatory bacterial sulfate reduction, its critical controls, detailed mechanisms, and isotopic effects are different and distinctive. Its major controls are (1) temperature, (2) sulfate availability (in solution or solid phase), (3) fluid pH (buffer systems), and (4) type and abundance of organic matter. Because TSR occurs on a geologic time-scale of millions of years, it is difficult to establish the optimum conditions, either from laboratory simulations or field observations. Nevertheless, there is increasing evidence that the rates are significant above 140-150 °C, and can account for natural gases with more than 5-10 % H_2S; in the 100-140 °C range, the rates are uncertain. In any case, TSR becomes important only in the catagenesis zone where kerogen is in the late stages of oil generation, and oil is cracking to gas. On a sedimentary volume basis, the rate of H_2S production from TSR probably is orders of magnitude lower than that from bacterial reduction during early diagenesis, even under optimum conditions (including high temperatures).

High-H_2S gas fields (>5-10% H_2S) are believed to receive a major H_2S input from TSR. Such sour-gas fields are mostly in hot (deep) carbonate reservoirs and often contain S^0 and polysulfides associated with the H_2S. It is likely that the H_2S/S^0 system reacts with organic matter under these conditions (see Figure 1) but the competing decomposition of organically bound sulfur keeps the level of organic sulfur low. Low-molecular-weight thiols (mercaptans) may become significant gas-phase components in this dynamic, steady-state system. In natural oil-gas reservoirs, distributions of reactants between gas, oil, water and solid phases, as well as mixing (mass transfer) processes, may have significant rate-limiting effects not easily modelled for subsurface conditions with spatial and temporal variations. Chapters 23-25, this volume, discuss recent research on the geochemical variables controlling thermochemical sulfate reduction, and efforts to model these processes.

Current Research

Current research on sulfur geochemistry is diverse, and includes research on diagenetic processes involving the interactions between sulfur and organic matter, the metabolic transformations of sulfur species, the formation of metal sulfides, and the application of sulfur-containing biomarkers as well as catagenetic transformations, such as thermochemical reduction of sulfate. How different depositional environments affect the pathways of sulfur transformations is of major interest. Incorporating sulfur into sedimentary organic matter received much attention with concerns about the relative reactivity of organic and inorganic reactants, the products formed, and how they change in structure and abundance during early diagenesis, and later, with deeper burial, when they are subjected to thermal processes. These aspects of the geochemistry of sedimentary organic sulfur are dealt with in Chapters 2-8 in this volume. Research on the geochemical formation of metal sulfides has emphasized an understanding of the molecular mechanisms involved (see chapters 9-13, this volume).

There is interest in sulfur isotopic studies because the isotopic composition of pyrite and sedimentary organic sulfur differs, although they are believed to be formed from a common pool of reduced sulfur, and the reasons are not clear (see chapters 21 and 22, this volume).

Analytical advances involving new and improved techniques, and modelling approaches generate more details of, and promise answers to, many longstanding questions. For example, x-ray absorption spectroscopy used to measure sulfur speciation in a variety of organic-rich sediments revealed sulfonates as a novel class of sedimentary organic sulfur compounds (79). Improved techniques for desulfurization, coupled with compound-specific carbon isotope analysis, adds new knowledge about sedimentary organic sulfur compounds (80; chapter 6 by Hefter et al., this volume). Although much has been learned about the types and abundance of sulfur functional groups, their thermal stability, and how they change with time during sediment burial, the models are still crude. Sequences of sulfur reactions in recent and older sediments, and in petroleum and natural gas deposits still pose partially answered questions. The rate of recent progress in sulfur geochemistry promises rapid advances as new information is assimilated and extended by further research. This book provides a framework for further research, and defines the current state of knowledge on the geochemistry of sedimentary sulfur.

Acknowledgements

We thank Avril Woodhead, Stephen Schwartz, and an anonymous reviewer for constructive comments on the manuscript. Vairavamurthy and Manowitz acknowledge U.S. Department of Energy, Division of Engineering and Geosciences of the Office of Basic Energy Sciences for support during the preparation of the manuscript (BNL Contract No. DE-AC02-76CH00016).

Literature Cited

1.   Andreae, M. O.; Raemdonck, H. *Science* **1983**, *221*, 744-747.
2.   Aneja V.P. J. *Air Waste Manage. Assoc.* **1990**, *40*, 469-476.
3.   Andreae, M. O., Barnard, W. R. *Mar. Chem.* **1984**, *14*, 267- 279.
4.   Vairavamurthy, A.; Andreae, M. O.; Iverson, R.L. *Limnol. Oceanogr.* **1985**, *30*, 59-70.
5.   Andreae, M. O. *Mar. Chem.* **1990**, *30*, 1-29.
6.   Bates, T. S.; Charlson R. J.; Gammon R. H. *Nature* **1987**, *329*, 319-321.
7.   Charlson, R. J.; Lovelock, J. E.; Andreae, M. O.; Warren, S. G. *Nature* **1987**, *326*,   655- 661.
8.   Schwartz, S. E. *Nature* **1988**, *336*, 441-445.
9.   *Biogenic Sulfur in the Environment*; Saltzman, E.S.; Cooper, W.J., Eds.; ACS Symposium Series 393; American Chemical Society: Washington DC, 1989.
10.   *Environmental Geochemistry of Sulfide Oxidation*; Alpers, C.N.; Blowes, D.W., Eds.; ACS Symposium Series 550; American Chemical Society: Washington, DC, 1994.
11.   Gransch, J. A.; Posthuma, J. In *Advances in Organic Geochemistry 1973*; B.

Tissot and F. Bienner, Eds.; Editions Technip: Paris, 1974; pp. 727-739.

12.   Orr, W.L.; Sinninghe Damsté, J.S. In Geochemistry of Sulfur in Fossil Fuels;
       Orr W.L; White, C.M. eds., ACS Symposium Series 429, American Chemical
       Society, Washington D.C., 1990, pp. 2-29.

13.   Orr, W. L. *Org. Geochem.* **1986**, *10*, 499-516.

14.   Tegelaar E.W.; de Leeuw J.W.; Derenne S.; Largeau C. *Geochim.
       Cosmochim. Acta*, **1989**, *53*, 3103-3106.

15.   Rall, H.T.; Thomson, C.J.; Coleman, H.J.; Hopkins, R.L. *Sulfur Compounds
       in Crude Oil*; U.S. Bureau of Mines Bulletin 659; US Government Printing
       Press: Washington DC, 1972, 187 pp.

16.   Tissot,B.P.; Welte, D.H. *Petroleum Formation and Occurrence (2nd Edition);*
       Springer-Verlag: Berlin, 1984.

17.   *Geochemistry of Sulfur in Fossil Fuels*; Orr, W.L.; White, C.M.; Eds.; ACS
       Symposium Series 429; American Chemical Society: Washington, DC, 1990.

18.   Garrels, R.M.; Perry, E.A. In *The Sea*; Goldberg, E., Ed.; Vol.5; John
       Wiley and Sons: New York, 1983; pp 303-336.

19.   Berner, R.A. *Am. J. Sci.* **1987**, *287*, 177-197.

20.   Berner, R.A.; Canfield, D.E. *Am. J. Sci.* **1989**, *289*, 333-361.

21   Berner, R.A. *Am. J. Sci.* 1982, **282**, 451-473.

22.   Berner, R.A. *Geochim. Cosmochim. Acta* 1984, *48*, 605-615.

23.   Raisewell, R.; Berner, R.A. *Am. J. Sci.* **1985**, *285*, 710-724.

24.   Goldhaber, M.B.; Kaplan, I.R. In *The Sea;* Goldberg, E.D., Ed.; Vol.5; John
       Wiley and Sons: New York, 1983; pp 569-656.

25.   Goldhaber, M.B.; Kaplan, I.R. *Mar. Chem.* **1980**, *9*, 95-143.

26.   Morse, J.W.; Millero, F.J.; Cornwell, J.C.; Rickard, D. *Earth Science
       Reviews* **1987**, 24, 1-42.

27.   Morse, J.W.; Emeis, K.C. *Am. J. Sci.* **1990**, *290*, 1117-1135.

28.   Luther, G.W. III *Geochim. Cosmochim. Acta* 1991, **55**, 2839-2849.

29.   Luther III, G.W.; Church, T.M. In *Sulfur Cycling on the Continents*;
       Howarth, R.W.; Stewart, J.W.B; Ivanov, M.V., Eds.; SCOPE 48; John
       Wiley and Sons: New York, 1992; pp 125-144.

30.   Canfield, D.E.; Raiswell, R. In *Releasing the Data Locked in the Fossil
       Record*; Allison, P.A.; Briggs, D.E.G., Eds.; Topic in Geobiology volume 9;
       Plenum: New York, 1991; pp 337-387.

31.   Schoonen, M.A.; Barnes, H.L. *Geochim. Cosmochim. Acta* **1991**, *55*, 1495-
       1504.

32.   Chambers, L.A.; Trudinger, P.A. *Geomicrobiol. J.* **1979**, *1*, 249-293.

33.   Jørgensen, B.B. *Nature* 1982, **296**, 643-645.

34.   Jørgensen, B.B. *Phil. Trans. R. Soc. Lond.* B. **1982**, *298*, 543-561.

35.   Trüper, H. G. In *Sulfur, its Significance for Chemistry, for the Geo-, Bio-, and
       Cosmosphere and Technology*; Müller, A.; Krebs, B., Eds.; Studies in
       Inorganic Chemistry, Volume 5; Elsevier: Amsterdam, 1984; pp 351-365.

36.   Jannasch, H.W. In The Major Biogeochemical Cycles and their Interactions;
       Bolin, B; Cook, R.B., Eds.; SCOPE 21; John Wiley and Sons: New York,
       1983; pp 517-525.

37.   Zobell, C.E. In *Organic Geochemistry*; Breger, I.A., Ed.; MacMillan: New
       York, 1963; 576-595.

38. Sinninghe Damsté, J.S.; de Leeuw, J.W. *Org. Geochem.* **1989**, *16*, 1077-1101.
39. Brassell, S.C.; Lewis, C.A.; de Leeuw J.W.; de Lange F.; Sinninghe Damste J.S. *Nature* **1986**, *320*, 643-645.
40. Raiswell, R.; Bottrell, S.H.; Al-Biatty, H.J.; Tan, M.D. *Am. J. Sci.*. **1993**, *293*, 569-596.
41. Eglinton, T.; Irvine, J.E.; Vairavamurthy, A.; Zhou, W.; Manowitz, B. *Org. Geochim.Cosmochim. Acta* **1994**, *22*, 781-799.
42. Krein, E. B. In *Supplement S: The chemistry of sulphur-containing functional groups*; S. Patai and Z. Rappoport, Eds.; John Wiley & Sons: New York, **1993**; pp 975-1032.
43. *The Global Biogeochemical Sulfur Cycle*; Ivanov, M. V.; Freney, J. R., Eds.; SCOPE 19; John Wiley and Sons: New York, 1983.
44. *Evolution of the Biogeochemical Sulfur Cycle*; Brimblecombe, P.; Lein,, A. Yu., Eds.; SCOPE 39; John Wiley and Sons: New York, 1989.
45. *Stable Isotopes: Natural and Anthropogenic Sulfur in the Environment*; Krouse, H.R.; Grinenko, V.A., Eds,; SCOPE 43; John Wiley and Sons: New York, 1991.
46. *Sulfur Cycling on the Continents*; Howarth, R.W.; Stewart, J.W.B; Ivanov, M.V., Eds.; SCOPE 48; John Wiley and Sons: New York, 1992; pp 125-144.
47. *Petroleum Geochemistry and Geology;* Hunt, J.M.; Springer-Verlag: Berlin, 1979.
48. Schidlowski, M. In *Evolution of the Biogeochem. Sulfur Cycle*; Brimblecombe, P.; Lein, A. Yu., Eds.; SCOPE 39; John Wiley and Sons: New York, 1989; pp 3-19.
49. Berner, R. A. In *The Changing Chemistry of the Oceans*; Dyrssen, D; Jagner, D., Eds.; Nobel Symposium 20; John Wiley and Sons: New York, 1989; pp 347-361.
50. Holser, W.T.; Maynard, J.B.; Cruikshank, K.M. In *Evolution of the Biogeo-chemical Sulfur Cycle*; Brimblecombe, P.; Lein, A. Yu., Eds.; SCOPE 39; John Wiley and Sons: New York, 1989; pp 21-56.
51. *Marine Evaporites*; Kirkland, D.W.; Evans, R. Eds.; Dowden, Hutchinson and Ross: Stroudsburg, 1973.
52. Hardie, L.A. *Am. J. Sci.* **1984**, *284*, 193-240.
53. Javor, B.; Springer: *Hypersaline Environments: Microbiology and Biogeo-chemistry*; Springer: Berlin, 1989.
54. *The Global Sulfur Cycle*; Sagan, D., Ed.; NASA Technical Memorandum 87570; NASA: Washington, DC, 1985.
55. Zehnder, A. J. B.; Zinder, S. H. In *The Handbook of Environmental Chemistry, Vol.1;* Hutzinger, O., Ed.; Springer: Berlin, 1980; pp 105-145.
56. Volkov, I.I.; Rozanov, A.G. In *The Global Biogeochemical Sulfur Cycle*; Ivanov, M. V.; Freney, J. R., Eds.; SCOPE 19; John Wiley and Sons: New York, 1983; pp 357-448.
57. Pingitore, N.E.; Meitzner, G.; Love, K.M. *Geochim. Cosmochim. Acta*, **1995**, *59*, 2477-2483.

58. Corliss, J.B. et al. *Science*, **1979**, *203*, 1073-1083.
59. Spiess, F.N. et al. *Science*, **1980**, *207*, 1421-1433.
60. Jannasch, H.W. In *Evolution of the Biogeochem. Sulfur Cycle*; Brimblecombe, P.; Lein, A. Yu., Eds.; SCOPE 39; John Wiley and Sons: New York, 1989; pp 181-190.
61. Arnold, M.; Sheppard, S.M.F. *Earth Planet. Sci. Lett.* **1981**, 56, 148-156.
62. Kerridge, J.K., Haymon, R.M., Kastner, M. *Earth Planet. Sci. Lett.* **1983**, *66*, 91-100.
63. Mottl, M.J.; Holland, H.D.; Corr, R.F. *Geochim. Cosmochim. Acta*, **1979**, *43*, 869-884.
64. Jannasch, H.W., Mottl, M.J. *Science*, **1985**, *229*, 717-725.
65. Anderson, D.F. In *The Biochemistry of Plants*, vol. 5; Stumpf, P.K.; Conn, E.E., Eds.; Academic Press: New York, 1980; 203-223.
66. Froelich, P.N.; Klinkhammer, G.P.; Bender, M.L. *Geochim. Cosmochim. Acta* **1979**, *43*, 1075-1090.
67. Widdel, F.; Hansen, T.A. In The Prokaryotes, 2nd Edition; Balows, A.; Truper, H.G.; Dworkin, M.; Harder, W.; Schleifer, K.H.; Eds.; Springer-Verlag: New York, 1992; pp 583-624.
68. Rees, C.E. *Geochim. Cosmochim. Acta* **1973**, *37*, 1141-1162.
69. Kaplan, I.R.; Rittenberg, S.C. *J. Gen. Microbiol.* **1964**, *34*, 195-212.
70. Jørgensen, B.B. *Limnol. Oceanogr.* 1977, **22**, ,814-832.
71. Thamdrup, Bo; Fossing, H.; Jørgensen, B.B. *Geochim. Cosmochim. Acta* **1994**, *58*, 5115-5129.
72. Mossmann, J.R.; Aplin, A.C.; Curtis, C.D.; Coleman, M.L *Geochim. Cosmochim. Acta* **1991**, *55*, 3581-3595.
73. Morse, J. W.; Berner, R. A. *Geochim. Cosmochim. Acta*, **1995**, *59*, 1073-1077
74. Kirkland, D. W.; Evans, R. *Bull. Am. Assoc. Pet. Geol.* **1981**, *65*, 181-190.
75. Evans, R.; Kirkland, D. W. In *Evaporities and Hydrocarbons*; Schrciber, D. C. Eds.;   Columbia Univ. Press: New York, 1988; 256-299.
76. Orr, W.L. *AAPG Bull.* **1974**, *58*, 263-276.
77. Krouse, H. R.; Viau, C.A.; Eliuk, L.S.; Ueda, A.; Halas, S. *Nature* **1988**, *333*,415-419.
78. Jørgensen, B.B; Isaksen, M.F.; Jannasch, H.W. *Science* **1992**, *258*,1756-1757.
79. Vairavamurthy, A.; Zhou, W.; Eglinton, T.; Manowitz, B. *Geochim. Cosmochim. Acta*   **1994**, *58*, 4681-4687.
80. Schouten, S.; Sinninghe Damste, J.S.; Kohnen, M.E.L.; de Leeuw, J.W. *Geochim. Cosmochim. Acta* **1995**, *59*, 1605-1609.

RECEIVED August 11, 1995

GEOCHEMISTRY OF ORGANIC SULFUR IN SEDIMENTARY SYSTEMS

Chapter 2

Role of Sulfur in the Transformations of Sedimentary Organic Matter: A Mechanistic Overview

Zeev Aizenshtat[1], Eitan B. Krein[1], Murthy A. Vairavamurthy[2], and Ted P. Goldstein[1]

[1]Department of Organic Chemistry and Casali Institute for Applied Chemistry, Energy Research Center, Hebrew University of Jerusalem, Jerusalem 91904, Israel
[2]Department of Applied Science, Geochemistry Program, Applied Physical Sciences Division, Brookhaven National Laboratory, Upton, NY 11973

Sulfur has a unique role in the various transformations of organic matter, from early diagenesis to the late stages of catagenesis. This activity of sulfur is facilitated by its ability to exist in many oxidation states, and to form catenated (polysulfidic) and mixed oxidation state species. This paper deals with mechanistic aspects of the transformations of sedimentary organic sulfur; in particular, it focuses on the transformations of polysulfidic sulfur. The nucleophilic introduction of these forms into organic matter generates thermally labile polysulfide cross-linked polymers during early diagenesis. The thermal processes occurring during later diagenesis, and catagenesis stabilize the sulfur in polymeric structures by forming new functionalities of sulfur (for example, thiophenic), and eliminating H_2S/S_8. At higher temperatures, the latter can reincorporate into organic matter or can aid in the oxidation of organic matter. While the introduction of sulfur into sedimentary organic matter occurs mainly during the early diagenesis, and is ionically controlled, the later catagenetic transformations probably are controlled by radical mechanisms.

Organically bound sulfur constitutes a significant fraction of the organic matter in Recent and ancient sediments, and in petroleums. The bulk of the sedimentary organic sulfur usually is present in the insoluble fraction of the organic matter, and probably most of it is associated with macromolecular structures as mono-, di-, and polysulfide linkages bridging carbon skeletons. The source of all the sedimentary organic carbon is the organic matter from living organisms; a fraction of this includes organically bound sulfur produced by the biosynthesis of autotrophic organisms (for example, phytoplankton). It is unlikely that organic sulfur of biosynthetic origin can

0097–6156/95/0612–0016$12.50/0

account for all the organically bound sulfur present in sedimentary and fossil organic matter because the abundance of sulfur in biomass usually is very low (usually $<2\%$ by weight) (*1*). The structures of organic sulfur compounds found in petroleums, kerogens, and sediments also are very different from those biosynthesized (*2*). Furthermore, the $^{34}S/^{32}S$ ratio of sulfur in fossil and sedimentary organic matter is enriched with lighter isotope by $>10\%$ relative to that in the biomass (*3-5*; see 6 for a detailed treatment of the isotopic issue). Thus, it is now generally accepted that the major fraction of the organically bound sulfur in fossil organic matter is formed through geochemical pathways, and not through assimilatory sulfate reduction. Because of the ubiquitous occurrence of organic sulfur in sedimentary systems, and the overwhelming evidence to suggest that most of this sulfur is of geochemical origin, it is thought that organic matter, like iron (which forms pyrite), is an important sink for sedimentary sulfur (*2,7,8*).

The geochemical pathways for incorporating sulfur into sedimentary organic matter, and the timing of such incorporation reactions are not completely understood, although it is becoming increasingly clear that reactions between functionalized organic compounds and reactive inorganic sulfur species, H_2S and its oxidation intermediates such as polysulfides, are mainly involved. In reducing marine sediments, H_2S is generated from the bacterial reduction of sulfate, known as dissimilatory sulfate reduction, in which sulfate is used as a terminal electron acceptor to support the respiratory metabolism of sedimentary bacteria when other oxidants are depleted (*9,10*). The abundant sulfate ion in seawater is chemically non-reactive under ambient sedimentary conditions, and hence, it is not considered to be a reactant in the geochemical incorporation of sulfur into organic matter. It is not known whether sulfate esters can be formed in marine sediments through bacterial mediation although such a process has been suggested in soils (*11*). Although H_2S (and its conjugate base HS^-) should be the main sedimentary sulfur nucleophile, there is compelling evidence from field observations and laboratory experiments that polysulfides play a dominant role in incorporating sulfur into organic matter (*12-19*). The role of other H_2S oxidation intermediates is unclear, particularly, that of oxygenated sulfur nucleophiles, such as sulfite and thiosulfate. However, Vairavamurthy et al. recently suggested that sulfite and thiosulfate may participate in incorporating sulfur into sedimentary organic matter, generating sulfonates (*20*). Elemental sulfur (S_8) cannot react directly under the mild conditions of early diagenesis, although it may be involved indirectly through the formation of polysulfides when H_2S is present.

Sulfur exhibits a unique chemical activity in its reactions with sedimentary organic matter, both in its ability to react with and be incorporated into organic matter (OM), and as a redox reagent. This activity mainly stems from its ability to undergo many, stepwise, redox reactions which produce a wide variety of sulfur species; furthermore, sulfur can catenate, and form, mixed-valent compounds (*21,22*). These species can react with organic matter over a wide range of temperature, and by different mechanisms, e.g. ionic-nucleophilic, electrophilic, and free radical (*23*). These characteristics facilitate a complex network of chemical reactions during the transformations of organic matter. In this paper, we provide a mechanistic framework for such transformations of organically bound sulfur in sediments, focusing particularly on the role of polysulfidic sulfur, from its low temperature reactions in early diagenesis to the thermal reactions during later diagenesis to catagenesis. In doing so, we review the literature, and include some

new ideas emanating from our recent work. In particular, the following concepts and issues are discussed:
    (a) What are the most probable mechanisms for incorporating sulfur in depositional environments ?
    (b) What is the fate of the labile sulfur released during thermal maturation of sedimentary organic matter ?
    (c) Can sulfur initiate redox transformations in organic molecules without becoming incorporated into the products ?
    (d) What are the active species involved in the transformations of sulfur-containing sedimentary organic matter?
    (e) Is there a temperature barrier between the ionic and the free-radical controlled chemical activity of sulfur in sedimentary organic matter?
    (f) Under what conditions is sulfur an active reactant, and under what others does it act as an initiator or catalyst to redox reactions ?

The Incorporation of Sulfur into Organic Matter During Early Diagenesis: Formation of Organic (Poly)sulfides

Despite the fact that mechanisms are known for reactions between sulfur nucleophiles and organic molecules in specific model systems, the mechanisms for the first formation of the C-S bond in natural sediments remain unclear. Figures 1 and 2 summarize some well- established reaction pathways for adding sulfur species to organic functional groups both under acidic (Figure 1) and basic (Figure 2) conditions (17-19, 24-30). Under acidic conditions, the reaction proceeds *via* the electrophilic addition mechanism in which a proton adds first to the unsaturated organic molecule, followed by the addition of sulfur species. Under basic conditions, the reaction proceeds *via* the nucleophilic addition mechanism that involves the initial addition of sulfur as a nucleophile in the rate limiting step (Figure 2); the organic molecule should have an activated unsaturated bond, as in α, β-unsaturated carbonyl compounds, for this reaction to proceed at kinetically significant rates under ambient conditions. Phase-transfer catalysis accelerates the rates of reactions between sulfur nucleophiles and organic electrophiles (17,18). Since most marine depositional environments are slightly basic, the nucleophilic pathway, also known as the Michael addition, is likely to occur there. Vairavamurthy and Mopper demonstrated this mechanism for the formation of the thiol 3-mercaptopropionate, detected in coastal sediments, from the reaction of HS$^-$ with acrylate ion, and suggested that it could represent a major pathway for incorporating sulfur into sedimentary organic matter during early diagenesis (24). The major findings of the studies on pathways of sulfur incorporation into organic matter, and the reactivities between sulfur nucleophiles and various functionalized organic molecules are summarized in the following points.
    1. Polysulfide linkages are present in numerous immature natural samples, either in the bulk OM (1,4,12), or in the structure of specific di- and tri-sulfide monomers and polymers (13,31-35) suggesting that polysulfide anions are incorporated into organic matter. For example, studies by Adam et al. (31,32) and Kohnen et al. (33-35) revealed mainly the presence of sulfur cross-linked macromolecules in sulfur-rich immature bitumens and petroleums; furthermore, most of the sulfur, according to these studies, cross-links aliphatic and alicyclic structures

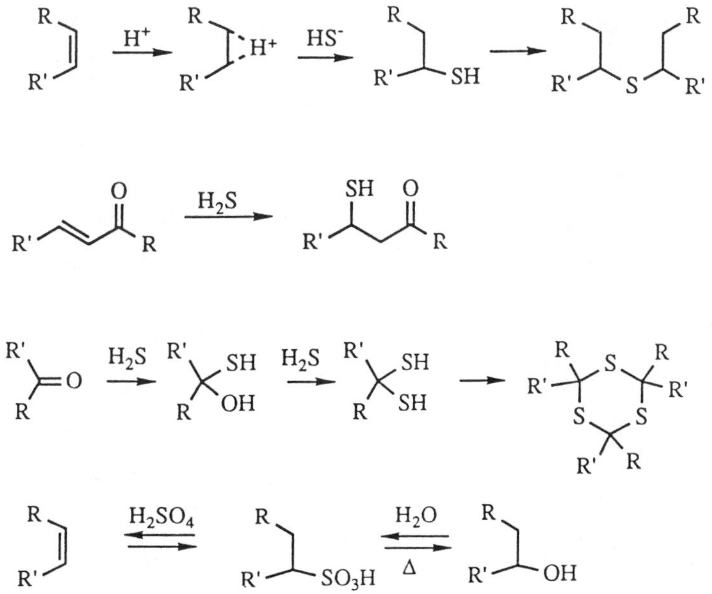

Figure 1. Acid-catalyzed electrophilic addition of sulfur species to organic functional groups.

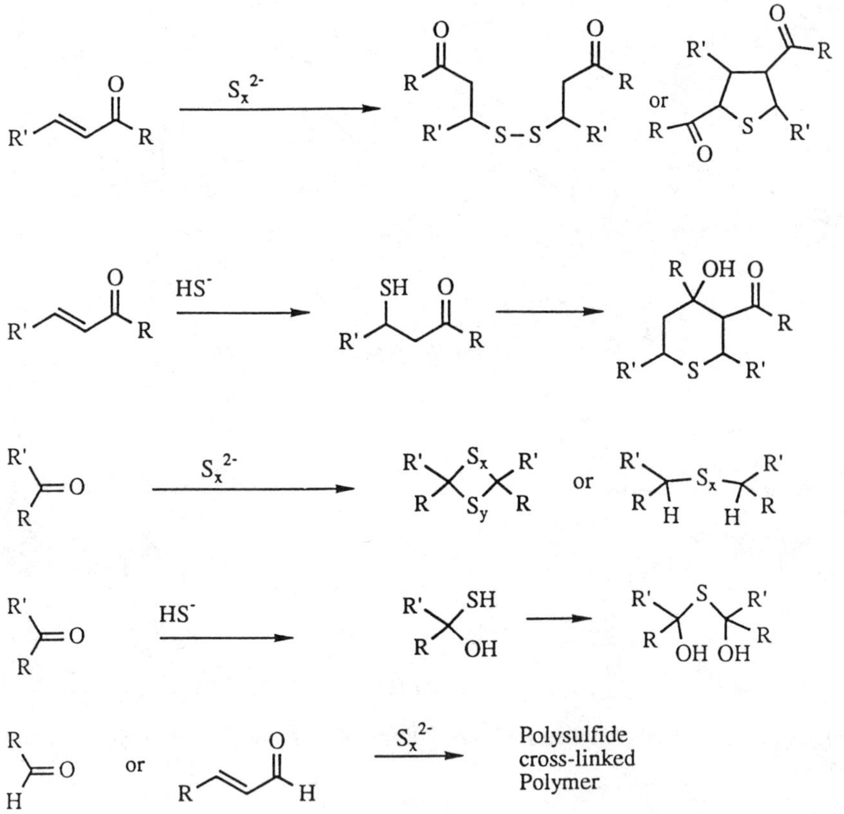

Figure 2. Base-catalyzed nucleophilic addition of sulfur species to organic functional groups .

generated from the lipids of the algae (e.g. isoprenoids, steroids, carotenoids).

2. Polysulfide anions are stronger nucleophiles than bisulfide anions, and the longer the polysulfidic chain, the stronger its nucleophilicity. Polysulfides are stable only at basic pH values, and as the pH increases, the length of the polysulfide chain decreases (*36*).

3. Electrophilic organic functional groups, such as activated-unsaturated double bonds, aldehydes, and ketones are the most active functional groups (*14-16,18-19, 24-26, 28, 37*). However, under certain experimental conditions (for example, with dimethylformamide as the reaction medium), dienes and monoenes also react with polysulfides (*17*), although higher temperatures or longer reaction times are required. Thus, natural aldehydes and ketones, whether they are of biogenic origin, or are formed by diagenetic transformation of other compounds (for example, by oxidation of alcohols, or dehydratation of polyols), will be among the rapidly reacting compounds with polysulfide anions. Because less electrophillic compounds, such as dienes, and monoenes react much slower, if at all, under typical sedimentary conditions, they can remain unchanged for longer periods. This difference in the reaction rate explains the co-occurrence of unsaturated compounds and sulfur-containing compounds in recent sediments (*38,39*), and the specificity of C-S bond positions at locations of double bonds in known precursors in mature sedimentary organic matter (*2,23*). (An alternative explanation for these phenomena is discussed later). Therefore, it is noteworthy that partial oxidation of OM, specially of alcohols, e.g. phytol to phytenal (*40*), will increase its capability to react with polysulfides. Also, the occurrence of one electrophilic functional group in a molecule, which serves as the initial reaction center, in some cases may enhance the reactivity of less reactive groups in the same molecule (*18*).

4. The products of low-temperature simulations experiments often are di- or polysulfide cross-linked dimers, oligomers or polymers (*17,19,25*). In some cases, monomeric mono- , di-, or poly-sulfide heterocycles are formed also (*14,18*). The formation of thiophenes is quite rare.

5. In general, reactive polysulfides are inorganic ions (S_n^{2-}) that usually are distributed in the aqueous phase. However, a major fraction of the reactive sedimentary polysulfides appears to be bound to the particle phase (*41*). It was suggested that such particle-bound polysulfides have one end of their chains linked to the particulate phase so that the other end is available for reactions (for example, with added organic molecules). Recently, Vairavamurthy et al. suggested that humic coatings on mineral particles could serve as attachment points for these particle-bound polysulfides (*42*). The presence of these particle-bound polysulfide anions essentially provides a phase-transfer mechanism for incorporating sulfur (nucleophiles) into organic matter.

Thus, a major consequence of incorporating sulfur into organic matter during early diagenesis is the formation of polysulfide cross-linked polymers. The formation of such structures implies that catenation of sulfur (S-S$_x$-S) has a major influence on the fate of both sulfur and carbon. In particular, catenation of sulfur seems to affect the properties of the molecules in the following ways:

(a)   Enhancing the nucleophilicity of sulfur (*14,16*).

(b)   Solubilizing and activating elemental sulfur (S_8), which is non-reactive to OM under ambient sedimentary conditions.

(c) Enabling different amounts of sulfur to be incorporated into SOM, independent of the structure of the OM, i.e. a specific, active, organic functional group may react similarly with mono-, di-, tri-, or poly-sulfide ions, thus enriching the SOM with as much sulfur as is available in the environment (the limit probably depends on its solubility).

(d) Lowering the thermal stability of the sulfur containing OM, since organic di- or poly-sulfides decompose at lower temperatures than the corresponding sulfides(43,44).

(e) Enabling the formation of transient mixed oxidation state species which can facilitate redox reactions (21); an example of a mixed oxidation state compound is thiosulfate, the outer sulfur with oxidation state -1, and the inner sulfur +5 (45).

Thermal Alteration of (Poly)sulfide Cross-Linked Polymers During Catagenesis

The main question at the late diagenesis and catagenesis stages is, what do these sulfur cross-linked polymeric structures yield upon thermal maturation ? Table 1 shows the bond dissociation energies (BDE) for various sulfur bonds; these values are distributed over a wide range due to the very strong influence of the environment and substituents (R-). As shown, organic polysulfides appear to be less stable thermally than the corresponding sulfides (see also 43,44). Thus, the thermal decomposition organic polysulfides is likely to proceed mainly through homolytic dissociation of S-S and/or C-S bonds, and therefore, the chemistry of polysulfidic sulfur at elevated temperatures is most likely to be controlled by free-radical mechanisms. As discussed elsewhere (19, 46), there is competition between C-S and S-S bond cleavage, controlled mainly by the stability of the alkyl or thiyl radicals (R• and RS•) that form after the bonds are cleaved. Also, stable thiyl radicals can be rearranged to produce stable heterocycles (thiophenes) (19, 46). However, the main reaction pattern for thermal decomposition of simple polysulfides is the elimination of hydrogen sulfide and elemental sulfur to form alkenes. Therefore, the next sections will deal with the following issues:

(a) Elimination and release of sulfur (as H_2S; elemental sulfur, or polysulfidic sulfur) causing unsaturation and aromatization of the residual kerogen.

(b) Rearrangement of sulfur-containing structures to form thermally more stable moieties (i.e. thiophenes).

(c) Re-incorporation of the released active sulfur species into the various fractions formed.

(d) Initiation of radical controlled reactions by sulfur radicals, which do not re-incorporate sulfur.

Elimination of Sulfur. During catagenesis, upon continued burial, the sulfur content of sedimentary organic matter gradually decreases, probably due to thermal degradation involving organic polysulfides. Heating experiments provide strong evidence that organic polysulfides behave this way during catagenesis. For example, heating protokerogen from Solar Lake (Golf of Eilat, Sinai) to 175°C decreased its sulfur content, eliminating it as elemental sulfur (12). Step-wise pyrolysis of Recent kerogens from Abu Dhabi also released elemental sulfur (39). Similar results were

Table 1. Bond dissociation energies (BDE)[1] of different types of sulfur bonds

Molecular type	Bond type	BDE (Kcal/mol)	Reference
SO_4^{2-}	S-O	> 120 ± ?	74,75,76,
$[O_3S\text{-}SO_3]^{2-}$	S-S	38-54 ± ?	74
SO_3^{2-}	S-O	35-45 ± ?	74
$R\text{-}CH_2\text{-}SH$	C-S	65-70 ± 10	44,74,75,77
$R\text{-}CH_2\text{-}S\text{-}CH_2\text{-}R'$	C-S	70-80 ± 5	44,74,75,77
R-S-S-R'	S-S	64 ± 10	44,74,75,77,78
Ph-S-S-Ph	S-S	53 ± 5	78
R-S-S-R'	C-S	57 ± 5	44,74,75
R-S-S-S-R'	S-S	54 ± 10	66,74
$R\text{-}S_x\text{-}R'$ (x>3)	S-S	36 ± 5	74,79
	C-S	64-75 ± 5	80
	C-S[2]	70-80 ± 5	80
	C-S[2]	85-90 ± 5	80

[1] Correspond to thermochemical values for ideal gas-phase molecules at 25 °C.
[2] Ring strain decreases the BDE values for C-S bonds in thiolane rings by about
2 Kcal/mol and increases these values in thiophene rings by about 16 Kcal/mol

obtained in the pyrolysis of synthetic polysulfide cross-linked polymers (*19*). Upon further maturation of the sedimentary organic sulfur, hydrogen sulfide also is eliminated along with elemental sulfur.

Using the relative abundance of 2,3-dimethylthiophene in pyrolysates of kerogens as a measure of the organic sulfur content, Eglinton et al. found that in natural maturity sequences, and in thermal experiments, this parameter decreased, indicating a decrease in organic sulfur with maturation (*47,48*). This finding agreed with the much earlier results of Gransch and Posthuma (*49*), which suggested that organically bound sulfur is eliminated during the early stages of oil production. Such elimination is probably the reason for the early formation of heavy, high sulfur-oil from Type-II-S kerogens because the loss of sulfur cross-linking is accompanied by the release of smaller sulfur-containing molecules (*50*). Molecular evidence for the elimination of sulfur was given by Kohnen et al. (*35*) who showed that shallow, less mature sediments from the Peru upwelling area (Pliocene sediments) contain cyclic trisulfides in higher concentrations than more mature, deeper sediments (Vena del Gesso - Upper Miocene) which contain mainly cyclic disulfides; many other sediments which are relatively more mature contain only monosulfide heterocycles (thiolanes and thianes).

Formation of Thermally Stable Compounds. Despite extensive work over the last ten years on the molecular identification of sulfur-containing compounds in bitumens, asphalts, and petroleum fractions, the factors controlling the formation of thiolanes and thiophenes are not well established, nor is the distribution between alkyl-thiophenes *vs.* benzo- and dibenzothiophenes fully understood.

Formation of Alkylthiophenes. Figure 3 summarizes the three main routes for the formation of alkylthiophenes in sedimentary organic matter. The first path is the direct formation of thiophenes by incorporation reactions during early diagenesis (path I). This route was suggested by several authors (*2,14,15,32,51,52*), usually from the reactions between sulfur species and dienes. However, most simulation reactions do not support this suggestion because polysulfide cross-linked polymers, rather than sulfur heterocycles, are the main products, regardless of the structure of their precursors (e.g. aldehydes and alkenes). Nevertheless, the occurrence of thiophenes in very recent sediments, and even in surface sediments, suggests that an unknown, direct mechanism of formation probably operates (*39,51,53,54*).

A second pathway is the aromatization of alkyl thiolanes (path II) (*2,52,55,56*), studied by Pyzant and co-workers (*57*) who examined the thermal stability of alkylthianes. These compounds, which are less thermally stable than the corresponding thiolanes, were transformed into the latter upon heating. The formed alkylthiolanes subsequently were aromatized to alkylthiophenes, along with isomerization. In all these transformations, there is a loss of sulfur (probably as H_2S) and formation of lower molecular weight compounds. However, this path does not explain the formation of thiolanes, and does not include the direct formation of thiophenes without thiolanes as intermediates. Since elemental sulfur is a known aromatization agent (transformed to H_2S) and is formed during thermal degradation of sulfur-rich sediments, it may catalyze this transformation.

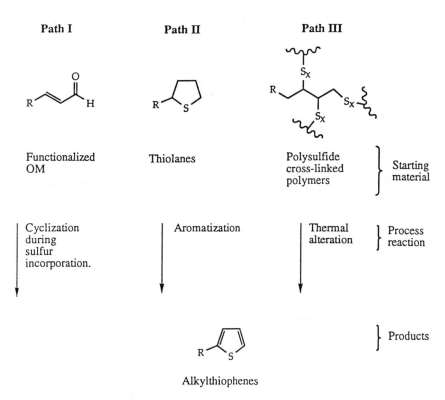

Figure 3. Suggested pathways for the formation of alkylthiophenes in sediments.

The third pathway for the formation of alkylthiophenes is during, and as a result of, the thermal decomposition of polysulfide cross-linked polymers (path III). Recently, Krein and Aizenshtat reported that polysulfide cross-linked macromolecules were generated from the reaction between ammonium polysulfide and phytenal (19). Thermal exposure of these polymers gave high yields of C_{20}-isoprenoid thiophenes. Recently, similar results were noted for the decomposition of polymers formed from the reaction between polysulfides and phytadienes(26). These studies and a more recent one (46) suggest that the formation of alkyl thiophenes from the pyrolysis of polysulfide cross-linked polymers is a general phenomenon. The structure of the polymers influence the yield and the isomer distribution of the thiophenes. In all of these studies, thiolanes are only a minor product, suggesting that, under these conditions, path II is not operative.

The three pathways are not exclusive; however, from the simulation experiments and maturation sequences, we prefer path III as the major route for the formation of alkyl thiophenes. Also, we suggest that alkylthiophenes may be formed from the reaction between sulfur species released from the decomposing polysulfide cross-linked polymers and dienes (further discussed in the following sections). It is noteworthy that the thermal formation of alkylthiophenes, as described for path III, was observed during GC and GCMS analysis (19); therefore, we suggest considering the formation and occurrence of thiophenes in recent sediments (as suggested in path I) only if their structure was established by analytical methods not involving thermal exposure.

Formation of Benzo- and Dibenzothiophenes. In petroleum, only a minority of sulfur-containing compounds are thiolanes, and there are even smaller amounts of thianes (58-62). Except for several sulfur-rich, heavy oils (e.g. Rosel Point), most petroleums are dominated by benzo[b]thiophenes or dibenzo[b,d]thiophenes and their methylated derivatives. However, the mechanism of their formation is unknown. These compounds were not detected at any of the simulation experiments described above, even when bithiophenes and thienothiophenes were formed (46). Therefore, none of the paths suggested for generating alkylthiophenes is applicable to the more aromatic compounds. From indirect evidence, we suggest four possible paths for the formation of these compounds; they are summarized in Figure 4.

One mechanism (pathway I) is the cyclization and aromatization of alkyl side-chains of long chain alkylthiophenes (55,63,64). The presence of 2,4 dialkyl-benzo[b]thiophenes in relatively high abundance in petroleum provides strong support for this mechanism (52). Intermediate non-aromatic thiatetralines also are found in some crude oils (23). The presence of elemental sulfur (S⁰) probably facilitates the transformation of these compounds to aromatic structures. The existence of this path is supported by comparing sulfur-containing compounds having highly branched isoprenoid carbon skeletons from different locations which form a "maturity sequence"(64). While the less mature samples contain thiolanes and thiophenes, only the more mature kerogens (Jurf ed Darwish, Jordan) contain highly branched isoprenoid benzo[b]thiophenes. Further support comes from pyrolysis experiments of a kerogen (Kimmeridge, U.K) pre-heated at different temperatures for 72h ("artificial maturation") (64). These experiments showed a systematic decrease of alkylthiophenes and increase of alkylbenzothiophenes as the pre-heating temperature

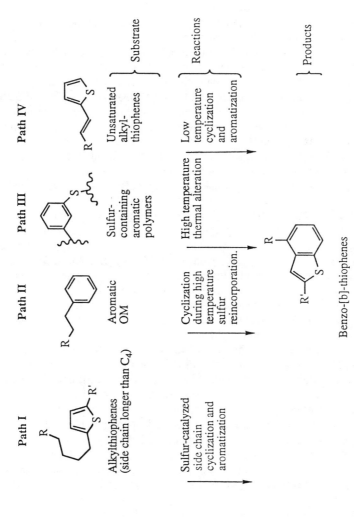

Figure 4. Suggested pathways for the formation of benzo- and dibenzo-thiophenes in sediments.

increased from 250 to 360°C, suggesting that such aromatizations occur inside the polymeric structure.

In a second pathway (II), condensed thiophenes are formed by the high-temperature incorporation of sulfur into aromatic, organic compounds. The elemental sulfur can originate from the catagenetic breakdown of polysulfide cross-linked polymers. Przewocki et al. (65) investigated the high temperature (>160 °C) reactions between elemental sulfur and toluene, dibenzyl, and z-stilbene. They showed that benzothiophenes form; however, most sulfur-containing compounds were poly-phenyl-substituted thiophenes and some thiols. The reaction between elemental sulfur and butylbenzene (>200 °C, in closed vessels) gave the expected 2-ethylbenzo[b]thiophene only as a very minor product, whereas the main low-molecular-weight product was 3-phenylthiophene (Figure 5) (Goldstein and Aizenshtat, unpublished results). These results suggest that the first point of S attack is at the α-position (benzylic carbon) of the alkyl side chain in the aromatic organic molecule. Almost 90% of the reacted butylbenzene formed a polymer with the empirical formula $C_{10}H_{10}S_2$. Similar results had been reported by Pryor (66), although it was not clear from that study whether the main product of the reaction was a polymer. Upon pyrolysis at higher temperatures, the polymer yielded benzo-thiophenes and thieno[2,3][b]benzothiophene; the latter also appear in some crude oils (69).

The third possible pathway (III) considers that dibenzothiophenes and benzothiophenes are generated from the pyrolysis of sulfur-containing aromatic polymers, where sulfur bridges form between two benzene rings or, at least, substitutes on the aromatic structure, as shown by Cohen and Aizenshtat (67,68). They also showed that pre-heating such polymers (e.g. polyphenylene sulfide) increased the yield of condensed thiophenes in comparison with a non-heated sample, indicating that thermal transformations from sulfidic to aromatic sulfur take place inside the polymer matrix. However, these reactions required high temperature, and hence, have high activation energies.

The fourth pathway for the formation of benzo- and dibenzo-thiophenes involves cyclization of polyunsaturated alkylthiophenes during early stages of diagenesis. In fact, 2,4-Dialkyl-benzothiophenes occur even in immature sediments (64), suggesting that their formation does not require severe thermal stress. Thus, some benzothiophenes may be formed through cyclization of polyunsaturated alkylthiophenes during early diagenesis by a mechanism comparable to the formation of benzohopanes, which also occurs in immature sediments.

Our evaluation of the evidence for each of the four suggested pathways favors the idea that condensed thiophenes are formed in nature by sulfur-catalyzed cyclization of structures with alkyl side chains (pathway I).

Reincorporation of the Active Sulfur Released from (Poly)sulfides Decomposition. The fate of the sulfur species released from thermally decomposing polymers is unclear. However, unlike oxygen, nitrogen, and carbon, which are released as relatively non-active low-molecular-weight compounds (e.g. CO_2, H_2O, NH_3, CH_4), sulfur is released as highly active species, such as H_2S, $S°$ and polysulfide ions and free radicals. Therefore, it is very likely that these species will be re-incorporated into the sedimentary organic matter by one of the many known high-temperature

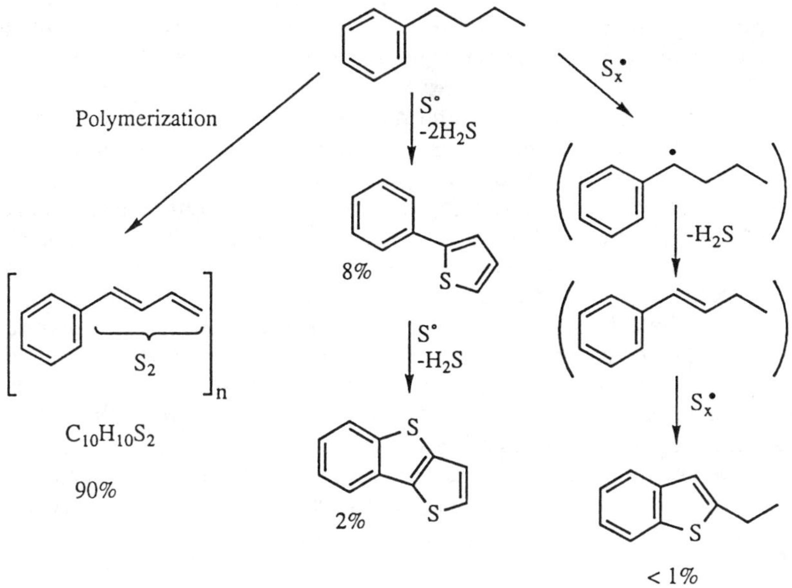

Figure 5.   The reaction products between elemental sulfur and butylbenzene (>200°C, autoclaved).

pathways (such as in Figure 5). This pathway may be regarded as a continuation of the formation of alkylthiophenes.

Heating a synthetic polysulfide cross-linked aliphatic polymer together with a monoene (1-tetradecene) at 160°C in the laboratory (46), resulted in the formation of a thiol and the incorporation of the monoene into the polymeric structure *via* one or two polysulfide linkages. Heating a similar polymer with phytadienes similarly generated C_{20}-isoprenoid thiophenes and thiols, and caused the incorporation of phytanyl moieties into the polymeric structure (Krein E.B. and Aizenshtat Z., unpublished results). Schmid (55) showed that elemental sulfur reacted with n-octadecane in a sealed glass tube between 200 and 250°C to produce a complex mixture of C_{18}-alkylthiophenes. Stoler (69) obtained similar results. These reactions, which are most likely radical chain reactions, cannot be accepted as the initial mechanism for sulfur enrichment, but can contribute to the content of sulfur compounds in kerogens and crude oils in reservoirs.

Hydrogen sulfide is another possible source for the re-incorporation of sulfur in oils. Ho and coworkers (58) described the occurrence of high thiol mature oils, and explained this phenomenon by H_2S incorporation. Orr (70) suggested three sources of hydrogen sulfide in oil reservoirs: (a) microbial reduction of sulfate in low temperature reservoirs (<50°C), (b) thermal cleavage from organic matter as described above, and (c) high-temperature thermochemical sulfate reduction (TSR). Orr also pointed out that the thermal maturation of oil in the absence of sulfate may cause a continued decrease in sulfur content, while thermal maturation in the presence of sulfate in high-temperature reservoirs (>80-120°C) may result in competing sulfurization and desulfurization in the oils, which can give them an abnormally high thiol content. Such processes can be monitored by the increase in the S/N atomic ratio because while nitrogen continues to decrease, sulfur may be maintained at the same level. The $\delta^{34}S$ values of these oils (and the related H_2S) will change towards the values of reservoir sulfate (because non-microbial sulfate reduction has low isotopic fractionation). In natural settings, the occurrence of metal ions (such as copper and iron) also may trap the released sulfur as sulfur ores (22,71).

Sulfur as a Mediator of Organic Redox Reactions

The previous sections dealt with reactions of sulfur which lead to its incorporation into organic molecules during diagenesis. However, sulfur also participates in the redox transformations of organic matter without becoming incorporated. A well-known example is thermochemical sulfate reduction (TSR) involving the oxidation of organic matter during late diagenesis (see last three chapters in this volume). Sulfate requires a very high activation energy to react as an oxidizing reagent. However, it is well known that sulfides accelerate this reaction. This mediation of redox conversions in organic molecules stems from the ability of sulfur to form a wide variety of species with different sulfur oxidation states; this ability is augmented due to its unique properties to catenate and "mix" different sulfur atoms of different oxidation states. The stepwise oxidation/reduction reactions between these species lowers the overall activation energy of the reduction of sulfate or oxidation of sulfide (21). Thus, sulfur can serve as an electron donor or acceptor for the reduction or oxidation of OM. Goldstein and Aizenshtat used this principle to suggest a chemical mechanism for TSR (22).

Several studies investigated the fate of the organic compounds reacting with sulfur species (see reference 22 for a recent review) at high temperatures. In most, benzylic compounds were used because the benzylic position is highly susceptible to attack by radical sulfur; thus, the latter can easily abstract hydrogens to form a benzylic radical (Figures 5 and 6). Early work by Toland (72) (S^0, NaOH-H_2O) showed p-toluic acid was oxidized to terphtalic acid, and toluene, xylene and other compounds to the corresponding aromatic acids. Pryor (66) showed that aqueous solutions of polysulfides (S_x^{2-}) above 175^0C caused similar oxidations. The reaction of butylbenzene (Figure 6) with aqueous ammonium polysulfides, unlike its reaction with elemental sulfur under dry conditions (Figure 5), formed benzylic acid along with acetic and propionic acids. Although the mechanism of this reaction is not clear, we think that sulfur and water play crucial roles, as presented in Figure 6. Earlier, organic acids, including low-molecular-weight aliphatic acids, were detected in the water fraction in hydrous pyrolysis experiments (73), but again the mechanism was not understood. We suggest that the organic acids in these experiments were generated by a mechanism similar to that proposed in Figure 6 involving sulfur-initiated reactions.

Conclusions

Sedimentary sulfur interacts with organic matter throughout all stages of diagenesis, from early diagenesis through catagenesis, to metagenesis. Because sulfur exists in many oxidation states, and forms catenated and mixed oxidation state species, it is involved in transforming both organic matter and minerals. The incorporation of sulfur into sedimentary organic matter occurs mainly during early diagenesis, and is ionically controlled; however, radical mechanisms appear to control the later transformations, through catagenesis to metagenesis (Figure 7).

The labile polysulfide cross-linked organic structures, formed during early diagenesis, change into more stable moieties during late diagenesis, such as thiophenes. These transformations could release H_2S/S^0, which then can react and reincorporate sulfur into SOM through either a free radical or even a carbocationic mechanism. In general, elemental sulfur becomes active at higher temperatures than organic polysulfidic sulfur.   Hence, polysulfide cross-linked polymers can be considered as a pool of active sulfur which is transferred by organic matter from the interstitial water of the depositional environment to the sediment, and stored during diagenesis in a relatively labile form, until higher temperature reactions take place during the late diagenesis. In some late diagenetic transformations, sulfur acts as a reactant, and is incorporated in the organic compounds, while in others, sulfur acts as an initiator or catalyst for reactions such as aromatization or oxidation and is not incorporated in the products. The end products formed are very much influenced by the whole sedimentary system; for instance, the presence of metal ions or excess water affects the type and distribution of products.

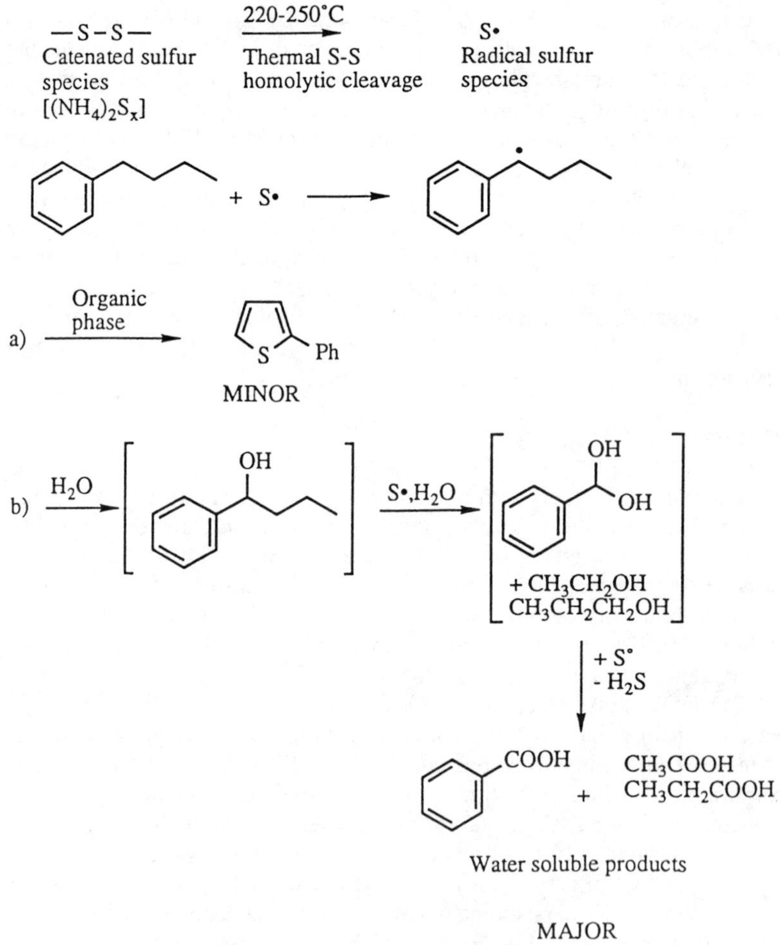

Figure 6. A suggested mechanism for the high-temperature oxidation of butylbenzene with ammonium polysulfide.

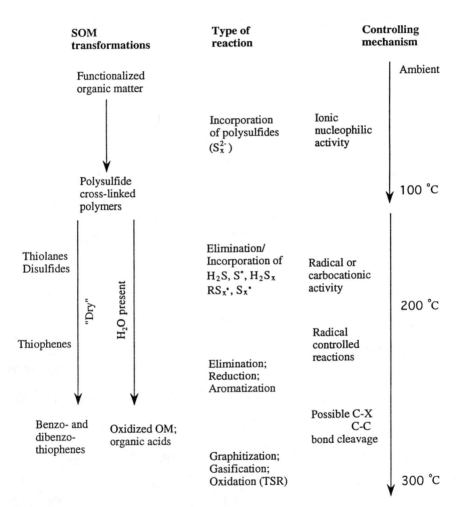

Figure 7. A diagrammatic presentation of sulfur transformations in the different thermal regimes of sediments.

Acknowledgements

This material is based upon work supported by the Moshe Shilo Center for Marine Biogeochemistry, Israel, and the U.S. Department of Energy, Division of Engineering and Geosciences of the Office of Basic Energy Sciences (Brookhaven National Laboratory Contract No. DE-AC02-76CH00016). We thank Tim Eglinton and Pierre Adam, and Wilson Orr for reviews which greatly benefited the revision.

Literature Cited

1. Francois, R. *Geochim. Cosmochim. Acta* **1987**, *51*, 17-27.
2. Sinninghe Damste, J.S.; de Leeuw, J.W. *Org. Geochem.* **1989**, *16*, 1077-1101.
3. Dinur, D.; Spiro, B.; Aizenshtat, Z. *Chem. Geol.* **1980**, *31*, 37-51.
4. Mossmann, J.R.; Aplin, A.C.; Curtis, C.D.; Coleman, M.L *Geochim. Cosmochim. Acta* **1991**, *55*, 3581-3595.
5. Raiswell, R.; Bottrell, S.H.; Al-Biatty, H.J.; Tan, M.D. *Am. J. Sci.*.**1993**, *293*, 569-596.
6. Anderson, T.F.; Pratt, L.M. Chapter: *Isotope evidence for the origin of organic sulfur and elemental sulfur in marine sediments*, this volume.
7. Vairavamurthy, A.; Wilson, W.L.; Manowitz, B. Chapter: *Geochemical transformations of sedimentary sulfur: an introduction*, this volume.
8. Brassell, S.C.; Lewis, C.A.; de Leeuw J.W.; de Lange F.; Sinninghe Damste J.S. *Nature* **1986**, *320*, 643-645.
9. Goldhaber, M. B.; Kaplan, I. R. In *The Sea, Vol 5*; E. D. Goldberg, Ed.; Wiley: New York, **1974**; Vol. 5; pp 659.
10. Jørgensen, B.B. *Nature* **1982**, *296*, 643-645.
11. Fitzgerald, J.W.; Jaru, T.A.; Strickland, T.C.; Swank, W.T. *Can. J. For. Res.* **1983**, *13*, 1077-1082.
12. Aizenshtat, Z.; Stoler, A.; Cohen, Y.; Nielsen, H. In *Advances in Organic Geochemistry*, Wiley, Chichester, **1983**, 279-288.
13. Kohnen, M.E.L.; Sinninghe, Damste J.S.; ten Haven, H.L.; de Leeuw, J.W. *Nature* **1989**, *341*, 640-641.
14. LaLonde, R.T.; Ferrara, L.M.; Hayes M.P. *Org. Geochem.* **1987**, *11*, 563-571.
15. LaLonde, R. T. In *Geochemistry of sulfur in fossil fuels*; W. L. Orr and C. M. White, Eds.; Am. Chem. Soc.: Washington, D.C., **1990**; Vol. 429; pp 68-82.
16. Vairavamurthy, A.; Mopper, K. In *Biogenic sulfur in the environment*, E.S. Saltzman and W.C. Cooper, Eds.; Am. Chem. Soc.: Washington, D.C., **1989**; Vol. 393; pp 231-242.
17. de Graaf, W.; Sinninghe, Damste J.S.; de Leeuw, J.W. *Geochim. Cosmochim. Acta* 1992, **56**, 4321-4328.
18. Krein, E.B.; Aizenshtat, Z. *J. Org. Chem.* **1993**, *58*, 6103-6108.
19. Krein, E. B.; Aizenshtat, Z. *Org. Geochem.* **1994**, *21*, 1015-1025.
20. Vairavamurthy, A.; Zhou, W.; Eglinton, T.; Manowitz, B. *Geochim. Cosmochim. Acta* **1994**, *21*, 4681-4687.
21. Williamson, M. A.; Rimstidt, J. D. *Geochim. Cosmochim. Acta* **1992**, *56*, 3867-3880.

22. Goldstein, T. P.; Aizenshtat, Z. *J. Therm. Anal.* **1994**, *42*, 241.
23. Krein, E. B. In *Supplement S: The chemistry of the sulphur-containing functional groups*; S. Patai and Z. Rappoport, Eds.; Wiley & Sons: New York, **1993**; pp 975-1032.
24. Vairavamurthy, A.; Mopper, K. *Nature* **1987**, *329*, 623-625.
25. Schouten, S.; van Driel, G. B.; Sinninghe Damsté, J. S.; de Leeuw, J. W. *Geochim. Cosmochim. Acta* **1994**, *58*, 5111-5116.
26. Schouten, S.; de Graaf, W.; Sinninghe Damsté, J. S.; van Briel, G. B.; de Leeuw, J. W. *Org. Geochem.* **1994**, *22*, 825-834.
27. March, J. *Advanced organic chemistry;* 4th ed.; Wiley & Sons: New York, 1992.
28. Magnusson, B. *Acta Chem. Scand.* **1959**, *13*, 1031-1035.
29. Magnusson, B. *Acta. Chem. Scand.* **1962**, *16*, 1536-1538.
30. Reinecke, M. G.; Morton, D. W.; Del Mazza, D. *Synthesis* **1983**, 160-161.
31. Adam, P.; Mycke, B.; Schmid, J. C.; Connan, J.; Albrecht, P. *Energy Fuels* **1992**, *6*, 533-559.
32. Adam, P.; Schmid, J. C.; Mycke, B.; Strazielle, C.; Connan, J.; Huc, A.; Riva, A.; Albrecht, P. *Geochim. Cosmochim. Acta* **1993**, *57*, 3395-3419.
33. Kohnen, M. E. L.; Sinninghe Damsté, J. S.; Kock van Dalen, A. C.; de Leeuw, J. W. *Geochim. Cosmochim. Acta* **1991**, *55*, 1375-1394.
34. Kohnen, M. E. L.; Sinninghe Damsté, J. S.; Baas, M.; Kock van Dalen, A. C.; de Leeuw, J. W. *Geochim. Cosmochim. Acta* **1993**, *57*, 2515-2528.
35. Kohnen, M. E. L.; Sinninghe Damsté, J. S.; ten Haven, H. L.; Kock-van Dalen, A. C.; Schouten, S.; de Leeuw, J. *Geochimica and Geochimica Acta* **1991**, *55*, 3685-3695.
36. Licht, S.; Hodes, G.; Manassen, J. *Inorg. Chem.* **1986**, *25*, 2486-2489.
37. Rowland, S. J.; Rockey, C.;Al-Lihaibi, S.S.; Wolff, G. A. *Org. Geochem.* **1993**, *20*, 1-5.
38. Kohnen, M. E. L.; Sinninghe Damsté, J. S.; Kock-van Dalen, A. C.; ten Haven, H. L.; Rullkötter, J.; de Leeuw, J. W. *Geochim. Cosmochim. Acta* **1990**, *54*, 3053-3063.
39. Kenig, F.; Huc, A. Y. In *Geochemistry of sulfur in fossil fuels*; W. L. Orr and C. M. White, Eds.; Am. Chem. Soc.: Washington, D.C., 1990; Vol. 429; pp 170-185.
40. Rontani, J.-F.; Giusti, G. *J. Photochem. Photobiol. A:chem.* **1988**, *42*, 347-355.
41. Vairavamurthy, A.; Mopper, K.; B.F., T. *Geophysical. Res. Lett.* **1992**, *19*, 2043-2046.
42. Vairavamurthy, A.; Zhou, W.; Manowitz, B. *Di-and Polysulfide Cross linking in the Formation of Humic Polymers in Marine Sediments.* Abstracts, Am. Chem. Soc. Fall National Meeting, Washington, D.C., 1994
43. Voronkov, M. G.; Deryagina, E. N. In *Chemisty of Organosulfur Compounds - General problems*; L. I. Belen'kii, Ed.; Ellis Horwood: Chichester, 1990; pp 48.
44. Martin, G. In *Supplement S: The chemistry of the sulphur-containing functional groups*; S. Patai and Z. Rappoport, Eds.; John Wiley & Sons: Chichester, 1993; pp 395-437.
45. Vairavamurthy, A; Manowitz, B; Luther, G. W. III; Jeon, Y. J. *Geochim. Cosmochim. Acta*, **1993**, *57*, 1619-1623.

46. Krein, E. B.; Aizenshtat, Z., this volume.
47. Eglinton, T. I.; Sinninghe Damsté, J. S.; Pool, W.; de Leeuw, J.; Eijkel, G.; Bonn, J. J. *Geochim. Cosmochim. Acta* **1992**, *546*, 1545-1560.
48. Eglinton, T. I.; Sinninghe Damsté, J. S.; Kohnen, M. E. L.; de Leeuw, J.; Larter, S. R.; Patience, R. L. In *Geochemistry of sulfur in fossil fuels*; W. L. Orr and C. M. White, Eds.; Am. Chem. Soc.: Washington, D.C., 1990; Vol. 429; pp 529-565.
49. Gransch, J. A.; Posthuma, J. In *Advances in Organic Geochemistry 1973*; B. Tissot and F. Bienner, Eds.; Editions Technip: Paris, 1974; pp 727-739.
50. Orr, W. L. *Org. Geochem.* **1986**, *10*, 499-516.
51. Fukushima, K.; Yasukawa, M.; Muto, N.; Uemura, H.; Ishiwatari, R. *Org. Geochem.* **1992**, *18*, 83-91.
52. Sinninghe Damsté, J. S.; de Leeuw, J. *Intern. J. Environ. Anal. Chem.* **1987**, *28*, 1-19.
53. Barbe, A.; Grimalt, J. O.; Pueyo, J. J.; Albaigés, J. *Org. Geochem.* **1990**, *16*, 815-828.
54. Grimalt, J. O.; Yruela, I.; Saiz-Jimenez, C.; Toja, J.; de Leeuw, J. W.; Albaigés, J. *Geochim. Cosmochim. Acta* **1991**, *55*, 2555-2577.
55. Schmid, J. C.; Connan, J.; Albrecht, P. *Nature* **1987**, *329*, 54-56.
56. Sinninghe Damsté, J. S.; de Leeuw, J.; Kock-van Dalen, A. C.; de Zeeuw, M. A.; de Lange, F.; Rijipstra, W. I. C.; Schenck, P. A. *Geochim. Cosmochim. Acta* **1987**, *51*, 2369-2391.
57. Payzant, J. D.; McIntyre, D. D.; Mojelsky, T. W.; Torres, M.; Montgomery, D. S.; Strausz, O. P. *Org. Geochem.* **1989**, *14*, 461-473.
58. Ho, T. Y.; Rogers, M. A.; Drushel, H. V.; Koons, C. B. *The American Association of petroleum Geologists Bulletin* **1974**, *58*, 2338-2348.
59. Bolshakov, G. F. *Sulfur Rep.* **1986**, *5*, 103-393.
60. Coleman, H. J.; Hopkins, R. L.; Thompson, C. J. *Int. J. Sulfur Chem. B* **1971**, *6*, 41-61.
61. Rall, H. T.; Thompson, C. J.; Coleman, H. J.; Hopkins, R. L. *Sulfur compounds in crude oil; US Bureau of Mines, Bulletin 659*; US government printing office: Washington D.C., 1986, pp 187.
62. Czogalla, C. D.; Boberg, F. *Sulfur Reports* **1983**, *3*, 121-167.
63. Eglinton, T. I.; Sinninghe Damsté, J. S.; Kohnen, M. E. L.; de Leeuw, J. *Fuel* **1990**, *69*, 1394.
64. Sinninghe Damsté, J. S.; Eglinton, T. I.; Rijpstra, W. I. C.; de Leeuw, J. In *Geochemistry of sulfur in fossil fuels*; W. L. Orr and C. M. White, Eds.; Am. Chem. Soc.: Washington, D.C., 1989; Vol. 429; pp 486.
65. Przewocki, K.; Malinski, E.; Szafranek, J. *Chem. Geol.* **1984**, *47*, 347.
66. Pryor, W. A. *Mechanisms of sulfur reactions*; McGraw-Hill: 1962.
67. Cohen, Y.; Aizenshtat, Z. *J. Anal. Appl. Pyrolysis* **1994**, *28*, 231.
68. Cohen, Y.; Aizenshtat, Z. *J. Anal. Appl. Pyrolysis* **1993**, *27*, 131.
69. Stoler, A. Thesis, Hebrew University of Jerusalem, 1990.
70. Orr, W. L. *Bull. Am. Assoc. Geol.* **1974**, *50*, 2295-2318.
71. Rospondek, M. J.; de Leeuw, J. W.; Baas, M.; van Bergen, P. M.; Leereveld, H. *Org. Geochem.* **1994**, *21*, 1181-1191.

72. Toland, W. G. *J. Am. Chem. Soc.* **1960**, *82*, 1911.
73. Aizenshtat, Z. In *Mineral/matrix effects in organic geochemistry - ACS 205th meeting*; Am. Chem. Soc.: Denver, 1993; Abstract 43.
74. Benson, S.W. *Chem. Rev.* **1978**, *78*, 23-35.
75. McMillan, D.F.; Golden, D.M. *Ann. Rev. Phys. Chem.* **1982**, *33*, 493-532.
76. Bond Dissociation Energies in Simple Molecules; Darwent, D. de B.; NSRDS-NBS, 31; National Bureau of Standards: Washington, D.C., 1970.
77. Nicovich, J.M.; Kreutter, K.D.; van Dijk, C.A.; Wine, P.H. *J. Phys. Chem.* **1992**, *96*, 2518-2528.
78. Griller, D.; Simöes, J.A.M.; Wayner, D.D.M. In *Sulfur Centered Reactive Intermediates in Chemistry and Biology*; Chatgilialogu, C; Asmus, K., Eds.; Plenum Press: New York, 1989; pp 37-52.
79. Hawari, J.A.; Griller, D.; Lossing, F.P. *J. Am. Chem. Soc.* **1986**, *108*, 3273-3275.
80. Lipschitz, A. The Hebrew University of Jerusalem, Jerusalem, personal communication, 1995.

RECEIVED July 7, 1995

Chapter 3

Sulfur Transformations in Early Diagenetic Sediments from the Bay of Concepcion, Off Chile

Murthy A. Vairavamurthy[1], Shengke Wang[1], Bandana Khandelwal[1], Bernard Manowitz[1], Timothy Ferdelman[2], and Henrik Fossing[2]

[1]Department of Applied Science, Geochemistry Program, Applied Physical Sciences Division, Brookhaven National Laboratory, Upton, NY 11973
[2]Department of Biogeochemistry, Max Planck Institute for Marine Microbiology, Fahrenheitstrasse 1, W–28359 Bremen, Germany

Despite the recognition that both organic sulfur and pyrite form during the very early stages of diagenesis, and that the amount of H_2S generated in bacterial sulfate reduction primarily limits their formation, the mechanisms and the active species involved still are not clear. In this study, we quantified the major forms of sulfur distributed in sediments to assess the geochemical mechanisms involved in these transformations. XANES spectroscopy, together with elemental analysis, were used to measure sulfur speciation in the organic-rich sediments from the Bay of Concepcion, Chile. Organic polysulfides constituted the major fraction of the organic sulfur, and occurred maximally just below the sediment surface (1-3 cm), where intermediates from H_2S oxidation were likely to be generated most abundantly. Sulfonates, which could be formed through the reactions of sulfite and thiosulfate, also showed a sub-surface maximum in the vicinity of the "oxic-anoxic interface". These results strongly suggest a geochemical origin for organic polysulfides and sulfonates, and illustrate that intermediates from H_2S oxidation play a dominant role in incorporating sulfur into organic matter. Pyrite was absent in the surficial layer, and first appeared just below the H_2S maximum, where organic polysulfides began to decrease in abundance. From these results, we argue, that an iron monosulfide precursor formed first from reactions with H_2S, and then reacts with organic polysulfides, completing the synthesis of pyrite in the sediment column.

Sulfur transformations are closely coupled to the diagenesis of organic matter in reducing organic-rich sediments. In the absence of oxygen, caused by its rapid consumption near the sediment-water interface by a large flux of organic matter into the sediments, sulfate becomes the dominant terminal electron acceptor to oxidize organic matter, producing hydrogen sulfide (1-3). The hydrogen sulfide so formed

0097–6156/95/0612–0038$12.25/0

can react in several ways, depending on environmental conditions, but there are three major pathways (I) oxidation, (II) formation of pyrite, and (III) formation of organic sulfur. Oxidation through chemical and biological pathways is prominent at oxic-anoxic interfaces where sulfide encounters oxidants, such as oxygen and oxidized metal ion species (for example, MnO_2) (4). This oxidation converts hydrogen sulfide back to sulfate, but also forms several partial oxidation products, such as elemental sulfur, polysulfides, sulfite, thiosulfate, and polythionates (5-8). Since most of these intermediates, particularly polysulfides, sulfite, and thiosulfate, are strong sulfur nucleophiles, they form an important part of the sedimentary pool of reactive sulfur species, in addition to hydrogen sulfide. In the presence of reactive iron, principally occurring as ferric oxide and hydroxide coatings on clay particles, the reduced sulfur species react with it to form unstable iron sulfides (greigite and mackinawite) first, and eventually pyrite (2,3,9). Similarly, organic sulfur compounds are formed when reduced inorganic sulfur species react with functionalized organic molecules (10-12). There is a competitive balance between the reactions of reduced sulfur species with either iron minerals or with reactive organic matter. The amount of sulfur in organic matter usually is enhanced in iron-poor sediments, such as those of carbonate platforms, and also in clastic-starved basins which accumulate pelagic materials such as algal (e.g. cocolithophores, diatoms) and radiolarian oozes. Thus, it is believed that reactive iron competes faster than organic matter for the available reduced sulfur (13). Because sulfate reduction occurs mainly in near-surface sediments, it is believed that reactive sulfur becomes locked up in pyrite and organic sulfur during the very early stages of diagenesis. However, the actual mechanisms and the active species involved in both pyrite and organic sulfur formation are unclear although there is compelling evidence from field observations and laboratory experiments that partial oxidation intermediates, such as polysulfides, are important (14-21). Early incorporation of sulfur into organic matter is of particular significance because sulfur binding can affect the subsequent susceptibility of organic molecules to microbial degradation, and, thus, can aid in preserving organic matter in sediments (22).

In this study, we focused on measuring sulfur speciation in the organic-rich sediments from the Bay of Concepcion, a highly productive area of approximately 120 km^2 adjacent to the seasonal upwelling front off Chile; our aim was to better understand the early diagenetic transformations involving organic sulfur. We used x-ray absorption near-edge structure (XANES) spectroscopy, in concert with elemental analysis, to obtain a detailed, down-core inventory of the major forms of sulfur. In general, the highly organic-rich sediments off the coast of Peru and Chile are considered ideal sites for obtaining important insights about early diagenetic transformations of reduced sulfur, and particularly, about sulfur-organic matter interactions (23). An important feature of these sediments is the presence of a dense populations of colorless sulfur bacteria, *Beggiatoa* and *Thioploca sp.*, that form thick mats covering the surface of sediments (24-25). These filamentous bacteria are characterized by rich accumulations of granules of elemental sulfur, formed by the oxidation of hydrogen sulfide in energy-related conversions by the bacteria (26). Polysulfides are likely to be formed under such conditions through the reactions of elemental sulfur and hydrogen sulfide, especially near the surface. Since polysulfides are implicated as active reactants in both pyrite formation and sulfur incorporation into organic matter, measurements of sulfur speciation in the sediments of the Bay of Concepcion might give us important insights about the partitioning of polysulfides between these pathways.

Samples and Methods of Analysis

Sediment Samples. In March/April 1994, scientists from the Max Plank Institute for Marine Microbiology, Bremen, Germany and the University of Concepcion, Concepcion, Chile organized a cruise onboard R/V Vidal Gormaz (The Thioploca Cruise 1994) to undertake a detailed, interdisciplinary study on the biology and chemistry of the sediments in the upwelling region off the coast of Chile. A sediment core obtained from Station 6 (36^0 36' 5' S and 73^0 00' 7' W) in the Bay of Concepcion was used in this study (Figure 1). The water depth at this Station was 34 m. The bottom water conditions at the time of collection were: temperature 11.4^0C; Salinity 34 $^o/_{oo}$; oxygen 14 μM. The core was taken using a composite Muticorer around 7 pm on March 16, 1994 and stored in incubator at 5 ^{0}C until noon the next day before sectioning. The sediments from Sta. 6 were covered on the surface by a dense mat of the filamentous bacteria, mainly *Beggiatoa sp.* interspersed with *Thioploca* sp. The bacterial mat was underlain by a soupy, black sedimentary ooze that was interwoven with bacterial filaments. Below ca. 4 cm, the sediment appeared grayish. Consolidated clay began around 18 cm. The core was sectioned in a nitrogen-filled glove-bag on the ship. From the surface to 10 cm down, the core was sectioned at one cm intervals. Below this depth, it was cut at two cm intervals down to the bottom at about 38 cm. The series of sediment sections were stored in air-tight bags and kept frozen until analysis.

The samples were water-washed, and dried before analysis. The samples were prepared in this way because (1) washing with water removes sulfate, the abundant form of sedimentary sulfur, that can cause errors in the XANES determination of the low levels of oxidized sulfur forms (e.g. sulfonates), (2) removing water concentrates the samples components, thus improving the accuracy of their measurement, and (3) drying the samples gives a more uniform basis for comparing the variations between sequences. Our previous studies indicated that washing with water does not affect the composition of major sinks for reduced sulfur, pyrite, and macromolecular organic sulfur. The samples were washed with water in a nitrogen glove-box by mixing them with de-aerated water which then was removed by centrifugation. The water-washed sediments were dried at 30^{0}C in a vacuum oven connected to a supply of nitrogen gas. Initially, the oven was flushed with nitrogen gas to ensure that air was completely removed. After drying, the samples were stored in a nitrogen atmosphere or in a freezer.

Analysis. Three major techniques were used to analyze the samples (1) elemental analysis for C, N, and S, (2) gas chromatography for elemental sulfur, and (3) XANES spectroscopy for quantifying the major forms of sulfur. By combining information from these analyses, we constructed a detailed down-core inventory of the major forms of sulfur.

Elemental Analysis. Total elemental composition for C, N, and S was determined using a Perkin Elmer Model 2400 CHNS elemental analyzer.

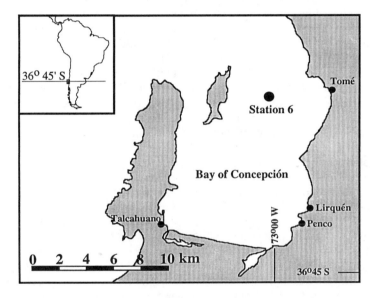

Figure 1. Site map showing the location of Station 6 of the "Thioploca Cruise 1994" in the Bay of Concepcion, off the coast of Chile.

Gas Chromatographic Analysis of Elemental sulfur. We used capillary gas chromatogra-phy and electron-capture detection for determining elemental sulfur. Essentially, we adapted the method by Chen et. al. (*27*). In this study, we used a 15 m DB-5 megabore column (JW Scientific) connected to a 2 m de-activated quartz megabore tubing for GC separation; the carrier gas was He, and the make-up gas was N_2. The GC conditions were: initial temperature and hold time: 140 ^{0}C, 2 min; rate: 8 ^{0}C/min; final temperature and hold: 250 ^{0}C, 4 min. Under these conditions, with a 1 μL volume of a toluene extract injected, we obtained linear response up to ca. 50 μM concentration. The samples were diluted to be within this range.

XANES spectroscopy. XANES (X-ray Absorption Near-Edge Structure) spectroscopy is a valuable tool for sulfur speciation because, the K-edge XANES spectra (produced by 1s electronic transitions) of the variety of sulfur forms each are richly endowed with characteristic features, including the edge energy, which facilitates qualitative recognition among various oxidation states and structures (see Figure 2). This technique is element specific, and non-destructive, and does not require cumbersome preparation of the sample before analysis. Thus, XANES spectroscopy recently emerged as an important tool for determining sulfur speciation in a variety of geochemical samples including coal, petroleum, and sedimentary rocks (*28-31*). For sulfur speciation in sediments, XANES is superior to other common techniques, such as chromatography, mainly because it provides (1) simultaneous information (both qualitative and quantitative) on all the sulfur forms present, and (2) the ability to analyze whole sediments.

The sulfur K-edge XANES spectra were collected as fluorescence excitation spectra at the National Synchrotron Light Source (NSLS) X-19A beam line at the Brookhaven National Laboratory. The spectra were recorded so that the scanning procedure yielded sufficient pre-edge and post-edge data for precisely determining the background, which is needed for analysis. The x-ray energy was calibrated using XANES spectra of elemental sulfur measured between sample runs, assigning 2472.7 eV to the peak of the spectrum for elemental sulfur. The uncertainty of the energy calibration was less than $\pm$ 0.15 eV, determined by comparing the spectra of model compounds obtained at different times. A non-linear least-squares fitting procedure, using linear combinations of normalized spectra of model compounds, gave quantitative information of the different sulfur species present (*7,28*).

Results

Elemental Analyses: Depth Profiles for Carbon, Nitrogen, and Sulfur. The depth distribution for carbon and nitrogen in water-washed sediments are very similar to each other (Figure 3). Their solid phase concentrations decrease rapidly at first in the top 5 cm of the sediment, then slowly from 5 to 10 cm depth, and somewhat flatten out below 10 cm depth. As indicated by the fit, the overall decrease appears to follow an exponential pattern of decay. The approximately constant level from 10 cm downwards suggests that remineralization of organic carbon and organic nitrogen mostly occurs within the top 10 cm of the sediment column, and that the organic matter remaining ca. 10 cm below surface is somewhat non-reactive and refractory. After correcting for carbon contained in $CaCO_3$ (as described below), we

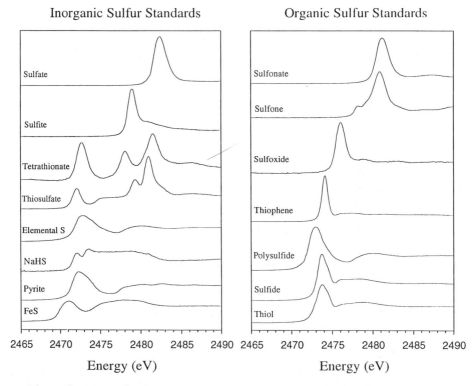

Figure 2. Normalized XANES spectra of various inorganic and organic sulfur compounds showing their characteristic features including edge energies.

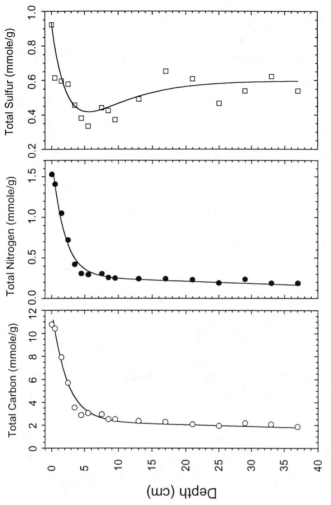

Figure 3. Down-core distribution of total carbon, nitrogen, and sulfur in water-washed and dried sediments from the Bay of Concepcion, off Chile.

estimated that about 13% of the initially deposited organic carbon is (or becomes) refractory; the nitrogen data also shows a similar distribution.

Figure 4 plots the total concentrations of carbon versus nitrogen, and shows the excellent correlation between their distribution, which is reflected by the correlation coefficient (r^2) of 0.997 for the linear regression line. The regression gives a positive intercept of 0.76 in the axis for carbon, suggesting that a fraction of the sedimentary carbon is independent of the nitrogen distribution. This is not surprising because a part of the sedimentary carbon usually is deposited as $CaCO_3$, derived from the exoskeletal tissues of organisms. We confirmed that $CaCO_3$ caused this intercept by determining the elemental values for some acid-washed (3M HCl) sediment samples that showed a regression passing through the origin. Therefore, to calculate the values for total organic carbon, we subtracted this amount (0.76 mole C g^{-1}) from the total C values.

The slope of the regression line, which gives the atomic ratio for C/N in the sedimentary organic matter, is 6.72. This value agrees excellently with the Redfield ratio of 6.6 (106/16) which is the ratio of these elements in the primary organic matter (phytoplankton) that source the sediments. Similar results also were reported for Cape Lookout Bight sediments by Martens et al. (*32*) who obtained an average value of 6.6 for C/N. However, Froelich et al. (*33*), who analyzed several sediment samples from the Peru margin, reported an atomic C/N value of 10.6 for all samples, including those in the upper few centimeters of the sediment; therefore, they speculated that the C/N ratio is somewhat reset a little from the planktonic value immediately upon incorporation into the sediment column. The reason for this discrepancy between Froelich et al.'s value and the data from Martens et al. and our work is unclear. The maintenance of the Redfield ratio for C/N in the solid phase throughout the entire depth of the sediment column suggests that processes controlling the diagenesis of organic matter in the Bay of Concepcion have not altered the relative abundance of carbon to nitrogen in the bulk insoluble organic matter. Furthermore, these results suggest that nitrogen diagenesis, like that of carbon, involves mainly degradation, and does not include the production of new organic nitrogen.

For total sulfur, the distribution in water-washed sediments is similar to those of carbon and nitrogen in the top 5 cm of the sediment column, showing a maximum at the surface, and a rapid decline in concentration from the surface to the 5 cm depth. However, below 5 cm, the sulfur profile differs markedly from those of the other two elements; the concentration increases again, indicating continued accretion of insoluble sulfur in deeper sediments. In sediments, pyrite usually remains stable after its formation. Since the concentration of total sulfur decreases within the top 5 cm , we think that pyrite was not formed appreciably in this top sediment column. Thus, organic sulfur and /or elemental sulfur should comprise the total sulfur in the top 5 cm of the sediment column. The diagenetic formation of pyrite might contribute to the increase in sulfur content below 5 cm depth.

Distribution of Elemental Sulfur. Measurements with gas chromatography revealed that elemental sulfur was present at relatively high concentrations in the surfacial layers of the sediment (Figure 5). Probably a major source of this sulfur is the filamentous sulfide-oxidizing bacteria, *Beggiatoa* and *Thioploca* sp, which form thick

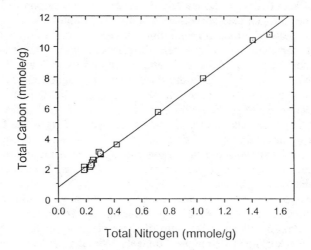

Figure 4. Plot of solid-phase total carbon versus total nitrogen for a series from the surface to 38 cm depth.

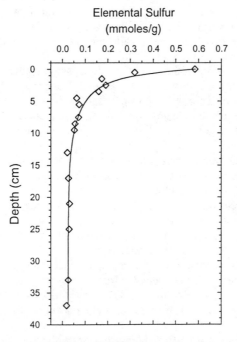

Figure 5. Down-core distribution of solid-phase elemental sulfur determined by gas chromatography and electron capture detection.

mats covering the sediments of the upwelling region off the coast of Peru and Chile (*24-25*). At Sta. 6, the mat was composed principally of *Beggiatoa sp.* These bacteria produce elemental sulfur by oxidizing hydrogen sulfide in a chemoautotrophic mode of energy metabolism. The sulfur then is stored internally within invaginations of the cell membrane and the cell wall (*26*). The burial of the sulfide oxidizing bacteria can contribute to a high content of elemental sulfur in the sediment. In fact, the surficial layer of the sediment column with bacterial mat contained the maximum amount of sulfur (0.6 mmole per g dried sample corresponding to 63 % of total sulfur), supporting the view that sulfide-oxidizing bacteria generate the elemental sulfur in these sediments. Similarly, Schimmelmann and Kastner suggested that *Beggiatoa sp.* directly contributed to the large amounts of sulfur preserved in the varved sediments of the Santa Barbara Basin, California (*34*). However, our results suggest that significant amounts of elemental sulfur were not formed in the sediments studied here. The depth distribution shows an exponential degradation of sulfur in the top sediment column (down to ca. 10 cm) that matches closely those of carbon and nitrogen. From ca. 10 cm downwards, the concentration remains at a steady state level of 0.1 % (w/w) .

We point out that there is some uncertainty in the elemental sulfur measurements due to the possibility of it being formed from the oxidation of labile sulfides (for example, FeS) during preparation of the samples. However, because we carefully processed the samples under anaerobic conditions in a glove box, we believe that such artifactual sulfur was minimal, if it at all occurred.

XANES Analysis: Speciation of Major Forms of Sulfur. The XANES spectra of the water washed samples for the station 6 depth sequences were very similar to those we reported previously for sediments from other organic-rich locations (*35*). As Figure 6 shows for some samples, the spectra consisted of two absorption bands, the first one in the energy range of 2471-2475 eV, corresponding to reduced forms of sulfur , and the second in the range of 2480-2485 eV corresponding to oxidized forms of sulfur. In general, our aim was to deconvolute the spectra to quantitatively estimate the major forms of sedimentary sulfur, such as pyrite, elemental sulfur, organic polysulfide, sulfonate, and sulfate. This analysis was carried out using a non-linear least squares fitting procedure in which the spectra of samples are fitted with various linear combinations of the spectra of appropriate model compounds, and the results of the best fit are taken to indicate the actual sulfur composition. The choice of fitting models was discussed in a previous study (*35*).

We obtained the best fits to the spectra of the samples with three models, pyrite, elemental sulfur, and an organic polysulfide (nonyl polysulfide) for the first absorption band, and two models, sulfonates and sulfate, for the second absorption band. The fitting is unambiguous for sulfonates and sulfate because these models have peaks that are reasonably well separated (Figure 2). However, a major concern here is the accuracy of the XANES estimates for elemental sulfur and organic polysulfides. These compounds have very similar spectra with peaks at almost the same energy position (Figure 2), hence they can affect each other mutually on fitting, causing one to increase and the other one to decrease simultaneously. Therefore, we totaled these values, and took their sum as a reliable estimate for organic polysulfides and elemental sulfur combined (total[organic+elemental]). Then, we calculated the

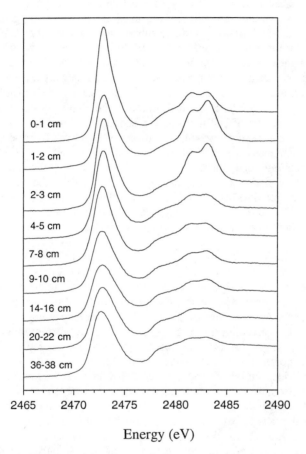

Figure 6. XANES spectra showing variations in spectral features for the depth series.

relative concentrations of organically bound sulfur by subtracting the experimental values for elemental sulfur from the total[organic+elemental] estimates. The fitting results for pyrite are expected to be reasonably accurate because its spectrum is further apart from the others used for fitting, and also it differs in shape. Although organic polysulfides accounted for all the organic sulfur in sediments in our XANES fitting, it is likely that these estimates also include other organic sulfur forms, such as thiols, and disulfides. It is difficult to estimate the abundance of minor organic constituents separately by XANES fitting mainly because most reduced organic sulfur forms have XANES peaks that are not widely separated. Nevertheless, our results suggest that organic polysulfides are the dominant forms of organic sulfur in early diagenetic sediments, which is in agreement with previous studies (20,36).

The relative depth distribution of organic polysulfides shows a maximum near the surface, in the top 5 cm of the sediment column (Figure 7). In the region below, the relative abundance of this polysulfidic sulfur decreases first down to a depth of ca. 10 cm, but then remains almost constant down to the bottom of the core. Pyrite is not present near the sediment-water interface, and appears first ca. 3 cm below surface, where the concentration of organic polysulfides begins to decrease. Below this level, pyrite builds up until a maximum is obtained around 10 cm depth, and after which the concentration remains constant with depth, reflecting steady-state conditions. In deeper sediments, pyrite represents the most abundant sulfur form accounting for almost 70 % of the total solid phase sedimentary sulfur; organic sulfur constitutes about 20 % of the total sulfur at corresponding depths.

Our XANES analysis of the water-washed sediments of the Bay of Concepcion also show sulfonates composing about 30 % of the total organic sulfur. As Figure 7 shows, the distribution of sulfonates reaches a sub-surface maximum about 2 cm below the sediment-water interface, and then remains nearly constant to the bottom (Figure 7c). These results agree with our recent study on sulfonates in which we described sulfonates as a novel class of sedimentary organic sulfur compounds, constituting about 20-40% of the total organic sulfur in near-surface marine sediments from various organic-rich locations (35). The sulfonates were associated mainly with the sedimentary solid phase, and were not detected in the sediment pore water. We also detected some sulfate in water-washed sediments at concentrations comparable to those of sulfonates. This sulfate is likely to be associated with $CaCO_3$ in sediments; several studies recently have demonstrated sulfate inclusions in sedimentary carbonates (37). Furthermore, we observed that in some samples treated with hydrochloric acid, the signal ascribed to sulfate disappeared completely.

Discussion

Sulfur Nucleophiles. The presence of sulfur nucleophiles in sediments is crucial for the diagenetic formation of organic sulfur and pyrite. The abundant sulfate ion in seawater is chemically non-reactive under ambient sedimentary conditions, and thus, is not involved in any diagenetic conversions other than dissimilatory sulfate reduction. It was suggested that, in soils, sulfate forms organic sulfate esters through bacterial mediation (38), but it is not certain whether such a process occurs in marine sediments. The primary, and the most abundant, sulfur reactant in sedimentary systems is hydrogen sulfide formed from sulfate reduction. The depth

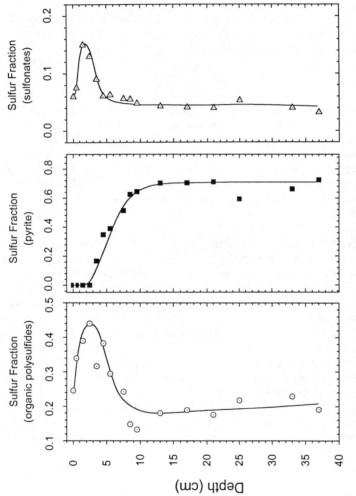

Figure 7. Depth variations in the relative amount of sulfur contained in organic polysulfides, pyrite, and sulfonates.

profiles for carbon and nitrogen showing a rapid decrease in the top 5 cm of the column (Figure 3) apparently suggest that bacterial sulfate reduction that generates H_2S should be at maximum in this zone of the sediment.   In fact, direct measurements of total sulfate reduction rates using the $^{35}SO_4^{2-}$ tracer injection method showed maximal rates just below the sediment surface between 0-2 cm (Figure 8). However, measurements of hydrogen sulfide concentrations in pore water showed a maximum (about 1.3 mM) at a depth of 3.5 cm (Figure 8).   (A detailed discussion of the sulfate reduction rates and the pore water H_2S concentrations will be published elsewhere; Ferdelman et al., manuscript in preparation).   Thus, it appears that the hydrogen sulfide generated from the reduction of sulfate by bacteria was transformed rapidly into other species in the top 0-3 cm of the sediment column.   A major pathway for H_2S transformations near the oxic-anoxic interface is oxidation through reactions with sedimentary metal (Fe, Mn) oxides or with molecular oxygen diffusing downwards from the water column just above the sediment-water interface(4).   As we noted earlier, oxygen was present at measurable levels in the water column above the sediment surface at Station 6.   The oxidation of hydrogen sulfide leads to the formation of several partial oxidation products, including polysulfides, sulfite, and thiosulfate, that also are active species.   In general, the type and abundance of H_2S oxidation   products formed are controlled largely by environmental variables, such as pH.   Under conditions typical of marine sediments (excess sulfide and limiting oxygen concentrations, and pH near neutral), thiosulfate is the dominant product; the formation of (inorganic) polysulfides was favored by catalysis of transition metal ions, such as Fe^{3+} and Ni^{2+} (7). (Hereafter we will refer to inorganic polysulfides as polysulfides, and organic molecules with polysulfide moieties as organic polysulfides).   Consequently, products of H_2S oxidation, including polysulfides, are likely to be at a maximum in the top layers of the sediment between the surface and the depth of sulfide maximum, about 3 cm below the sediment surface.

The polysulfides formed from sulfide oxidation can generate elemental sulfur and HS^- through an equilibrium reaction (equation 1).   Because of this

$$S_n^{2-} + H^+ \quad <------> \quad HS^- + (n-1)/8 \; S_8 \qquad (1)$$

equilibrium, polysulfides also are generated when elemental sulfur is formed in sediments by either microbial or chemical oxidation of hydrogen sulfide.   This latter pathway could be important for forming polysulfides in the sediment core studied here because, as discussed, significant amounts of elemental sulfur was formed in the surficial layers of the sediment by the filamentous sulfide-oxidizing bacteria living at the interface.   Consequently, the maximum level of polysulfides generated in this pathway   might occur   ca. 2 cm   below the sediment-water interface because, it is expected usually to lie between the maxima of elemental sulfur and H_2S.   Thus, polysulfides formed from this elemental sulfur pathway would be at nearly the same depth as those formed directly from the oxidation of H_2S.

Diagenetic Formation of Organic Sulfur and Pyrite: Mechanisms and Timing.
Using the data on the down-core distribution of the various sulfur forms, we assessed (1) the relative timing of   sulfur incorporation into organic matter vs. pyrite formation, and (2) which sulfur species were important in these processes.   To put our inferences in the proper perspective, we first give a brief background about sulfur incorporation into organic matter and pyrite formation.

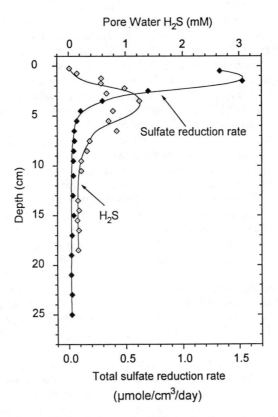

Figure 8. Variation with depth in total sulfate reduction rate and porewater H_2S concentration. $^{35}SO_4^{2-}$ - tracer injection method and the methylene blue spectrophotometric method were used for determining the sulfate reduction rate and the pore water H_2S concentration respectively; experimental details are as described elsewhere (53,54).

Mechanisms of sulfur incorporation into organic matter: background. Sulfur is incorporated into organic matter when appropriate functionalized organic molecules come into contact with sulfur nucleophiles. Organic molecules containing activated unsaturated bonds (for example, unsaturated carbonyl compounds) have been considered important because reactions between them and sulfur nucleophiles can proceed under ambient sedimentary conditions via the Michael addition mechanism (*10*). Laboratory simulations indicated that compounds containing non-activated, but conjugated multiple double bonds and those with carbonyl groups react with sulfur nucleophiles under phase-transfer conditions, which probably exist in microenviron-ments in the sedimentary solid phase (*16-18*).

Since HS^- is the most abundant sulfur nucleophile in sedimentary systems, it is expected to play an important role in organic sulfur formation in sediments. Thiols are the initial products from the reactions of HS^- with organic molecules (10). However, there is growing evidence that H_2S oxidation intermediates, particularly polysulfides, also are important. Because organic polysulfides, and thiophenic compounds formed from their thermal transformations, are the dominant organic sulfur forms in sedimentary systems, reactions involving polysulfides were considered even more important than those of HS^- (*14-18,36,39-41*) Aizenshtat et al. were among the first to suggest that polysulfides are involved in secondary sulfur enrichment, based on studies in the Solar-Lake sediments (*36*). Kohnen et al. identified cyclic di- and tri-sulfides with C_{20} isoprenoid and linear carbon skeletons in Quaternary sediments (*20*). Laboratory kinetic studies of the reactions of tetrasulfide (S_4^{2-}) and bisulfide with simple molecules containing activated unsaturated bonds (acrylic acid, acrylonitrile) clearly indicated that polysulfide ions are more reactive than bisulfide in forming organosulfur compounds under conditions typical of marine sediments (*14*). Also, Francois observed that polysulfides were more reactive than bisulfide ions in incorporating sulfur into humic substances (*19*). Recently, laboratory simulations provided strong support for the involvement of polysulfides in generating many organic sulfur compounds, including organic polysulfides and thiophenes, identified in various sulfur-rich sediments and oils (*14-18*). For example, the reaction of polysulfides into phytenal (or even phytol) was shown to be a route for the formation C_{20} isoprenoid polysulfides (*16,17*), and thiophenes (*41*).

Although polysulfides are considered important for sulfur incorporation into organic molecules, the role of other H_2S oxidation intermediates is unclear, particularly, oxygenated sulfur nucleophiles such as sulfite and thiosulfate. However, Vairavamurthy et al. recently suggested that sulfite and thiosulfate may participate in incorporating of sulfur into sedimentary organic matter, generating sulfonates (*35*). Elemental sulfur cannot react directly, although it may be involved indirectly through the formation of polysulfides when H_2S is present.

Pyrite formation: background. Because pyritic sulfur usually has the same isotopic composition as the hydrogen sulfide in pore-water, the latter is believed to be the only sulfur source for pyrite (*23,42,43*). The reaction between HS^- and Fe^{2+} should involve an oxidation step because the pyritic sulfur is at -1 oxidation state, whereas the sulfur in H_2S is at -2 state. Various pathways were proposed for this redox change. In one pathway, Fe^{2+} reacts with H_2S to form FeS or a soluble bisulfide complex (for example, $Fe(SH)^+$) first, which then reacts with H_2S oxidation products, elemental sulfur or polysulfides, to form pyrite (9,44-46). A second pathway considers the direct formation of pyrite from the reaction of polysulfides (or elemental sulfur and HS^-) with Fe^{2+} in the presence of HS^- (21,45,47). However, laboratory studies suggest that pyrite never forms directly from the reaction of polysulfides with Fe^{2+}. An iron monosulfide precursor (FeS and/or iron sulfide complex ($Fe(SH)^+$) always forms first, which then reacts again with polysulfides to form pyrite (21,46). It was suggested that the second step proceeds through a cyclic intermediate (21). In both pathways, the presence of an oxidized form of sulfur, the polysulfides (or elemental sulfur and HS^-), is a prerequisite for forming pyrite. However, a third pathway was described which does not require elemental sulfur or polysulfides; instead, an iron-bisulfide complex first forms, and then undergoes an internal redox change to form FeS_2, accompanied by the production of molecular hydrogen (48).

Inferences from this study. Our present results, and earlier ones indicate that organic polysulfides are the dominant fraction of organic sulfur in early diagenetic sediments. If organic polysulfides form by the geochemical pathway involving the reaction of (inorganic) polysulfides with organic molecules, then they must occur maximally in the very top of the sediment column where polysulfides are generated. In fact, the distribution of organic polysulfides show maximum around 2.5 cm depth, where rates of H_2S oxidation are expected to be at maximum, that strongly suggests they were formed by this geochemical pathway. Similarly, the subsurface maximum for sulfonates appearing just above that of organic polysulfides also indicates their geochemical origin. It was suggested that sulfonates could be generated by reactions involving sulfite and thiosulfate, although they also could have a biological origin, derived from sinking particles or benthic biomass (35). However, in the latter case, the maximum would be expected to span right from the surface, rather than to be present as a subsurface peak. Thus, sulfonates appear to be formed by the geochemical pathway in the sediment core we studied here. These results, strongly support the view that the partial oxidation products of H_2S play a dominant role in incorporating sulfur into sedimentary organic matter. Since these oxidation products are generated maximally at the oxic-anoxic interface, which usually lies above the hydrogen sulfide maximum, the oxidation intermediates will have the first chance to react with reactive molecules present in the deposited organic matter rather than hydrogen sulfide. This hypothesis may explain the dominance of organic polysulfides in geochemical systems.

According to our XANES results, pyrite first appears at ca. 3 cm depth, where organic polysulfides start to decrease in abundance. As we argued previously, the depth profile for total sulfur also shows the same trend, adding confidence to the XANES data. The appearance of pyrite after organic polysulfides is intriguing because pyrite is expected to be formed before organic sulfur as iron usually outcompetes functionalized organic molecules for available sulfur nucleophiles (13). For sediments from the Miocene Monterey Fm., it was suggested that pyrite preceded the formation of organic sulfur based on sulfur isotope and elemental distribution (C,S) (43). As discussed, the polysulfides that we presumed to have reacted with organic molecules to form organic polysulfides also are known to be the most important reactants for forming pyrite through reactions with iron. We do not think that iron limited the formation of pyrite in the top sediment column because pyrite rapidly builds up below 4 cm.   A possible short-term explanation for the formation of organic polysulfides before pyrite is that reactive organic molecules provide a more stable sink than iron for reactions with polysulfides in the sediments studied here. Indeed, laboratory studies suggest that pyrite is not formed rapidly from polysulfide reactions.   For example, Luther observed that 3-4 months were required for near complete (95%) pyrite formation from the reaction of equimolar quantities of Fe(II) with either S_2^{2-} or S_5^{2-} (*21*).   Thus, organic molecules may trap polysulfides faster than iron.   To test this hypothesis, we need not only the kinetic data of the relevant reactions, but also a detailed knowledge of (1) the reaction mechanisms, and (2) the chemical nature of polysulfides, functionalized organic molecules, and reactive iron present in sediments.

The view that polysulfides react with organic matter first to form organic polysulfides is consistent with the suggestion by Mossman et al. for Peru margin sediments, who argued, from considerations of sulfur isotopes and mass balance, that organic polysulfides are formed in the uppermost surfacial layer of the sediment (*23*). The presence of abundant organic polysulfides in the top sediment column is also in agreement with an earlier study by Vairavamurthy et al. who suggested that a major fraction of reactive polysulfides in near-surface sediments is bound to the particle phase (*15*).   They believed that a significant fraction of such particle-bound polysulfides have one end of their chains linked to the particulate phase so that the other end is available for reactions (for example, with added organic molecules). Recently, it was suggested that humic coatings on mineral particles could serve as attachment points for these particle-bound polysulfides (*49*).   Similarly, the organic polysulfides in the top part of the sediments studied here could also be present as particle-bound polysulfides attached to humic coatings on mineral particles.   In agreement, studies by Francois (*19*) and Ferdelman et al. (*50*) showed that humic substances extracted from near-surface coastal and saltmarsh sediments were highly enriched in sulfur, with maximum in the top sediment column in the region of the oxic-anoxic interface.   In fact, Francois ascribed reactions involving polysulfides as mainly responsible for this sulfur enrichment (*19*).

In the sediment core we studied, pyrite formation probably proceeded through reaction with H_2S, first forming an iron monosulfide precursor because (1) pyrite only began to increase just below the depth of hydrogen sulfide maximum, and (2) pyrite was not detected in the region of polysulfide maximum.   As we pointed out earlier, after iron monosulfides form, polysulfides (or elemental sulfur and HS^-) are required to complete pyrite synthesis, unless the iron monosulfide precursor undergoes an internal redox change.   Because the build up of pyrite appears to correlate with a decline in organic polysulfides, the latter may provide the sulfur to generate pyrite from this precursor.   A problem in this explanation is the inability of the organic polysulfides to react with FeS if both of them exist in the sedimentary solid phase as particle-bound organic polysulfides, and solid-phase FeS.   However, because FeS exists in equilibrium with soluble complexes, such as $Fe(SH)^+$, the reaction can proceed between solid-phase organic polysulfides and iron sulfide complexes.   Hence, the particle-bound organic polysulfides also can serve as nuclei for the formation of pyrite.   It was suggested earlier that if a hydrophilic molecule (for example, acrylate) reacted with a particle-bound polysulfide (containing a reactive free end), then some of the organic polysulfides so formed (perhaps only a small fraction), could break from the solid phase and become distributed in the pore water (15). Therefore, even if the solid-phase FeS was the reactant, the hydrophilic organic polysulfides equilibrated between the solid phase and the pore water might be the way by which sulfur is transferred from organic polysulfides to pyrite.

The idea that organically bound sulfur could be used to form pyritic sulfur was originally proposed by Altschuler et al. who described such transfer as the major source of sulfur for the pyrite formed in peat from the Florida Everglades (51). Through bacterial conversions, both reducible organic sulfur (for example, ester sulfates), and reduced organic sulfur (for example, organic sulfides) would provide the nucleophilic sulfur species required for forming pyrite. This study suggests that organic polysulfides also are an important sulfur source for pyrite in marine sediments. We envisage this transformation as a direct reaction which does not require bacterial mediation. However, the outcome would not be different even with bacterial involvement because degradation of the organic part in organic polysulfides would release the inorganic polysulfides which then could react with iron-monosulfide precursors to form the pyrite. We do not know whether sulfur is isotopically fractionated when the sulfur in organic polysulfides is converted into the sulfur in pyrite. If the lighter isotope of sulfur (^{32}S) is preferentially involved in this conversion, then the organic sulfur will be enriched in the heavier sulfur isotope (^{34}S) compared to pyrite. The view that organic polysulfides form first and then help to generate pyrite, together with the idea that sulfur isotopic fractionation occurs in transferring sulfur from organic polysulfides to pyrite, may explain the isotopic disparity between organic sulfur and pyrite in sediments, although there also are other possible explanations for the latter (52).

Acknowledgements

This material is based upon work supported by the U.S. Department of Energy, Division of Engineering and Geosciences of the Office of Basic Energy Sciences (Brookhaven National Laboratory Contract No. DE-AC02-76CH00016), the Max Plank Society, and the University of Concepcion. We thank Tim Eglinton, Math Kohnen and George Luther for reviews which helped to improve the manuscript, and Martin Schoonen for discussions. We also thank Captain Javier Boto, his crew, and technicians from the Hydrographic and Oceanographic Service of the Chilean Navy for their help and cooperation onboard RV Vidal Gormaz during the "Thioploca Cruise 1994".

Literature Cited

1. Jørgensen, B.B. *Nature* **1982**, *296*, 643-645.
2. Berner, R.A. *Am. J. Sci.* **1982**, *282*, 451-4731.
3. Goldhaber, M.B.; Kaplan, I.R. In The Sea; Goldberg, E.D. ed.; Vol. 5; John Wiley and Sons: New York, 1974; pp 569-655.
4. Jørgensen, B.B; *Limnol. Oceanogr.* **1977**, *22*, ,814-832.
5. Chen, K.Y.; Morris, J.C. *Environ. Sci. Technol.* **1972**, *6*, 529-537.

6. Zhang, J-Z.; Millero, F. *Geochim. Cosmochim. Acta* **1993**, *57*, 1705-1718.
7. Vairavamurthy, A.; Manowitz, B.; Zhou, W.; Jeon, Y.; In *The Environmental Geochemistry of Sulfide Oxidation*; Alpers, C. and Blowes, D. eds.; ACS Symposium Series 550; American Chemical Society, Washington D.C., 1994, pp 412-430.
8. Steudel, R.; Holdt, G.; Gobel T.; Hazeu W. *Angew. Chem. Int. Ed. Engl.* **1987**, *26*, 151-153.
9. Berner, R.A. *Geochim. Cosmochim. Acta* **1984**, *48*, 605-615.
10. Vairavamurthy, A.; Mopper, K. *Nature* **1987**, *329*, 623-625.
11. Sinninghe Damste, J.S.; de Leeuw, J.W. Org. Geochem. **1989**, *16*, 1077-1101.
12. Orr, W.L.; Sinninghe Damste, J.S. In *Geochemistry of Sulfur in Fossil Fuels*; Orr W.L; White, C.M. eds., ACS Symposium Series 429; American Chemical Society: Washington D.C., 1990; pp 2-29.
13. Gransch, J.A.; Posthuma, J. In *Advances in Organic Geochemistry*; Editions Technip: Paris, 1974; pp 727-739.
14. Vairavamurthy, A.; Mopper, K. In *Biogenic sulfur in the environment*; Saltzman, E.S.; Cooper, W.J., Eds.; ACS Symposium Series No. 393; American Chemical Society: Washington DC, 1989; pp 231-242.
15. Vairavamurthy, A.; Mopper, K.; Taylor, B.F. *Geophys. Res. Lett.* **1992**, *19*, 2043-2046.
16. de Graaf, W.; Sinninghe, Damste J.S.; de Leeuw, J.W. *Geochim. Cosmochim. Acta* 1992, *56*, 4321-4328.
17. Krein, E.B.; Aizenshtat, Z. *J. Org. Chem.* **1993**, *58*, 6103-6108.
18. Schouten, S.; van Driel, G.B.; Sinninghe, Damste J.S.; de Leeuw, J.W. *Geochim. Cosmochim. Acta.* **1994**, *45*, 5111-5116.
19. Francois, R. *Geochim. Cosmochim. Acta* **1987**, *51*, 17-27.
20. Kohnen, M.E.L.; Sinninghe, Damste J.S.; ten Haven, H.L.; de Leeuw, J.W. *Nature* **1989**, *341*, 640-641.
21. Luther, G.W. III *Geochim. Cosmochim. Acta* **1991**, *55*, 2839-2849.
22. Sinninghe Damste, J.S.; Irene, W.; Rijpstra, C.; Kock-van Dalen, A.C.; de Leeuw, J.W.; Schenck, P.A. *Geochim. Cosmochim. Acta* **1989**, *53*, 1343-1355.
23. Mossmann, J.R.; Aplin, A.C.; Curtis, C.D.; Coleman, M.L *Geochim. Cosmochim. Acta* **1991**, *55*, 3581-3595.
24. Gallardo, V.A. *Nature* **1977**, *26*, 331-332.
25. Fossing, H. et al. *Nature* **1995**, *374*, 713-715.
26. Nelson, D.C.; Castenholz, H.W. *J. Bacteriol.* **1981**, *147*, 140-154.
27. Chen, K.Y.; Moussavi, M.; Sycip, A. *Environ. Sci. Technol.* **1973**, *7*, 948-951.
28. Waldo, G.S.; Carlson, R.M.K.; Moldowan, J.M.; Peters, K.E.; Penner-Hahn, J.E. *Geochim. Cosmochim. Acta* **1991**, *55*, 801-814.
29. Huffman, G.S.; Mitra, S.; Huggins, F.E.; Shah, N.; Vaidya, S.; Lu F. *Energy & Fuels* **1991**, *5*, 574-581.
30. Spiro, C.L.; Wong, J.; Lytle, F.W.; Geegor, R.B.; Maylotte, D.H.; Lamson, S.H. *Science* **1984**, *226*, 48-50.
31. George, G.N.; Gorbaty, M.L *J. Am. Chem. Soc.* **1989**, *111*, 3182-3186.

32. Martens, C.S.; Haddad, R.I.; Chanton, J.P. In *Organic Matter: Productivity, Accumulation, and Preservation in Recent and Ancient Sediments.* Whelan, J.K.; Farrington, J.W., Eds.; Columbia University Press: New York, 1992; pp 82-98.
33. Froelich, P.N.; Arthur, M.A; Burnet, W.C; Deakin, M.; Hensley, V.; Jahnke, R.; Kaul, L.; Kim, K.H.; Roe, K.; Soutar, A.; Vathakanon, C. *Mar. Geol.* **1988**, *80*, 309-343.
34. Schimmelmann, A.; Kastner, M. *Geochim. Cosmochim. Acta* **1993**, *57*, 67-78.
35. Vairavamurthy, A.; Zhou, W.; Eglinton, T.; Manowitz, B. Geochim. Cosmochim. Acta **1994**, *21*, 4681-4687.
36. Aizenshtat, Z.; Stoler, A.; Cohen, Y.; Nielsen, H. In Advances in Organic Geochemistry; Wiley: Chichester, 1983; pp 279-288.
37. Staudt, W.J.; Schoonen, M.A.A. *Sulfate incorporation into sedimentary carbonates*, this volume.
38. Fitzgerald, J.W.; Jaru, T.A.; Strickland, T.C.; Swank, W.T. *Can. J. For. Res.* **1983**, *13*, 1077-1082.
39. LaLonde, R.T.; Ferrara, L.M.; Haye,s M.P. *Org. Geochem.* **1987**, *11*, 563-571.
40. LaLonde, R.T. In *Geochemistry of Sulfur in Fossil Fuels;* Orr, W.L.; White, C.M., Eds.; ACS Symposium Series 429; American Chemical Society: Washington D.C., 1990, pp 69-82,
41. Krein, E.B.; Aizenshtat, Z. *Org. Geochem.* **1994**, *21*, 1015-1026.
42. Raiswell, R.; Bottrell, S.H.; Al-Biatty, H.J.; Tan, M.MD. *Am. J. Sci.* **1993**, *293*, 569-596.
43. Zaback, D.A.; Pratt, L.M. *Geochim. Cosmochim. Acta* **1992**, *56*, 763-774.
44. Schoonen, M.A.; Barnes, H.L. *Geochim. Cosmochim. Acta* **1991**, *55*, 1495-1504.
45. Rickard, D.T. *Am. J. Sci.* **1975**, *275*, 636-652.
46. Luther, G.W. III; Ferdelman, T.G. *Environ. Sci. Technol.* **1993**, *27*, 1154-1163.
47. Giblin, A.E.; Howarth, R.E. *Limnol. Oceaogr.* **1984**, *29*, 47-63.
48. Drobner, E.; Huber H.; Wachtershauser, G.; Rose, D.; Stetter, K.O. *Nature* **1990**, *346*, 742-744
49. Vairavamurthy, A.; Zhou, W.; Manowitz, B. *Di-and Polysulfide Cross Linking in the Formation of Humic Polymers in Marine Sediments*; Abstracts, Am. Chem. Soc. Fall National Meeting, Washington, D.C., 1994
50. Ferdelman, T.G.; Church, T.M.; Luther, G.W. *Geochim. Cosmochim. Acta* **1991**, *55*, 979-988.
51. Altschuler, Z.S.; Schnepfe, M.M.; Silber, C.C.; Simon, F.O. *Science* **1983**, *221*, 221-227.
52. Anderson, T.F.; Pratt, L.M. *Isotope evidence for the origin of organic sulfur and elemental sulfur in marine sediments*, this volume.
53. Fossing, H. *Continental Shelf Res.* **1990**, *10*, 355-367.
54. Cline, J.D. *Limnol. Oceanogr.* **1969**, *14*, 454-458.

RECEIVED May 11, 1995

Chapter 4

Organic Geochemistry of Sulfur-Rich Surface Sediments of Meromictic Lake Cadagno, Swiss Alps

Anke Putschew, Barbara M. Scholz-Böttcher, and Jürgen Rullkötter

Institute of Chemistry and Biology of the Marine Environment,
Carl von Ossietzky University of Oldenburg, P.O. Box 2503,
D–26111 Oldenburg, Germany

Samples of Recent sediment from Lake Cadagno in the Swiss Alps were examined to study organic facies in a restricted setting and sulfur incorporation into organic matter at a very early stage of diagenesis. Lake Cadagno represents an uncommon lacustrine depositional environment due to a permanently anoxic bottom water column and a constant inflow of sulfate-rich groundwater near the bottom of the lake. The sediment is rich in organic carbon and sulfur. The extractable bitumen contains distinct molecular markers reflecting at least part of the floral and microbial community in the lake and allochthonous higher plant contribution. No low-molecular-weight organo-sulfur compounds were detected. Desulfurization of the heterocomponent and asphaltene fractions in the bitumen with nickel boride, however, yielded hydrocarbons with a strong dominance of phytane and minor concentrations of n-alkanes, steranes and squalane. Carbon isotope ratio mass spectrometry of individual compounds distinguishes phytane of obviously microbial origin in the high-molecular-weight organo-sulfur compounds from free phytane in the bitumen derived from algae or higher plants.

The analysis of organo-sulfur compounds (OSC) in fossil organic matter has become a subject of increasing interest in organic geochemistry. One major reason is the enhanced preservation of biological markers bound into macromolecular sedimentary organic matter *via* sulfur bonds and thus their potential availability, e.g. for paleoenvironmental reconstruction even at advanced stages of organic matter diagenesis and catagenesis or in crude oils (*1-2*).

Analysis of geological samples suggests that the formation of OSC occurs by intra- or intermolecular addition of reduced inorganic sulfur species into functionalised biogenic lipids (*3*) leading to compounds of different molecular size.

The sulfur found in OSC comprises single sulfur-carbon as well as polysulfide bonds. Hydrogen sulfide, polysulfides, and elemental sulfur have been suggested as the reactive inorganic sulfur species. Accordingly, a number of laboratory experiments have been carried out in which diagenetic sulfur incorporation has been simulated in a variety of low-molecular-weight organic compounds, e.g. phytol (*4-5*), phytadienes (*4*), hop-17(21)-ene (*6*) and a series of aldehydes and ketones (*7*). Particularly the successful addition of hydrogen polysulfide to isolated double bonds under mild conditions (*4*) was supporting evidence of the view that sulfur incorporation can occur at a very early stage of diagenesis. In contrast to this, elemental sulfur probably reacts with sedimentary organic constituents only at elevated temperatures, i.e. during catagenesis (*8*).

The first OSC identified in geological samples mostly had carbon skeletons smaller than C_{15} and did not provide much information about their origin (e.g. *9-12*). In 1984, Valisolalao *et al.* (*13*) identified OSC (e.g. a C_{35} hopanoid containing a thiophene ring) which were structurally related to known biochemical precursors. Since that time a great number of novel low-molecular-weight OSC with structures similar to geologically occurring hydrocarbons and their precursor compounds were identified in bitumens extracted from sediments and in crude oils (*14-18*).

Advanced analytical procedures have been developed which allow the study of organically-bound sulfur compounds in macromolecules and provided new information related to the understanding of OSC formation in sediments. GC-pyrolysis of sulfur-rich kerogens using sulfur-specific detectors revealed evidence of the presence of sulfur containing moieties (*19-20*). Desulfurization of soluble macromolecular material using Raney nickel and nickel boride (Ni_2B) released low-molecular-weight compounds and lead to information concerning the structure of these macromolecules (*21-22*). The hydrocarbons obtained in this way had structures identical to known biological marker hydrocarbons, but in most cases both the compound distributions and the relative concentrations of stereoisomers differed dramatically from those in the free hydrocarbon fractions indicating that progress of diagenesis - as common for other macromolecularly bound moieties (*23*) - was slower for the sulfur-bound moieties in the macromolecules than for the free hydrocarbons (*24-25*). Selective cleavage of di- and polysulfide bonds by MeLi/MeI revealed different modes of carbon-sulfur cross-linking and thus provided clear evidence that polysulfidic reduced inorganic sulfur species are involved in the diagenetic reaction of functionalised lipids (*26*).

Previous investigations used sediments older than 10,000 years to examine the formation of OSC. In the present study fossil organic material not older than 100 years was analysed in an attempt at elucidating the earliest processes of sulfur incorporation into fossil organic matter.

Lake Cadagno

Lake Cadagno is located south of the St.Gotthard massif in the Ticino Alps of Switzerland (Figure 1) at an altitude of 1923 m above sea level. It is situated in a

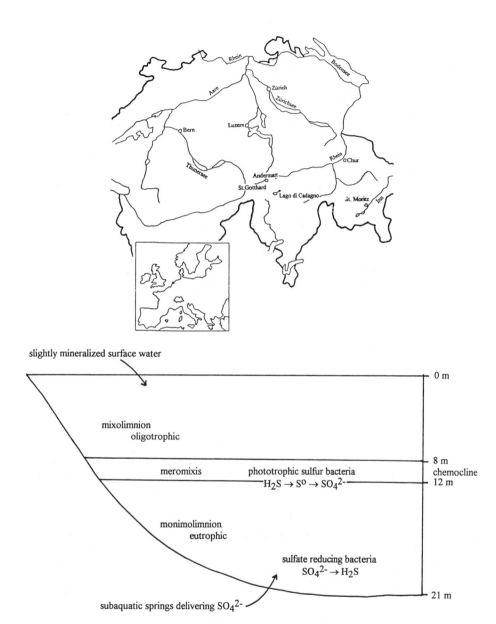

Figure 1: Geographic location of Lake Cadagno in the Swiss Alps and schematic cross section of the water body of the lake.

cirque created by the action of glaciers on pre-Triassic gneisses and mica shists of central alpine crystalline rocks. The southern lake shore follows the contact of the Triassic and early Jurassic rocks folded into the so-called Piora syncline. The rocks comprise tectonized dolomite limestone and gypsum deposits which favor mineral dissolution by groundwater in the fractured subsurface (karst hydrology). The lake surface is about 0.27 km^2, the maximal depth is 21 m (Figure 1). The upper part of the water column regularly turned over by weather forces, i.e. the mixolimnion, is oxic and fed by slightly mineralized surface water. This supply is controlled by seasonal changes, e.g. melting waters in the spring. The lower part of the water column not affected by wind stress and other turn-over processes, the monimolimnion, is anoxic. Subaquatic springs highly concentrated in sulfate (2-4 mmol/l) feed this zone *(27-28)*. The density difference between the upper and the lower parts of the water body leads to a stable chemocline at a water depth between 8 and 12 m, which inhibits a vertical mixture of the water column across this boundary. Due to the high sulfate concentration, bacterial sulfate reduction is the main biological process in the monimolimnion. The boundary layer (meromixis) is dominantly populated by phototrophic sulfur bacteria, especially *Chromatium* (sulfide-oxidizing purple bacteria) *(27,29)*. These bacteria are enriched in the pigment okenone which causes the red color in the water column at the chemocline. Chemotrophic sulfide oxidizing bacteria thrive in the upper part of the meromixis where oxygen is present.

Materials and Methods

A gravity core of Lake Cadagno sediment, 36 cm long, 7 cm wide, and immediately frozen at the sampling site, was provided by A. Losher (ETH Zürich). Based on the work of Züllig *(29)* it represents the sedimentary history of the lake over the last 80-100 years which corresponds to a sedimentation rate of about 4 mm/a, a value common for moderately to highly productive lacustrine systems with a significant supply of detrital material. The sediments were homogeneously black and had a strong odor of hydrogen sulfide and possibly other volatile sulfur compounds; this odor ceased considerably below a depth of 32 cm. The core was separated into sections of 2 cm thickness which were freeze-dried and ground. Aliquots of the ground samples were used to determine total carbon (TC) and total sulfur (TS) contents with a LECO SC-444 instrument. Before total organic carbon (TOC) determination with the same instrument, the samples were treated with hydrochloric acid to remove carbonates. Carbonate contents were calculated by difference and expressed as percent calcium carbonate.

For molecular organic geochemical analysis the samples were ultrasonically extracted with dichloromethane. The solvent was evaporated and the extract suspended in a small volume of *n*-hexane. After addition of internal standards (squalane, thianthrene, 5α(H)-androstan-17-one) asphaltenes (used here and throughout this paper as an operational term and comprise humic substances etc.) were precipitated with an approximately 100 fold excess of *n*-hexane. The compounds soluble in *n*-hexane were separated on a column filled with silica deactivated with 5% water. The aliphatic hydrocarbon fraction was eluted with *n*-hexane, the aromatic hydrocarbon fraction with *n*-hexane/dichloromethane (9:1 by volume) and the hetero(NSO)-compounds with dichloromethane/methanol (9:1 by volume). For the investigation of extractable bound fatty acids, alcohols and sterols, aliquots of the NSO fractions were saponified with 3 ml 5% KOH in methanol/water (8:2 by volume) for 2 h at 80°C. Aliquots of the asphaltene and NSO fractions were desulfurized with Ni_2B following a published precedure (*22*). The reaction products were separated by column chromatography on silica gel into an apolar and a polar fraction. The apolar fraction was eluted with *n*-hexane/dichloromethane (9:1 by volume) and analysed before and after hydrogenation; the polar fraction was eluted with dichloromethane/methanol (1:1 by volume) but not further investigated. For the analysis of compounds bound to the insoluble organic matter, extract residues were saponified with 20 ml 5% KOH in methanol/water (8:2 by volume) for 24 h under reflux. Where appropriate, the extract fractions were derivatized with diazomethane and MSTFA (*n*-methyl-*n*-trimethylsilyl-trifluoroacetamide) before analysis.

GC analysis was carried out on a Hewlett Packard 5890 series II instrument equipped with a temperature programmable injector system (Gerstel KAS 3) and a flame ionization detector (FID) or a sulfur-selective chemoluminescence detector (Sievers SCD). The detection level for sulfur was 17 pg S/µl. A DB-5 (J&W) fused silica capillary column (30 m x 0.25 mm i.d., df = 0.25 µm) was used with helium as carrier gas. The temperature of the GC oven was programmed from 60°C (1 min isothermal) to 300°C (50 min isothermal) at 3°C/min. The injector temperature was programmed from 60°C (5 s hold time) to 300°C (60 s hold time) at 8°C/s.

GC/MS measurements were performed with the same type of GC system under the conditions described above. The gas chromatograph was coupled to a Finnigan SSQ 710 B mass spectrometer operated at 70 eV with a scan range of *m/z* 50 to 600 and a scan time of 1 scan/s. Carbon isotope ratios of individual hydrocarbons were determinded using a Finnigan MAT 252 isotope-ratio monitoring mass spectrometer coupled with a Varian gas chromatograph. An Ultrix 2 (Hewlett-Packard) capillary column (50 m x 0.32 i.d., df = 0.17 µm) was used with helium as carrier gas. The temperature program was identical to the one described before.

Structural Assignment of a Major Steroid Olefin in the Aliphatic Hydrocarbon Fractions. The mass spectrum of the most abundant compound, apart from *n*-alkanes, in most of the aliphatic hydrocarbon fractions of the Lake Cadagno sediments is characterized by a molecular ion at *m/z* 394 and significant fragment ions at *m/z* 69, *m/z* 255 and *m/z* 257. These data suggest that the compound may be a C_{29} steratriene with two double bonds in the ring system and one double bond in the side chain. The mass spectrum of 24R-ethylcholesta-3,5,22-triene, a commercially available standard (Chiron), is similar to that of the hydrocarbon in the Lake Cadagno sediments but the standard has a slightly longer gas chromatographic retention time. Also, the fragment ion at *m/z* 69 is subordinate in the mass spectrum of the standard. This fragment is typical of side chain-unsaturated steroids with a double bond postion at C-22 and a 23,24-dimethyl substitution pattern. It is generated by the formation of an isopentyl ion after double hydrogen transfer and cleavage of the C-23/C-24 bond. This principal fragmentation behavior has already been described for dinosterol, i.e. 4α,22,23-trimethylcholest-22-en-3β-ol (*30*), and 23,24-dimethylcholesta-5,22-dien-3β-ol (*31*). According to the molecular ion and the fragment at *m/z* 255, the unknown compound does not carry additional methyl substituents in the ring system. Thus, the steratriene in the Lake Cadagno sediments appears to be 23,24 dimethylcholesta-3,5,22-triene. After hydrogenation of an aliphatic hydrocarbon fraction of a Lake Cadagno sediment and the standard steratriene, the two resulting steranes coelute. This is in accordance with the interpretation of the steratriene characteristics, because gas chromatographic coelution of 23,24-dimethyl- and 24-ethylcholestanols is known (*32*).

Results and Discussion

Bulk parameters and extract fractionation. The anoxic sediment of Lake Cadagno is rich in organic carbon and sulfur (Figure 2). The highest TOC value (13.5%) was measured at the sediment surface, the lowest value of 2.9% corresponds to a sample at a depth of 20 cm. The sulfur content shows a trend which at the the first glance seems to grossly covary with that of the TOC values. TS reaches a maximum at a depth of 10 cm (3.3%) and a minimum at a depth of 24 cm (1.3%). The relative variation of the total sulfur content is much smaller, however, than that of the organic carbon which leads to lower TOC/S ratios in those sediment layers which are less enriched in organic carbon (Fig ure 2). This means that the rate of microbial sulfate reduction was not limited by the supply of organic matter.

While graded turbidite sequences, caused by episodic inflow of sandy material, have been observed in some parts of Lake Cadagno (*28*), no sedimentological evidence (e.g. color change) for dilution of organic matter was observed in the core. Thus, diagenetic processes as well as fluctuations in the supply of organic matter, both qualitatively and quantitatively, are more likely to have affected the distribution of organic matter, bulk sulfur and carbonate concentrations than physical processes.

The amount of extractable organic matter is around 10%, normalized to TOC, down to a depth of 16 cm. Below that depth the extract yield drops to about 5% (Table I). This indicates that there may be two sections characterised by differences in the bulk composition/type of the organic matter above and below the boundary of about 16 cm depth despite the fairly uniform relative proportions of the gross extract fractions (Table I). All extract yields are high in view of the early stage of diagenesis of the organic matter, but are typical of sediments rich in sulfur (e.g. *33*). As expected, the NSO compounds represent the most important extract fraction (Table I). The relative proportion of the polar fractions is even higher than displayed in Table I, because the aliphatic hydrocarbon fractions include elemental sulfur although this has not been determined quantitatively in all samples. The amount of elemental sulfur decreases with increasing depth. The sample from 6-8 cm depth contains 2.7% elemental sulfur (3.1% TS), whereas the sediment at 30-32 cm depth has an elemental sulfur content of only 0.2% (2.8% TS).

Aliphatic Hydrocarbon Fraction. The aliphatic hydrocarbon fractions are dominated by n-alkanes with a strong odd over even carbon number predominance and a maximum at n-C_{29} or n-C_{31} (Figure 3). This type of distribution is characteristic for land plant-derived organic matter (*34*). The most likely source is the grass growing around the lake.

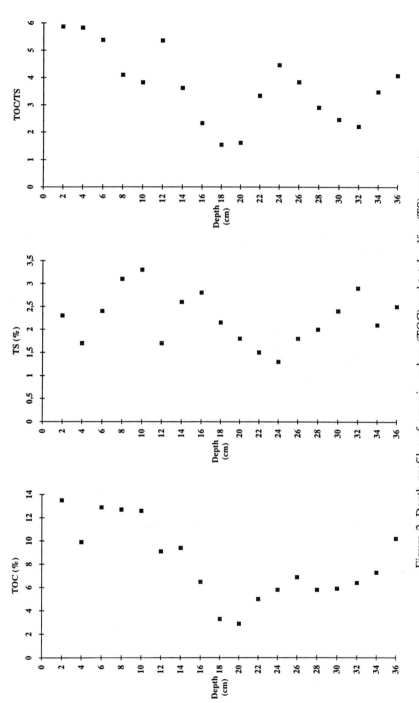

Figure 2. Depth profiles of organic carbon (TOC) and total sulfur (TS) contents in the top 36 cm of lake Cadagno sediments and of the TOC/TS ratio.

Table I. Extract yields and composition of total extracts of Lake Cadagno sediments

Depth (cm)	Extract yield (mg/g C_{org})	NSO fraction (%)	Aromatic fraction (%)	Aliphatic fraction (%)	Asphaltenes(%)
0-2	92	55.3	0.6	34.8	7.4
2-4	114	36.1	4.1	35.1	7.2
4-6	153	64.8	0.9	34.7	6.5
6-8	109	49.9	0.7	55.8	13.3
8-10	116	42.9	0.3	45.4	9.6
10-12	66	78.7	1.6	16.1	14.6
12-14	92	46.7	1.7	35.1	11.7
14-16	94	43.6	1.1	27.1	16.8
16-18	43	47.8	0.9	37.4	10.4
18-20	63	30.7	2.0	20.8	4.9
20-22	36	64.1	2.8	30.3	11.3
22-24	43	55.4	2.8	26.3	10.3
24-26	52	46.4	0.2	24.8	14.4
26-28	42	77.7	0.8	15.7	10.7
28-30	47	63.6	2.4	18.4	12.8
30-32	56	69.4	18.9	9.7	39.4
32-34	41	40.0	2.7	50.0	8.7
34-36	46	74.5	0.5	2.2	17.4

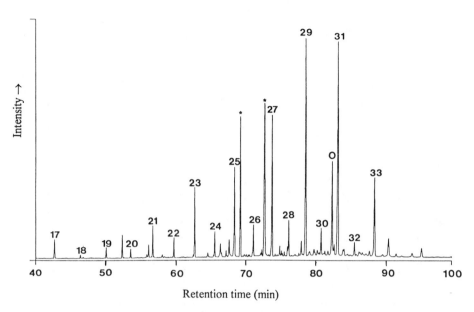

Figure 3. Gas chromatogram of the aliphatic hydrocarbons of the sample from 28-30 cm depth. *n*-Alkanes are indicated by their carbon numbers. O = steratriene (see Table II), * = internal standards.

Hopanoid hydrocarbons of microbial origin (*35*) are the second-most important group of constituents of the aliphatic hydrocarbon fractions. Hopanes with the thermodynamically least stable, biogenic 17β(H),21β(H) configuration and a carbon number range of C_{27}-C_{31} occur together with high relative amounts of hopenes like 22,29,30-*trinor*-hop-17(21)-ene, hop-17(21)-ene, *neo*-hop-13(18)-ene and hop-22(29)-ene (Figure 4 and Table II). This compounds distribution is typical of immature organic matter and common in Recent sediments, yet it demonstrates the rapid conversion of functionalized precursors into hydrocarbons and their concurrent or subsequent transformation into a mixture of isomers, at least for the unsaturated hopenes.

Sterenes are present in low concentrations with the exception of one compound which elutes just before the C_{31} *n*-alkane (Figure 3) and is conspicuously abundant. It was tentatively identified as 23,24-dimethylcholesta-3,5,22-triene, based on comparison with the mass spectrum and relative retention time of a commercially available 24R-ethylcholesta-3,5,22-triene standard and published mass spectra of 4α,22,23-trimethylcholest-22-en-3β-ol (*30*) and 23,24-dimethylcholesta-3,5,22-dien-3β-ol (*31*) as outlined in the experimental section. Another aspect in favor of the structure of 23,24 dimethylcholesta-3,5,22-triene is the occurrence of a potential precursor compound, namely 23,24-dimethylcholesta-5,22-dien-3β-ol in the NSO fraction of the Lake Cagano sediment extracts. This sterol is known to be characteristic of dinoflagellates (*36-37*), and dinoflagellates are known to occur in Lake Cadagno (K. Hanselmann, private communication 1994). The amount of the

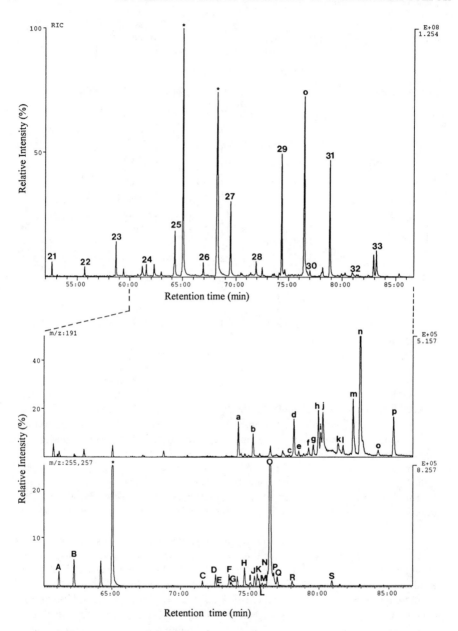

Figure 4. Partial reconstructed ion chromatogram RIC (top) and mass chromatograms (m/z 191, middle; Σm/z (255+257), bottom) of the aliphatic hydrocarbon fraction of the sample from 26-28 cm depth. n-Alkanes are indicated by their carbon numbers (RIC). For compounds identified in the mass chromatograms see Table II. * = internal standards. Note: Retention times of cyclic hydrocarbons relative to n-alkanes differ between GC analysis (see peak O in Figure 3) and GC-MS analysis (see peak O here) due to the effect of the mass spectrometer vacuum on the GC column.

Table II. Compounds identified in the aliphatic hydrocarbon fractions of Lake Cadagno sediments (see Figure 4)

Symbol/Compound	Symbol/Compound
a 22,29,30-*trinor*-hop-17(21)-ene	A des-A-oleanadiene
b 17β(H)-22,29,30-*trinor*-hopane	B des-A-oleanadiene
c triterpene (m/z 410,395,109)	C cholesta-5,22-diene
d hop-17(21)-ene	D 5α(H)-cholest-2-ene
e 17β(H),21α(H)-30-*nor*-hopane	E sterene (m/z 370,355,316,257, 215,147,108)
f 17β(H),21α(H)-hopane	F 24-methylcholesta-5,22-diene
g *neo*-hop-13(18)-ene	G unknown (m/z 368,353,247,213, 159,147)
h triterpene (m/z 410, 395,191, 189)	H 24-methylcholestatriene
i 17β(H),21β(H)-30-*nor*-hopane	I 24-methyl-5α(H)-cholest-2-ene
j hop-21-ene	J C_{29}-steratriene
k triterpene (m/z 410,218,205,191, 175)	K C_{29}-steratriene
l 17α(H),21β(H)-*homo*-hopane	L C_{29}-steratriene
m 17β(H),21β(H)-hopane	M unknown (m/z 394/392,380,253)
n hop-22(29)-ene	N C_{29}-steratriene
o *homo*-hop-22(29)-ene	O 23,24-dimethylcholesta-3,5,22-triene
p 17β(H),21β(H)-*homo*-hopane	P C_{29}-steratriene
	Q 24-ethyl-5α(H)-cholest-2-ene
	R unknown (m/z 396,381,275,255, 213,160,147
	S C_{30}-steratriene

steratriene, normalized to organic carbon, generally appears to increase with depth although fluctuations were observed in this trend (Table III). The increase is attributed to a progress in diagenesis, i.e. transformation of a functionalized precursor into the triunsaturated hydrocarbon.

In the surface sample of the lake (0-2 cm), the highly branched isoprenoid 2,6,10-trimethyl-7-(3-methylbutyl)-dodecane was detected. Its concentration decreases rapidly with depth. 2,6,10-Trimethyl-7-(3-methylbutyl)-dodecane is typically used as a marker for the green alga *Enteromorpha prolifera* (*38*), although more recently highly-branched isoprenoids have been predominantly related to diatoms, at least in the marine environment (*39-41*). *Chlorophyceae* and *Cryptophyceae* were described as the main phytoplankton species in Lake Cadagno (*29*), but no natural product survey for isoprenoids (or sterols) has yet been performed for these lake communities.

Aromatic Hydrocarbon Fraction. Aromatic hydrocarbons are the least abundant fraction of all sediment extracts from Lake Cadagno as common for Recent sediments. The only compound found in higher concentrations was squalene which is a component widespread in many organisms. In these fraction the concentrations of OSC such as thiophenes or thiolanes are below the detection level of the sulfur-selective detector.

NSO Fraction. The GC-amenable portions of the NSO fractions are dominated by free *n*-fatty acids with an even over odd carbon number predominance. *iso*-Hexadecanoic, *n*-hexadecanoic, *n*-octadecanoic acid and fatty acids with higher carbon numbers (n-C_{24}, n-C_{26}, n-C_{28}) are the most abundant compounds (Figure 5a). After hydrolysis of the total NSO fraction, which yields the sum of the free and bound extractable acids, the fatty acid distribution pattern changes and is characterized by *n*-hexadecanoic and *n*-octadecanoic acid together with their monounsaturated analogs. The concentration of extractable bound fatty acids with carbon numbers higher than eighteen is very low (Figure. 5b; extractable bound fatty acids represent the calculated difference between total extractable fatty acids after hydrolysis and free extractable fatty acids). A distribution pattern similar to that of the extractable bound acids was found for the kerogen-bound fatty acids after saponification of the extract residues (Figure 5c). The total amount of fatty acids bound to kerogen and - to a lesser extent - of the bound extractable acids is much higher than the amount of free fatty acids. The bound fatty acids apparently are mainly derived from algal and bacterial biomass (*42*). Only the small fraction of the free fatty acids contains a significant relative proportion of long-chain acids of terrestrial higher plant origin. Unsaturated fatty acids, *iso*- and *anteiso*-acids which are typical of microbial biomass (*42*) are present in a free as well as in a bound form.

The main sterols identified in the NSO fractions are cholest-5-en-3β-ol, 24-ethylcholest-5,22-dien-3β-ol, 23,24-dimethylcholest-22-en-3β-ol, 24-ethylcholest-5-en-3β-ol, and 23,24-dimethylcholest-5-en-3β-ol (Figure 6 and Table IV). *n*-Alcohols were found in the range of C_{16} to C_{28} with an even over odd carbon number predominance. Sterols and alcohols were mainly found as free compounds indicating that the major part of the bound fatty acids released by hydrolysis are not derived from wax or steryl esters. The abundance of sterols relative to the concentrations of steroid hydrocarbons is typical of the early stage of diagenesis. Only the

Table III. Concentrations (μg/g C_{org}) of free phytane and 23,24-dimethylcholesta-3,5,22-triene in the aliphatic hydrocarbon fractions and of phytane released by desulfurization

Depth (cm)	Aliphatic hydrocarbon fraction		After desulfurization	
	Free phytane	23,24 Dimethyl-cholesta-3,5,22-triene	NSO fraction	Asphaltenes
0-2	17.0	7.0	1624.0	169.0
2-4	30.0	11.0	799.0	
4-6	16.0	11.0	1568.0	154.0
6-8	4.5	6.0	911.0	35.0
8-10	12.0	7.0	1419.0	750.0
10-12	6.0	69.0	1098.0	238.0
12-14	3.0	32.0	926.0	374.0
14-16	6.0	38.0	1203.0	371.0
16-18	<1.0	14.0	669.0	134.0
18-20	<1.0	31.0	174.0	76.0
20-22	<1.0	55.0	991.0	151.0
22-24	<1.0	64.0	358.0	108.0
24-26	<1.0	50.0	795.0	480.0
26-28	<1.0	133.0	1888.0	35.0
28-30	<1.0	56.0	618.5	445.0
30-32	<1.0	89.0	839.0	576.0
32-34	<1.0	113.0	1107.0	38.0
34-36	<1.0	129.0	869.0	81.0

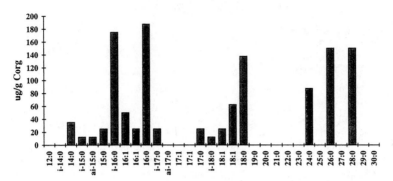

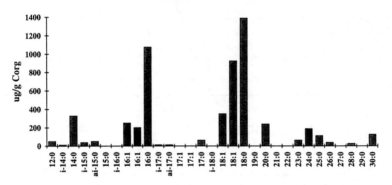

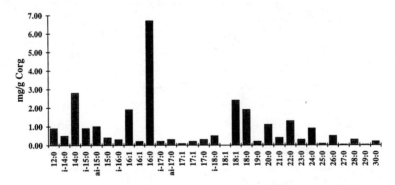

Figure 5. Distribution of extractable free fatty acids in the NSO fraction, of bound fatty acids in the NSO fraction after hydrolysis (calculated difference of total acids after hydrolysis and free acids), and of kerogen bound acids after hydrolysis for the Lake Cadagno sediment from 6-8 cm depth.

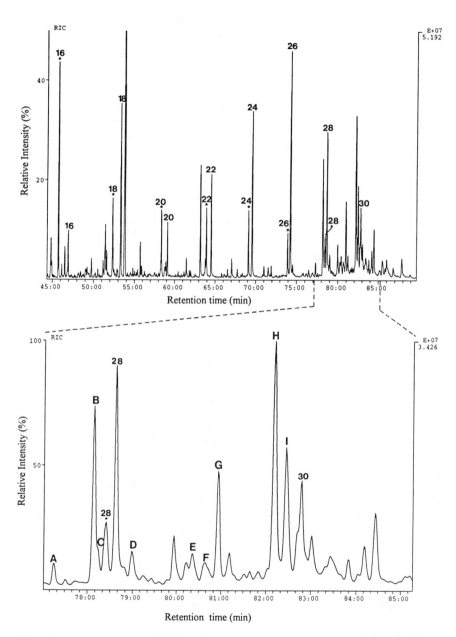

Figure 6. Reconstructed ion chromatogram of the NSO fraction (methylated and trimethylsilylated derivatives) of the sample from 30-32 cm depth. Numbers with asterisk = *n*-fatty acids, numbers = *n*-alcohols. See Table IV for sterol identification (letter symbols).

concentration of the dominant steratriene in the aliphatic hydrocarbon fractions is close to that of the corresponding steroid alcohol. This may indicate that dehydration has started early after decay of the precursor organism already in the water column.

Table IV. Sterols in the NSO fractions of Lake Cadagno sediments (see Figure 6)

Symbol	Compound
A	cholesta-5,22-dien-3β-ol
B	cholesta-5-en-3β-ol
C	cholestan-3β-ol
D	24-methylcholesta-5,22-dien-3β-ol
E	24-methylcholesta-5-en-3β-ol
F	24-methylcholestan-3β-ol + 23,24-dimethylcholesta-5,22-dien-3β-ol
G	24-ethylcholesta-5,22-dien-3β-ol + 23,24-dimethylcholesta-22-en-3β-ol
H	24-ethylcholesta-5-en-3β-ol + 23,24-dimethylcholestan-3β-ol
I	24-ethylcholestan-3β-ol

Desulfurization of Asphaltenes and NSO Fractions. Desulfurization of NSO fractions as well as asphaltenes yielded phytane as the by far most abundant hydrocarbon (Figure 7, Table III). The phytane concentrations obtained this way are higher by two orders of magnitude or more in most of the sediments. While there is no clear downhole trend in the amount of bound phytane, free phytane concentrations drop significantly below 16 cm depth, i.e. at the level where total organic carbon concentrations indicated a change in organic matter supply. n-Alkanes were released only in minor concentrations. In contrast to the n-alkanes in the aliphatic hydrocarbon fractions, the n-alkanes in the desulfurization products show an even over odd carbon number predominance although this is not strongly pronounced (Figure 7). This predominance suggests a potential relationship to n-alcohols or n-fatty acids but this could not yet be substantiated. Besides phytane and n-alkanes, squalane, cholestane and 23,24-dimethylcholestane are additional compounds which were liberated by the desulfurization procedure.

Stable Carbon Isotope (δ^{13}C) Ratios for Selected Compounds. The stable carbon isotope compositions of n-alkanes, phytane, and 23,24-dimethylcholesta-3,5,22-triene were determined in a selected aliphatic fraction and an apolar fraction obtained after desulfurization of an NSO fraction. The results are compiled in Table V. There is relatively little variation in the carbon isotopic composition of the free n-alkanes (average around -31‰), but there is a tendency of a slight ^{13}C enrichment with decreasing carbon number. In the bound n-alkanes this is particularly noticeable for the C_{21} to C_{23} n-alkanes. In general, the n-alkanes of the free aliphatic hydrocarbon fraction are slightly more depleted in ^{13}C than the same compounds released after desulfurization.

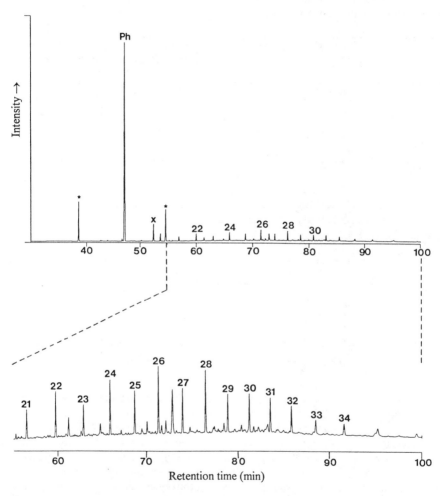

Figure 7. Gas chromatogram of hydrocarbons released by desulfurization of the NSO fraction of the sample from 28-30 cm depth. *n*-Alkanes are indicated by their carbon numbers. Ph = phytane, * = standards.

Table V. δ^{13}C-values for hydrocarbons in the aliphytic hydrocarbon fraction of an extract of a sample from 20-22 cm depth and in the desulfurization products of asphaltenes extracted from a sediment from 28-30 cm depth. So far, these two samples are the only ones investigated for molecular isotope composition. Taking the simularities in compound distributions, the values are considered representative for the entire sediment interval studied.

	Free aliphatic hydrocarbons (20-22 cm)	Hydrocarbons released after desulfurization of asphaltenes (28-30 cm)
Compound	δ^{13}C ($^{o}/_{oo}$)	δ^{13}C ($^{o}/_{oo}$)
n-C_{18}	-29.5	
Phytane	-24.1	-42.5
n-C_{20}	-31.0	
n-C_{21}	-29.9	-26.4
n-C_{22}	-31.1	-27.8
n-C_{23}	-29.7	-29.1
n-C_{24}	-30.2	-31.0
n-C_{25}	-30.9	-30.6
n-C_{26}	-31.5	-31.5
n-C_{27}	-32.0	-31.0
n-C_{28}	-34,1	-33.6
n-C_{29}	-32.8	-30.4
n-C_{30}	-33.1	-31.9
Steratriene	-30.4	
n-C_{31}	-33.4	-32.4
n-C_{32}	-32.0	
n-C_{33}	-33.1	

The carbon isotope values for phytane in the free and bound fractions are remarkably different (Table V). Phytane released after desulfurization is depleted in ^{13}C by almost 20‰ relative to phytane in the aliphatic hydrocarbon fraction which strongly suggests that phytane incorporated into the macromolecular fraction is of a different origin to the free phytane. The sulfur-bound phytane in the NSO fraction due to its strongly negative carbon isotope value, common for microbial isotope fractionation, is likely to be related to the phototrophic sulfur bacteria which are abundant at the chemocline of the lake. In contrast to this, the quantitatively much less important free phytane probably originates from the transformation of chlorophyll a-derived phytol, either from the lacustrine phytoplankton or from higher plants in the vicinity of the lake. Because the carbon isotope ratio of free phytane is neither in accordance with those of the long-chain n-alkanes assumed to be of higher plant origin nor with that of 23,24-dimethylcholesta-3,5,22-triene supposed to be a biomarker of the dinoflagellates in the lake, a source assignment for the free phytane cannot be made as long as no information is available for the living biota in the lake and from its vicinity.

Conclusions

The anoxic sediment of Lake Cadagno is rich in organic carbon and sulfur. Organic geochemical analysis revealed that the organic matter deposited in the sediments is mainly derived from bacteria and phytoplankton, whereas the terrigenous contribution of organic material is very small and primarily reflected by the n-alkane distribution.

In the examined extract fractions no soluble low-molecular-weight organosulfur compounds were detectable. Sulfur compounds were only found indirectly in the macromolecular material of the asphaltenes and of the NSO fractions. The removal of sulfur revealed that hydrocarbons with a phytane, n-alkane, squalane, and sterane skeleton were bound *via* sulfur into the macromolecular network. The results of this investigation are in agreement with the idea that formation of organic sulfur compounds takes place in an early stage of diagenesis.

The isotopic data show that phytane released after desulfurization has an origin which is different from that of the other hydrocarbons released by desulfurization and also different from that of free phytane in the aliphatic hydrocarbon fraction. The bound phytane apparently stems from the phototrophic bacteria living near the chemocline of Lake Cadagno. For a rigid interpretation of the isotopic data and a clear identification of the source organisms, biological material from the water column at different depths and from the higher plants on the shore will have to be analysed. This work is in progress.

Acknowledgments

We thank Dr. A. Losher (ETH Zürich) for providing us with a sediment core from Lake Cadagno and Finnigan MAT, Bremen, for the measurement of stable carbon isotopes. We are grateful to Dr. K. Hanselmann (University of Zürich) for the support in retrieving additional core material and helpful discussions. The thorough review of the manuscript by Dr. Linda Stalker (University of Oklahoma, Norman) was highly appreciated and greatly helped to improve the quality of the publication. For financial support we thank the Deutsche Forschungsgemeinschaft (DFG grant no. Ru 458/4).

References

1. Sinninghe Damsté, J.S.; Rijpstra, W.I.C.; de Leeuw, J. W.; Schenck, P.A. *Geochim. Cosmochim. Acta* **1989**, *53*, 1323-41.
2. de Leeuw, J.W.; Sinninghe Damsté, J.S. In *Geochemistry of Sulfur in Fossil Fuels*; Orr, W.L; White, C.M., Eds.; ACS Symposium Series 429; American Chemical Society: Washington, D.C., **1990**; pp. 417-43.
3. Kohnen, M.E.L.; Sinninghe Damsté, J.S.; Baas, M.; Kock-van-Dalen, A.C.; de Leeuw, J.W. *Geochim. Cosmochim. Acta* **1993**, *57*, 2515-28.
4. de Graf, W.; Sinninghe Damsté, J.S.; de Leeuw, J.W. *Geochim. Cosmochim. Acta* **1992**, *57*, 4321-28.
5. Rowland, S.; Rockey, C.; Al-Lihaibi, S.S.; Wolff, G.A. *Org. Geochem.* **1993**, *20*, 1-5.

6.   Bisseret, P.; Rohmer, M. *Tetrahedron Lett.* **1993**, *34*, 5295-98.
7.   Schouten, S.; van Driel, G.B.; Sinninghe Damsté, J.S.; de Leeuw, J.W. *Geochim. Cosmochim. Acta* **1994**, *58*, 5111-16.
8.   See Aizenshtat et al., this volume.
9.   Dean, R.A; Whitehead, E.V. In *Proc. 7th World Petroleum Congress*: Panel Discussion 23; Paper 7, **1967**.
10.   Gal'pern, G.D. *Int. J. Sulfur Chem.* **1971**, *6*, 115.
11.   Thompson, C.J. In *Organic Sulfur Chemistry*; Freidlina, R. Kh.; Skarova, A.E., Eds.; Pergamon Press: Oxford, **1981**; pp. 189-208.
12.   Aksenov, V.S.; Kamyanov, V.F. In *Organic Sulfur Chemistry*; Freidlina, R. Kh.; Skarova, A.E., Eds.; Pergamon Press: Oxford, **1981**; pp. 1-13.
13.   Valisolalao, J.; Perakis, N.; Chappe, B.; Albrecht, P. *Tetrahedron Lett.* **1984**, *25*, 1183-86.
14.   Sinninghe Damsté, J.S.; ten Haven, H.L.; de Leeuw, J.W.; Schenck, P.A. *Org. Geochem.* **1986**, *10*, 791-805.
15.   Sinninghe Damsté, J.S.; de Leeuw, J.W.; Kock-van Dalen, A.C.; de Zeeuw, M.A.; de Lange, F.; Rijpstra, W.I.C.; Schenck, P.A. *Geochim. Cosmochim. Acta* **1987**, *51*, 2369-91.
16.   Sinninghe Damsté, J.S.; de Leeuw, J.W. *Int. J. Environ. Anal. Chem.* **1987**, *28*, 1-19.
17.   Sinninghe Damsté, J.S.; Kock-van Dalen, A.C.; de Leeuw, J.W.; Schenck, P.A. *Tetrahedron Lett.* **1987**, *28*, 957-60.
18.   Sinninghe Damsté, J.S.; Rijpstra, W.I.C.; de Leeuw, J. W.; Schenck, P.A. *Geochim. Cosmochim. Acta* **1987**, *51*, 2369-91.
19.   Eglinton, T.I.; Sinninghe Damsté, J.S.; Pool, W.; de Leeuw, J.W.; Eijkel, G.; Boon, J.J. *Geochim. Cosmochim. Acta* **1992**, *56*, 1545-60.
20.   Sinninghe Damsté, J.S.; Kock-van Dalen, A.C.; de Leeuw, J.W.; Schenck, P.A. *J. Chromatogr.* **1988**, *435*, 435-52.
21.   Sinninghe Damsté, J.S.; Eglinton, T.I.; Rijpstra, W.I.C.; de Leeuw, J.W. In *Geochemistry of Sulfur in Fossil Fuels*; Orr, W.L.; White, C.M., Eds.; ACS Symposium Series 429; American Chemical Society: Washington, D.C., **1990**, pp. 486-528.
22.   Schouten, S.; Pavlovic, D.; Sinninghe Damsté, J.S.; de Leeuw, J.W. *Org. Geochem.* **1993**, 20, 901-9.
23.   Seifert, W.K. *Geochim. Cosmochim. Acta* **1978**, *42*, 473-84.
24.   Sinninghe Damsté, J.S.; Rijpstra, W.I.C.; Kock-van Dalen, A.C.; de Leeuw, J.W.; Schenck, P.A. *Geochim. Cosmochim. Acta* **1989**, *53*, 1343-55.
25.   Rullkötter, J.; Michaelis W. *Org. Geochem.* **1990**, 16, 829-52.
26.   Kohnen, M.E.L.; Sinninghe Damste, J.S.; Kock-van Dalen, A.C.; de Leeuw, J.W. *Geochim. Cosmochim. Acta* **1991**, *55*, 1375-94.
27.   Wagener, S.; Schulz, S.; Hanselmann, K. *FEMS Microbiol. Ecol.* **1990**, *74*, 39-48.
28.   Losher, A. *Ph.D. Thesis* **1989**, ETH Zürich.
29.   Züllig, H. *Schweiz. Z. Hydrol.* **1985**, *47*, 87-125.
30.   Volkman, J.K.; Kearney, P.; Jeffrey, S.W. *Org. Geochem.* **1990**, *15*, 489-97.
31.   Wardroper, A.M.K. *Ph.D. Thesis* **1979**, University of Bristol.
32.   Robinson, N. *Ph.D. Thesis* **1984**, University of Bristol.

33. Rullkötter, J.; Littke, R.; Schaefer, R.G. In *Geochemistry of Sulfur in Fossil Fuels*; Orr, W.L.; White, C.M., Eds.; ACS Symposium Series 429; American Chemical Society: Washington, D.C., **1990**, pp. 149-69.
34. Eglinton, G.; Hamilton, R.J. In *Chemical Plant Taxonomy*; Swain, T., Ed.; Academic Press: London, **1963**, pp. 187-217.
35. Ourisson, G.; Albrecht, P.; Rohmer, M. *Pure Appl. Chem.* **1979**, *51*, 709-29.
36. Volkman, J.K. *Org. Geochem.* **1986**, *9*, 83-101.
37. Thomas, J. Ph.D. Thesis 1990, University of Bristol.
38. Rowland, S.J.; Yon, D.A.; Lewis, C.A.; Maxwell, J.R. Org. Geochem. 1985, 8, 207-13.
38. Nichols, P.D.; Volkman, J.K.; Palmisano, A.; Smith, G.A.; White, D.C. J. Phycol. 1988, 24, 90-6.
40. Summons, R.E.; Barrow, R.A.; Capon R.J.; Hope, J.M.; Stranger, C. Aust. J. Chem. 1993, 46, 907-15.
41. Volkman, J.K.; Barrett, S.M.; Dunstan, G.A. Org. Geochem. 1994, 21, 407-13.
42. Meyers, P.A.; Ishiwatari, R. Org. Geochem. 1993, 20, 867-900.

RECEIVED June 27, 1995

Chapter 5

Influence of Sulphur Cross-linking on the Molecular-Size Distribution of Sulphur-Rich Macromolecules in Bitumen

Stefan Schouten[1], Timothy I. Eglinton[2], Jaap S. Sinninghe Damsté[1], and Jan W. de Leeuw[1]

[1]Division of Marine Biogeochemistry, Netherlands Institute for Sea Research (NIOZ), P.O. Box 59, 1790 AB Den Burg, Texel, Netherlands
[2]Department of Marine Chemistry and Geochemistry, Woods Hole Oceanographic Institution, Woods Hole, MA 02543

A sulphur-rich bitumen was separated by gel permeation chromatography (GPC) and the fractions obtained were analyzed for free and sulphur-bound carbon skeletons. Free and low-molecular-weight sulphur compounds are mainly present in GPC-fractions representing $M_w < 800$. Desulphurization with deuteriated nickel boride of the polar fractions of the GPC-fractions showed that fractions of increasing molecular weight have increasing amounts of deuterium incorporated into the released carbon skeletons. This indicates that sulphur-rich geomacromolecules were separated on basis of molecular weight through gel permeation chromatography. These results further support the idea that sulphur-crosslinking is an important factor in the formation of high-molecular-weight substances present in bitumens.

The incorporation of sulphur into organic matter in the subsurface is believed to occur during very early diagenesis (e.g. *1-3*). Hydrogen sulphide (e.g. *2,4*), polysulphides (e.g. *5-7*) and elemental sulphur (e.g. *8*) are suggested as inorganic sulphur species which react with functionalized lipids. Depending on the number and positions of the functionalities in the organic substrate the sulphur may react either intermolecularly or intramolecularly. In the case of the latter reaction, cyclic (poly)sulphides are formed which are subsequently converted to thiophenes during increased diagenesis. Intermolecular reactions of sulphur with functionalised lipids results in the formation of dimers or, when multiple functionalities are present, in a complex mixture of oligo- and polymers. Several studies (e.g. *8-14*) have shown that this extensive crosslinking can lead to high-molecular-weight compounds present in polar and asphaltene fractions and in kerogens. Sinninghe Damsté *et al.* (*10*) and Kohnen *et al.* (*11*) suggested that with increasing molecular size (i.e. with increasing crosslinking) the solubility of the macromolecular aggregates decreases. Thus, kerogens may contain sulphur-rich macromolecules with a high degree of crosslinking whilst asphaltenes and polar fractions may contain less cross-linked macromolecules which are still soluble in organic solvents.

To test this hypothesis, a sulphur-rich bitumen from the Monterey Formation was subjected to gel permeation chromatography (GPC), a technique commonly used in polymer sciences to separate polymers into molecular size fractions. Eglinton *et al.* (*15*) already reported on the X-ray absorption and pyrolysis-mass spectrometry analyses of these fractions. The data indicated that the thiophene/sulphide ratio, a parameter thought to be related to the degree of polymerization, increases with GPC-fractions of decreasing molecular weight. Here we report on the presence of free and sulphur-bound hydrocarbons in the total bitumen and its GPC-fractions.

Experimental

Sample Preparation. An immature sediment sample from the Monterey Formation at Naples Beach (California) was taken from a 10 cm interval at the base of Unit 315 approximately 9 m below the lowest phosphorite horizon (*15*) and was comprised of a lenticularly laminated claystone. It was ultrasonically extracted with $CH_3OH/CHCl_3$ and the resulting extract was washed with distilled water and dried using sodium carbonate.

Gel Permeation Chromatography. The gel permeation chromatography (GPC) of the Monterey bitumen has been described by Eglinton *et al.* (*15*). Briefly, the bitumen was separated into nine fractions by GPC using three Shodex styrene-divinylbenzene columns (KF-801, KF-802 and KF-802.5; column dimensions 300 mm x 8 mm) and $CHCl_3$ as eluent (0.5 ml/min). The column effluent was monitored by a UV/VIS (254 nm) detector (Hitachi L-4200) and a refractive index detector (Shodex RI-71). Approximate molecular sizes of GPC fractions were determined by comparison with retention times for polystyrene standards (Polymer Labs Inc.)

Analysis of Bitumen and GPC-fractions. The bitumen and its GPC-fractions were analyzed as shown in Figure 1. Apolar fractions were isolated by column chromatography using Al_2O_3 as stationary phase and a mixture of hexane/dichloromethane (9:1 v/v) as eluent. The residual polar material left on the column was ultrasonically extracted from the stationary phase using a CH_3OH/CH_2Cl_2 (1:1 v/v) mixture. The polar material was desulphurized using deuterated nickel boride (*16*) and the released hydrocarbons were isolated using column chromatography (Al_2O_3; eluent hexane/dichloromethane 9:1 v/v). A known amount of an internal standard [2,3-dimethyl-5-(1',1'-d$_2$-hexadecyl)thiophene] was added prior to isolation of the apolar fraction and the desulphurization of the residue. Before quantification of the hydrocarbons released by desulphurization of the residue the mixture was hydrogenated using H_2/PtO_2.

Gas Chromatography. GC was performed using a Carlo Erba 5300 instrument, equipped with an on-column injector. A fused silica capillary column (25 m x 0.32 mm) coated with CP Sil-5 (film thickness 0.12 μm) was used with helium as carrier gas. For detection a flame ionization detector (FID) was used. The samples (dissolved in ethyl acetate) were injected at 75 °C and subsequently the oven was programmed to 130 °C at 20 °C/min and then at 6 °C/min to 320 °C at which it was held for 10 min.

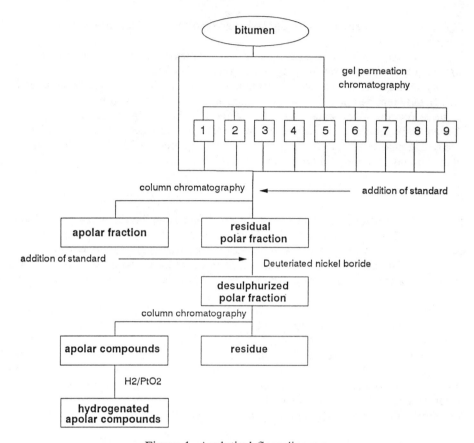

Figure 1. Analytical flow diagram.

Gas Chromatography-Mass Spectrometry. GC-MS was performed on a Hewlett-Packard 5890 gas chromatograph interfaced to a VG Autospec Ultima mass spectrometer operated at 70 eV with a mass range m/z 40-800 and a cycle time 1.8 s (resolution 1000). The gas chromatograph was equipped with a fused silica capillary column (25 m x 0.32 mm) coated with CP Sil-5 (film thickness = 0.2 μm). The carrier gas was helium. The samples were on-column injected at 50 °C and subsequently the oven was programmed to 130 °C at 20 °C/min and then at 4 °C/min to 300 °C at which it was held for 10 min.

Results

A sulphur-rich bitumen, obtained by extraction of a sediment sample from the Monterey Formation, was subjected to gel permeation chromatography to separate sulphur-rich macromolecules on bases of molecular-weight (Figure 2). Nine fractions were obtained (*15*) of which the two highest molecular weight fractions (1 and 2) were combined since they represented very low amounts of material. The total bitumen and the GPC-fractions were subjected to column chromatography to obtain the apolar fraction. In this way apolar low-molecular-weight compounds were removed from the more polar high-molecular-weight compounds. The residual polar fraction was desulphurized using deuteriated nickel boride to asses the amount of sulphur-linkages of the sulphur-rich geomacromolecules (Figure 1).

Free Hydrocarbons. The gas chromatogram of the apolar fraction of the total bitumen revealed that it predominantly consists of a complex mixture of sterenes, aromatic steroids, phytane, lycopane, hopanoids, thiophenes and a C_{13} alkene (Figure 3).

GC-analysis of the apolar fractions of the 8 GPC-fractions revealed that only GPC-fractions 7 to 9 contained significant amounts of GC-amenable material. Other apolar fractions only contained internal standards and contaminants as GC-amenable components. Eglinton *et al.* (*15*) determined, using polystyrene standards, that fractions 7 to 9 mainly contain compounds with a molecular weight lower than 800. This is in agreement with the presence of low-molecular-weight apolar compounds in these fractions since apolar fractions are generally assumed to contain compounds with a molecular weight smaller than 800 (*11*). Most of the sterenes, aromatic steroids and thiophenes are present in GPC-fraction 8 whilst in fraction 9 naphthalenes, phenanthrenes and dibenzothiophenes are dominant. Fraction 9 thus contains compounds with a similar or even higher polarity as those in GPC-fraction 8 (dibenzothiophenes *versus* thiophenes and sterenes) but generally with lower molecular weights. This suggests, at least for the compounds in the apolar fractions, that the GPC-separation is based on molecular-weight rather than polarity.

Sulphur-Bound Hydrocarbons. Desulphurization with deuteriated nickel boride of the residual polar material (Figure 1) of the total bitumen resulted in the formation of several deuteriated carbon skeletons (Figure 4). Phytane is present in significant amounts and possesses 0-2 deuterium atoms. Steranes range in carbon number from C_{27} to C_{29} and have predominantly 0-3 deuterium atoms incorporated. The dominant hopane is the C_{35} 17ß,21ß(H)-hopane which possesses 0 to 8 deuterium atoms. N-alkanes are present in relatively high amounts but, due to the low intensity of molecular ions in their mass spectra, the extent of deuterium labelling could not be determined.

Desulphurization of the residual polar fraction of the GPC-fractions only yielded deuteriated hydrocarbons for fractions 4 to 9. The GPC-fractions 1+2 and 3 may nevertheless contain sulphur-bound compounds but due to the high-molecular-weight nature of the fractions the nickel boride reagent may be less effective because of steric hindrance. Quantification of the cholestane and C_{35} hopane carbon skeletons released upon desulphurization show that the low-molecular-weight fraction 8 contains the highest amounts (Figure 5).

Importantly, the pattern of deuteriation changes with GPC-fraction. Cholestane possesses 0 to 4 deuterium atoms in GPC-fraction 5 (1 deuterium atom dominant) but this changes to 0-1 deuterium atoms in GPC-9 (0 deuterium atom dominant; Figure 6). The C_{35} hopane has incorporated 0-9 deuterium atoms in GPC-fraction 5 (7 deuterium atoms dominant) but 0-2 deuterium atoms in GPC-fraction 9 (0 deuterium atom dominant; Figure 7). Using these data in combination with the quantitative data presented in Figure 5, it is possible to calculate the average deuteriation of cholestane and the C_{35} hopane in the desulphurized bitumen: 0.7 deuterium atoms for cholestane and 2.4 deuterium atoms for the C_{35} hopane. These average numbers of deuterium atoms compare very well with those obtained by desulphurizing the total bitumen: 0.8

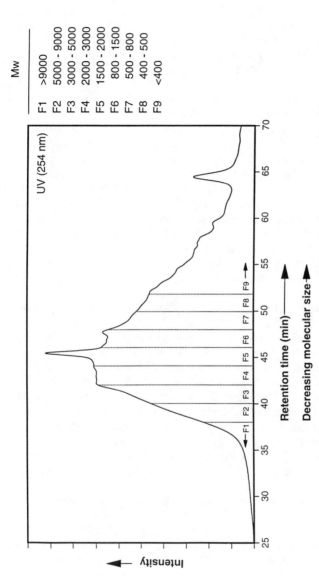

Figure 2. GPC chromatogram of Monterey Formation bitumen. F1-F9 denote the time zones used to collect molecular size-fractions. Approximate molecular size ranges for each fraction based on calibration with polystyrene standards.

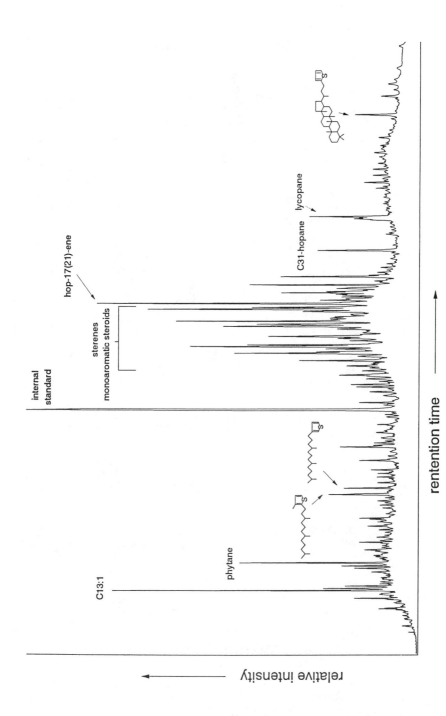

Figure 3. Gas chromatogram of free hydrocarbons present in the bitumen.

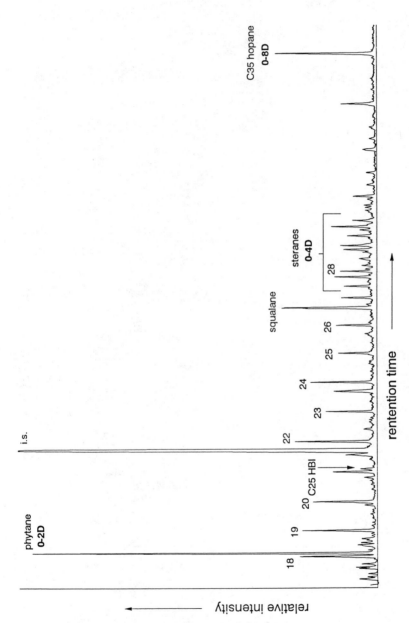

Figure 4. Gas chromatogram of hydrocarbons released upon desulphurization of the residual polar fraction of bitumen using deuteriated nickel boride. Numbers indicate total number of carbon atoms of n-alkanes. C_{25} HBI = highly branched isoprenoid.

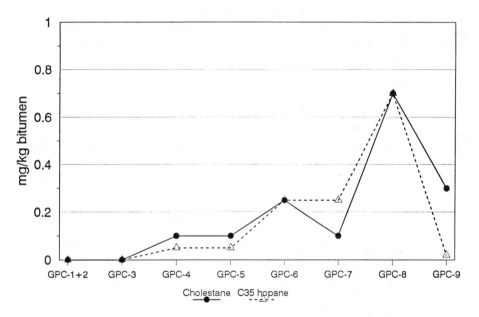

Figure 5. Amount of cholestane and C_{35} hopane released upon desulphurization of the GPC-fractions.

for cholestane and 2.3 for the C_{35} hopane. This indicates that the sulphur-rich geomacromolecules present in the GPC-fractions are representative of the bitumen as a whole.

The occurrence of carbon skeletons without deuterium atoms seem rather peculiar but has been noted before using other deuteriated desulphurization reagents like Nickelocene (12) and Raney Nickel (13) and is attributed to impurities in the reagents and the faster reaction rate of hydrogen compared to deuterium and exchange reactions. It must also be noted that the high amounts of cholestane and C_{35} hopane without deuterium atoms in GPC-fraction 9 may have been due to an incomplete removal of the apolar fraction due to some problems with the column chromatography at that time.

Using the results from desulphurization of model compounds by deuteriated nickel boride (16) it is possible to reconstruct the type of sulphur compound from the deuteration pattern. Compounds with one (poly)-sulphide bond yield compounds with predominantly one deuterium atom. Desulphurization of thiolanes yield compounds with two deuterium atoms although small amounts of 1 and 3 deuterium atoms can be detected. Desulphurization of a thiophene yields a more complex pattern which is dominated by 6 deuterium atoms, but also with substantial amounts of 7 and 8 deuterium atoms probably due to exchange with the allylic hydrogen atoms.

Released cholestane possesses up to 4 deuterium atoms. This deuterium pattern can be obtained by desulphurizing steroidal thiolanes and/or thiols. Such compounds, however, can be excluded in the residual polar material since steroidal thiolanes and thiols elute in the apolar fraction (8-9). The deuterium pattern of cholestane is therefore explained by the presence of cholestane with predominantly 1 and 2 intermolecular sulphur-linkages. The small amounts of cholestane with three and 4 deuterium atoms may be derived from desulphurisation of steroidal thiolanes linked by a sulphur-bond to the macromolecular network or, less likely, from cholestane with 3 intermolecular sulphur-linkages.

The deuteriation pattern of the C_{35} hopane is dominated by 2 and 7 deuterium atoms. The hopane with mainly 2 deuterium atoms may result from desulphurization of a hopanoid thiolane. As in previous example, this is considered unlikely, since hopanoid thiolanes are eluting in the apolar fraction. Thus the hopane carbon skeletons with mainly two deuterium atoms are derived from hopanes attached to the macromolecular network by two sulphur atoms. The hopane carbon skeletons with mainly 7 deuterium atoms are probably derived from a hopanoid thiophene linked to the macromolecular network by one sulphur atom. A similar complex deuteration pattern is visible for the C_{35} hopane released by desulphurization of GPC-fractions 4, 5 and 6 as for standard hopanoid thiophenes, indicating that it is indeed derived from a hopanoid thiophene linked by one sulphur. The question remains why the intermolecularly linked hopanoid thiophene, present in small amounts (Figure 5), is predominant in the high-molecular-weight fractions. A separation based on polarity instead of molecular weight can be excluded since hopanoid thiophenes linked by one sulphur atom are less polar then the C_{35} hopanes linked by two sulphur atoms which are present in the other fractions. One might reason that this C_{35} hopanoid thiophene moiety is only linked to the macromolecular network by one sulphur-atom, so that it can not act as a cross-linking agent but is only attached at the fringes of the crosslinked macromolecule. The nickel boride reagent may still be capable of desulphurizing peripheral parts of the macromolecular network but fails to desulphurize the core of the network due to steric hindrance.

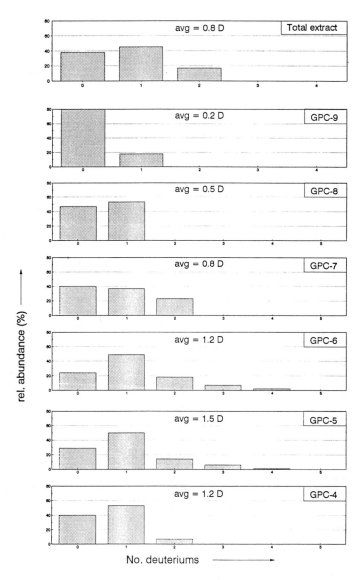

Figure 6. Extent of deuterium labelling of cholestane released upon desulphurization of the bitumen and its GPC-fractions.

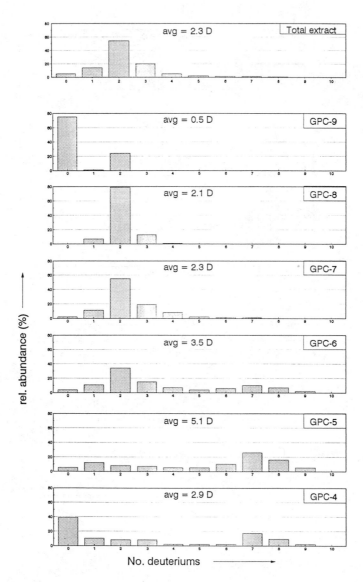

Figure 7. Extent of deuterium labelling of pentakishomohopane released upon desulphurization of the residual polar fraction of the bitumen and its GPC-fractions.

Excluding this peripheral moiety, the overall pattern is thus characterized by an increasing average amount of deuteriation for GPC-fractions of increasing molecular weight. This supports the concept that geomacromolecules in fractions with increasing higher molecular weight are more crosslinked and thus have more C-S bonds per carbon skeleton. XANES- and pyrolysis-data indicated this already for the Monterey bitumen (Eglinton *et al.*, 1994) but these analysis were performed on the whole fractions without prior removal of low-molecular-weight apolar compounds. In addition these prior studies did not yield direct information on the type and number of sulphur linkages for discrete compounds that the chemolysis results have provided.

Conclusions

Free and low-molecular-weight sulphur compounds are mainly present in GPC-fractions representing $M_w < 800$. Desulphurization of the polar material of the GPC-fractions of increasing molecular weight showed that carbon skeletons released have an increasing amount of deuterium labelling. This indicates that sulphur-rich geomacromolecules in fractions of higher molecular weight are indeed more crosslinked and thus have more C-S bonds per carbon skeleton.

Acknowledgements

This study was partly supported by a PIONIER-grant to JSSD from the Netherlands Organization for Scientific Research (NWO) and a grant to TIE from the Basic Energy Science division of the US Department of Energy (DE-FG02-92ER14232). Dr. M. Lewan (USGS, Denver) is thanked for provision of the Monterey sediment sample and Ms. J.E. Irvine is thanked for technical assistance. This is Division of Marine Biogeochemistry Contribution no. 178.

Literature Cited

1.	Vairavamurthy, A.; Mopper, K. *Nature* **1987**, 329, pp. 623-625.
2.	Sinninghe Damsté, J.S.; Rijpstra, W.I.C.; Kock-van Dalen, A.C.; de Leeuw, J.W.; Schenck, P.A. *Geochim. Cosmochim. Acta* **1989**, 53, pp. 1443-1455.
3.	Kohnen, M.E.L.; Sinninghe Damsté, J.S.; Kock-van Dalen, A.C.; ten Haven, H.L.; Rullkötter, J.; de Leeuw J.W. *Geochim. Cosmochim. Acta* **1990**, 54, pp. 3053-3063.
4.	Fukushima, K.; Yasukawa, M.; Muto, N.; Uemura, H.; Ishiwatari, R. *Org. Geochem.* **1992**, 18, pp. 83-91.
5.	Kohnen, M.E.L.; Sinninghe Damsté, J.S.; ten Haven, H.L.; de Leeuw, J.W. *Nature* **1989**, 341, pp. 640-641.
6.	de Graaf, W.; Sinninghe Damsté, J.S; de Leeuw, J.W. *Geochim. Cosmochim. Acta* **1992**, 56, pp. 4321-4328.
7.	Krein, E.B.; Aizenshtat Z. *Org. Geochem.* **1994**, 21, pp. 1015-1025.
8.	Schmid, J.C. *Ph.D. dissertation* **1986**, University of Strasbourg, pp. 263.
9.	Sinninghe Damsté, J.S.; Rijpstra, W.I.C.; de Leeuw, J.W.; Schenck, P.A.; In *Advances in Organic Geochemistry 1987* (Matevelli, L.; Novelli,M., Eds.), *Org. Geochem.* **1988**, 13, pp. 593-606.

10. Sinninghe Damsté, J.S.; Eglinton, T.I.; Rijpstra, W.I.C.; de Leeuw, J.W. In: *Geochemistry of Sulphur Compounds in Fossil Fuels* (Orr, W.L.;White, C.M., Eds.), ACS Symp. Ser. **1989**, Washington DC, pp. 486-528.

11. Kohnen, M.E.L.; Sinninghe Damsté, J.S.; Kock-van Dalen, A.C.; de Leeuw, J.W. *Geochim. Cosmochim. Acta* **1991**, 55, pp. 1375-1394.

12. Richnow, H.H.; Jenisch, A.; Michaelis, W. In: *Advances in Organic Geochemistry 1991* (Eckhardt, C.B. *et al.*, Eds.), *Org. Geochem.* **1992**, 19, pp. 351-370.

13. Adam, P.; Schmid, J.C.; Mycke, B.; Strazielle, C.; Connan, J.; Huc, A.; Riva, A.; Albrecht, P. *Geochim. Cosmochim. Acta* **1993**, 57, pp. 3395-3419.

14. Hoffman, I.C.; Hutchison, J.; Robson, J.N.; Chicarelli, M.I.; Maxwell, J.R. In: *Advances in Organic Geochemistry 1991* (Eckhardt, C.B. *et al.*, Eds.), *Org. Geochem.* **1992**, 19, pp. 371-388.

15. Eglinton, T.I; Irvine, J.E.; Vairavamurthy, A.; Zhou, W.; Manowitz, B. In: *Advances in Organic Geochemistry 1993* (Telnaes, N.; van Graas, G.; Øygard, K., Eds.), *Org. Geochem.* **1994**, 21, pp 781-800.

16. Schouten, S.; Pavlovic, D.; Sinninghe Damsté, J.S.; de Leeuw, J.W. *Org. Geochem.* **1993**, 20, pp. 901-909.

RECEIVED May 11, 1995

Chapter 6

Alkanoic Subunits in Sulfur-Rich Geomacromolecules

Information from Stepwise Chemical Degradation and Compound-Specific Isotopic Analysis

J. Hefter, V. Hauke, H. H. Richnow, and W. Michaelis[1]

Insitute für Biogeochemie und Meereschemie, Universität Hamburg, Bundesstrasse 55, 20146 Hamburg, Germany

Alkanoic subunits (*n*-alkanes and carboxylic acids) are released from sulfur-rich geomacromolecules using a stepwise chemical degradation sequence. The sites of former sulfur bonds were labelled with deuterium during the degradation reaction to provide evidence of sulfur linking and cross-linking. The extent of deuteriation increases with the chain length of the degraded alkanoic structural subunits. GC-MS analyses confirm the presence of sulfur-bound carboxylic acids (*n*-C_{13} to *n*-C_{29}) and additional sulfur functionalities in alkyl-aryl substructures of high-sulfur macromolecules. Measurements with isotope ratio monitoring gas chromatography - mass spectrometry (IrmGC-MS) suggest that different chemical bonds (e.g. C-S-bonds, alkyl-aryl-bonds) are involved in the selective preservation of biogenic material in macromolecular structures. Carbon isotope fractionation due to macromolecular organic matter formation is discussed.

The geochemical transformation of sedimentary sulfur and its interaction with organic matter is well documented, and is reflected in the high abundance of this element (up to 14%) in heavy petroleum and kerogen (*1*). At least 90 % of sedimentary organic matter is present as macromolecular material. Thus, the chemical characterization of sulfur-rich geomacromolecules is a necessity and should provide information concerning the role of sulfur - organic matter interaction in sedimentary environments.

The main approaches used to elucidate the structure of sulfur species in macro-molecules can be divided into two classes: (I) non-destructive spectroscopic methods (e.g. IR, NMR, XANES) revealing bulk information about the abundance of aromatic and aliphatic carbon atoms and functional groups (including sulfur speciation); (II) chemical degradation of the macromolecular structure using selective cleavage reactions of distinct bonds followed by characterization of the resultant low-molecular-weight products with conventional analytical techniques, such as GC and GC-MS.

Here, we have combined two selective chemical degradation methods with com-pound-specific carbon isotopic analysis (IrmGC-MS). This approach permits both the characterization of inter- and intramolecular linking of sulfur or other structural subunits with organic compounds and the identification of different organic precursors

---

[1]Corresponding author

0097–6156/95/0612–0093$12.00/0

contributing to the formation of the macromolecules. Chemolysis of organic sulfur compounds combined with compound specific carbon isotope measurements have already been proven as a valuable approach to provide information on sulfurization mechanisms and the original depositional environment (2-4).

Materials and Methods

Basic data concerning the samples studied are summarized in Table I. The Monterey oils and sediments are part of the Cooperative Monterey Organic Geochemistry Study. From all samples, macromolecular fractions of different polarity, such as resins, asphaltenes and kerogens were prepared. The elemental composition and the abundance of the respective fractions in the investigated sediments and oils are shown in Table II.

Table I. Samples

Sample	Type	Origin	Age	Lithology
Monterey KG4	Sediment	California	Middle Miocene	calcareous shales and mudstones
Monterey #4	Oil	California	Miocene	silicates and carbonates
Monterey KG8	Sediment	California	Late Miocene	siliceous mudstones (carbonate poor)
Monterey #8	Oil	California	Miocene	silicates and carbonates
Sicilian shale	Sediment	Italy	Upper Triassic	laminated black shales
Boscan	Oil	Venezuela	Cretaceous	carbonates
Bati Raman	Oil	Turkey	Cretaceous	carbonates

Table II. Amount and elementary composition of studied fractions

Sample	Fraction	[wt% of extract or oil]	C [wt%]	S [wt%]	C/S-ratio
Monterey KG4	LMW[a]	3.4	n.d.[b]	n.d.	n.d.
	Resin	42.0	49.3	5.2	9.5
	Asphaltene	45.1	56.0	7.0	8.0
	Kerogen	n.d.	38.6	6.9	5.6
Monterey #4	LMW	52.3	n.d.	n.d.	n.d.
	Resin	24.9	74.6	8.0	9.3
	Asphaltene	18.6	79.0	8.0	9.9
Monterey KG8	Kerogen	n.d.	36.7	6.9	5.3
Monterey #8	LMW	18.9	n.d.	n.d.	n.d.
	Resin	26.2	76.9	9.4	8.2
	Asphaltene	18.0	78.8	9.5	8.3
Sicilian shale	LMW	1.1	n.d.	n.d.	n.d.
	Resin	23.4	49.6	12.7	3.9
	Asphaltene	59.9	48.8	10.6	4.6
Boscan	LMW	49.3	n.d.	n.d.	n.d.
	Resin	25.9	61.7	7.0	8.8
	Asphaltene	18.3	82.1	6.4	12.8
Bati Raman	LMW	45.1	n.d.	n.d.	n.d.
	Resin	15.9	73.9	8.0	9.2
	Asphaltene	25.4	78.3	10.1	7.8

[a]LMW = extractable low-molecular-weight compounds
[b]n.d. = not determined

Preparation of Low- and High-Molecular-Weight Fractions. Sediments were ultrasonically extracted several times with a mixture of dichloromethane/methanol (3:1, v/v) until the solvent was colourless. Asphaltenes were precipitated from the sediment extracts or the crude oils by addition of a fifty-fold excess of *n*-heptane. Chromatographic separation of the *n*-heptane soluble fraction was performed by silica column chromatography. Elution of the column with solvents of increasing polarity yielded the following fractions: hydrocarbons (*n*-hexane), aromatic and sulfur compounds (*n*-hexane/10-15 % diethyl ether), acids (dichloromethane) and resins (dichloromethane/methanol/water, 70:25:5, v/v/v). Kerogen isolates were prepared using conventional techniques (*5*). To achieve a better precision in the IrmGC-MS measurements, the *n*-alkanes were further separated from branched and cyclic alkanes by molecular sieve (0.5 nm, Merck) addition. The *n*-alkanes were recovered by dissolution of the molecular sieve with hydrofluoric acid.

Stepwise Chemical Degradation. A sequence of two degradation steps was applied to the macromolecular fractions. The initial desulfurization step provided labelling of the former sites of sulfur bonds with deuterium. After chromatographic separation of the resulting low-molecular-weight products, the remaining high-molecular-weight material was subjected to RuO_4 oxidation. The degradation steps, fractionation procedures and analytical techniques are summarized in Figure 1.

Desulfurization. Desulfurization of the respective macromolecular fractions was achieved with nickel boride, which has been shown to be an effective reducing agent for carbon-sulfur bonds (*6*, *7*). Recently, it has been demonstrated that this reagent can be successfully applied for the desulfurization of macromolecular organic matter of geological origin (*8*, *9*). In our degradation experiments, labelling of the desulfurization products was performed by using deuteriated nickel boride prepared *in situ* by the reaction of $NiCl_2$ with $NaBD_4$ in a molar ratio of 1:3. A more detailed description of the method is given elsewhere (*9*). For comparison, selected samples were additionally desulfurized with Raney nickel (*10-12*) and nickel(0)cene (*13*, *14*).

Ruthenium Tetroxide Oxidation. RuO_4 oxidation is a well-established method for the cleavage of alkyl-aryl moieties yielding carboxylic acids (*13-17*). A 15- to 20-fold molar excess of periodic acid (H_5IO_6), calculated on the organic carbon content of the sample, was used as co-oxidant. After mixing the sample with the co-oxidant in dichloromethane/acetonitrile/water (2:2:3, v/v/v), 20 mg of RuO_2 was added as a catalyst. The reaction mixture was stirred for 15h at 35°C. Organic acids obtained by the reaction were esterified with diazomethane and purified by column chromatography.

Analytical Techniques

Elemental Composition. Sulfur and carbon contents were analysed on a *Carlo Erba* 1500 elemental analyzer.

Gas Chromatography. Analyses of low molecular weight hydrocarbons and carboxylic acids (as methyl esters) were carried out using a *Carlo Erba* 4160 GC equipped with a fused silica capillary column (DB-5, 30 m x 0.25 mm; J&W Scientific). The temperature program was: 80 °C, 3 min isothermal; 80-300 °C, 3 °C/min; 300 °C, 30 min isothermal. Injection mode: on column; carrier gas: H_2.

Gas Chromatography-Mass Spectrometry. GC-MS analyses were performed on a *Carlo Erba* 4160 instrument interfaced to a *Varian* CH7A mass spectrometer. MS conditions: 70 eV ionization energy; source temperature 250 °C, mass range *m/z* 50-

800; resolution 1000. The temperature program was: 80 °C, 5 min isothermal; 80-300 °C, 3 °C/min; 300 °C, 30 min isothermal. Injection mode: on column; carrier gas: He.

Isotope Ratio Monitoring Gas Chromatography-Mass Spectrometry. The IrmGC-MS system used was a Finnigan MAT 252 controlled by the ISODAT software. It includes a gas chromatograph interfaced *via* a combustion and water-removal assembly to an isotope-ratio mass spectrometer. A detailed description of the instrumental set-up and the methodology is given elsewhere (*18-20*). Isotope ratios reported in this paper are given as δ-values in parts per thousand (δ^{13}C °/oo) calculated relative to the Pee Dee Belemnite standard (PDB), Equation 1:

$$\delta^{13}C_S = 1000\left(\frac{R_S}{R_{PDB}} - 1\right) \tag{1}$$

R = ^{13}C/^{12}C ratio, subscripts S and PBD refer to sample and standard Pee Dee Belemnite, respectively.

Values were calibrated against an external CO$_2$ standard and the performance of the instrument was checked daily with a standard solution of hydrocarbons with known δ^{13}C values. All samples were measured at least twice, but generally three times. Replicate measurements produced differences usually lower than 0.8 °/oo.

The analysis of the carboxylic acids requires an esterification for GC/MS and IrmGC-MS measurement. The addition of a methyl group from the derivatization reagent to the carboxylic acids needs a correction for the measured methyl ester carbon isotope values. Equation 2 was used to calculate the isotopic composition of a fatty acid (FA) with *n* carbon atoms from the measured δ^{13}C value of the corresponding fatty acid methyl ester (FAME).

$$\delta^{13}C_{FA} = \frac{(n+1)\,\delta^{13}C_{FAME} - \delta^{13}C_{H2CN2}}{n} \tag{2}$$

The δ^{13}C value of diazomethane used in equation 2 was experimentally determined by measuring the δ^{13}C values of several fatty acid reference compounds prior to and following derivatization. Based on these measurements, the δ^{13}C value of the used diazomethane was calculated to be -73.0 ± 0.8 °/oo. The application of the mini-mum (-73.8 °/oo) and maximum (-72.2 °/oo) value for the diazomethane derivatization to equation 2 showed that the uncertainty introduced for the calculated isotopic values of fatty acids was ± 0.05 °/oo. Therefore all calculations were based on a δ^{13}C value of -73 °/oo for diazomethane.

Results and Discussion

Desulfurization - Efficacy and Specificity. The efficiency of the nickel boride reagent was first of all compared to previously used methods (*13, 14*) applied to the same samples (Monterey #4 oil resin and asphaltene, Monterey KG4 and KG8 kerogens). A desulfurization is obtained with all reagents, as indicated by increasing C/S ratios of the desulfurized residues (Figure 2).

Different trends are observed for the degradability of the various polar fractions as well as for the efficacy of the applied reagents. In general, the resin fractions were desulfurized with the highest efficiency, followed by the asphaltene and kerogen fraction. This decreasing degradability is most probably related to the solubility of the

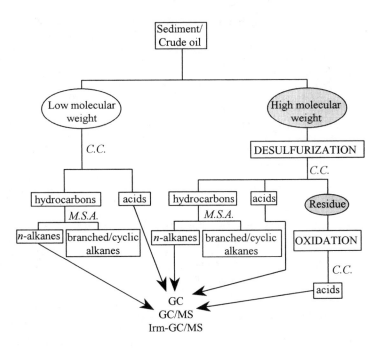

Figure 1. Generalyzed flow chart of sample fractionation, degradation and analysis. *C.C.* = column chromatography, *M.S.A.* = molecular sieve addition. Details for the preparation of macromolecular fractions depending on the sample type are given in the text.

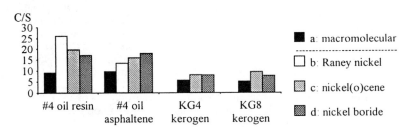

Figure 2. C/S ratios of Monterey macromolecular fractions prior (a) and after desulfurization with different reagents (b-d).

respective fractions, their accessibility to the applied reagents, or, in case of the kerogen, to the insolubility of the reactant.

Resin fractions were desulfurized most effectively by Raney nickel, followed by nickel(0)cene and nickel boride. In contrast, the highest degree of desulfurization was obtained for the asphaltene fraction with the nickel boride reagent. Confirming earlier work (8), this is probably due to an enhanced solubility of asphaltenes in the solvents used for this reaction (THF and CH_3OD). However, it should be mentioned that none of these reagents is capable of removing sulfur quantitatively.

Several synthetic reference compounds of different functionality were used to test the selectivity and reactivity of the nickel boride reagent (Figure 3 A, B). Major emphasis was placed on the investigation of the labelling specificity of this reagent during deuteriation. Usually, high degrees of sulfur conversion (> 90 %) were observed. Aliphatic thioethers and dithioethers as well as thiolanes and thiophenes were reduced to n-alkanes. The number of deuterium atoms incorporated allows to distinguish between these different types of sulfur species. Indeed 78 % of the alkanes released by the desulfurization of thioethers are labelled with one deuterium atom (Figure 3 C), whereas 82 % of the alkanes deriving from thiolanes incorporated two deuterium atoms. Desulfurization of the thiophene reference compound afforded alkanes of which 68 % were labelled by six deuterium atoms.

Sulfur-bound Alkyl Subunits. Desulfurization of geomacromolecules results in products, which can comprise a broad range of known hydrocarbon biomarkers (n-alkanes, isoprenoids, steroids, hopanoids, etc., 11-14). In earlier studies we investigated the biomarker content, composition, distribution and their selective mode of binding in sulfur-rich macromolecular fractions (13, 14). The present study focuses on aliphatic degradation products with a linear carbon skeleton. These compounds represent, at least for the samples investigated, significant sulfur-bound macromolecular building blocks. In the following sections, different types of inter- and intramolecular sulfur speciation of alkyl subunits are discussed in relation to sulfur content, diagenetic alteration and source of the biologically derived material.

Sulfur Content and Modes of Sulfur Incorporation. Resin fractions of different sulfur content (Monterey KG4: 5.2 wt.%S; Sicilian shale: 12.7 wt.%S), but comparable maturity, were analyzed to investigate the relationship between the sulfur content of the macromolecular substrate and the mode of n-alkane incorporation via C-S bonds.

Figure 4 shows the relative abundance of deuterium atoms in n-C_{18} and n-C_{22} alkanes obtained by nickel boride desulfurization of the respective resin fractions. The uptake of two deuterium atoms predominates in both samples (Monterey resin: n-C_{18}: 37.6 %, n-C_{22}: 36.9 %; Sicilian resin: n-C_{18}: 33.2 %, n-C_{22}: 24.5 %), inferring a significant contribution of dithioether structures independent of the sulfur content. Significant contributions of free or trapped thiolane structures which could also yield two-fold deuteriated products seem unlikely, as low molecular weight thiolanes were removed during the resin preparation. It has to be considered, however, that a small amount of desulfurization products could originate from polar components (sulfoxides) trapped in the resin fraction.

A significant difference was observed between both samples for the multideuteriated components of the respective n-alkanes (Figure 4). Compounds released upon desulfurization of the Monterey KG4 resin with 5.2 wt. % S possess at most 8 deuterium atoms, whereas the range of multideuteriation is extended up to 20 deuterium atoms in alkanes derived from the Sicilian shale resin with 12.7 wt. % S. The sum of components with more than 8 deuterium atoms represents a considerable contribution to the n-alkanes desulfurized from the sulfur-rich Sicilian shale resin.

Obviously the sulfur content of the macromolecular substrate mirrors the type of

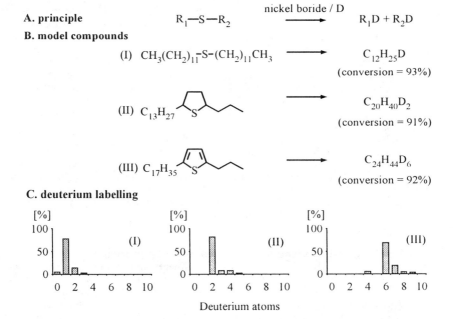

Figure 3. Chemical degradation of C-S bonds by nickel boride/deuterium (A, B) and deuterium labelling of reference compounds (C): desulfurized thioether (I), thiolane (II) and thiophene (III). Data are shown as relative contributions to the molecular ion(s).

compounds incorporated *via* C-S bonds. An increase in the number of deuterium atoms found in the desulfurization products can for example be related to the amounts of double bonds of biological precursor molecules. Double bonds are favoured sites for an attack of active sulfur species whereby organic compounds become linked *via* sulfur bonds to the macromolecular network, or to form intramolecularly bound sulfur compounds (e.g. in thiolanes, thiophenes) (*21-23*).

Chain Length Related Effects. Figure 4 shows a chain length related dependence on the amount of multideuteriation in *n*-alkanes obtained by desulfurization of the Sicilian shale resin. About 18 % of the n-C_{18} compound was labelled with 8-20 deuterium atoms, while the respective amount for the n-C_{22} alkane was nearly 30 %. This trend is shown even more by comparing the relative amounts of C_{14}-C_{30} *n*-alkanes containing 0 to 7 with those having 8 or more deuterium atoms (Figure 5). The contribution of polydeuteriated compounds ($\geq$ 8 deuterium atoms) clearly increases with increasing carbon atom number, reaching more than 40 % in the long chain homologues. Similar results were obtained from the desulfurization of the Bati Raman resin fraction (*9*).

Polydeuteriation with more than 7 deuterium atoms cannot be sufficiently explained by the presence of multiple thioether linkages between alkanoic subunits in the macromolecules. Sulfur cross-linking of this type would require, in some instances, more than 20 linkages on one alkanoic unit. However, the observed polydeuteriation could result from the former presence of thiolane and thiophene structures connected through thioether bonds to the macromolecule.

The number of sites suitable for attack of an active sulfur species forming inter- and/or intramolecular sulfur functionalities statistically increase with increasing carbon atom number of the biological precursor molecule. This may possibly explain the observed relation between chain length of the cleaved individual components and the deuterium uptake. However, the natural product precursors for these long chain structural subunits are not quite clear (*12*). One possible explanation is that they arise from polyenes or polyunsaturated functionalized compounds present in algal or bacterial biomass (*12, and ref. therein*).

Tracing Diagenetic History and Modes of Sulfur Incorporation. The structure of sulfur-rich macromolecular fractions present in sediments and crude oils should trace significantly the biological input in addition to the diagenetic history. To investigate the influence of the latter, a preliminary study on resin fractions with comparable sulfur contents but different diagenetic history was initiated, analyzing Monterey sediment KG4 (5.2 wt. %S), Monterey oil #4 (8.0 wt. %S) and Bati Raman oil (8.0 wt. %S).

The deuterium signature of *n*-octadecanes obtained by the desulfurization of the respective resin fractions revealed only slight differences in deuteriation for octadecanes labelled with 2 to 7 deuterium atoms. Generally, about 37 % of the n-C_{18} compounds have incorporated two deuterium atoms (Figure 6).

In contrast, a significant diversity was observed for the polydeuteriated homologues containing 8 to 17 deuterium atoms. They are absent from the sediment derived resin (KG4), but constitute 3.6 % (Monterey #4) and 11.5 % (Bati Raman) in the oil derived resins. Additionally, monodeuteriated species are more abundant in the relatively immature Monterey samples (14-17 %) than in the Bati Raman sample (5 %).

These observations imply that immature resins are characterized by high relative amounts of aliphatic moieties linked *via* one or two C-S bridges to the macrostructure, whereas C-S bound aliphatics in more mature resins are characterized by additional intramolecular sulfur structures such as thiolanes and thiophenes. This difference may be related to rearrangement processes during diagenesis within the macromolecules resulting in an enhanced cross-linking of aliphatics *via* sulfur or the formation of sulfur-bound thiolanes and thiophenes. Results obtained by simulated sulfur incorporation and thermal transformation (*23*) support the suggested diagenetic transformation of

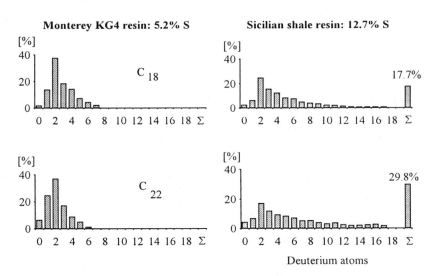

Figure 4. Relative distribution of deuterium atoms in n-C_{18} (above) and n-C_{22} (below) alkanes from the desulfurization of macromolecular fractions with different sulfur content. Σ = sum of 8-20 deuterium atoms, as a percentage of the total.

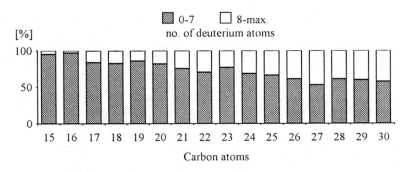

Figure 5. Relative distribution of deuterium atoms in n-alkanes obtained by nickel boride desulfurization of the Sicilian shale resin.

polysulfides to alkylthiophenes. Moreover, the proportion of thiophene sulfur in oils has been shown to increase with thermal maturity, which is probably caused by the conversion of more labile sulfide moieties to thiophene containing compounds (24). Artificial maturation studies confirmed that increasing thermal maturity results in a selective loss of thermally labile sulfide accompanied by a relative enrichment of thiophene moieties (24). Early oil generation from sulfur-rich kerogens has been related to the abundance of aliphatic sulfide and disulfide groups, which are expected to be thermally more labile than thiophenic type ring systems (25). Therefore, oil generation by thermal degradation and release of thioether bound alkanoic subunits should also lead to a residual macromolecule relatively enriched in thiophenic moieties. The relatively low amounts of alkyl moieties labelled by only one deuterium atom, accompanied by relatively high amounts of multideuteriated alkanes in the desulfurization products of the mature Bati Raman resin provides further evidence for an influence of diagenesis on the loss of sulfide moieties and the enrichment of thiophenic structures. However, more detailed studies are needed to rule out the possibility that the decrease in sulphide containing substructures could be caused by transfer to another macromolecular fraction.

Precursor Related Sulfurization Effects. The deuterium distribution in products released by desulfurization of the various macromolecular fractions provided information on the chemical structure of a biological precursor compound and the mode of its incorporation *via* sulfur. Moreover, it should be possible to distinguish different biological sources of the entrapped organic material by measuring the $\delta^{13}C$ isotopic values of the products released. The relationship between the chemical structure of a precursor and its mode of sulfurization was examined on the Monterey #4 oil resin. The overall distribution of *n*-alkanes obtained by chemical degradation of this sample suggested a multiple palaeobiological origin of sulfur-bound alkyl subunits present in the macromolecule.

The GC trace can be broadly subdivided into three sections (A, B and C, see Figure 7). Inserts show the deuteriation pattern of one individual *n*-alkane from each of the sections. In addition, the range and the numerical average of $\delta^{13}C$ values have been determined for all the components of the respective sections.

Section A includes alkanes with 15 to 24 carbon atoms and shows an even over odd predominance. The *n*-C_{24} alkane is the most abundant alkane in the aliphatic hydrocarbon fraction. The deuteriation pattern observed in *n*-C_{24} (Figure 7) is typical for *n*-alkanes in Section A. It clearly maximizes at two deuterium atoms indicating that significant amounts (~35 %) of the *n*-C_{15} to *n*-C_{24} alkanes have been linked to the macromolecular network *via* two C-S bonds. An average $\delta^{13}C$ value of -27.0 ‰ was determined for these hydrocarbons.

Section B comprises *n*-alkanes with 25 to 36 carbon atoms, showing no even or odd carbon number preference. *n*-C_{27} Alkane slightly predominates in this range. There is a clear difference in the deuterium signature from the *n*-alkanes in Section A. The relative amount of *n*-alkanes labelled with two deuterium atoms is minor, instead a shift towards an uptake of more deuterium (>3) can be observed. The average $\delta^{13}C$ value for *n*-alkanes within this group was found to be -29.6 ‰.

The *n*-C_{37} to *n*-C_{39} homologues, defined as Section C, are distinctive from Section B by a sharp increase in the relative concentration of the *n*-C_{37} alkane. The deuteriation pattern strongly differs from those of Sections A and B hydrocarbons. Significant amounts (~50 %) of the *n*-C_{37} and the *n*-C_{38} alkanes are labelled with 5 to 10 deuterium atoms. Carbon isotopic measurements on the *n*-C_{37} and *n*-C_{38} alkanes resulted in an average $\delta^{13}C$ value of -26.1 ‰.

The observed variations of the deuterium signature and the carbon isotopic composition provide evidence that depending on the type of precursor, different species of C-S bonds are present in the alkyl subunits. For example, the *n*-C_{37} and *n*-C_{38} alkanes released by desulfurization of the Monterey #4 oil resin can be assigned

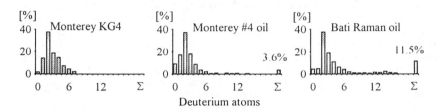

Figure 6. Relative deuterium distribution in *n*-octadecanes from desulfurization of resin fractions. Maturity increases from the left to the right side. Σ = sum of 8 to 17 deuterium atoms.

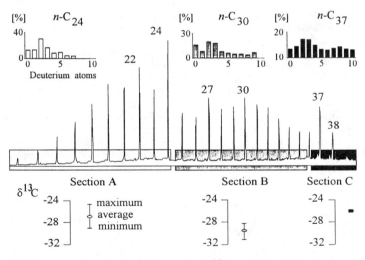

Figure 7. Monterey #4 oil resin: GC profile, δ^{13}C values and deuteriation pattern of distinct *n*-alkanes obtained after nickel boride desulfurization.

from their specific deuteriation pattern as well as from the $\delta^{13}C$ values as derivatives of long chain unsaturated (LCU) ketones and or alkenes. These compounds, indicators for the presence of marine prymnesiophyte algae, are abundant in the widespread coccolithophorid *Emiliana huxleyi* (26) and in some other species of prymnesiophytes (27). Alkenones are usually composed of 37 to 39 carbon atoms with a predominance of the C_{37} homologue and comprise between two and four double bonds. The degree of unsaturation is believed to depend on the growth temperature (28, 29).

The double bonds of these LCU ketones and alkenes provide preferred sites for a reaction with sulfur species, leading to the formation of intramolecularly sulfur-bound constituents of the macromolecular structure (30). Recently it has been shown that inorganic sulfur species can interact with double bonds and ketone functionalities to form organic sulfur compounds (31). The amount of double bonds present in the original alkenones (2-4) or alkenes may therefore result in at least 4 to 8 C-S linkages within the macromolecular network. Figure 7 clearly demonstrates that the distribution of deuterium atoms found in n-C_{37} and n-C_{38} alkanes obtained by desulfurization of the Monterey oil #4 resins is best understood in terms of this consideration. The $\delta^{13}C$ values of these hydrocarbons (n-C_{37} = -25.9 $^o/oo$; n-C_{38} = -26.4 $^o/oo$) support further evidence for a palaeobiological origin from prymnesiophyte related algae. The data fit well into the carbon isotopic range reported for algal biomass components in Monterey sediment samples (4).

Sulfur-bound Carboxylic Acids. Desulfurization of the Sicilian shale resin yielded considerable amounts of n-carboxylic acids, not detected in the degradation products of the other samples investigated. The contribution of these carboxylic acids to the macromolecular organic matter (85.6 mg/g_resin) is twice the amount of the total hydrocarbons (42.5 mg/g_resin) released upon desulfurization.

The carbon atom number distribution of the free extractable acids and the corresponding homologues released by desulfurization are depicted in Figure 8 together with the corresponding $\delta^{13}C$ values. The two fractions are quite similar regarding the relative distribution of individual fatty acids. This may suggest a common source. Both series are characterized by a general predominance of the short chain compounds with a concentration maximum at n-hexadecanoic acid. Long chain fatty acids (n-C_{19} to n-C_{29}) are present in minor, but nevertheless significant amounts, where their relative contribution is slightly enhanced in the sulfur-bound fraction. Both free and bound fractions reveal a remarkable predominance of even numbered compounds.

Table III. Carbon isotopic compositions, $\delta^{13}C$ ‰, of extractable and desulfurized resin n-carboxylic acids for the Sicilian shale sample

C^a	free	sulfur bound (resin)	Δ^b	C	free	sulfur bound (resin)	Δ
13	-25.6 ±0.4[c]	-26.5 ±0.0	-0.9	22	-27.2 ±0.3	-24.8 ±0.5	+2.4
14	-26.6 ±0.1	-25.3 ±0.1	+1.3	23	-28.8 ±0.5	-26.0 ±0.0	+2.8
15	-27.1 ±0.3	-24.0 ±0.3	+3.1	24	-27.6 ±0.5	-24.6 ±0.9	+3.0
16	-26.6 ±0.3	-25.2 ±0.4	+1.4	25	-29.1 ±0.7	-26.2 ±0.7	+2.9
17	-27.1 ±0.3	-23.9 ±0.7	+3.2	26	-29.2 ±0.4	-27.4 ±0.4	+1.8
18	-27.4 ±0.3	-23.4 ±0.7	+4.0	27	-29.6 ±0.6	-25.6 ±0.1[d]	+4.0
19	-27.9 ±0.4	-23.2 ±0.9	+4.7	28	-29.2 ±0.5	-26.6 ±0.6	+2.6
20	-27.3 ±0.3	-23.9 ±0.8	+3.4	29	-29.8 ±1.0	-28.6 ±0.4	+1.2
21	-27.3 ±0.2	-25.5 ±0.6	+1.8				

[a] Number of carbon atoms. [b] $\delta^{13}C$ difference of free and sulfur bound fatty acid. [c] Indicated uncertainities are standard errors of means of 3 or 4 measurements. [d] Co-elution of two compounds as indicated by *m/z* 45/44 ratio fluctuations.

The $\delta^{13}C$ values for the extractable fatty acids ranged from -25.6 to -29.8 ^{o}/oo (average -27.8 ^{o}/oo), while sulfur-bound fatty acids range from -23.2 to -28.6 ^{o}/oo (average -25.3 ^{o}/oo; Table III), that is the sulfur-bound acids are enriched by 2.5 ^{o}/oo (average) in ^{13}C relative to the free fatty acids.

This enrichment may be an indication of different biological sources, diagenetic isotopic fractionation during the formation of the macromolecular organic matter, or a mixture of both. However, the assumption that the observed isotopic shift is caused by two different organic sources is unlikely. In particular, the nearly constant difference in the carbon isotopic composition for the entire carbon atom number range indicates a close biogenetic relation of both acid fractions. Therefore, we propose an isotopic fractionation process has occurred in the sulfurization step of fatty acids during early diagenesis. Support for this hypothesis might be drawn from the fact that an analogous, yet weaker isotopic shift is observed for the free and sulfur-bound hydrocarbons of this sample (data not shown here).

The deuterium signature of the fatty acids obtained by desulfurization revealed a preferred linkage to the macromolecular substrate by one and two C-S bonds. However, as described above for the *n*-alkanes released upon desulfurization, multideuteriation of components also increases with the chain length of the *n*-carboxylic acids, reaching up to 20 deuterium atoms in the long chain homologues. Significant amounts of naturally occurring fatty acids are unsaturated homologues or possess additional functionalities (hydroxy acids), thus providing reactive sites for an early diagenetic sulfur incorporation and subsequent formation of macromolecular organic matter.

Alkanoic Structural Subunits Released by Ruthenium Tetroxide Oxidation of Previously Desulfurized Macromolecular Organic Matter. In order to obtain a more complete picture of structural moieties not exclusively bound by sulfur linkages, ruthenium tetroxide oxidation was applied subsequent to the desulfurization step. This reagent induces the chemical degradation of alkyl-aryl units in high yields (*33*). Furthermore, the reagent oxidizes ethers to the corresponding esters, which can be hydrolyzed under acidic conditions. Other functional groups, such as primary alcohols and olefins are oxidized to fatty acids (*34, 35*). Oxidation of alkyl-aryl units produces aliphatic acids bearing an additional carbon atom from the oxidation of the aryl structure. Similarly, aryl-alkyl-aryl groups are oxidized to aliphatic α,ω-dicarboxylic acids.

Figure 9 shows the GC trace of the acidic fraction obtained by RuO_4 oxidation of a previously desulfurized resin fraction (Sicilian shale). The oxidation afforded 90.1 mg acids per gram desulfurized resin. The main degradation products are aliphatic monocarboxylic acids ranging from n-C_7 to n-C_{27}, with maxima at n-C_9 and n-C_{16} and α,ω-dicarboxylic acids from n-C_8 to n-C_{27} maximizing at n-C_{17}. GC-MS investigation revealed the monoacids as well as the diacids to be a mixture of non deuteriated and deuteriated compounds. Usually, ~50-70 % of the molecular ions are deuterium free species, thus representing sulfur free alkyl-aryl units in the macromolecular organic matter. The remaining portions contain 1 to 7 deuterium atoms, which only can result from the former desulfurization step. Therefore, significant amounts of alkyl units in the parent macromolecular network must have been linked simultaneously *via* C-S and alkyl-aryl bonds or, in case of multideuteriated products, may arise from aryl-substituents possessing a former thiolane or thiophene structure.

The monocarboxylic acids and α,ω-dicarboxylic acids released upon RuO_4 oxidation cannot exclusively be assigned to former aryl-alkyl and aryl-alkyl-aryl structures, respectively. A certain proportion may derive from the oxidation of olefinic substructures or alcohol functionalities present in the macromolecular structure. Furthermore, α,ω-dicarboxylic acids can also derive from alkyl groups which were linked at one end to an aromatic ring, whereas at the opposite terminal position there

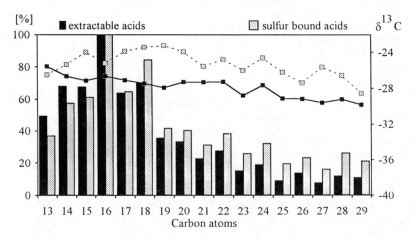

Figure 8. Normalized carbon atom number distribution of extractable fatty acids (black bars) and fatty acids (dotted bars) obtained by nickel boride degradation of the Sicilian shale resin. Lines indicate carbon isotopic composition.

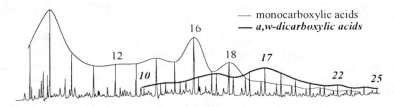

Figure 9. GC profile of carboxylic acids obtained by RuO_4 oxidation of the desulfurized Sicilian shale resin.

could have been a free carboxylic group, an olefinic moiety or alcohol structure. However, all of these possible precursors represent alkanoic building blocks of the parent macromolecule.

Conclusions

Selective stepwise chemical degradations combined with carbon isotope determinations were successfully used to characterize n-alkanoic structural elements of sulfur-rich macromolecular fractions.

Resin fractions with low sulfur content are characterized by the abundance of thioether and dithioether bound alkyl moieties, whereas sulfur-rich resins are typified by multiple C-S linked compounds and sulfur-bound alkylthiolanes and alkylthiophenes.

The detailed investigation of n-alkanes released by desulfurization of the Sicilian shale resin revealed a positive correlation between the number of carbon atoms and the amount of deuterium incorporation during sulfur reduction. This indicates that the number of possible sites of sulfur incorporation increase with increasing chain length.

A distinct difference in the deuteration pattern was observed for n-alkanes of identical carbon skeletons released from different resin fractions. This could be due to maturity related processes. Immature macromolecular organic matter is characterized by relative high amounts of alkanoic subunits linked *via* one or two C-S bridges to the macrostructure, whereas in mature high molecular weight material, C-S bound alkylthiolanes and -thiophenes are more common.

Straight chain substructures obtained by desulfurization of the Monterey #4 oil resin could be assigned to at least three different palaeobiologic sources, each characterized by a distinct mode of sulfurization and specific carbon isotopic values. For example, the deuteration pattern as well as the $\delta^{13}C$ values determined for n-C_{37} and n-C_{38} alkanes released upon desulfurization allowed us to relate these compounds to alkenones of a prymnesiophyte algae origin.

For the first time, to our knowledge, sulfur-bound n-carboxylic acids, mainly attached by one and two C-S linkages, were detected as structural elements of macromolecular organic matter. Their stable carbon isotope composition showed an enrichment of ^{13}C (2.5 ^{o}/oo) compared to their extractable homologues, independent of the carbon atom number. The corresponding relative distributions and $\delta^{13}C$ variations within the extractable acids and the fatty acids released upon desulfurization most probably indicate a common source of both fractions. Therefore we propose that the observed isotopic shift is caused by an isotopic fractionation during sulfurization and incorporation of fatty acids into sulfur-rich macromolecular organic matter.

Ruthenium tetroxide oxidation was applied to investigate alkyl structural subunits not exclusively sulfur-bound. Considerable amounts of the resulting degradation products were deuteriated, which only could arise from a former cross-linking of alkyl units to aromatic structures and simultaneously *via* sulfur to the macromolecules.

Our results suggest that combining stepwise selective chemical degradation techniques and stable carbon isotopic measurements of the individual degradation products is an effective approach for providing information on the chemical structure of geomacromolecules, the presence of cross-linking alkanoic compounds, and the palaeobiological origin of distinct macromolecular network constituents.

Acknowledgements

Isotopic measurements were performed at the Laboratoire de Geochimie Organique, Université Louis Pasteur, Strasbourg, France. We thank Dr. P. Albrecht for providing instrumental and laboratory facilities, P. Wehrung for helpful discussions and assistance in the IrmGC-MS measurements and Dr. R. Seifert (Hamburg) for an initial review of this manuscript. We are indebted to Dr. L. Stalker (School of Geology and Geophysics, University of Oklahoma) and two anonymous reviewers for their help in improving the quality of the paper. This work was supported by the Deutsche Forschungsgemeinschaft (Mi 157/7-3). The Monterey samples are part of the Cooperative Monterey Organic Geochemistry Study. We thank Agip, S. Donato Milanese - Milano for permission to publish and for providing the Sicilian shale sample.

Literature Cited

1. Sinninghe Damsté, J.S.,de Leeuw, J.W. *Org. Geochem.* **1990**, *16*, 1077-1101.
2. Kohnen, M. E. L., Schouten, S., Sinninghe Damsté, J. S., de Leeuw, J. W., Merritt, D. A., Hayes, J. M. *Science* **1992**, *256*, 358-362.
3. Sinninghe-Damsté, J. S., Wakeham, S. G., Kohnen, M. E. L., Hayes, J. M., de Leeuw, J. W. *Nature* **1993**, *362*, 827-829.
4. Schoell, M. Schouten, S., Sinninghe-Damsté, J.S., de Leeuw, J. W., Summons, R. E. *Science* **1994**, *263*, 1122-1125.
5. Orr, W. L. *Org. Geochem.* **1986**, *10*, 499-516.
6. Back, T. G., Yang K., Krouse, H. R. *J. Org. Chem.* **1992**, *57*, 1986-1900.
7. Back, T. G., Baron D. L., Yang K. *J. Org. Chem.* **1993**, *58*, 2407-2413.
8. Schouten, S., Pavlovic, D., Sinninghe Damsté, J.S., de Leeuw, J. W. *Org. Geochem.* **1993**, *7*, 901-909.
9. Hefter, J. Richnow, H. H. Seifert, R. Michaelis, W. In *Composition, Geochemistry and Conversion of Oil Shales*; Snape, C. E., Ed.; NATO ASI Series C; Kluwer Academic Publishers: Dordrecht, The Netherlands, 1995, Vol. 455, pp. 395-406.
10. Schmid, J. C, Ph.D thesis, **1986**, Université Louis Pasteur, Strasbourg, France.
11. Sinninghe Damsté, J. S., Rijpstra, I. W. C., de Leeuw, J. W., Schenck, P. A. *Org. Geochem.* **1988**, *13*, 593-606.
12. Adam, P., Schmid, J. C., Mycke, B., Strazielle, C., Connan, J., Huc, A., Riva, A., Albrecht, P. *Geochim. Cosmochim. Acta* **1993**, *57*, 3395-3419.
13. Richnow, H. H., Jenisch, A., Michaelis, W. *Org. Geochem.* **1992**, *19*, 351-370.
14. Richnow, H. H., Jenisch, A., Michaelis, W. *Geochim. Cosmochim. Acta* **1993**, *57*, 2767-2780.
15. Standen, G., Patience, R. L., Boucher, R. J., Eglinton, G. In *Organic Geochemistry -Advances and Applications in the Natural Environment*; Manning, D.A.C. Ed.; University Press: Manchester, U.K., 1991, pp. 454-457.
16. Standen, G., Boucher, R. J., Rafalska-Bloch, J., Eglinton, G. *Chem. Geol.* **1991**, *91*, 297-313.
17. Trifilieff, S., Sieskind, O., Albrecht, P. In *Biological Markers in Sediments and Petroleum - A Tribute to Wolfgang K. Seifert*; Moldowan, J. M., Albrecht, P., Philp, R. P. Ed.; Prentice Hall: Englewood Cliffs, USA, 1992, pp. 350-369.
18. Hayes, J. M., Freeman, K. H., Popp, B. P. In *Advances in Organic Geochemistry 1989*, Durand, B., Béhar, F. Ed.; *Org. Geochem. 16*, Pergamon Press, Oxford, 1990, 1115-1128.
19. Ricci, M. P., Merritt, D. A., Freeman, K. H., Hayes, J. M. *Org. Geochem.* **1994**, *21*, 561-571.
20. Merritt, D. A., Brand, W. A., Hayes, J. M. *Org. Geochem.* **1994**, *21*, 573-583.
21. Fukushima, K., Yasukawa, M., Muto, N., Uemura, H., Ishiwatary, R. *Org. Geochem.* **1992**, *18*, 83-91.
22. De Graaf, W., Sinninghe Damsté, J. S., de Leeuw, J. W. *Geochim. Cosmochim. Acta* **1992**, *56*, 4321-4328.
23. Krein, E. B., Aizenshtat, Z. *Org. Geochem.* **1994**, *21*, 1015-1025.
24. Waldo, G. S., Carlson, R. M. K., Moldowan, M., Peters, K. E., Penner-Hahn, J. E. *Geochim. Cosmochim. Acta* **1991**, *55*, 801-814.

25. Orr, W. L. In *Advances in Organic Geochemistry* 1985, Leythaeuser, D., Rullkötter, J. Ed.; *Org. Geochem.*, *10*, 1986, Pergamon, Exeter, U.K., Part I, pp. 499-516.
26. Volkman, J. K., Eglinton, G., Corner, E. D. S., Forsberg, T. E. V. *Phytochem.* **1980,** *19*, 2619-2622.
27. Marlowe, I. T., Brassel, S. C., Eglinton, G., Green, J. C. *Org. Geochem.* **1984,** *6*, 135-141.
28. Brassell, S. C., Eglinton, G., Marlowe, I. T., Pflauman, U., Sarntheim, M. *Nature* **1986,** *320*, 129-133.
29. Prahl, F. G., Wakeham, S. G. *Nature* **1986,** *320*, 367-369.
30. Sinninghe-Damsté, J. S., Rijpstra, W. I., Kock-van Dalen, A. C., de Leeuw, J. W., Schenck, P. A. *Geochim. Cosmochim. Acta* **1989,** *53*, 1343-1355.
31. Schouten, S., van Driel, G. B., Sinninghe Damsté, J. S., de Leeuw, J. W., *Geochim. Cosmochim. Acta* **1994,** *58*, 5111-5116.
32. Freeman, K. H., Wakeham, S. G., Hayes, J. M. *Org. Geochem.* **1994,** *6/7*, 629.644.
33. Stock, L. M., Tse, K. *Fuel*, **1983,** *62*, 974-975.
34. Carlsen, P. H. J., Katsukiti, T., Sharpless, K. B. *J. Org. Chem.* **1981,** *46*, 3936-3938.
35. Choi, C.-Y., Wang, S.-H., Stock, L. M. *Energy & Fuels* **1988,** *2*, 37-48.

RECEIVED May 11, 1995

Chapter 7

Proposed Thermal Pathways for Sulfur Transformations in Organic Macromolecules: Laboratory Simulation Experiments

Eitan B. Krein and Zeev Aizenshtat

Department of Organic Chemistry and Casali Institute for Applied Chemistry, Energy Research Center, Hebrew University of Jerusalem, Jerusalem 91904, Israel

We have investigated the possible role of thermal degradation of aliphatic polysulfide cross-linked polymers and oligomers, similar to those identified in low-maturity bitumens and oils, in the diagenesis and catagenesis of sedimentary organic sulfur. A number of polymers containing n-octyl building blocks were prepared and subjected to thermal alteration under different experimental conditions (160 and 300°C). At relatively low temperatures (160°C) the following processes were observed: (1) changes in the polymer structure; (2) formation of monomeric cyclic mono- and polysulfides; (3) incorporation of alkenes into the polymeric structure; and (4) elimination of hydrogen sulfide and elemental sulfur. At higher temperatures (300°C) elimination of H_2S and S_8 to form alkenes compete with the formation of thiophenes and low amounts of heavier aromatic sulfur heterocycles. The distribution of products depends on the structure of the polymers. These results may explain some of the reactions that take place in nature, and may provide a framework for the understanding of the chemistry that control the transformation of organic sulfur in sediments. The results also demonstrate the dynamic role of sulfur as an active element during diagenesis.

In recent years several investigations showed that immature sulfur-rich organic sediments contain high amounts of polysulfide cross-linked polymers and oligomers(1-5). These polymers consist of aliphatic biomarkers such as n-alkanes, isoprenoids, steranes and hopanes. The aliphatic building blocks are connected to each other by (poly)sulfide cross-linkages at specific locations in the aliphatic moieties. This specificity in the C-S bond positions supports the idea that sulfur incorporation takes place by low-temperature reactions between polysulfide anions and functionalized organic matter. This assumption is further supported by several recent studies which demonstrated the reactivity of polysulfide anions with different electrophilic functional groups such as activated double bonds, aldehydes and ketones, and under special conditions, even with dienes and monoenes (6-8,11,12). In most of the simulation

0097–6156/95/0612–0110$14.00/0

experiments polysulfide cross-linked oligomers or polymers are the major products. These polymers are very similar to the oligomers that occur in immature sediments.

Recently, we reported (*8*) that thermal exposure can transform polysulfide cross-linked oligomers formed from α,β-unsaturated aldehydes (phytenal and citral) into isoprenoid thiophenes. We proposed this reaction as a possible pathway for the geochemically controlled formation of naturally occurring alkyl isoprenoid thiophenes.

The objective of this study is to further investigate of the possible role of polysulfide cross-linked polymers as intermediates between sulfur incorporation during early diagenesis and the formation of stable sulfur heterocycles. We studied the effects of the initial polymer structure, and of the technique and temperature of pyrolysis, on the changes in the structure of the polymer and on the distribution of monomeric products. Several polysulfide cross-linked polymers were prepared from a number of n-octane derivatives by polymerization with ammonium polysulfide, under phase transfer catalysis (PTC) conditions. The starting materials were chosen so that the resulting polymer had a well defined structure, and not to mimic natural sulfurization. The choice of the positions of sulfidic cross-linkage was made to see the difference in the thermal behavior of the polymers. The polymers were subjected to thermal degradation by two different methods involving autoclaving at 160°C or treatment in a fluidized bed pyrolyzer at 300°C. This report presents the results of these experiments and the possible geochemical significance of the results.

Experimental

Materials. Polymers I-V were prepared under PTC conditions as described in Scheme 1. 1-Octene(**1**), 2-octene (cis+trans)(**2**), 1-octen-3-ol(**3**), 2-octenal(**4**) and 1-tetradecene(**5**) were purchased from Aldrich. Bromination with Br_2 was performed by the dropwise addition of a solution of bromine in CCl_4 (50 ml) to an ice-bath cooled solution of an equivalent amount of the alkene in CCl_4 (50 ml) over a period of 30 min until the resonance absorption peaks of the vinylic protons could no longer be detected in the ^{1}H-NMR spectrum. The solution was washed twice with water, dried over $MgSO_4$, and the solvent evaporated under reduced pressure at 40° C. Bromination with PBr_3 was performed by stirring a solution of 1-octen-3-ol in 50 ml of CCl_4 with an equal molar amount of PBr_3 in an ice-bath for 20 min, then the solution was washed and dried as described above. ^{1}H-NMR (400 MHz) analysis showed that a mixture of 3-bromo-1-octene and 1-bromo-2-octene (3:2) was formed. Octadienes (**6**) were prepared by steam distillation of 1-octen-3-ol (**3**) with a catalytic amount of H_3PO_4. The identity and purity of the products were confirmed by ^{1}H-NMR analysis, GC and GC/MS.

The reactions of the bromides and 2-octenal (**4**) with ammonium polysulfide and the depolymerization of the oligomers with MeLi and MeI were performed as described elsewhere(*8*)

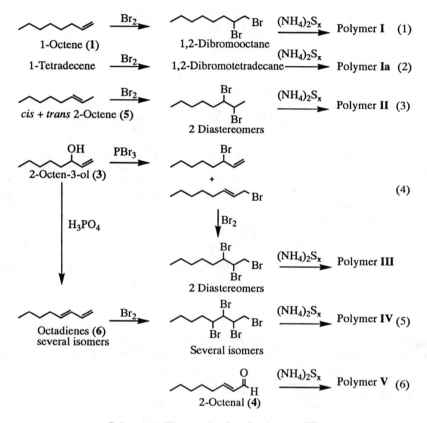

Scheme 1. The synthesis of polymers IV

Chromatography and Spectroscopy. Previously described methodes were used for GC, GC/MS and ^{1}H-NMR analyses (*8*), High resolution mass spectroscopy (resolution 3000) was carried out with ZAB-2F (VG Analytical) (*9*).

Thermal Decomposition: The general experimental procedure for the thermal decomposition experiments and for the analysis of the products are described in Scheme 2. In the method involving autoclaving, a sample (500 mg) of the polysulfide cross-linked polymer in a 25 ml glass vial was placed in a 25 ml autoclave (Parr), flushed thoroughly with N_2, and placed in an oil bath at 160°C. After 48h the released gas was passed through a solution of $AgNO_3$ in order to detect and collect hydrogen sulfide as Ag_2S. The liquid and solid residues were chromatographed on a short (1x10 cm) SiO_2 (70-230 mesh) column into two fractions as described in Scheme 2. Fluidized bed pyrolysis experiments were conducted at 300°C by a procedure described elsewhere(*8*). The products were extracted with hexane and analyzed by GC. Then the hexane solution was concentrated by evaporation under reduced pressure at 40°C in order to remove some of the light octenes and octadienes. The remaining, heavier, sulfur-containing compounds were analyzed by GC and GC/MS.

Results and Discussion

The Initial Structure of the Polymers. The initial structure of the polymers was established by elemental analysis, MeLi/MeI depolymerization and ^{1}H-NMR spectroscopy. The results of the elemental analysis are presented in Table I. The treatment with MeLi followed by MeI, is a chemical degradation method, which selectively cleaves S-S bonds(*10*), and has been used to analyse the structure of artificial(*8,11,12*) and natural(*3,4,13*) cyclic and cross-linked polysulfides.
In the present research this method was used to determine the degree of the substitution of bromides by (poly)sulfides. The distribution of the resulting methyl thioethers shows that elimination of HBr competes with the substitution reaction. For example, Figure 1a shows the gas chromatogram of the products from MeLi/MeI depolymerization of polymer **III** which was prepared from 1,2,3-tribromooctane. The mixture contains ca. 65% of the fully substituted 1,2,3-tri(methylthio)octane(**7**), and ca. 35% of unsaturated 3-(methylthio)-1-octene(**8**) and very small amounts of several di(methylthio)octanes. Therefore, the mean number of C-S bonds in the polymer is 2.3 for each octyl unit instead of the expected 3. Scheme 3a shows the reconstructed structure of polymer **III**. The presence of double bonds in the polymeric structure is also confirmed by the wide vinylic absorbances between 5.3 and 6.1 ppm in the ^{1}H-NMR spectrum of polymer **III** (as well as in all other similar polymers). Polymer **V**, which was prepared from 2-octenal, decomposed by MeLi/MeI to mainly 1,1,3-tri(methylthio)octane(**7a**) and, therefore, its structure is similar to the polymers formed from other α,β-unsaturated aldehydes described previously (*8*). Minor amounts of **7** and several di(methylthio)octanes were also formed.
Table I shows the distribution of the methylthio derivatives of the polymers, and the mean number of C-S bonds for each octyl unit. These results show that all polymers contain a mixture of saturated and unsaturated octyl units connected to each other by at least one polysulfide cross linking. These structures are very similar to the polymers obtained by the simulation reactions between polysulfides and alkenes(*11*), and to naturally occurring immature geopolymers. The polymers may also contain polysulfide cyclic moieties, however, there was no evidence for the formation of monomeric cyclic compounds during polymerization.

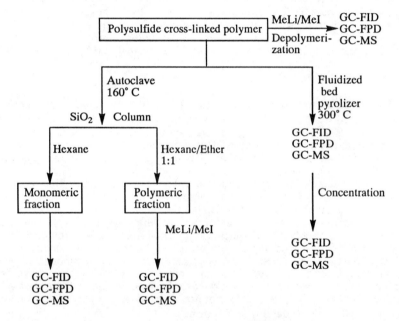

Scheme 2. Experimental procedures flow diagram for the thermal decomposition experiments, and for the analysis of the products.

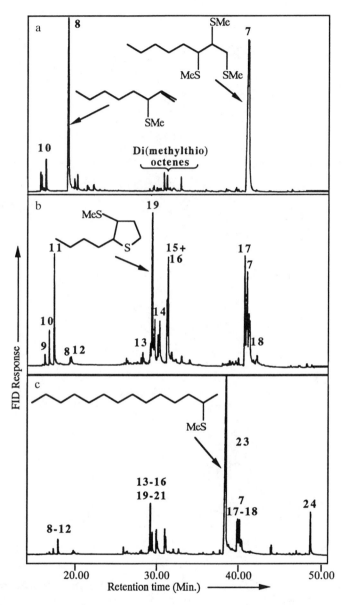

Figure 1. Gas chromatograms (FID) of the resulting products from the MeLi/MeI depolymerization of polymer **III**, (a) before heating, (b) after 48h in an autoclave at 160°C, (c) after 48h in an autoclave at 160°C with 1-tetradecene (**5**).

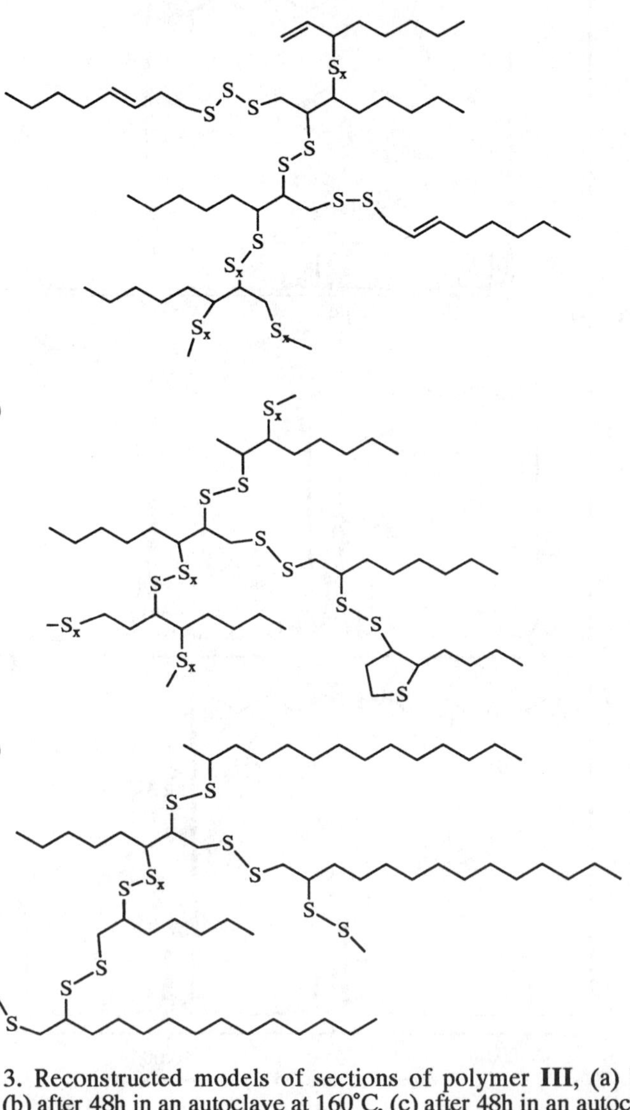

Scheme 3. Reconstructed models of sections of polymer **III**, (a) before heating, (b) after 48h in an autoclave at 160°C, (c) after 48h in an autoclave at 160°C with 1-tetradecene (**5**), based on the thioethers shown in Figure 1.

Table I: Elemental and chemical analyses of polymers I-V

Polymer	Br atoms position	Elemental Analysis C(%)	H(%)	S(%)	S/C	(Methylthio) octanes Mono-	Di-	Tri-	Tetra-	# C-S Bonds[a]
I	1,2	50.32	8.42	41	0.31	100[d]	57	0	0	1.4
Ia	1,2					100	54	0	0	1.4
II[b]	2,3	37.10	6.23	56	0.57	59	100	31	0	1.8
III	1,2,3	40.96	6.65	52	0.48	53	0	100	0	2.3
IV	1,2,3,4	35.54	5.42	59	0.62	9	100	12	3	2.1
V	2-Octenal	50.20[c]	8.06	38	0.28	0	2	98	0	3

[a]Average nubmer of C-S bonds [b] Prepared in ethanol. [c] Contains also 2.84% N.
[d](%)Normalized to most abundant derivative.

Autoclave Experiments (160°C).
Polymers **III** and **V** were treated at 160° for 48h under an inert atmosphere (N_2) in an autoclave. The released gas was passed through a solution of silver nitrate, and in both cases the precipitation of silver sulfide was observed, indicating the release of hydrogen sulfide during pyrolysis. The remaining material contained a mixture of a pale yellow liquid, soluble in hexane, and a brown tar, soluble in benzene or ether, but only partly in hexane. The mixture (liquid+tar) was subseqently separated on a short silica column into two fractions. The hexane fraction (25-30% by weight) contained monomeric products, and the hexane/ether fraction (70-75%) contained a brown-red polymeric substance. The latter was treated with MeLi/MeI in the same manner as the initial polymer (40-60% yield).

Figure 1b shows the gas chromatogram (FID) of the resulting methyl thioethers from the thermally treated polymer **III** and Scheme 3b shows the reconstruction of its structure. The identity of the methylthioethers (**7-18**) was established by GC/MS analysis based on the known fragmentation patterns of these compounds upon electron impact (EI) ionization in the mass spectrometer (*3,4,8,11*) Several alkyl-(methylthio) thiolanes were tentatively identified. All isomers of these compounds have a molecular ion at m/z 190. Exact mass measurements confirmed the expected molecular formula - $C_9H_{18}S_2$ (see appendix). A main fragment in the mass spectrum of compound **19** at m/z 133 (Figure 2) corresponds to the α–cleavage of an alkyl (methylthio)thiolane. The minor fragments at m/z 142 and 143 represent the elimination of MeSH and MeS respectively. The ions at m/z 85 and 100 (base peak) are assumed to be daughter ions of two possible structure of m/z 142. The ion at m/z 85 is the α–cleavage product of 2-butyl-2,5-dihydrothiophene and m/z 100 is formed by β–cleavage (and hydrogen transfer) of 2-butyl-4,5-dihydrothiophene. This fragmentation was observed also in the mass spectra of these two compounds (see following sections) and fit to a methylthio group at carbon 3. Thus, compound **19** was identified as 2-butyl-3-(methylthio)thiolane. Similar considerations were applied to identify compounds **20** and **21** as isomers of 2-(1-(methylthio)butyl)-thiolane, m/z 87 fits to a thiolane ring and m/z 113 fits to an ion formed after β–cleavage induced by a side chain double bond after elimination of MeSH.

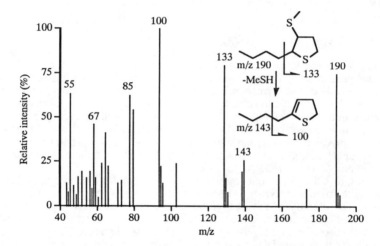

Figure 2. Mass-spectrum of 2-butyl-3-(methylthio)thiolane **(19)**.

The structure and distribution of the thioethers indicate that the thermal exposure changed the structure of polymer **III** quite significantly. The main changes are as follows:

- Most of the thioethers are saturated, for instance the unsaturated 3-(methylthio)-1-octene(**8**), which is a major component in the depolymerization products of the original polymer, disappeared almost completely, while saturated mono (methylthio)-octanes (**9-12**) were formed.

- The main components of the thioethers are di(methylthio)- (**13-16**) and tri(methylthio)-octanes (**8,17-18**) in contrast to mono- and tri(methylthio)-octanes (**7** and **8**) which were obtained from the polymer before heating.

- Four different isomers of tri(methylthio)-octane were formed from the thermally altered polymer while only one isomer, 1,2,3-Tri(methylthio)-octane (**8**), was formed from the initial polymer.

- The presence of the tentatively identified (methylthio)-thiolanes (**19-21**) is indicative for the presence of heterocyclic moieties connected to the polymeric framework as shown in Scheme 3b. No such evidence was found in the initial polymers.

Polymer **V** behaved very similarly, giving rise to 1,3-di-(methylthio)-octane (**16**) as the major product.

In order to understand the mechanisms that control the structural changes of the polymers, polymer **III** (250 mg) was treated exactly as above in the presence of 1-tetradecene (300 mg). As above, the resulting dark brown liquid which was soluble in hexane was separated on a SiO_2 column into a monomeric fraction (40%), eluting with hexane, and a polymeric fraction (60%), eluting with a mixture of ether/hexane (1:1). The polymeric fraction was treated with MeLi/MeI, both the monomeric and the MeLi/MeI treated fraction were analysed by GC and CG/MS. The results of these analyses are as follows:

- The major sulfur containing monomeric product (ca. 50%) is 2-thiotetradecane (**22**).

- The main monomeric compounds which do not contain sulfur (ca. 50%) are 1-tetradecene (starting material) and *cis+trans* 2-tetradecene. The distribution of the tetradecenes is almost identical to the distribution of 1- and 2-tetradecenes formed after fluidized bed pyrolysis of polymer **Ia** (see following text).

- The gas chromatogram of the MeLi/MeI treated fraction is presented in Figure 1c. This fraction contained all of the products which were obtained from pure polymer **III** and were described above. However, the main C_{14} products in this fraction are 2-(methylthio)tetradecane (**23**) and 1,2-di(methylthio)-tetradecane (**24**). Scheme 3c presents the reconstruction of the structure of this fraction.

Figure 3 shows the gas chromatograms of the monomeric fraction of the autoclave products of polymers **III** (3a) and **V** (3b). Both fractions contain sulfur-containing heterocycles as their main products. 3-Pentyl-1,2-dithiolane (**25**) and 3-pentyl-1,2,3-trithiane (**26**) are the dominant peaks in both cases (molecular ions at m/z 176 and 208 respectively, see appendix). The pyrolysis products of polymer **III** also contain 1,2,3,4-tetrathiepane (**27**, m/z 240) and dihydrothiophenes (**28-30**, m/z 142). One of the main characteristics in the monomeric products distribution from polymer **III** is that the cyclic di- or polysulfide compounds have several, less abundant, isomers in addition to the main isomers. This is in contrast to the pyrolysis products of polymer **V** which contain only one isomer of compounds 25 and **26**. The pyrolysis products of polymer **V** also contain 2-butylthiophene (**31**) and a very small amount of 2-butylthiolane (**32**) which are not present in the autoclave pyrolysis products of polymer **III**.

From the results of the autoclave experiments presented above it is clear that aliphatic polysulfide cross-linked polymers which are formed at low temperature are quite thermolabile at 160°C. The main processes that take place at this temperature are:

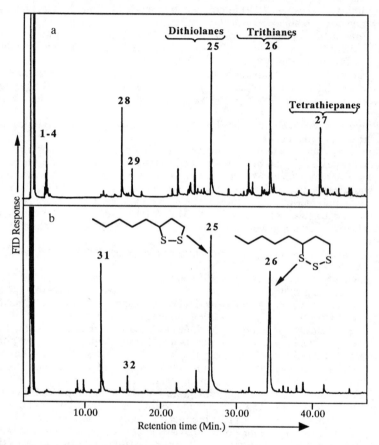

Figure 3. Gas chromatograms (FID) of the monomeric products formed after 48h in an autoclave at 160°C of polymers **III** (a) and **V** (b)

(1) changes in the polymer structure; (2) formation of monomeric cyclic mono- and polysulfides; (3) incorporation of alkenes into the polymeric structure; (4) elimination of hydrogen sulfide and elemental sulfur.

Mechanisms of Structural Changes. The processes which were described above may be explained by a general mechanism of elimination followed by addition of sulfur species. Di- or polysulfides are known to decompose at lower temperatures than the corresponding sulfides to form alkenes and hydrogen di- or polysulfide($14,15$). The latter can decompose to hydrogen sulfide and elemental sulfur:

$$R\text{-}CH_2\text{-}CH_2\text{-}S\text{-}S\text{-}CH_2\text{-}CH_2\text{-}R' \longrightarrow R\text{-}CH=CH_2 + R'\text{-}CH=CH_2 + H_2S_2$$

This process may proceed either by a multistep, radical or by a one-step, concerted mechanism. In a closed system such as an autoclave, this process or each of its steps may be reversible, i.e. the addition of the active sulfur species to double bonds which exist or form during pyrolysis. Such a process can explain all of the changes in the polymeric structure which were described above. The following equations show a few examples of such processes. S^* represents the active sulfur species (their identification will be discussed later), E represents elimination reaction and A addition reaction.

Disappearance of double bonds and of unsaturated octyl units connected with one C-S bond:

Rearrangement of C-S bond positions:

Cyclization:

Incorporation of alkenes into the polymeric framework, formation of thiols and isomerization of the alkenes:

The nature and the exact structure of the active sulfur species cannot be elucidated from the results described above. However, the incorporation of tetradecene into the polymers may provide some indirect evidence. The major amount of C_{14} units which are connected to a sulfur atom (compounds **22** and **23**) are bound at carbon 2. This specificity fits the Markovnikov orientation, and may indicate a carbocation intermediate, probably by a high-temperature addition of hydrogen sulfide and/or hydrogen polysulfide. On the other hand, the formation of 1,2-di(methylthio)tetradecane (**24**) may indicate a free radical mechanism. Two experiments were conducted in order to solve this problem. In the first experiment, 250 mg of 1-tetradecene (**5**) were heated under the same conditions as described above, together with 100 mg of elemental sulfur. Only 4% of the starting material polymerized, giving, after treatment with MeLi/MeI a mixture of **23** and **24** (1:2), indicating that in the pyrolysis experiments the latter compound is formed by reaction with released elemental sulfur. The non-polymeric fraction contained the starting material (without isomerization) and traces (<1%) of **22**. In a second experiment the same compounds were heated under hydrogen sulfide atmosphere (1 atm), but only traces of **22** were detected. These results show that the presence of elemental sulfur and/or hydrogen sulfide cannot explain the pyrolysis results, and that the sulfur species formed during the thermal decomposition of polysulfide cross-linked polymers are much more active. Our assumption is that a mixture of radical and hydrogenated sulfur species ($S_x^\bullet, HS_x^\bullet, H_2S_x, RS_x^\bullet$) which are released during the thermal decomposition of polysulfides via the mechanisms described in the literature(*14*), is responsible for the reactions described above.

Fluidized Bed Pyrolysis Experiments (300°C). In a previous study (*8*) we described the thermal formation of C_{10} and C_{20} isoprenoid thiophenes (see introduction) from polysulfide cross-linked polymers, formed from the corresponding α,β-unsaturated aldehydes, after fluidized bed pyrolysis at 300°C. In this investigation we studied the generality of this process in order to answer the following questions: Is the formation of alkyl thiophenes limited to polymers formed from α,β-unsaturated aldehydes which have a very distinctive structure? What is the influence of the polymeric structure (number and position of C-S bonds, sulfur content and S/C ratio) on the formation of thiophenes ? Do other products than thiophenes form, and what is their identity and distribution?

Table II presents the distributions of the main products from the 300°C fluidized bed pyrolysis of the six polysulfide cross-linked polymers which were described above. All six polymers produced alkyl thiophenes. This proves that the formation of alkyl thiophenes can be regarded as a general thermal decomposition pathway for all alkyl polysulfide cross-linked macromolecules. However, the structure of the polymer plays an important role in controlling the yield and the distribution of the alkyl thiophene isomers. In this regard, the polymers can be divided into two groups.

Table II: Fluidized bed pyrolysis results of polymers I-V

Polymer	# C-S Bonds[a]	Alk.[b] to Thio.Ratio	Alkane and Alkenes Distribution					Thiophenes Ratio[c]
			alkane	1-ene	2-ene	diene	triene	5-Me:2-Alk
I	1.4	100:12[d]	26[d]	61	89	100	0	21:100[d]
Ia	1.4	100:8	6	93	100	55	0	26:100
II	1.8	100:7	23	4	100	79	0	76:100
III	2.3	100:22	7	6	35	100	1	25:100
IV	2.1	100:28	0.5	0.8	2	100	5	32:100
V	3.0	0:100	0	0	0	0	0	00:100

[a]Average nubmer of C-S bonds. [b]Alk= alkanes+alkenes [c]2-Alkyl-5-methyl-thiophene to 2-alkyl-thiophene.[d](%)Normalized to most abundant derivative.

One group contains only polymer **V**. This polymer is formed from 2-octenal (**4**), and its thermal behavior is completely analogous to the behavior of the polymers formed from other α,β-unsaturated aldehydes, which were studied before(*8*) and therefore can be considered as members of this group. The polymers in this group decompose to give <u>only</u> alkyl thiophenes, without "free" alkanes or alkenes. Polymer **V** gave one isomer, 2-butylthiophene (**31**), which is the "end of chain" isomer. The isoprenoid thiophenes formed from phytenal and citral(*8*), are also two pairs of the "end of chain" isomers possible in isoprenoid systems.

The second group contains all other polymers which were investigated in the present study (**I-IV**). The polymers of this group form alkanes and alkenes as the main pyrolytic products (78-91%). Dienes are the most abundant products, especially from the more substituted polymers **III** and **IV**. However, mono-enes and even saturated alkanes are found in all cases. The identity of these compounds was established by comparison to GC retention times of authentic compounds. The yield of alkyl thiophenes is mainly controlled by the structure of the polymer, i.e. the number and position of C-S bonds, and not by the sulfur content. For example, polymer **II** has almost the same sulfur content as polymer **IV**, but the thiophene content in its pyro-products is only about one third of the content in the products of polymer **IV**. This point is even clearer in comparison to polymer **V**, which has the lowest sulfur content, but the highest thiophene content in the products. On the other hand, polymers **I** and **Ia**, which have the same average number of C-S bonds, in the same positions, gave very similar pyrolytic product distributions. This demonstrates also that the length of the alkyl chain had very little influence, if any, on the pyrolysis results. All five polymers, in this group, produced two isomers of alkyl thiophenes - 2-alkylthiophene (**31,33**) (the "end of chain isomer") and 2-alkyl-5-methylthiophene (**34,35**) (the "methyl isomer"). The "end of chain isomer" is the dominant isomer in all cases, indicating its higher thermodynamic stability. The ratio between the two isomers is probably controlled by the position of the C-S bonds in the alkyl chain. Polymer **II**, which is the only polymer that does not have a sulfur atom connected to carbon 1, gave the highest relative amounts of the "methyl isomer" (**34**), since the formation of the "end of chain" isomer (**31**) requires C-S bond rearrangement (this may also explain why polymer **II** gave the lowest yield of both isomers). The polymers, which have a C-S bond on carbon 1 (**II-IV**), have similar isomer ratios in their pyrolysis products, with a tendency to higher relative amounts of the "methyl isomer" in the more substituted polymers. Polymer **V**, which has two C-S bonds on carbon 1, gave only the "end of chain isomer"(**31**). It is noteworthy that no 2-alkyl-5-ethylthiophenes were found, even in polymers which contain C-S bonds on carbon 3.

Figure 4 shows the gas chromatograms of the fluidized bed pyrolytic products of polymers **III** (4a) and **IV** (4b) after concentration and removing most of the volatile compounds (octane and octenes), thus enabling the analysis of the less abundant, heavier, sulfur containing compounds. The content of these compounds usually does not exceed 1% of the pyrolytic products. The compounds were tentatively identified by GC/MS, and their molecular formulae were confirmed by high resolution MS as shown in the appendix. These compounds compose cyclic di- and trisulfides (**25,26**) and dihydrothiophenes (**28-30,36,37**) similar to those identified in the lower-temperature autoclave experiments. In addition to these compounds, unsaturated alkyl thiophenes (**38,39**,m/z 138) 2,2'-bithiophene (**40**, m/z 166) two isomers of condensed dithiophenes (2-ethylthieno[3,2-b]thiophene,**41**, 2,5-dimethylthieno[3,2-b]thiophene,**42**, m/z 168) and a condensed trithiophene (dithieno[3,2-b,d]thiophene,**43**, m/z 196) were also detected.

Mechanisms of Fluidized Bed Pyrolysis. In our previous work(*8*) we presented a detailed mechanism for the formation of alkyl thiophenes as the result of the pyrolysis of polysulfide cross-linked polymers formed from α,β-unsaturated aldehydes. We also discussed the energetic considerations that may explain this mechanism. The results described above are indicating a dramatic difference between the pyrolysis products of polymer **V** and those from other polymers supports the proposed mechanism. Therefore, we adopted the same energetic considerations to propose a mechanism for the pyrolysis of the second group of polymers. Scheme 4 shows the general mechanisms for the formation of thiophenes in the pyrolysis of a polymer similar to polymer **III** and of polymer **V**. Both mechanisms are series of eliminations of polysulfides to form double bonds, and in both cases thiophenes are formed by a cyclization of dienic intermediates that contain a vinylic sulfur atom (**44,45**). In the case of polymer **III**, where only one C-S bond is positioned on each carbon atom, two such intermediates can be formed (**44,45**), leading to the two thiophene isomers (**31,34**). Similar sequences can be adopted for the formation of octenes and octadienes. Disproportionation of alkyl radicals may explain the formation of saturated alkanes. In polymers like polymer **V**, the two C-S bonds on carbon 1 enable only one dienic intermediate (**44**), hence the formation of the "end of chain" thiophene (**31**) only. It is noteworthy that while **31** is also a major constituent of the

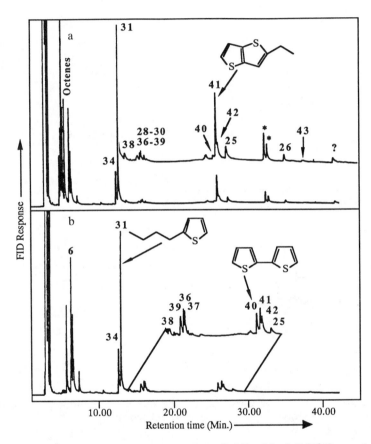

Figure 4. Gas chromatograms (FID) of the fluidized bed (300°C) pyrolytic products of polymers **III** (a) and **IV** (b). (*-unreacted tribromooctanes)

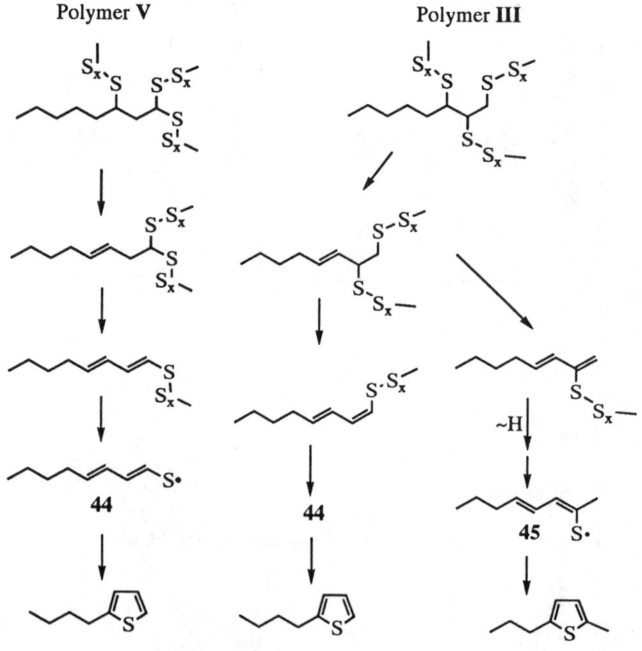

Scheme 4. Proposed mechanisms for the formation of thiophenes in the fluidized bed pyrolysis (300°) of polymers **III** and **V**.

monomeric fraction of the 160°C autoclave pyrolysate of polymer **V**, alkyl thiophenes were not identified in the corresponding fractions of the other polymers. This observation may indicate that polymers formed from α,β-unsaturated aldehydes (or similar sites in heterogeneous polymers) can yield alkyl thiophenes at lower temperatures than other polysulfide cross-linked polymers/moieties.

Geochemical Significance

It is now widely accepted that polysulfide cross-linked macromolecules, termed by Adam et al. (*2*) "nonpolar macromolecular fraction" (NPMF), occur as major constituents in relatively immature sulfur-rich bitumens and crude oils (*1,2,4,5,16*). Several theories were put forward in order to explain the formation of these macromolecules, either based on the analysis of their structure or on laboratory simulation studies. We suggested (*6*) the nucleophilic addition of polysulfide anions to electrophilic organic compounds such as aldehydes and α,β-unsaturated carbonyl compounds as a possible mechanism for cross-linking. Schouten et al. (*12*) also demonstrated the importance of aldehydes and ketones as precursors of NPMF, while de Graaf et al. (*11*) showed that in basic organic solvents, such as dimethylformamide, even non-electrophilic compounds such as dienes and monoenes can react with polysulfide anions. All of these reports proved that low temperature, anionic reactions of polysulfides can explain the early incorporation of sulfur into organic matter in nature, and the formation of cross-linked products. In many of the cases described in these reports the reactants lose their original functional groups (e.g aldehyde or keto-groups) and give products which contain alkyl units. Adam et. al. (*2*) preferred a radical mechanism for the formation of such compounds, but it seems that the formation of radical species requires somewhat elevated temperatures (*17,18*). Furthermore, most of these mechanisms could not explain the formation of sulfur heterocycles, especially aromatic thiophenes which are the main constituents of more mature bitumens and crude oils. The results of this research offer a possible explanation for the transformations of sedimentary polysulfide cross-linked polymers to aromatic sulfur-containing compounds.

Since the mechanism of formation of polysulfide cross-linked polymers was not the goal of this study, the polymers which were studied here were prepared by known synthetic methods. The main advantage of the use of such synthetic polymers is the possibility to tailor a defined structure for each polymer in order to check the influence of structural parameters on the results. Nevertheless, the structures of the polymers are very similar to polymers formed in simulations of natural sulfurization. For example, the elimination of HBr that takes place during the polymerization of polybrominated alkanes gives a polymer very similar to the polymers formed by the reaction of polysulfides with dienes in N,N'-dimethylformamide (DMF)(*11*). Therefore, we believe that such polymers can serve as suitable models for the study of thermal alteration in nature.

The main conclusion from the present research is that thermal exposure of polysulfide cross-linked polymers can produce sulfur-containing heterocycles regardless of their structure. This conclusion is a generalization of our previous suggestion for the formation of isoprenoid thiophenes. Therefore, we suggest that natural polysulfide cross-linked macromolecules can be regarded as an intermediate stage in the maturation of sulfur containing organic matter. During the maturation processes, the macromolecules rearrange and change their structure while eliminating sulfur species (H_2S or $S°$) and sulfur heterocycles. The autoclave experiments described here may represent such rearrangements in nature, resulting in a red-brown nonpolar macromolecular fraction very similar in structure to the "red-band" described by Adam et. al. (*1*). The thermally altered polymers are much more similar to natural

NPMFs than the original polymers formed after low temperature sulfurization as described here and elsewhere (11). For example the distribution of mono(methylthio)octanes (see Figure. 1b) is almost identical to the distribution of mono(methylthio)eicosanes which were formed after MeLi/MeI depolymerization of a polar fraction from the Vena del Gesso bitumen (3). Another example is cyclization inside the polymeric structure, which is a process identified in our experiments and in macromolecular natural fractions (4,19). This high similarity may suggest that the naturally occurring NPMF samples are the result of mild thermal exposure of the products of sulfur incorporation during early diagenesis .

The results also demonstrate that thiophenes are not the only heterocycles that can be obtained by the thermal decomposition of polysulfide cross-linked polymers. The main heterocycles that result from the autoclave experiments are cyclic di- and trisulfides. These are very similar (as lower homologues) to cyclic n-alkyl-disulfides identified in nature (for example in the Vena del Gesso sample)(13,20). Cyclic di- and trisulfides with a phytanyl chain were also found in nature. When the polymer formed from phytenal (8) was heated in an autoclave at 160°C we indeed recovered cyclic C_{20} isoprenoid di- and trisulfides (46,47, Figure 5) similar to those described by Kohnen et. al.(13). The proposed mechanism, based on our results, may provide an alternative explanation to the one suggested by Kohnen et al. (13) for the formation of cyclic alkyl polysulfides in nature. It is noteworthy that these compounds were identified (12) in the products of a simulation reaction between polysulfides and phytenal by on-column GC/MS analysis. We, however did not observe these compounds before thermal exposure.

Another observation from the autoclave experiments, which may have geochemical significance is the high reactivity of the sulfur species formed during thermal alteration of the polymers, i.e. incorporation of alkenes into the polymeric structure, and the formation of new sulfur containing compounds from these alkenes, show that sulfur must be regarded as an active element in all stages of organic matter transformation. It seems that during early stages of diagenesis anionic species of sulfur (especially catenated forms such as polysulfides) are the most important sulfurization agents. At this stage electrophilic organic compounds (such as aldehydes and ketones) will be the most active compounds and the first to react with sulfur species. However, at later stages of diagenesis, and during oil formation, radical and hydrogenated sulfur species may act as major sulfurizing agents, thus enabling the incorporation of sulfur into organic compounds that did not react with anionic species at earlier stages. Olefins and other organic compounds that have radical "sensitive" functional groups (e.g. benzylic carbons) may be the most important reactants during late diagenesis. At the late diagenesis and early catagenesis stages the mechanism presented by Adam et al. (2) for the formation of NPMF seems valid. It is important to note that in our experiments the polysulfide cross-linked polymers are much more active (towards alkenes) than elemental sulfur, hence such polymers can be regarded as a storage of "active sulfur".

At high temperatures (300°C), all of the polymers studied produced alkyl thiophenes. This result strengthens our suggestion that polysulfide cross-linked polymers can be the precursors of alkyl thiophenes in nature. We believe that this thermal stabilization mechanism rather than cyclization during the sulfurization of functionalized organic compounds (2,21) can better explain the high abundance of thiophenic compounds in mature organic matter, either as monomeric compounds or as a part of kerogen. The results also show that in addition to alkyl thiophenes, thermal decomposition of these polymers can give even more complicated sulfur compounds such as bithiophenes (40) and thienothiophenes (41-43) which are also found in nature(4,22,23).

The difference between the distribution of the pyrolytic products from polymers formed from unsaturated aldehydes (such as 2-octenal and phytenal), which give "end

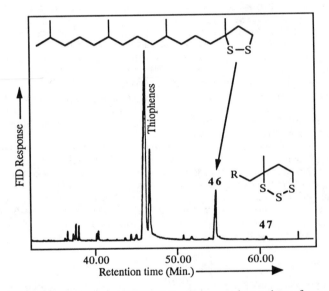

Figure 5. Gas chromatogram (FID) of the monomeric products formed after 48h in an autoclave at 160°C of the polymer formed from phytenal (*8*)

of chain" thiophenes, and the products from other polymers, which give a mixture of "end of chain" and "methyl" thiophenes, may serve as an indicator for the precursor of these compounds in nature. This phenomenon was also observed in several immature sediments for C_{20} isoprenoid thiophenes (24-26). Our results may suggest that in sediments, where this specific distribution is observed, phytenal is the precursor. This suggestion needs further investigation by more studies.

In the past, several authors have suggested that monomeric sulfur-containing compounds are formed as a result of intramolecular sulfur incorporation, while intermolecular reactions give sulfur cross-linked polymers (2,21). Our results show that the sulfur cycle is much more dynamic. In a previous study on polysulfide anion addition mechanisms (6) we have shown that these reactions are reversible at low temperatures (0-70°C), thus dimeric disulfides could be converted into cyclic monomers. The results presented here show that this reversibility of sulfur incorporation continues at higher temperatures, and that the monomeric and polymeric paths interchange, probably until thermodynamically stable structures such as thiophenes are formed. Scheme 5 shows a suggested general pathway for the formation and thermal alteration of the sulfur-containing macromolecules in nature.

The transformation of phytol is a good example for this scheme since sulfur-containing phytanyl derivatives which fit all of the stages were identified in nature. Another example of a specific compound which follows the transformations outlined in Scheme 5 is the C_{35} bacteriohopane tetrol (48) (Scheme 6). Since alcohols are not reactive towards polysulfides (as was found by others(12) and in our experiments), some diagenetic transformations must take place in order to give an electrophilic derivative. Elimination of two molecules of water will give diketones such as 49, however, since such compounds were not yet identified in nature other transformations can be valid as well. Such compounds may react with polysulfides and be incorporated in polymers. C_{35} hopanoid units indeed were found in NPMFs to be connected to the macromolecular structure by more than one C-S bond positioned on the side chain as in structure 50 (2). Relatively low thermal exposure of such structures, as in our autoclave experiments, can give hopanoid thiolane (51) and hopanoid 1,2-dithiolane or dithiane (52). Such compounds were indeed identified in immature sediments (5,13). Further thermal transformation will lead to hopanoid thiophene 53 , which was described in immature sediments as well (27).

Conclusions

In this chapter we present the results of laboratory experiments that were aimed to provide some answers on the thermal transformation of polysulfide cross-linked macromolecules which can be a major constituent in sulfur-rich bitumens and petroleums such as the Rozel Point oil , Monterey oils, Vena del Gesso bitumens and others. The general mechanism involving thermal decomposition of sulfur cross-linked polymers can explain the formation of all sulfur heterocycles that were identified in sulfur rich immature sediments.

The main conclusions of these results are as follows:

    - Polysulfide cross-linked polymers can be regarded as thermolabile intermediates in the formation of sulfur-containing heterocycles.

    - Low thermal exposure of polysulfide cross-linked polymers yields cyclic di- and polysulfides, as a product of internal restructuring of the polymers, which also include the formation of cyclic moieties inside the polymeric structure.

    - High thermal exposure (or longer periods of times ?) yield thiophenic (or bithiophenic) compounds.

    - The number and the positions of the C-S bonds in the alkyl units control the yield and the structure of the heterocyclic compounds.

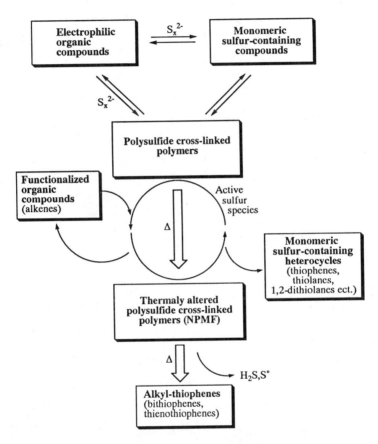

Scheme 5. A suggested general pathway for the diagenetic formation and for the thermal alteration of polysulfide cross-linked macromolecules in the geosphere

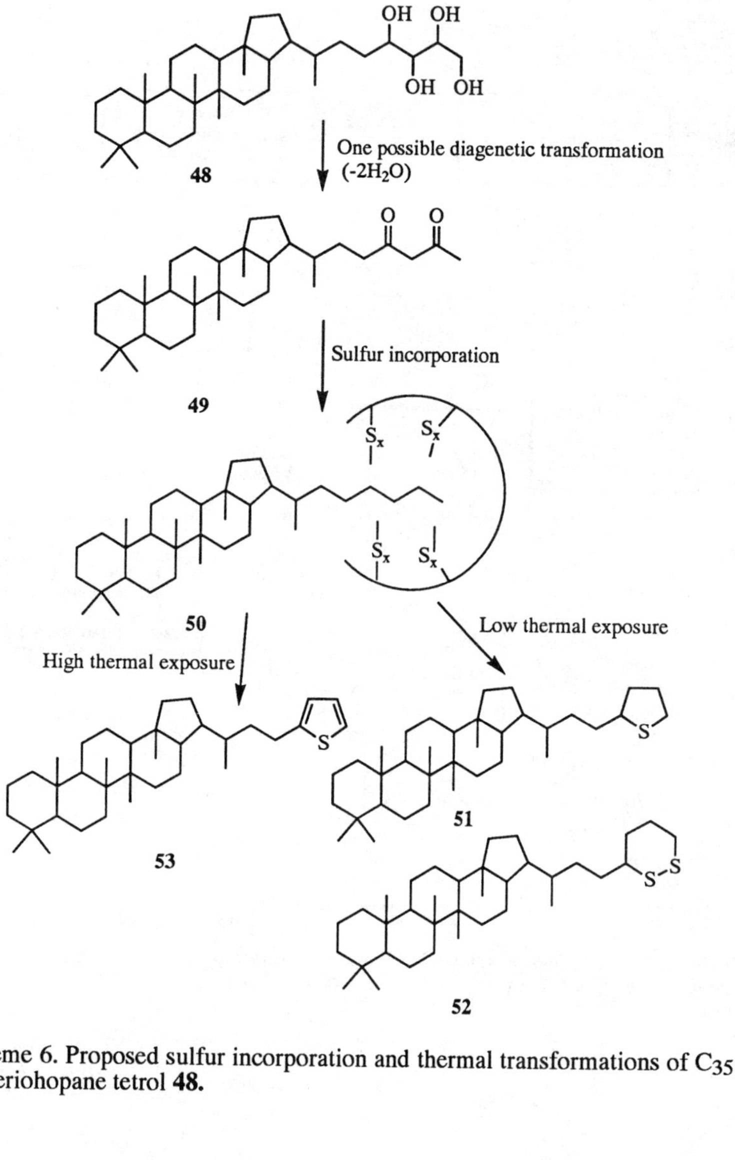

Scheme 6. Proposed sulfur incorporation and thermal transformations of C$_{35}$ bacteriohopane tetrol **48**.

Appendix-Identification and mass spectrometric data of the sulfur-containing compounds discussed in the text.

	Name	M° (Calc.)(Intensity-%)	Mol. formula[a]	Base peak	Characteristic peaks (Intensity-%)
7	1,2,3-Tri(methylthio)octane	252.1041(252.1040)(10)	$C_{11}H_{24}S$	61[CH_2SMe]	121(21)[CH(SMe)CH$_2$SMe],131(74)[C$_6$H$_{12}$SMe], 157(23)[204-MeS],204(9)[M-MeSH]
7a	1,1,3-Tri(methylthio)octane	252(24)		131[$C_6H_{12}SMe$]	61(36)[CH$_2$SMe],107(9)[CH(SMe)$_2$,157(13)[205-MeSH],205(6)[M-MeS]
8	3-(Methylthio)-1-octene	158(0)		69[C_5H_9]	110(68)[M-MeSH]
9	4-(Methylthio)octane[b]				
10	3-(Methylthio)octane	160(32)		57[C_4H_9]	89(94)[C$_3$H$_6$SMe],131(64)[C$_6$H$_{12}$SMe]
11	2-(Methylthio)octane	160.1286(160.1286)(32)	$C_9H_{20}S$	75[C_2H_6SMe]	112(20)[M-MeSH],145(41)[C$_7$H$_{14}$SMe]
12	1-(Methylthio)octane	160(65)		61[CH_2SMe]	112(8)[M-MeSH],145(38)[C$_7$H$_{14}$SMe]
13	3,4-Di(Methylthio)octane	206(40)		61[CH_2SMe]	89(76)[C$_3$H$_6$SMe],117(65)[C$_5$H$_{10}$SMe]
14	2,3-Di(Methylthio)octane	206.1170(206.1163)(54)	$C_{10}H_{22}S_2$	131[$C_6H_{12}SMe$]	61(49)[CH$_2$SMe],75(38), 159(6)[M-MeS]
15	1,2-Di(Methylthio)octane	206.1152(206.1163)(56)	$C_{10}H_{22}S_2$	145[$C_7H_{14}SMe$]	61(45)[CH$_2$SMe],159(5)[M-MeS]
16	1,3-Di(Methylthio)octane	206(60)		61[CH_2SMe]	87(47)[135-MeSH], 131(10)[C$_6$H$_{12}$SMe], 135(3)[C$_3$H$_5$+2SMe], 191(57)[M-Me]
17	1,3,4-Tri(methylthio)octane	252(4)		117[$C_5H_{10}SMe$]	61(78)[CH$_2$SMe],87(20)[C$_3$H$_4$SMe, 157(24)[204-MeS],189(16)[204-MeSH]
18	1,2,4-Tri(methylthio)octane (two isomers)	252(40)		61[CH_2SMe]	87(18)[C$_3$H$_4$SMe],117(8)[C$_5$H$_{10}$SMe], 157(42)[204-MeS], 237(18)[M-Me]
19	2-Butyl-3-(methylthio)thiolane	190.0894(190.0850)(74)	$C_9H_{18}S_2$	100[C_5H_8S]	85(63)[C$_4$H$_5$S],133(79)[M-butyl]

Continued on next page

Appendix. Continued

No.	Compound	Molecular ion	Formula	Base ion	Fragment ions
20	2-[1-(methylthio)butyl]thiolane	190(51)		$113[C_6H_9S]$	$87(83)[C_4H_7S], 142(19)[M\text{-}MeSH], 175(31)[M\text{-}Me]$
21	Isomer of 20	190(35)		$113[C_6H_9S]$	$87(89)[C_4H_7S], 142(17)[M\text{-}MeSH], 175(27)[M\text{-}Me]$
22	2-Thiotetradecane	230(13)		$57[C_4H_9]$	$61(40)[C_2H_4SH], 196(26)[M\text{-}H_2S]$
23	2-(Methylthio)tetradecane	244(25)		$75[C_2H_6SMe]$	$196(11)[M\text{-}MeSH], 229(16)[C_{13}H_{26}SMe]$
24	1,2-Di(methylthio)tetradecane	290(18)		$229[C_{13}H_{26}SMe]$	$61(49)[CH_2SMe]$
25	3-Pentyl-1,2-dithiolane	$176.0699(176.0693)(36)$	$C_8H_{16}S_2$	$69[C_5H_9]$	$87(27)[C_4H_7S], 101(6)[C_5H_9S], 105(8)[C_3H_5S_2], 143(21)[M\text{-}SH]$
26	3-Pentyl-1,2,3-trithiane	$208.0403(208.0414)(52)$	$C_8H_{16}S_3$	$87[C_4H_7S]$	$101(33)[C_5H_9S], 143(34)[M\text{-}HS_2], 175(5)[M\text{-}SH]$
27	3-Pentyl-1,2,3,4-tetrathiepane	240(27)		$69[C_5H_9]$	$87(54)[C_4H_7S], 101(20)[C_5H_9S], 105(15)[C_3H_5S_2], 143(32)[M\text{-}HS_3], 176(26)[M\text{-}S_2], 208(12)[M\text{-}S]$
28	2-Butyl-4,5-dihydrothiophene	$142.0805(142.0816)(27)$	$C_8H_{14}S$	$100[C_5H_8S]$	$85(20)[C_4H_5S], 97(9)[C_5H_5S], 113(4)[C_6H_9S]$
29	2-Propyl-5-methyl-4,5-dihydrothiophene	$142.0802(142.0816)(26)$	$C_8H_{14}S$	$113[C_6H_9S]$	$85(9)[C_4H_5S], 100(4)[C_5H_8S]$
30	Isomer of 29	142(31)		$113[C_6H_9S]$	$85(5)[C_4H_5S]$
31	2-Butylthiophene	$140.0678(140.0660)(29)$	$C_8H_{12}S$	$97[C_5H_5S]$	$111(17)[C_6H_7S]$
32	2-Butylthiolane	144(22)		$87[C_4H_7S]$	$101(11)[C_5H_9S]$
33	2-Decylthiophene	224(20)		$97[C_5H_5S]$	
34	2-Propyl-5-methylthiophene	$140.0678(140.0660)(34)$	$C_8H_{12}S$	$111[C_6H_7S]$	$98(46)[C_5H_6S], 111(15)[C_6H_7S]$
35	2-Nonyl-5-methylthiophene	224(11)		$111[C_6H_7S]$	
36	2-Butyl-2,5-dihydrothiophene	142(22)		$85[C_4H_5S]$	
37	2-Propyl-5-methyl-2,3-dihydrothiophene	$142.0793(142.0816)(36)$	$C_8H_{14}S$	$99[C_5H_7S]$	$85(21)[C_4H_5S]$
38	2-(1-Butenyl)thiophene	$138.0490(138.0503)(43)$	$C_8H_{10}S$	$123[M\text{-}Me]$	$97(9)[C_5H_5S]$
39	Isomer of 38	$138.0459(138.0503)(92)$	$C_8H_{10}S$	$123[M\text{-}Me]$	$97(33)[C_5H_5S], 111(17)[C_6H_7S]$

No.	Compound	Formula	m/z (HRMS)		Fragment peaks
40	2,2'-Bithiophene	$C_8H_6S_2$	166(100) 165.9901(165.9911)	166[M]	121(30)[M-HCS],134(11)[M-S]
41	2-Ethylthieno[3,2-b]thiophene	$C_8H_8S_2$	168.0049(168.0067)(52)	153[M-Me]	103(52)[M-HS$_2$]
42	2,5-Dimethylthieno[3,2-b]thiophene	$C_8H_8S_2$	168.0055(168.0067)(78)	167[M-H]	153(10)[M-Me],134(9)M-S]
43	Dithieno[3,4-b,d]thiophene	$C_8H_4S_3$	196(100) 195.9488(195.9475)	196[M]	152(16)
46	3-Methyl-3-(4,8,12-trimethyltridecyl)-1,2-dithiolane	$C_{20}H_{40}S_2$	344.2576(344.2571)(24)	55[C$_4$H$_7$]	311(9)[M-SH],alkenic peaks
47	4-Methyl-4-(4,8,12-trimethyltridecyl)-1,2,3-trithiepane		376(8)	55[C$_4$H$_7$]	311(57)[M-HS$_2$],344(14)[M-S],alkenic peaks

a Calculated from HRMS b Identified by retention time after ref. 3

- Organic sulfur transformations in sediments are continuous and reversible. Consequenly both dissociation of macromolecules to form sulfur species and unsaturated compounds and association of such molecules into polymers occur during thermal exposure. Hence, sulfur is an active element through all stages of organic matter transformation from early diagenesis to catagenesis.

Acknowledgements

We wish to thank the Moshe Shilo Centre for Marine Biogeochemistry for financial support.

Literature Cited

(1)   Adam, P.; Mycke, B.; Schmid, J. C.; Connan, J.; Albrecht, P. *Energy Fuels* **1992**, 533-559.
(2)   Adam, P.; Schmid, J. C.; Mycke, B.; Strazielle, C.; Connan, J.; Huc, A.; Riva, A.; Albrecht, P. *Geochim. Cosmochim. Acta* **1993**, *57*, 3395-3419.
(3)   Kohnen, M. E. L.; Sinninghe Damsté, J. S.; Kock van Dalen, A. C.; de Leeuw, J. W. *Geochim. Cosmochim. Acta* **1991**, *55*, 1375-1394.
(4)   Kohnen, M. E. L.; Sinninghe Damsté, J. S.; Baas, M.; Kock van Dalen, A. C.; de Leeuw, J. W. *Geochim. Cosmochim. Acta* **1993**, *57*, 2515-2528.
(5)   Schmid, J. C. Thesis Thesis, University Louis Pasteur, Strasbourg, 1986.
(6)   Krein, E. B.; Aizenshtat, Z. *J. Org. Chem.* **1993**, *58*, 6103-6108.
(7)   Krein, E. B.; Aizenshtat, Z. In *Organic Geochemistry: Poster sessions from the 16th international meeting on organic geochemistry, Stavanger 1993*; K. Øygard, Ed.; Falch Hurtigtrykk: Oslo, 1993; pp 596-602.
(8)   Krein, E. B.; Aizenshtat, Z. *Org. Geochem.* **1994**, *21*, 1015-1025.
(9)   Heller-Kallai, L.; Miloslavski, I.; Aizenshtat, Z. *Clays Clay Miner.* **1989**, *37*, 446-450.
(10)   Eliel, E. L.; Hutchins, R. D.; Mebane, R.; Willer, R. L. *J. Org. Chem.* **1976**, *41*, 1052-1057.
(11)   de Graaf, W.; Sinninghe Damsté, J. S.; de Leeuw, J. W. *Geochim. Cosmochim. Acta* **1992**, *56*, 4321-4328.
(12)   Schouten, S.; van Driel, G. B.; Sinninghe Damsté, J. S.; de Leeuw, J. W. *Geochim. Cosmochim. Acta* **1994**, *58*, 5111-5116.
(13)   Kohnen, M. E. L.; Sinninghe Damsté, J. S.; ten Haven, H. L.; Kock-van Dalen, A. C.; Schouten, S.; de Leeuw, J. *Geochimica and Geochimica Acta* **1991**, *55*, 3685-3695.
(14)   Martin, G. In *Supplement S: The chemistry of the sulphur-containing f unctional groups*; S. Patai and Z. Rappoport, Eds.; John Wiley & Sons: Chichester, 1993; pp 395-437.
(15)   Voronkov, M. G.; Deryagina, E. N. In *Chemisty of Organosulfur Compounds - General problems*; L. I. Belen'kii, Ed.; Ellis Horwood: Chichester, 1990; pp 48.
(16)   Hofmann, I. C.; Hutchison, J.; Robson, J. N.; Chicarelli, M. I.; Maxwell, J. R. *Org. Geochem.* **1992**, *19*, 351-370.
(17)   Porter, M. In *Organic chemistry of sulfur*; S. Oae, Ed.; Plenum press: New York, 1977; pp 71-118.
(18)   Mayer, R. In *Organic chemistry of sulfur*; S. Oae, Ed.; Plenum press: New York, 1977; pp 33-69.
(19)   Adam, P.; Kanz, C.; Huc, A.; J., C.; Albrecht, P. In *Organic Geochemistry, poster sessions from the 16th international meeting on organic geochemistry, Stavanger, 1993*; K. Øygard, Ed.; Falch Hurtogtrykk: Oslo, 1993; pp 502-505.

(20)   Kohnen, M. E. L.; Sinninghe Damsté, J. P.; ten Haven, H. L.; de Leeuw, J. *Nature* **1989**, *341*, 640-641.
(21)   Sinninghe Damsté, J. S.; de Leeuw, J. *Org. Geochem.* **1990**, *16*, 1077-1101.
(22)   Sinninghe Damsté, J. S.; de Leeuw, J. *Intern. J. Environ. Anal. Chem.* **1987**, *28*, 1-19.
(23)   Czogalla, C. D.; Boberg, F. *Sulfur Reports* **1983**, *3*, 121-167.
(24)   Kenig, F.; Huc, A. Y. In *Geochemistry of sulfur in fossil fuels*; W. L. Orr and C. M. White, Eds.; Am. Chem. Soc.: Washington, D.C., 1990; Vol. 429; pp 170-185.
(25)   Barbe, A.; Grimalt, J. O.; Pueyo, J. J.; Albaigés, J. *Org. Geochem.* **1990**, *16*, 815-828.
(26)   Grimalt, J. O.; Yruela, I.; Saiz-Jimenez, C.; Toja, J.; de Leeuw, J. W.; Albaigés, J. *Geochim. Cosmochim. Acta* **1991**, *55*, 2555-2577.
(27)   Valisolalao, J.; Perakis, N.; Chappe, B.; Albrecht, P. *Tetrahedron Lett.* **1984**, *25*, 1183-1186.

RECEIVED June 26, 1995

Chapter 8

Transformations in Organic Sulfur Speciation During Maturation of Monterey Shale: Constraints from Laboratory Experiments

Bryan C. Nelson[1], Timothy I. Eglinton[1], Jeffrey S. Seewald[1], Murthy A. Vairavamurthy[2], and Francis P. Miknis[3]

[1]Department of Marine Chemistry and Geochemistry, Woods Hole Oceanographic Institution, Woods Hole, MA 02543
[2]Department of Applied Science, Geochemistry Program, Applied Physical Sciences Division, Brookhaven National Laboratory, Upton, NY 11973
[3]Western Research Institute, Laramie, WY 82071

A series of hydrous pyrolysis experiments were conducted at temperatures ranging from 125 to 360°C at 350 bars pressure to examine variations in sulfur speciation during thermal maturation of Monterey shale. The total sediment, kerogen and bitumen from each experiment in addition to unheated representatives were analyzed via x-ray absorption spectroscopy, pyrolysis-gas chromatography, ^{13}C NMR spectrometry, elemental analysis, thin-layer chromatography and reflected light microscopy.

Based on these measurements, it was possible to recognize three distinct temperature regimes, within which the type and amount of sulfur in the analyzed fractions underwent transformations: (i) between 150 and 225°C, a significant proportion of kerogen-bound sulfur is lost probably due to the collapse of polysulfide bridges; (ii) between 225 and 275°C, cleavage of -S-S- and -S-C- linkages within the kerogen is believed to occur, resulting in substantial production of polar sulfur-rich bitumen; (iii) above 275°C total bitumen yields as well as the proportion of bitumen sulfur decrease, while C-C bond scission leads to increased yields of saturated and aromatic hydrocarbons.

The results from this study clearly and quantitatively establish a link between organically-bound sulfur, and more specifically, organic polysulfides, and the low-temperature evolution of soluble petroleum-like products (bitumen) from sulfur-rich source rocks.

There have been numerous investigations of the effects of heteroatoms (atoms other than carbon and hydrogen) in kerogen on the timing and thermal stress required for petroleum generation (*1-3*). In particular, the role of sulfur during petroleum

0097–6156/95/0612–0138$14.25/0

generation has received a great deal of attention (*3-8*). These studies typically attributed early petroleum generation from sulfur-rich kerogens to the inherent weakness of sulfur-sulfur and sulfur-carbon bonds relative to carbon-carbon bonds (*9*). Studies of the role of sulfur in the sedimentary formation of kerogen (*10-12*) suggest that sulfur linkages must be ubiquitous in sulfur-rich Type II-S kerogens ($S_{org}/C > 0.04$; (*5*)). These inferences implicating sulfur in maturation, however, are based largely on indirect and circumstantial evidence and the connection between the chemistry of sulfur-rich kerogens and the physical processes leading to petroleum generation remain unclear (*13*).

The goal of this study was to gain further insight into the chemical changes involving sulfur during source rock maturation. In contrast to previous studies, emphasis was placed on direct assessment of the changes in sulfur species in the organic matter itself. A series of hydrous-pyrolysis experiments were conducted with sulfur-rich Monterey shale at temperatures ranging from 125 to 360°C to provide constraints on the speciation of sulfur in kerogen and generated products as a function of temperature. Although, temperatures higher than those found in natural settings are employed during hydrous-pyrolysis (in order to enhance reaction rates and allow organic transformations to be observed on a laboratory time scale), these experiments have proven an effective means to simulate the generation of hydrocarbons in the laboratory. Previous studies have shown that the pyrolysis-products are similar to products of geological maturation (*3,14*). In addition, the processes leading to the generation and expulsion of oil during hydrous-pyrolysis experiments appear to be the same as those occurring in natural petroleum formation (*14*).

An analytical scheme was designed to facilitate a mass-balance approach involving gaseous, liquid (bitumen) and solid (altered sediment and kerogen) phases that allowed a detailed characterization of the fate of organic sulfur during thermal maturation. Analyses were conducted using x-ray absorption near edge structure (XANES) spectroscopy, pyrolysis-gas chromatography (Py-GC), solid-state ^{13}C nuclear magnetic resonance (NMR) spectrometry, thin-layer chromatography with flame ionization detection (TLC-FID), elemental analyses and reflected light microscopy. This blend of analyses enabled the determination of gross and molecular-level transformations in sulfur speciation, as well as the manner by which they lead to the early conversion of kerogen to oil.

Experimental

Sample Description and Preparation. A thermally immature ($R_o = 0.25\%$) consolidated sediment sample from the Miocene Monterey Fm. (ML91-17) was obtained from an outcrop at Naples Beach, CA. The sample was removed from a 10 cm interval at the base of Unit 315, approximately 9 m below the lowest phosphorite horizon (M.D. Lewan, personal communication). The sample comprised a lenticularly laminated claystone and visually appeared fresh (i.e. unweathered) and blocky. The surface of the sample was scraped prior to disc-mill pulverization to expose pristine material and remove possible contaminants.

Figure 1 illustrates the experimental scheme used in this study. Our primary goal was to examine the fate of kerogen-bound sulfur during artificial maturation. Consequently, after pulverization, the sediment sample was sequentially extracted

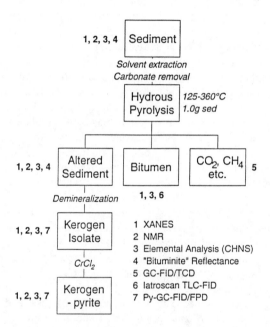

Figure 1. Outline of analytical scheme for separation and characterization of hydrous-pyrolysis products.

by sonic disruption using $CH_3OH/CHCl_3$ to remove indigenous bitumen. Sedimentary carbonate was removed from the sample by treating with 10% HCl at 40°C for 2 hrs. For each experiment 1.0 g of solvent-extracted, carbonate-free sediment was used as the starting material except for the experiments at 225 and 275°C, in which carbonate was not removed from the starting material. The sediment sample was loaded into a 20 mL (30.5 x 0.9 cm i.d.) pipe-bomb, sealed and placed into a 62 x 12.0 cm i.d. furnace. The pipe-bomb was evacuated and then partially filled with argon-purged distilled water to ensure that the sediment was in contact with water during heating. Once at the experimental temperature, the pipe-bomb was filled completely with distilled water and pressurized to 350 bars so that only a single liquid phase was present (i.e., no head-space). Temperature was monitored (± 2°C) with a thermocouple at each end of the pipe-bomb to ensure there were no thermal gradients. All experiments were 168 hrs in duration.

During an experiment, fluid samples were withdrawn from the pipe-bomb through a 10 µm filter into glass gas-tight syringes after approximately 24, 72 and 168 hrs and analyzed for the gas concentrations of dissolved gases. Pressure was maintained during the sampling process by pumping argon-purged distilled water into one end of the pipe-bomb while the fluid sample was removed from the other. The sampling process was performed rapidly (1 to 2 min) to avoid dilution of the fluid samples with the freshly injected water. At the end of the experiment the pipe-bomb was cooled and all liquid contents were removed by sequentially pumping 30 mL of CH_3OH and CH_2Cl_2 through the pipe-bomb. Subsequently, the pipe-bomb was opened and the solid residue was removed by rinsing with additional solvent. The combined CH_3OH/CH_2Cl_2 extract was centrifuged and the liquid products decanted. After drying at 40°C in air, the solid residue was sequentially extracted by sonic disruption in CH_3OH/CH_2Cl_2 to ensure complete recovery of the generated bitumen. This extract was combined with the pipe-bomb rinses, back-extracted with distilled H_2O, dried over anhydrous Na_2SO_4 and filtered (0.45 µm) to remove residual suspended material yielding the generated bitumen. A split of the altered sediment (~350 mg) was demineralized through HCl and HF acid digestions at room temperature (*15*) to obtain a kerogen isolate. An aliquot (~50 mg) of the kerogen isolate was treated with $CrCl_2$ and concentrated HCl to remove inorganic sulfur (*16,17*).

Gas Analysis. Dissolved CO_2 and light (C_1-C_6) hydrocarbons in the fluid samples from all experiments were analyzed using a purge and trap apparatus interfaced to a gas chromatograph. Dissolved H_2S concentrations were monitored to obtain an estimate of the amount of sulfur, both inorganic and organic, removed from the system as H_2S during sediment maturation. The absolute concentration of dissolved H_2S was determined gravimetrically by precipitation as Ag_2S in a 3 wt% $AgNO_3$ solution.

Elemental Analysis. Elemental analyses were performed in order to monitor fluctuations in the distribution of sulfur and carbon among the various fractions resulting from maturation of the Monterey shale. Weight percent C, H, N and S were determined for all fractions including those obtained from unheated, solvent-extracted Monterey shale. Elemental analyses were conducted using a Leco 932 elemental analyzer.

X-ray Absorption Spectroscopy (XANES). X-ray Absorption Near Edge Structure (XANES) Spectroscopy was performed at the National Synchrotron Light Source at Brookhaven National Laboratory on Beamline X-19A (*18,19*). XANES was conducted on the unheated and altered sediments, bitumen, and $CrCl_2$-treated kerogens to observe thermally-induced changes in sulfur speciation within and between the various fractions. Solid samples were prepared as boric acid pellets, whereas liquid samples were taken to near dryness and adsorbed onto pre-combusted GF/F filters under a N_2 atmosphere, heat-sealed in air-tight pouches and kept frozen prior to analysis. Solid and liquid samples were analyzed while mounted in a He-purged sample chamber. XANES spectra of all fractions were acquired from 2465 to 2900 eV using a Lytle fluorescence detector. A monochromator step function was chosen to provide a resolution of 0.5 eV at the near-edge region.

Quantitative deconvolution of the XANES spectra was accomplished using a computer algorithm developed by Waldo et al. (*20*). After corrections to account for background and self-absorption, the proportions of different sulfur species were calculated by fitting the normalized spectra (least-squares procedure) with up to six reference-compound spectra. For this study, FeS_2, elemental sulfur, cysteic acid, sodium sulfate, benzyl sulfide, dibenzyl trisulfide, dibenzothiophene and dibenzylsulfoxide were used as reference compounds. The accuracy of these results are estimated to be $\pm 10\%$. Optimal curve-fitting of the XANES spectra from analyses of altered sediments and kerogens prior to $CrCl_2$ treatment was difficult due to the presence of inorganic sulfur compounds (pyrite). Consequently, sulfur speciation determinations are reported only for the bitumen and $CrCl_2$-treated kerogen.

Analytical Pyrolysis (Py-GC). Analytical pyrolysis was conducted to determine the relative distribution of volatile thiophenic and hydrocarbon pyrolysis-products from the kerogen and pyrite-free kerogen samples from the unheated and matured Monterey Shale. Pyrolysis-Gas Chromatography was performed using a FOM-3LX Curie-point pyrolysis unit (controlled by a Horizon RF generator) interfaced to a Hewlett-Packard 5890 Series II GC (*18*). Kerogen samples were loaded onto Fe/Ni wires with a Curie temperature of 610°C. Samples were pyrolyzed for 5 s and the pyrolysis interface temperature was set at 200°C. Helium was used as the carrier gas. Separation of the pyrolysis products was achieved on a Restek R_{TX-1} column (50 m x 0.32 mm i.d.; film thickness 0.5 μm) using a temperature program from 30°C (5 min initial time) to 320°C (15 min final time) at a rate of 3°C min^{-1}. The GC effluent was split and simultaneously monitored by a flame ionization detector (FID) and a sulfur-selective flame photometric detector (FPD). Flash pyrolysis of the kerogen and pyrite-free kerogen fractions yielded near-identical results, indicating that the $CrCl_2$-treatment did not affect the relative distribution of GC-amenable (i.e. volatile) products analyzed.

Nuclear Magnetic Resonance (NMR) Spectrometry. Solid-state ^{13}C NMR spectrometry was used to determine changes in the carbon structure of the altered sediments, kerogens and $CrCl_2$-treated kerogens that occurred as a result of thermal maturation. Solid-state ^{13}C NMR measurements were made using the technique of cross polarization (CP) with magic-angle spinning (MAS) at a carbon frequency of 25 MHz using a ceramic probe and a 7.5 mm o.d. zirconia pencil rotor. As a consequence of the small amounts of sample material available for the NMR measurements (~20 mg for the kerogens and ~100 mg for the altered sediments), data were collected over periods of 15 to 18 hrs (54,000 and 64,800 transients). A pulse delay of 1 s, a contact time of 1 ms, a 5.0 μs pulse width and a sweep width of 16 kHz were used to acquire data. Sample spinning speeds were ~4.5 kHz. A 50 Hz exponential multiplier was applied to the free induction decay of each ^{13}C spectrum before integration.

The NMR spectra were integrated between 90 and 260 ppm for the "aromatic" region and -40 to 90 ppm for the "aliphatic" region. The spinning rates were sufficiently high so that contributions to the aliphatic integrals from high field aromatic carbon spinning sidebands were negligible and were not included in the aliphatic carbon integrals. The carbon aromaticity values can contain contributions from carbonyl (~210 ppm) and carboxyl carbons (~180 ppm), if present.

Bituminite Reflectance. Microscopic examination of the initial (unheated) sediment sample revealed that the majority of organic matter was comprised of an amorphous groundmass termed "bituminite" (*21*), with only trace amounts of recognizable vitrinite. Bituminite, which is believed to derive from degraded algal and bacterial matter, is thus considered to be a major kerogen constituent. For this reason, and because vitrinite is so scarce, direct measurements of reflectance were made on bituminite in order to assess the extent of thermal alteration.

Thin-layer Chromatography. Quantification of the saturated, aromatic and polar compound classes in the bitumen was performed by thin-layer chromatography with flame ionization detection (TLC-FID) according to the methods of Karlsen and Larter (*22*) using an Iatroscan TH-10 Mark III instrument. 1.5-2.0 μL of each sample (dissolved in CH_2Cl_2) was applied to silica rods which were sequentially developed in C_6H_{14} (10 cm), C_7H_8 (5 cm) and a 95:5 solution of CH_2Cl_2:CH_3OH (2 cm). A standard mixture comprised of *n*-eicosane, dibenzothiophene and 2,6-dimethoxyphenol was used for calibration of response factors for aliphatic, aromatic and polar compounds, respectively.

Results

Gaseous Phases. Dissolved H_2S concentrations were measured during five of the hydrous-pyrolysis experiments (Table 1). During these experiments the concentration of dissolved H_2S was likely controlled by the solubility of iron sulfides such as pyrite and/or pyrrhotite owing to the rapid precipitation/dissolution kinetics for these phases under hydrothermal conditions (23). Pyrite was abundant in the unheated sediment and persisted along with newly formed pyrrhotite in the thermally altered sediments. At temperatures below 225°C dissolved H_2S concentrations were too low to be determined gravimetrically, but H_2S was detectable by odor in all experiments. Sources of dissolved H_2S during the experiments include diagenetic pyrite and organic sulfur. It is not possible to directly determine the relative contributions from these two sources based on dissolved concentrations alone. However, at 350 and 360°C 12.7 and 18.8 mg H_2S/g rock was released to solution respectively. These amounts exceed the amount of sulfur present as diagenetic pyrite in the initial sediment (10.5 mg/g sediment), and thereby provide direct evidence for the release of organically bound sulfur to solution. Because dissolved sulfide is cycled through the fluid into sulfide minerals, the calculated contributions of organic sulfur to solution represent an absolute minimum.

Generated Bitumen. The total bitumen extract, saturated, aromatic and polar compound yields from each experiment and the indigeneous material removed from the sample prior to heating, are listed in Table 1 and illustrated in Figure 2. The initial bitumen content of the unheated sediment was approximately 11 mg/g rock. We estimate an extraction efficiency of >70% in removing this bitumen from the starting material, and hence any contribution from indigenous bitumen to extractable yields after the experiments is likely to be minor. In all cases, polar compounds comprised the majority of each bitumen fraction and, as a result, the total extract largely mirrored polar compound yields. The yield of polars increased with temperature to a maximum in the 275°C experiment (62.0 mg/g rock), above which concentrations decreased. The yield of saturate and aromatic compounds were similar to each other with maxima in the experiments at temperatures above 275°C.

The results of the elemental analyses (C, H, N and S) of the bitumen are listed in Table 2. Above 250°C the atomic ratio of hydrogen to carbon (H/C) decreased while the atomic ratio of nitrogen to carbon (N/C) remained relatively constant. Above 150°C the bitumen initially exhibited an increase in the sulfur to carbon (S/C) ratio with temperature which reached a maximum at 250°C followed by a decrease (Figure 3).

Normalized (K-edge) XANES spectra of the bitumen are illustrated in Figure 4. Reduced sulfur species (sulfides) are represented by peaks that occur towards the left side of each spectra, whereas peaks for more oxidized species (sulfoxides) are found at higher energies. Fitting of the XANES spectra reveal that organic sulfides and polysulfides were the primary sulfur forms contained in the bitumen and

Table 1. Total extract, saturated, aromatic and polar compound yields. Concentrations of dissolved CO_2 + C_1-C_3 (light hydrocarbon) and H_2S.

Temperature (°C)	Total (mg/g rock)	Saturates (mg/g rock)	Aromatics (mg/g rock)	Polars (mg/g rock)	CO_2 + C_1-C_3* (mg/g rock)	H_2S* (mg/g rock)
Unheated	10.7	0.0	0.0	10.7	n.d.	n.d.
125	1.23	0.0	0.0	1.23	6.01	n.d.
150	1.72	0.0	0.0	1.72	8.68	n.d.
175	7.26	0.58	0.0	6.68	13.3	n.d.
200	22.4	0.11	0.27	22.0	20.7	n.d.
225	10.5	0.64	0.87	9.01	n.d.	n.d.
250	36.9	0.54	2.08	34.2	29.8	6.9
275	68.4	2.69	3.70	62.0	n.d.	n.d.
300	29.7	1.60	1.45	26.6	47.9	8.4
325	29.2	3.25	4.39	21.5	50.3	9.3
350	22.7	1.87	3.26	17.5	69.3	12.7
360	33.1	1.21	4.81	27.1	77.4	18.8

n.d. not determined.
* calculated from the concentrations of these species dissolved in the aqueous phase.

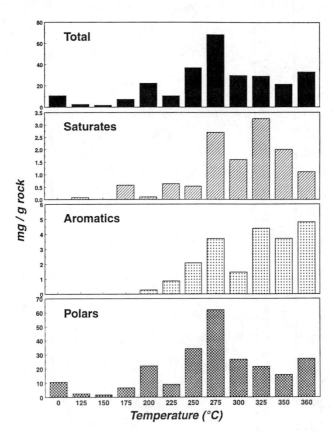

Figure 2. Bar charts showing yields of total extract, saturated, aromatic and polar compound classes for the soluble (bitumen) products from each hydrous-pyrolysis experiment.

Table 2. Elemental Analysis of bulk sediments, CrCl$_2$-treated kerogens and solvent (bitumen) extracts

Temp. (°C)	Bulk Sediment				CrCl$_2$-treated Kerogen					Bitumen				
	%OC	%N	%S	S/C	%OC	%H	%N	%S	S/C	%OC	%H	%N	%S	S/C
Unheated	20.58	1.24	3.86	0.070	60.44	6.52	3.22	8.25	0.051	68.86	8.34	2.05	7.25	0.039
125	22.60	1.36	3.67	0.061	59.83	6.41	3.06	7.83	0.049	56.40	5.67	2.44	5.12	0.034
150	23.54	1.42	3.67	0.058	59.57	6.86	2.91	7.50	0.047	67.68	7.02	2.23	6.92	0.038
175	22.39	1.34	3.50	0.059	60.79	6.22	3.11	7.39	0.046	69.62	8.39	1.56	8.30	0.045
200	22.62	1.23	3.02	0.050	64.56	6.45	3.14	6.92	0.040	n.d.	n.d.	n.d.	n.d.	n.d.
225*	18.64	1.08	2.64	0.053	62.88	6.28	3.24	6.04	0.036	55.35	4.69	1.76	6.79	0.046
250	20.32	1.21	3.06	0.056	60.81	5.99	2.93	5.21	0.032	68.66	7.59	1.68	10.8	0.059
275*	15.58	0.94	2.36	0.057	69.13	5.74	3.60	4.49	0.024	71.98	8.25	2.27	10.4	0.054
300	18.31	1.12	2.24	0.046	72.55	5.66	3.70	3.91	0.020	55.95	6.72	1.35	6.61	0.044
325	16.71	0.97	2.21	0.050	75.26	5.03	3.95	3.40	0.017	72.55	7.92	1.62	9.64	0.050
350	15.08	0.90	1.81	0.045	65.22	4.39	3.54	2.94	0.017	n.d.	n.d.	n.d.	n.d.	n.d.
360	14.58	0.98	2.44	0.063	74.79	4.23	4.01	3.48	0.017	76.48	6.86	2.06	7.61	0.037

* Sediment samples were not decarbonated prior to hydrous-pyrolysis.

n.d. not determined.

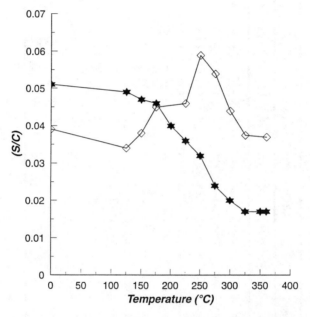

Figure 3. Variation in the atomic S/C ratio of the bitumen (open diamonds) and CrCl$_2$-treated kerogens (closed stars) with temperature.

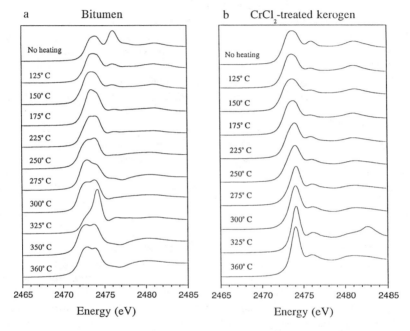

Figure 4. Normalized XANES spectra of (a) bitumen from the unheated, 125°C, 150°C, 175°C, 225°C, 250°C, 275°C, 300°C, 325°C, 350°C and 360°C hydrous-pyrolysis experiments, and (b) $CrCl_2$-treated kerogens from the unheated, 125°C, 150°C,175°C, 225°C, 250°C, 275°C, 300°C, 325°C and 360°C experiments.

typically accounted for over 50% of the sulfur present (Figure 5). The remaining sulfur was in the form of thiophenes and sulfoxides. Thiophenes and sulfoxides exhibited relatively little change with temperature, while organic sulfides and polysulfides increased significantly from 3% to a maximum of 8% of the bitumen at 275°C. Above this temperature the weight percent of sulfides and polysulfides in the bitumen decreased.

Solid Phases. In the altered sediments, weight percent organic carbon, nitrogen and sulfur as well as the S/C atomic ratio, decreased with increasing temperature (Table 2). In the kerogens the H/C atomic ratio decreased at higher temperatures (>250°C) while the N/C atomic ratio remained fairly constant. In contrast to the bitumen, the S/C ratio of the kerogen initially decreased with increasing temperature (Figure 3). This decrease was first apparent in kerogens from experiments above 175°C, with the S/C ratio changing from ~0.05 to <0.02 in kerogen from experiments above 300°C. Between 325 and 360°C, the S/C ratio remained constant.

Bituminite reflectance measurements of the altered sediments are listed in Table 3 and illustrated in Figure 6. The reflectance value of the bituminite present in the altered sediment were low (0.25%), relative to vitrinite of the same sample (0.39%), due to the hydrogen-rich nature of the former (24,25). Maturation of the altered sediments, however, was clearly indicated by increased reflectance with temperature, especially above 275°C (Figure 6), and R_o reached a maximum of 0.95% in the 360°C altered sediment samples.

Results from CP/MAS ^{13}C NMR measurements of the altered sediments and kerogens are listed in Table 3 and illustrated in Figure 6. ^{13}C NMR spectra of the unheated and artificially matured kerogens are shown in Figure 7. The major band of the right portion of each spectrum (0-60 ppm) represents carbon in aliphatic structures including structures with sulfide bonds. The major band to the left of each spectrum represents carbon in aromatic (90-160 ppm) structures (including thiophenes), and in any carboxylate (~180 ppm) and carbonyl (~210 ppm) structures that may be present. Very little variation was observed in the ^{13}C NMR results from the altered sediment, kerogen and $CrCl_2$-treated kerogen from the same experiments (Table 3, Figure 6). Comparison of the ^{13}C NMR spectra from each experiment, however, clearly show a preferential loss of aliphatic carbons relative to aromatic carbons with increasing temperature (Figure 7). Examination of Figure 6 reveals an exponential relationship between temperature and aromaticity with a major steepening in the curve above 275°C.

Partial FID and FPD chromatograms from flash pyrolysis (610°C) of the unheated, 250, 300 and 360°C kerogens are shown in Figure 8. FPD peak assignments were made by comparison of relative retention times to earlier studies (26,18), and are listed in Table 4 with the inferred carbon skeletons of the bound precursors (10). The major peaks in the FID chromatograms are due to n-alkanes and n-alkenes, alkylbenzenes and thiophenes (Figure 8). The Py-GC results indicate a preferential loss of aromatic and especially thiophenic compounds relative to n-hydrocarbons with increasing temperature. The generation potential for n-alkanes and n-alkenes remains high until approximately 300°C suggesting that C-C cracking has not occurred to a significant extent below this temperature.

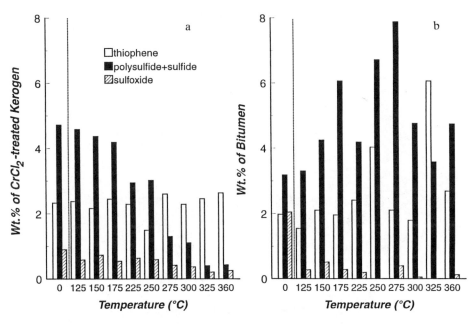

Figure 5. XANES-based sulfur speciation in the (a) CrCl$_2$-treated kerogen and (b) bitumen fractions.

Table 3. NMR results of bulk sediments, isolated kerogens and CrCl$_2$-treated kerogens and "bituminite" reflectance of bulk sediments

| Temperature (°C) | Aromaticity* | | | %R$_o$ |
	Bulk Sediment	Kerogen	CrCl$_2$-treated Kerogen	Bulk Sediment
Unheated	0.35	0.34	n.d.	0.25
125	0.32	0.35	0.32	0.24
150	0.36	0.32	0.33	0.28
175	0.41	0.37	0.36	0.31
200	n.d.	n.d.	n.d.	0.30
225	0.44	n.d.	n.d.	0.29
250	0.45	0.49	0.49	0.36
275	0.57	n.d.	n.d.	0.33
300	0.64	0.59	0.63	0.42
325	0.67	n.d.	n.d.	0.59
350	n.d.	n.d.	n.d.	0.85
360	0.88	0.83	0.83	0.95

* may include any contributions from carboxylate and carbonyl carbons if present.
n.d. not determined.

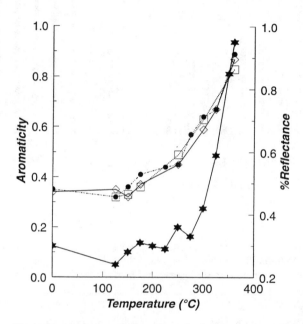

Figure 6. Variation in aromaticity (may include any contributions from carboxylate carbons at ~175 ppm) of kerogen, CrCl$_2$-treated kerogen and altered sediment fractions as well as variation in reflectance of altered sediments with temperature. Symbols represent: (Aromaticity) open diamonds kerogens, open squares pyrite-free kerogen, closed circles altered sediments; (%R$_o$) closed stars altered sediments.

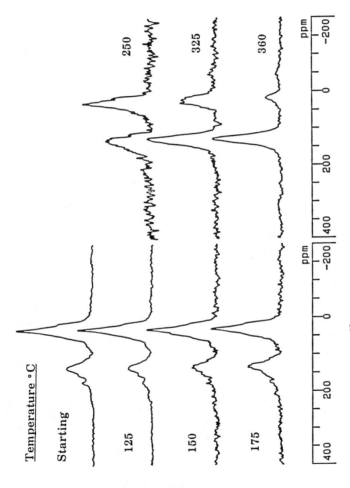

Figure 7. Solid-state ^{13}C CP-MAS NMR spectra of the kerogen isolates from the unheated, 125°C, 150°C, 175°C, 250°C, 325°C and 360°C experiments.

a

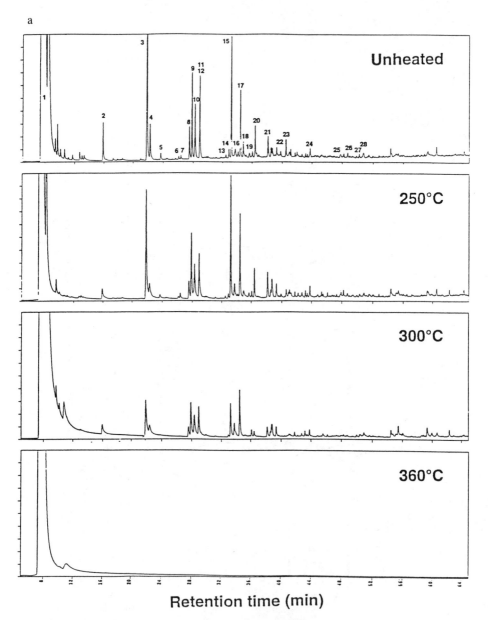

Retention time (min)

Figure 8.   Partial (a) FID and (b) FPD chromatograms from Py-GC of the kerogen isolates from the unheated, 250°C, 300°C and 360°C hydrous-pyrolysis experiments.   In (a) numbers represent *n*-hydrocarbon homologs, symbols denote: closed circles alkylbenzenes, open circles thiophenes. Peak assignments for (b) are listed in Table 4.

b

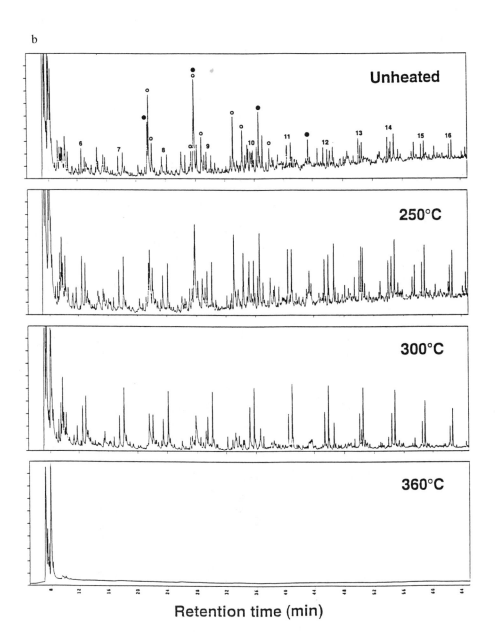

Retention time (min)

Figure 8. Continued.

Table 4. Peak identifications for FPD chromatograms

Peak	Compound	Origin*	Peak	Compound	Origin*
1	Hydrogen sulfide	?	16	2-ethyl-4-methylthiophene	B
2	Thiophene	?	17	2,3,5-trimethylthiophene	I,B
3	2-Methylthiophene	L	18	2-methyl-5-ethenylthiophene	L
4	3-Methylthiophene	I	19	2,3,4-trimethylthiophene	B,S
5	Thiolane	?	20	3-isopropyl-2-methylthiophene	S
6	Methyldihydrothiophene	L	21	2-methyl-5-propylthiophene	L
7	2-methylthiolane	L		+ 2,5-diethylthiophene	L
8	2-ethylthiolane	L	22	5-ethyl-2,3-dimethylthiophene	B
9	2,5-dimethylthiophene	L	23	Unknown	S?
10	2,4-dimethylthiophene	B,S	24	2-butyl-5-methylthiophene	L
11	2-ethenylthiophene	I	25	2-ethyl-5-butylthiophene	L
12	2,3-dimethylthiophene	I,B	26	2-methyl-5-pentylthiophene	L
13	2-ethylthiolane	L	27	2-methylbenzo[β]thiophene	L
14	2-propylthiophene	L	28	4-methylbenzo[β]thiophene	L
15	2-ethyl-5-methylthiophene	L		+ 3-methylbenzo[β]thiophene	B

* Inferred carbon skeleton of bound precursor: L=linear, B=branched, I=isoprenoid, S=steroid side chain.

Discussion

Compositional Characteristics of Unheated Monterey Shale Kerogen.   The Above 325°C essentially only gaseous products (CH_4, C_2H_6, C_2H_4, H_2S) were detected in the flash pyrolyzates.

Estimates of the relative abundance of volatile thiophenic versus hydrocarbon pyrolysis products were made from the ratio of FID peak areas for 2-methylthiophene (2MT) to the sum of toluene (Tol) and n-C_7 alkene (C_7H_{14}) (Figure 9).   The 2MT/(Tol+C_7H_{14}) ratio decreased slightly with increased temperature up to 250°C. At temperatures greater than 250°C, the ratio decreased sharply and approached zero above 325°C.   Although kerogens from the experiments below 250°C exhibited very little change in the 2MT/(Tol+C_7H_{14}) ratio, the S/C ratio decreased significantly (from 0.051 to 0.032) relative to the unheated sample (Figure 9).   Conversely, a sharper decrease in the 2MT/(Tol+C_7H_{14}) ratio above 250°C coincided with only a moderate change in the S/C ratio (Figure 9).

The predominant sulfur-containing pyrolysis products in all FPD chromatograms were H_2S and low molecular weight thiophenes (Figure 8). H_2S was likely derived from the thermal decomposition of pyrite, present in the kerogen, as well as from the cleavage of organically-bound thiols, aliphatic sulfides (thiolanes, thianes) or sulfide bridges (*18*).   Volatile thiophenic pyrolysis products were observed in the kerogens from all the experimental temperatures except for the 350 and 360°C experiments (Figure 8).   At temperatures <300°C no marked changes in the internal distribution of thiophenic products were observed.   At 300°C and above, however, the relative contribution of 2-methylthiophene, 2,5-dimethylthiophene, 2-ethyl-5-methylthiophene and 2-methyl-5-propylthiophene decreased more dramatically than other thiophenic products (Figure 8).

Normalized (K-edge) XANES spectra of the $CrCl_2$-treated kerogens are illustrated in Figure 4.   With increased temperature a sharpening in the shape of the major peak occurred, as well as a slight (~ 1 eV) shift towards higher energy.   This shift is detectable at 175°C and clearly evident by 225°C (Figure 4).   The percentages of the major forms of sulfur (normalized to %S) in the kerogen determined by fitting of the XANES spectra are illustrated in Figure 5.   The primary sulfur forms present in the unheated kerogen were organic sulfides and polysulfides. Together these two species account for ~50% of the sulfur while, thiophenes and sulfoxides comprised the majority of the remaining sulfur.   Systematic changes in sulfur speciation as a function of temperature are clearly evident (Figure 5).   Similar to the bitumen, thiophenes and sulfoxides present in the kerogens remained relatively constant with increased temperature.   In contrast, organic sulfides and polysulfides decreased with temperature from 4.7 to 0.4% in the unheated and 360°C pyrite-free kerogens, respectively (Figure 5) with two marked decreases occurring above 175°C and 250°C.   Above 250°C   the relative proportion of thiophenic sulfur exceeded that present as organic sulfides.

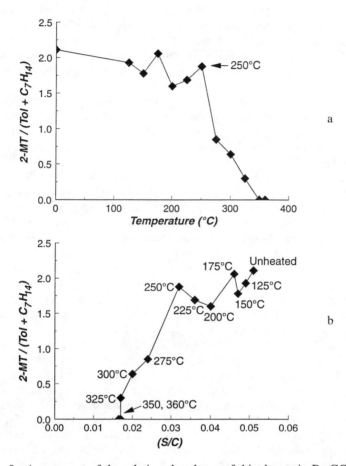

Figure 9. Assessment of the relative abundance of thiophenes in Py-GC (FID) traces of CrCl$_2$-treated kerogens from each hydrous-pyrolysis experiment (expressed as the ratio of [2-methylthiophene/(toluene+C$_7$H$_{14}$)]) with (a) temperature and (b) atomic S/C ratio.

values obtained from bituminite reflectance (%R_o = 0.25), solid-state CP/MAS ^{13}C NMR spectroscopy (aromaticity = 0.34) and Rock-Eval (T_{max} = 386°C) measurements all indicate that the unheated Monterey kerogen was relatively immature.   In addition, with an atomic S_{org}/C ratio of 0.051 the sulfur-rich Monterey kerogen (Wt.%S = 9.45) can be classified as "Type II-S" (5).

The homologous series of *n*-alkanes and *n*-alkenes present in the partial Py-GC-FID chromatogram of the unheated Monterey kerogen sample is typical of a wide range of oil-prone, immature kerogens (27). The distinguishing feature of Monterey shale and other sulfur-rich kerogens, however, is the high relative abundance of thiophenes released during pyrolysis (26,27). The thiophenic pyrolysis products detected by FPD can provide direct information regarding the chemical nature of macromolecularly bound sulfur (10,26). When pyrolysis is conducted using wires with a Curie temperature of 610°C, S-S, C-S and C-C bonds are cleaved and sulfur-bound or sulfur-containing moieties present in the kerogen will generate a variety of products through β-cleavage, γ-hydrogen rearrangement and by γ-cleavage (10). Based on the substitution patterns of the thiophenes and benzothiophenes yielded upon pyrolysis, carbon skeletons of different sulfur-containing moieties originally present in the kerogen can be distinguished (10). Similar to previous studies of Monterey shale and other immature Type II-S kerogens (18,26), a significant proportion of the sulfur-containing pyrolysis products from the unheated Monterey kerogen are likely derived from isoprenoid and/or steroidal carbon skeletons (Figure 8, Table 4).

Complementary information regarding the sulfur-containing structures present in the kerogen can be obtained through x-ray absorption (XANES) spectroscopy. In addition, XANES spectroscopy allows all forms of sulfur present in the sample to be "seen", whereas pyrolysis only liberates volatile species. In contrast to the flash-pyrolysis approach, XANES spectroscopy reveals the electronic environment of the sulfur atom and does not provide information on the configuration of carbon atoms, other than those in close proximity to sulfur. The XANES data reveal that organic sulfides and polysulfides, thiophenes and sulfoxides comprise 4.7, 2.3 and 0.9 weight percent of the unheated, CrCl$_2$-treated kerogen, respectively (Figure 5). Thus, sulfides and polysulfides account for >50% of the total sulfur in the unheated kerogen. The low percentage of sulfoxides indicates that oxidation of sulfur during sample manipulation was minor. K-edge XANES spectra do not allow for the discrimination between cyclic (thiolanes, thianes) and acyclic sulfides.   By assuming, however, that the relative proportion of cyclic and acyclic sulfides is approximately equivalent, the ratio of thiophenes (a cyclic form of sulfur) to total sulfides can be used as an estimate of the extent of *intra*- versus *inter*-molecular S-bonding (18). Based on this premise, XANES spectroscopic analysis of the unheated Monterey kerogen would indicate that sulfur cross-linking is extensive and that there are abundant, potentially weak S-linkages where thermally-induced cleavage might occur.

Transformations During Laboratory Maturation. In order to constrain sources and sinks of sulfur as a function of temperature, we have calculated the fraction of the total sedimentary sulfur present in three major phases: kerogen, bitumen and total inorganic sulfur (Figure 10). The latter pool is calculated by difference assuming a closed system and an initial total sedimentary S content (3.86%; Table 2) and is predominantly composed of H_2S, pyrite and pyrrhotite. For comparison, the distribution of carbon associated with gaseous species (C_1-C_3 hydrocarbons, CO_2), bitumen and kerogen are also shown in Figure 10. Plotting the data in this way shows several interesting trends which reveal systematic transformations in the concentrations and speciation of sulfur in each of the phases analyzed. These transformations also reveal the interplay between each phase during the maturation process.

In the unheated sediment the dominant portion (ca. 75%) of the sedimentary S is associated with kerogen, pyritic sulfur representing the remainder. Essentially no sulfur is associated with indigenous bitumen since extractable material was removed from the immature starting material (it is also assumed there is initially a negligible amount of adsorbed H_2S). This condition prevails at temperatures up to 150°C, above which the proportion of kerogen-bound sulfur decreases markedly from 3.0 to 1.8% at 225°C. Only a minor increase in solvent extractable sulfur was observed over this temperature interval, and accordingly the loss of kerogen sulfur is balanced by the production of inorganic sulfur. Above 250°C a further decrease in kerogen sulfur is apparent, and this is accompanied by an increase in bitumen sulfur, which peaked at 275°C. This maximum also corresponds with the temperature of maximum bitumen generation (Figure 2) and, in particular, polar compound evolution. Above 275°C, S in both bitumen and kerogen decreases resulting in an increase in inorganic sulfur. The total decrease of kerogen sulfur from the unheated sample to the 360°C residue was approximately a factor of 4 (i.e. 2.8 to 0.7%).

Based on this mass balance information, and the compositional transformations observed, we can construct an overview of the likely fate of organically-bound sulfur during laboratory maturation of the Monterey shale. The earliest transformations occur at very low (≤ 175°C) temperatures (in terms of artificial maturation experiments). This change has not been studied in detail previously, since it occurs well below the temperature at which the primary generation of hydrocarbon-like products takes place during laboratory heating experiments (3,7). The decrease in kerogen sulfur implies that a substantial degree of internal rearrangement takes place within the kerogen macromolecular network, even under mild thermal stress. These rearrangements do not result in significant amounts of soluble (petroleum-like) products, but H_2S is generated. We postulate that the H_2S is primarily an elimination product from polysulfide bridges within the kerogen.

Cleavage of bridges containing 2 or more S-atoms would yield H_2S and the bridge may subsequently reform, the result being negligible release of soluble organic products. Alternatively, sulfide bridges could be broken, yielding H_2S, but insufficient bridges are broken to release soluble, carbon-containing moieties from

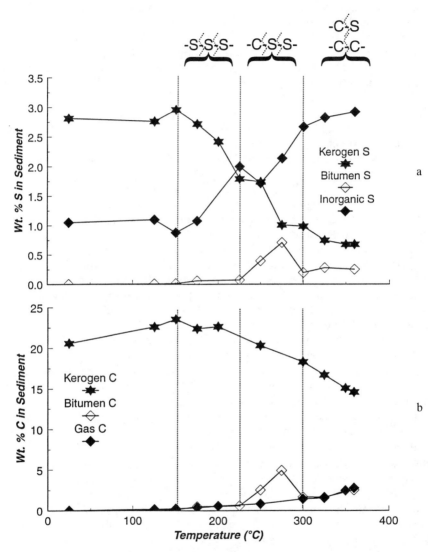

Figure 10.   Variation in total sedimentary (a) sulfur and (b) carbon with temperature.

the kerogen. Whatever the case, the net result is a loss of sulfur from the kerogen with no concomitant carbon loss. Based on the XANES analyses it is clear that the sulfur is removed from the kerogen during these "low temperature" experiments primarily as sulfides or polysulfides (Figure 5). This is also evident from the pyrolysis-based thiophene ratios, which show no decrease over this temperature interval (Figure 9). Both these observations are consistent with (poly)sulfide bridges as the reactive sulfur species. Unfortunately K-edge XANES does not allow unequivocal distinction of mono-sulfide from polysulfide species so this inference cannot be verified quantitatively. Nevertheless, qualitative support for the low-temperature reactivity of polysulfides is apparent in the normalized (K-edge) XANES spectra of the kerogens (Figure 4). Above 175°C there is a ~1 eV shift in the maximum for the lowest energy peak towards higher energy. Because organic polysulfides exhibit slightly lower K-edge energies than corresponding mono-sulfides this shift suggests preferential removal of polysulfides over mono-sulfides at lower temperatures.

Between 225°C and 300°C a significant proportion of the kerogen is converted to polar-rich bitumen (Figure 2). Polar compounds (resins and asphaltenes) are considered to be large, soluble macromolecules with a strong genetic link to kerogen (i.e. soluble kerogen moieties) (28). The bitumen generated at 275°C is enriched in sulfur compared to both lower and higher temperature soluble products and, based on the XANES data, is dominated by (poly)sulfide sulfur. We interpret these data to imply that the bitumen derived from these experiments is predominantly the result of cleavage of sulfide links in the kerogen, liberating lower molecular weight (and therefore soluble) kerogen sub-fragments which retain abundant sulfide linkages within their infrastructure. Since sulfur in the newly generated bitumen cannot account for all of the sulfur lost from the kerogen, continued H_2S formation at these temperatures is also implied.

At higher temperatures(>300°C) bitumen formation via cleavage of sulfur links gives way to hydrocarbon generation through C-C bond scission. Presumably hydrocarbons are generated from both the kerogen and bitumen. These conditions result in soluble products which contain increasing proportions of aliphatic and aromatic hydrocarbons (including aromatized sulfur compounds) with maximum yields observed at 325°C for the former and 360°C for the latter (Figure 2). The temperature regime between 275 and 360°C is also where the most marked changes in the carbon structure of the kerogen occurred, as indicated by bituminite reflectance and NMR spectrometry, which suggest increasing aromaticity. Also of note is the dramatic reduction in abundance of (poly)sulfide sulfur in the kerogen (Figure 5). In terms of relative proportions, thiophenic sulfur exceeds that of sulfidic sulfur above 250°C in the kerogen.

The distribution of carbon in the various pools resulting from the artificial maturation of Monterey shale (Figure 10) is consistent with the above interpretation of the fate of organic sulfur during maturation. It is expected that the relative concentration of organic carbon associated with the kerogen C would decrease as bitumen and gaseous products are generated. Although this held true for experiments at 250°C and above, an increase in kerogen C over that of the unheated sample was observed at 125-175°C. This was likely due to the low-temperature dissolution of non-carbon containing minerals originally present in the Monterey shale (e.g. refractory phosphates).   Overall, kerogen C decreased from approximately 21 to 15% in the unheated and 360°C samples, respectively (Figure 10), while gaseous organic carbon, in particular CH_4 and CO_2, increased with increasing temperature. Very little (<1%) organic carbon was associated with the bitumen C until 250°C, with a maximum in bitumen C at 275°C (Figure 10). This is consistent with the thermally induced cleavage of sulfide links in the kerogen freeing soluble, carbon-containing structures to the bitumen.   At higher temperatures C-C bonds in these soluble structures are likely broken leading to hydrocarbon and ultimately gas generation.

The generation of petroleum-like products as a function of temperature during the experiments presented here is consistent with previous experimental studies on sulfur-rich kerogens. The formation of expelled oil has been modeled as a two step process involving the decomposition of generated bitumen to form a compositionally distinct oil phase (*3,7*). Although we cannot distinguish between generated bitumen and expelled oil in our experiments, compositional variations in the total extractable bitumen support such a model. Bitumen, as defined by Lewan (*14*), is rich in polar compounds relative to expelled oil, which contains substantially greater proportions of saturate and aromatic hydrocarbons. During our experiments we observed a clear offset between peak generation of polar compounds at (275°C) and aliphatic and aromatic compounds (325-360°C), consistent with early bitumen generation which in turn decomposes to produce a more oil-like substance. These temperatures for peak bitumen and inferred oil formation are significantly lower than those from hydrous-pyrolysis (72 hrs) of the relatively sulfur-poor Woodford shale (330 and 350°C, respectively) and the Phosphoria shale (300 and 350°C), characterized by an intermediate organic sulfur content (*3*). Baskin and Peters (*7*) conducted hydrous-pyrolysis experiments for 72 hrs utilizing the Monterey shale and observed peak bitumen and oil formation at 280 and 330°C, respectively. These temperatures are almost identical to those observed during this study, the difference in heating time notwithstanding. Taken collectively, the experimental results provide strong evidence for the early generation of bitumen and oil from sulfur-rich kerogens, relative to sulfur-poor kerogen, owing to the preferential cleavage of weak sulfur linkages.

Interpreting the above results within the context of geological maturation of sulfur-rich source rocks such as the Monterey Fm., we can make several inferences. The first concerns the general phenomenon of low temperature petroleum generation. We have observed relatively low temperatures (experimentally-speaking) for maximum bitumen generation (275°C) during our experiments. This maximum reflects the liberation (or formation) of sulfur-rich high molecular weight heterocompounds (polar compounds) which clearly indicates the transfer of a significant proportion of sulfur from insoluble kerogen to soluble bitumen. Even before this event, however, partial elimination of sulfur from the kerogen takes place under mild thermal stress (175°C), and there is every reason to believe that this also occurs in the natural system within a late diagenetic/early maturation timeframe. For both of these phases of sulfur removal from the kerogen, geochemical evidence suggests that the sulfur species responsible are sulfide linkages. Initially, polysulfide bridges may collapse, liberating H_2S, but little or no bitumen (Figure 10). These bridges are likely to be the most thermally labile. Subsequently, C-S bonds in mono-sulfide bridges may undergo scission in concert with S-S bond cleavage, releasing polar-rich bitumen. Thus (poly)sulfide linkages are the species which display the greatest reactivity at lower temperatures and likely play a key role in dictating the evolution of petroleum-like products. At higher temperatures, reactions dominated by sulfur bond-breakage give way to those involving carbon-carbon bond scission (Figure 10). Here we envisage that sulfur plays a subordinate role owing to the fact that most of the labile sulfur is already eliminated from the kerogen at lower temperatures and condensed cyclic forms of sulfur, which are thermodynamically more stable and do not participate significantly in hydrocarbon generation, are more prevalent.

An interesting observation stemming from the present study is the non-linear relationship between the Py-GC based thiophene ratio and S_{org}/C ratio as a function of maturation temperature. A near linear relationship has previously been observed between these parameters for immature kerogens (27). Eglinton et al. (18) also observed this relationship for unconsolidated organic sulfur-rich sediments from the Peru margin, noting that the relation held despite the fact that XANES analysis indicated only a small proportion of the sulfur was thiophenic. These authors postulated that the ratio must reflect total sulfur because of a constancy in the relative abundance of sulfur species over the diagenetic interval studied. Indeed, for Peru margin kerogens, XANES-based speciation was remarkably constant, with sulfides dominating throughout. Like the Peru kerogens, the kerogens from the unheated Monterey shale showed a similar dominance of sulfides, but unlike the former there was a significant shift in the relative proportions of sulfur species as a function of temperature. Consequently, at lower temperatures (<250°C), the thiophene ratio remained constant while the S/C ratio decreased (Figure 9). This reflects removal of S from polysulfide bridges (as H_2S), while sulfur more intimately associated with carbon structures remains intact. At higher temperatures, the thiophene ratio shows a more precipitous reduction compared to the S/C ratio, where removal of thiophenic species becomes prevalent. The result of these thermally disparate events is a "dog-leg" relationship between the thiophene ratio and the S/C ratio. Although systematic decreases in the thiophene ratio as a function of maturity have been previously observed (29), this more complex relationship with respect to the S/C ratio has not been reported. It is clear, therefore, that caution should be exercized when using the Py-GC approach for estimating organic sulfur contents of kerogens spanning a wide maturity range.

Conclusions

Hydrous-pyrolysis experiments, when combined with an analytical scheme involving XANES spectroscopy, Py-GC, CP/MAS ^{13}C NMR spectrometry, elemental analysis, TLC-FID and relected light microscopy, represents an effective means to determine the speciation of sulfur during maturation of the Monterey shale.   The following conclusions regarding gross and molecular-level transformations in sulfur speciation and the manner by which they lead to the early conversion of kerogen to bitumen and oil are made:

1. The high sulfur content of Monterey kerogen leads to relatively low temperature (polar) bitumen (275°C) and (saturate and aromatic hydrocarbons) oil (325-360°C) generation as compared to sulfur-poor kerogens during experiments.

2. XANES spectroscopic analyses indicate that sulfides and polysulfides are the major forms of organic sulfur initially present in the Monterey Shale kerogen. Removal of these sulfur species from the kerogen begins at very low temperatures (150°C), with qualitative evidence indicating preferential elimination of polysulfides. The relative amount of organic sulfides and polysulfides increases in the bitumen fraction to a maximum at 275°C, whereas the relative proportion of total thiophenes in both the kerogen and bitumen fractions show comparatively little change with temperature. The early generation of bitumen is attributed to the inherent weakness of sulfur linkages within kerogen-bound sulfides and, in particular, polysulfides.

3. The fate of organic sulfur during laboratory thermal maturation of Monterey kerogen can be described in terms of three temperature regimes: (a) 150-225°C: (Poly)sulfide bridges collapse, leading to the formation of H_2S. (b) 225-275°C: Scission of sufficient sulfide linkages resulting in the release of soluble sulfur-rich fragments (bitumen) from the kerogen and continued H_2S production. (c) >275°C: Cleavage of S-S and C-S bonds gives way to C-C bond scission and the generation of saturated and aromatic hydrocarbons. At higher temperatures (360°C) sulfur may no longer be mechanistically important as most of the labile forms have been removed.

Acknowledgments

The authors would like to thank M. D. Lewan (USGS, Denver, CO) for providing the Monterey Shale sample, S. Wang and F. Lo (BNL, Upton, NY) for assisting with XANES analysis and curve fitting, L. B. Eglinton and N. L. Parmentier for various analyses, and E. Bailey for help with manuscript preparation. This research was funded by U.S. Department of Energy grant #'s DE-FGO2-86ER13466 (J. K. Whelan, J.S.S.), DE-FGO2-92ER14232 (T.I.E.), DE-AC02-76CH00016 (A.V.) and DE-FC21-93MC30127 (F.P.M.) This is Woods Hole Oceanographic Institution contribution number 8948.

Literature Cited

1. Tissot B. *Revue de l'Institut Francais du Petrol.* **1984**, *39*, 561-572.
2. Tannenbaum E. and Aizenshtat Z. *Org. Geochem.* **1984**, *6*, 503-511.
3. Lewan M.D. *Phil. Trans. R. Soc. Lond.* *A315*, **1985**, 123-134.
4. Gransch J.A. and Posthuma J. In: *Advances in Organic Geochemistry*; Tissot, B. and Bienner, F., Eds.; Paris Editions Technip: Paris, 1974; pp 727-739.
5. Orr W.L. *Org. Geochem.* **1986**, *10*, 499-516.

6. Hunt J.M., Lewan M.D. and Hennet R.J-C. *Bull. Amer. Assoc. Petrol. Geol.* **1991**, *75*, 795-807.
7. Baskin D.K. and Peters K.E. *Bull. Amer. Assoc. Petrol. Geol.* **1992**, *76*, 1-13.
8. Patience R.L, Mann A.L. and Poplett I.J.F. *Geochim. Cosmochim. Acta* **1992**, *56*, 2725-2742.
9. Lovering E.G. and Laidler K.J. *Canadian Jour. Chem.* **1960**, *38*, 2367.
10. Sinninghe Damsté J.S., Eglinton T.I., de Leeuw J.W. and Schenck P.A. *Geochim. Cosmochim. Acta* **1989**, *53*, 873-889.
11. Sinninghe Damsté J.S., Rijpstra W.I.C., Kock-van Dalen A.C., de Leeuw J.W. and Schenck P.A. *Geochim. Cosmochim. Acta* **1989**, *53*, 1343-1355.
12. Orr W.L. and Sinninghe Damste J.S. In: *Geochemistry of Sulfur in Fossil Fuels*; Orr, W.L. and White, C.M., Eds.; ACS Symposium Series 429, 1990; pp. 2-24.
13. Claxton M.J., Patience R.L. and Park P.J.D. In: *Poster sessions from the 16ᵗʰ International meeting on Organic Geochemistry*; Stavanger: 1993; pp. 198-201.
14. Lewan M.D. In: *Organic Geochemistry Principles and Applications*; Engel, M.H. and Macko, S.A., Eds; Plenum Press: New York, 1993; pp. 419-442,.
15. Eglinton T.I. and Douglas A.G. *Energy and Fuels* **1988**, *2*, 81-88.
16. Canfield D.E., Raiswell R., Westrich J.T., Reaves C.M., Berner R.A. *Chem. Geol.* **1986**, *54*, 149-155.
17. Acholla F.V. and Orr W.L. *Energy and Fuels* **1993**, *7*, 406-410.
18. Eglinton T.I., Irvine J.E., Vairavamurthy A., Zhou W. and Manowitz B. *Org. Geochem.* **1994**, *22*, 781-799.
19. Vairavamurthy A., Manowitz B., Zhou W. and Jeon Y In: *Environmental Geochemistry of Sulfide Oxidation*; Alpers, C.N. and Blowes, D.W. Eds.; ACS Symposium Series 550; 1994; pp. 412-430.
20. Waldo G.S., Carlson R.M.K., Moldowan J.M., Peters K.E. and Penner-Hahn J.E. *Geochim. Cosmochim. Acta* **1991**, *55*, 801-814.
21. Teichmuller M. *Org. Geochem.* **1986**, *10*, 581-599.
22. Karlsen D.A. and Larter S.R. *Org. Geochem.* **1991**, *17*, 603-617.
23. Seewald J.S. and Seyfried W.E. Earth Planet. Sci. Lett. **1990**, *101*, 388-403.
24. Robert P. In: *Organic Metamorphism and Geothermal History*; D. Reidel Publishing Co.: Holland, 1988; pp. 61-129,.
25. Lo H.B. *Org. Geochem.* **1993**, *20*, 653-657.
26. Eglinton T.I., Sinninghe Damsté J.S., Pool W., de Leeuw J.W., Eijkel G. and Boon J.J.. *Geochim. Cosmochim. Acta* **1992**, *56*, 1545-1560.
27. Eglinton T.I., Sinninghe Damsté J.S., Kohnen M.E.L. and de Leeuw J.W. *Fuel* **1990**, 69, 1394-1404.
28. Sinninghe Damsté J.S. and de Leeuw J.W. *Org. Geochem.* **1990**, *16*, 1077-1101.
29. Eglinton T.I., Sinninghe-Damste J.S., Kohnen M.E.L., de Leeuw J.W., Larter S.R. and Patience R.L. In: *Geochemistry of Sulfur in Fossil Fuels*; Orr, W.L. and White, C.M. Eds.; ACS Symposium Series 429; 1990; pp. 529-565.

RECEIVED June 30, 1995

GEOCHEMISTRY OF IRON SULFIDES IN SEDIMENTARY SYSTEMS

Chapter 9

Chemistry of Iron Sulfides in Sedimentary Environments

David Rickard[1], Martin A. A. Schoonen[2], and G. W. Luther III[3]

[1]Department of Earth Sciences, University of Wales, Cardiff CF1 3YE,
Wales, United Kingdom
[2]Department of Earth and Space Sciences, State University of New York,
Stony Brook, NY 11794–2100
[3]College of Marine Studies, University of Delaware, Lewes, DE 19958

Recent advances in understanding the chemistry of iron sulfides in sedimentary environments are beginning to shed more light on the processes involved in the global sulfur cycle. Pyrite may be formed via at least three routes including the reaction of precursor sulfides with polysulfides, the progressive solid-state oxidation of precursor iron sulfides and the oxidation of iron sulfides by hydrogen sulfide. The kinetics and mechanism of the polysulfide pathway are established and those of the H_2S oxidation pathway are being investigated. Preliminary considerations suggest that the relative rates of the three pathways are H_2S oxidation > polysulfide pathway >> solid-state oxidation. The kinetics and mechanisms of iron(II) monosulfide formation suggest the involvement of iron bisulfide complexes in the pathway and iron bisulfide complexes have now been identified by voltammetry and their stabililty constants measured. The framboidal texture commonly displayed by sedimentary pyrite appears to be an extreme example of mosaicity in crystal growth. Framboidal pyrite is produced through the H_2S oxidation reaction. Frontier molecular orbital calculations are beginning to provide theoretical underpinning of the reaction mechanisms. Recent progress in understanding iron sulfide chemistry is leading to questions regarding the degree of involvement of precursor iron sulfides in the formation of pyrite in sediments. Spin-offs from the work are addressing problems relating to the involvement of iron sulfides in the origin of life, the nature of metastability, the mechanism of precipitation reactions and the use of iron sulfides in advanced materials.

The importance of iron sulfides in the global environment is becoming increasingly apparent. Iron sulfides are the major sink for sulfur in the oceans and provide a key reservoir in the linked global cycles of O, C, N and S *(1,2,3,4)*. Jørgensen *(5)*, has estimated that as much as 50% of all organic matter in coastal sediments is metabolised by sulfate-reducing bacteria. Much of the sulfide produced by these bacteria ultimately ends up as pyrite fixed in sediments. There is considerable evidence, particularly isotopic, that this process has been important throughout geologic time, since the earliest sediments were formed on earth in excess of 3.5 Ga ago *(6)*. The processes leading to the formation of pyrite are thus key to understanding how the global atmosphere-ocean system works and how it has affected the Earth throughout its history.

The discovery in 1979 *(7)* of sulfide laden hydrothermal vents on the deep ocean floor added a new dimension to the global iron sulfide system. Since that time more than 150 sites have been discovered and the scale of this process is beginning to be appreciated. In 1994 the largest hydrothermal field on the ocean floor was discovered on the Atlantic ridge south of Iceland *(8)*. This discovery is significant since it suggests that these deep sea hydrothermal vents are even more abundant than hitherto suspected. Apart from the vast amounts of iron sulfides being produced annually from these systems, they also generate much biomass in the deep ocean. These isolated ecosystems appear to depend for their primary energy source on the sulfide contained in the hydrothermal fluids. Evidence for these deep ocean hydrothermal iron sulfide-rich systems is present throughout the geological record. The oldest known appears to be in excess of 3Ga in age and thus have been a key factor in the coupled global geochemical - biological systems since the earliest times on the planet.

The abundance of iron sulfides in time and space on the planet and their association with ecosystems where primary photosynthesis is precluded, has led to suggestions that pyrite may have played a crucial role in the origin of life on earth *(9,10,11,12)*. This suggestion is based both on the biochemistry of possible precursor moieties and the possibilities provided as templates by various pyrite crystalline forms.

Parkes et al. *(13)* identified sulfate reducing bacteria actively producing sulfide in rocks and sediments from more than 500m below the Pacific ocean floor. The five dispersed sites investigated all showed this deep activity. On the basis of these observations, they estimate that this deep microbial ecosystem constitutes more than 10% of the total global biosphere. Since only known microbial metabolisms have been included in these estimates, this figure is a minimum. The implications of these observations regarding the extent of the global sulfide system are significant. These ideas are also discussed by Deming and Baross *(14)*.

Iron sulfide formation has been documented in a wide variety of planetary environments apart from the oceans. These include lakes *(15,16,17,18,19,20)*, moors *(21)*, swamps *(22)* and aquifers *(23)*. The most recent thorough review of sedimentary iron sulfide chemistry was published in 1987 and covered the literature up to 1985 *(24)*. The purpose of this chapter is to provide a review of advances in studies relevant to the formation of iron sulfides during low-temperature, early diagenesis of marine sediments since 1985. Some of the recent advances have been made outside the natural environmental sciences as the formation of iron sulfides is becoming of industrial importance. For example, there is interest in synthesizing pyrite for use in solar cells *(25,26)*, high-energy batteries *(27)* and semiconductor applications *(28)*. Corrosion science is another important area of active research because iron sulfides are often formed as a result of corrosion of iron alloys under anoxic condition in industrial equipment, such as reactors and pipes *(29, 30,31)*.

The H$_2$S System

A knowledge of the dissociation constants of the H$_2$S system is important for understanding the interactions of the latter with the iron system. There is substantial agreement on the value of pK$_1$ for H$_2$S *(32,33)* in waters of varying salinity which confirms the values reported in Morse et al *(24)*. The value of pK$_2$ is still uncertain because of the ease of oxidation of sulfide to polysulfide at high pH. Schoonen and Barnes *(34)* re-evaluated pK$_2$ through an extrapolation of polysulfide pK$_2$ values as a reciprocal of the chain length. They estimated pK$_2$ (H$_2$S) at 18.5. Meyer et al *(35)* provided an independent estimate of 19.0. These values are 4-5 orders of magnitude greater than those values listed in Morse et al *(24)*. The larger values are more in-line with earlier observational reports suggesting that the value had to be > 17 *(35)*. These newer values indicate that estimates for metal sulfide complexes are much lower than reported previously *(36)*.

The Chemistry of Iron Monosulfides

Iron Monosulfide Complexes. It is generally observed that in stable sulfide environments the ion activity product (IAP) is undersaturated with respect to solid iron monosulfide phases *(37)*. However, the concentration of dissolved iron in sulfide-bearing waters which are subject to oxygen [or Fe(III) and Mn(III,IV) phases] infusion is analytically greater than that suggested by the equilibrium solubility of iron sulfide phases. Thus, there are kinetic limitations to phase formation. Even if the solubility controlling phase is a metastable high energy form such as colloidal iron (II) sulfide, the equilibrium solubility of iron (II) in solution appears high. There is of course an analytical problem in removing particulate iron sulfides in size ranges in the nanometer range. Even allowing for these systematic errors, the amount of iron in solution in sulfide-bearing waters appears high. Dyrrsen *(38)* and Elliot *(39)* used linear free energy relationship techniques to estimate the stability complexes of metal sulfide complexes in the absence of experimental values. These theoretical extrapolations also suggested the existence of relatively stable metal sulfide complexes in natural marine and freshwater systems.

There is further evidence for the existence of relatively stable iron(II) sulfide complexes in the apparent mobility of iron in anoxic sediments. Mass balance considerations on pyrite formation, for example, suggest strongly that iron must be transported to the site of precipitation *(40)* and thus it must be in a soluble form in sulfidic environments. Furthermore, the occurrence of iron sulfides within apparently closed microenvironments, such as plant cells *(41)* suggests that significant amounts of iron may be maintained in solution as sulfide complexes. It could be argued that sulfide is generated within an individual cell and iron could diffuse through non-sulfidic systems outside the cell. However, it seems improbable that, in a mass of cells, this type of non-sulfidic extra-cellular environment could be maintained.

Nanomolar levels of sulfide have been determined in oxic seawater *(42,43)* extending interest in iron sulfide complexation. Elliott et al. *(44)* suggested that carbonyl sulfide (COS) hydrolysis could release sulfide into oxygenated natural waters. Walsh et al. *(45)* showed that some phytoplankton can produce H_2S in culture when the trace metal content of the medium is high. Since the rate of (bi)sulfide oxidation in laboratory studies *(46)* and in field samples *(47,48,43,49,50)* occurs at reasonable rates, the occurrence of sulfide in oxic environments must be related to complex formation. The complexes involved may include metal complexes which are either kinetically inert or with high stability constants.

Luther and Ferdelman *(32)* identified $FeSH^+$ species in salt marsh creek and porewaters and Davison et al *(51)* suggested possible Fe_2S_2 species in anoxic lake waters. Rickard *(52,53)* proposed $FeSH^+$ and $Fe(SH)_2$ as kinetic intermediaries in the formation of iron (II) monosulfide from the reaction between iron(II) and sulfide in solution.   Luther and Ferdelman *(32)* and Zhang and Millero *(54)* used voltammetry at mercury electrodes to determine stability constants for metal sulfide complexes. The titration of Fe(II) with μM levels of sulfide in seawater at pH = 8.1, gave $FeSH^+$ ($\log \beta$ = 5.5 *(32)*; $\log \beta$ = 5.3 *(54)*) and $[Fe_2(SH)]^{3+}$ ($\log \beta$ = 11.1 *(32)*; not found *(54)*). Below pH 7, an $Fe(H_2S)$ complex forms which dissociates *(33)*. Luther et al. *(55)* demonstrated experimentally that in sulfidic-sulfur complexes with Mn, Fe, Co and Ni the sulfidic-sulfur is present as HS^-, whereas in complexes with Cu and Zn the ligand is S^{2-}. Both mononuclear and binuclear complexes exist at seawater pH. Luther et al. *(56)* determined polysulfide complexes of Mn, Fe, Co, Ni, Cu and Zn with one-to-one stoichiometry.

Iron Monosulfide Minerals. Using Molecular Orbital (MO) considerations, Tossell and Vaughan *(57,58)* reviewed the structures of iron sulfide minerals. A common element in most iron monosulfide structures is the presence of $Fe(II)S_6$ octahedra (e.g. troilite, pyrrhotite and greigite).   However, the iron in mackinawite is in tetrahedral coordination.   The six d electrons around the Fe(II) can result in both high spin and low spin cases in octahedral geometry. Only high spin is possible in the tetrahedral geometry. These unique electronic characteristics which affect the reactivities of the iron sulfide minerals, result from the sulfur species ligands which bind to the Fe(II). We can define a generalized FeS solid with octahedral geometry around Fe(II) to represent all monosulfide forms.   Although there are other geometries for FeS *(59)* essentially all forms of FeS have the same high spin electronic characteristics for Fe(II). For FeS minerals, S^{2-} is a weak field ligand which induces the high spin electron configuration for the d orbitals ($t_{2g}^4 e_g^2$) of the Fe(II).

Iron Monosulfide Reaction Chemistry. The highest occupied orbitals for S^{2-} are the 3s and 3p orbitals (Fig 1). Bonding to Fe(II) or any other atom occurs with the 3p orbitals. Fig. 2 shows a molecular orbital diagram for high spin octahedral Fe(II) bound to sulfide. On the surface of FeS, one of the S^{2-} 3p orbitals is bound to Fe(II) whereas the two other 3p orbitals are perpendicular to the bond axis in FeS and can bond with Lewis acids which attach to the surface of the solid. Through these 3p orbitals, the divalent ions of Mn, Co, Ni, Cd, Cu and Zn adsorb to the surface of mackinawite *(60,61)*. These orbitals impart the low zero point of charge, which makes the surface negatively charged at pH > 2 for greigite and pyrrhotite *(62)*. The 3p orbitals are of the same symmetry as the d_{xy}, d_{xz} and d_{yz} (or t_{2g}) orbitals of Fe(II) and can interact to give the high spin configuration for Fe(II) (Fig 2).

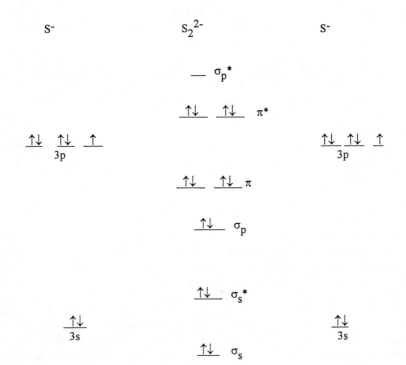

Figure 1. The molecular orbital combination of two S$^-$ anions to form the polysulfide anion, S$_2^{2-}$. The relative ordering of energy levels for S$^-$ is the same for S^{2-}.

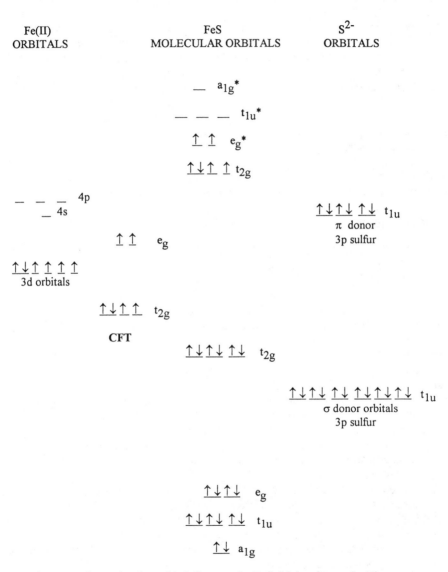

Figure 2. The molecular orbital diagram for FeS (high spin case). The combination of filled 3p orbitals from S^{2-} with the 3d, 4s and 4p orbitals of Fe(II). CFT indicates initial crystal field splitting of the metal d orbitals.

The initial splitting of the Fe(II) d orbitals is a crystal field or electrostatic interaction with the ligand prior to significant covalent bonding of the ligand and metal orbitals (Fig. 2). The t_{2g} and t_{1u} orbitals for both the metal and ligand respectively are π-type orbitals. For S^{2-} these orbitals are p orbitals which are filled and not involved in   bonding with Fe(II). Combination of these orbitals with those of the Fe(II) results in a set of t_{2g} molecular orbitals which are π-like and higher in energy than the original t_{2g} orbitals. Thus the pairing energy is prohibitively larger than the energy gap between the t_{2g} and e_g orbitals. In this case the combination results in the S^{2-} ligand donating electrons through the σ and π systems to the Fe(II) and stabilizing the high spin configuration. This configuration is kinetically labile and is the primary reason why FeS dissolves readily in even dilute mineral acids and can be measured quantitatively.

Synthesis of Iron(II) Monosulfides and Reaction Mechanisms.   Rickard *(63)* reported that iron (II) monosulfide formed in solution through two competing mechanisms:

(a) *the hydrogen sulfide pathway*, which involves the direct precipitation of Fe(II) monosulfide :

$Fe^{2+} + H_2S \rightarrow FeS + 2H^+$

This can be described by the pseudo first order rate law

$-d[H_2S]]/dt = k_1'[H_2S]$

where k_1' is the observed first order rate constant with a value of $9010s^{-1}$, $[H_2S]$ is the concentration of dissolved H_2S in moles and t is time in seconds.

(b) *The bisulfide pathway*, which involves the formation of the complexes $FeSH^+$ and the solid $Fe(SH)_2$:

$Fe^{2+} + 2HS^- \rightarrow Fe(HS)_2$

The rate of this reaction may be described by a law of the form

$-d[aHS^-]/dt = k_2'[aHS^-]^2$

where $[aHS^-]$ is the bisulfide activity and k_2' is the observed second order rate constant with value of 1.3×10^7 $mole^{-1} \cdot s^{-1}$ at $25^\circ C$. The result is consistent with an Eigen model of the process and suggests that nucleation in these systems is not rate-limiting.   The second stage of the reaction involves the condensation of $Fe(SH)_2$ to FeS with the release of dissolved sulfide back to solution:

$Fe(SH)_2 \rightarrow FeS + H_2S$

The net result of these competing reactions is complex behavior of the system with respect to total dissolved sulfide concentrations (S) and pH. At 25^0C the bisulfide pathway is faster at neutral to alkaline pH and at $S > 10^{-3}M$; the hydrogen sulfide pathway dominates in acid pH and at $S < 10^{-3}M$. Rickard *(64)* suggests that the competing reactions have variable temperature responses and that the hydrogen sulfide pathway becomes more important at temperatures less than $25^\circ C$.

Quantum-sized Particles.   The lack of any nucleation stage in Rickard's *(52,64)* kinetics suggests that the nucleation rates for precipitation of $Fe(SH)_2$ and FeS from aqueous solution are not rate-limiting.   This raises questions regarding the physical nature of these nuclei or small clusters. Rickard *(52,64)* showed that the original $Fe(SH)_2$ precipitate underwent a slower condensation reaction after 0.4s to form FeS with the release of H_2S back into solution. Buffle et al. *(65)* proposed a similar reaction on the basis of their field measurements.

The problem centers on the behavior of particles ranging in size from 1-10nm. These particles are quantum sized and research in this area is in its infancy. The processes involved link colloidal science, solid state physics and molecular chemistry. There are a host of metal sulfide species which yield clusters including ZnS, CdS and PbS *(66,67,68,69)*.

Some evidence exists for the existence of Fe(II) (bi)sulfide clusters. Luther*(70)* noted that the reaction of equimolar amounts of Fe(II) and S_x^{2-} (x=2,4,5) leads to fine particles and only a small quantity of FeS complex was measurable by voltammetry. On further addition of S_x^{2-}, all the FeS dissolved with the formation of a $[FeSH(S_x)]^-$ complex which could be measured by voltammetry. This suggests that the conversion between solid and solution species in this system is facile and not rate-determining.

The surface energy of precipitated Fe(II)S is unknown. By analogy with maximum and minimum values measured in other solids *(71)*, it probably lies between the limits of 10^{-5} J.cm^{-2} and 2 x 10^{-4} J.cm^{-2}. The molar volume is ca 20cm^3 by analogy with crystalline mackinawite. Applying the Kelvin equation, the critical nucleus size for precipitated Fe(II)S is constrained to between 11 and 220nm for supersaturations approaching unity. Such low supersaturations are suggested by the lack of evidence for a rate-controlling nucleation step in the precipitation reaction. The actual size of the tetragonal mackinawite unit cell is more than a magnitude less than these estimates *(72)*.   Note also that the estimated critical nucleus size is greater than the unit cell for cubic FeS *(73)*. These estimates suggest that amorphous FeS is a less ordered variant of the first crystalline phases rather than merely extremely fine-grained particles.

These estimates are consistent with observations on the first iron (II) sulfide precipitate. Vaughan et al. *(74)* reported x-ray absorption spectroscopic (EXAFS) evidence that the initial local structure was not mackinawite-like but that this developed rapidly with Fe in tetrahedral coordination with sulfur. Parise et al. *(75)* also found rapid mackinawite-like structure developing after 24 hours. Rickard *(76)* described the development of the mackinawite strongest x-ray line, from the 001 basal reflection, developing after 1 hour aging at 25°C in water. The line is very broad and the absence of other reflections suggests the initial lack of short range ordering. Rickard *(52)* postulated that the first formed precipitate is $Fe(SH)_2$ which subsequently condenses to Fe(II)S in a matter of hours.

The Chemistry of Iron Disulfides

FeS$_2$ Chemistry. The molecular orbitals for S_2^{2-} are combined from the 3s and 3p orbitals of two S atoms (Fig 1). The highest occupied MOs for S_2^{2-} are the two antibonding orbitals (π^*). One set of these orbitals donate electrons to Fe(II) and form angular bonds in FeS$_2$ (Fe-S-S). On the surface of pyrite, the other set of these orbitals is also available to bond with Lewis acids which can attach to the surface. Kornicker and Morse *(77)* have demonstrated in a systematic manner that the divalent ions of Ca, Mn, Co, Ni, Cd and Zn adsorb on pyrite. For the combination of S_2^{2-} and Fe(II) atomic orbitals, the highest occupied S_2^{2-}orbitals are used in bonding as above. However, the lowest unoccupied S_2^{2-} π orbitals are empty d orbitals that are similar in symmetry to the Fe(II) t_{2g} orbitals (Fig. 3). The combination of these metal and ligand "d" orbitals results in a lowering of the metal t_{2g} orbitals' energy and in the low spin configuration (t_{2g}^6) since the energy gap between the t_{2g} and e_g orbitals is larger than the pairing energy for two electrons in the same orbital. The extra stabilization resulting from this configuration is the reason for the thermodynamic and kinetic stability of pyrite with respect to non-oxidizing acids relative to other iron sulfides.

Kinetics and Mechanism of Pyrite Formation. Pyrite formation must focus on how the disulfide ligand can be formed and induce Fe(II) to change from high spin to low spin in the transition state, since solution and solid phase FeS or $FeSH^+$ precursors are high spin *(70,57)*

Pyrite is the stable iron sulfide in most sedimentary environments. Reports of pyrrhotite formation are scarce in modern low temperature systems. The reasons for this are not entirely obvious. Equilibrium thermodynamics suggests a large stability region for the pyrrhotites at low Eh or fO_2 values in neutral to alkaline solutions and at low S concentrations. All these conditions appear to be observed commonly in marine and freshwater environments - indeed they may be rather more abundant than the oxidized/reduced boundary conditions in which pyrite is stable and forms rapidly. The situation is made even more puzzling in that the metastable iron(II) monosulfides occur and may be variously involved in pyrite formation. It may be that microbiological activity encourages pyrite formation at the expense of pyrrhotite and/or that the transformation of metastable iron(II) monosulfides to pyrrhotite is kinetically hindered at low temperatures. Schoonen and Barnes *(78)* observed experimentally that pyrite nucleation was kinetically hindered at low temperatures leading to large potential supersaturations in solution. The reaction between polysulfides and the hexaquoiron (II) does not appear to lead to FeS_2 nucleation from solution, although it may result in pyrite growth *(70,78)*. The experimental problem is that, if the nucleation stage is not rate-limiting, it will not be discernible in the experimental kinetics. FeS_2 does not nucleate from FeS-undersaturated solutions; however, FeS_2 can grow from FeS-undersaturated solutions. This is consistent with early experimental work by Allen et al. *(79)*.

Pyrite is by far the most stable iron sulfide in marine and freshwater environments. Effectively it appears to occupy a great energy well into which all the iron and sulfide species will ultimately sink. This in turn implies that there are multiple competing pathways for its formation since, whatever the nature of the reactants or the details of the environments, the ultimate sedimentary iron sulfide will probably be pyrite.

(a) The Polysulfide Pathway. Luther *(70)* confirmed and refined the Rickard mechanism *(80)* for pyrite formation through the reaction between precursor Fe(II) monosulfide and polysulfide. Based on the molecular orbital diagram of FeS (Fig 1) and voltammetric data indicating that a complex of the form $[Fe(SH)S_x]$- exists in solution, Luther suggested that FeS (or an $FeSH^+$ complex) could donate a pair of electrons to the central and more positive S atoms in S_4^{2-} and S_5^{2-} thereby reducing the polysulfide ion. The $[Fe(SH)S_x]^-$ complex then breaks down to form FeS_2, H^+

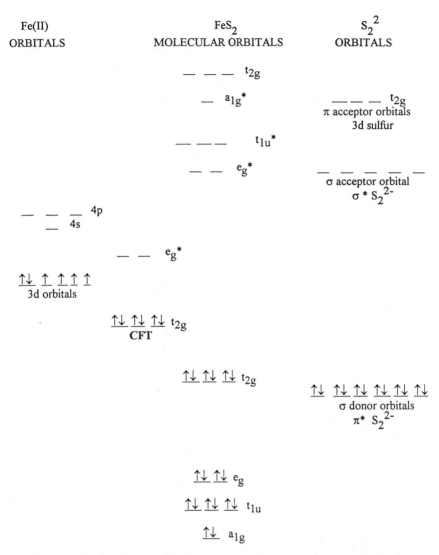

Figure 3. The molecular orbital diagram for FeS_2 (low spin case). The combination of filled π^* orbitals from S_2^{2-} with the 3d, 4s and 4p orbitals of Fe(II). The σ^* orbitals of S_2^{2-} do not interact with Fe(II) and are electron acceptor orbitals in Cr(II) reduction of pyrite. CFT indicates initial crystal field splitting of the metal d orbitals.

and S_{x-1}^{2-} with the pyrite sulfur atoms coming from the polysulfide ion in agreement with the Rickard mechanism (80). The significance of this mechanism is that (i) it does not necessarily depend on FeS as a precursor: $FeSH^+$ (or $Fe(SH)_2$) might do as well (52,63), (ii) that it is not a solid-solid reaction, but involves a dissolved stage, and (iii) the Fe(II) is converted from high to low spin so dissociation of disulfide is not possible. Luther (70) noted the possibility of polysulfide directly attacking FeS forming on the surface of FeOOH. Schoonen and Barnes (81) also observed this reaction and noted that it occurred under more reducing conditions than that involving greigite as an intermediate.

(b) The FeS Oxidation Pathway. The progressive oxidation mechanism has been confirmed by the careful experimental programme of Schoonen and Barnes (81). They found that amorphous FeS ($Fe_{1.11}S$-$Fe_{1.09}S$) aged to mackinawite $FeS_{0.94}$. Cubic FeS ($Fe_{1.028}S$-$Fe_{1.102}S$) may be an intermediate form in the process with a short half-life (82). Under slightly oxidizing conditions mackinawite transformed to greigite (Fe_3S_4) and then to pyrite. Individual steps in this complex reaction sequence are still not entirely understood. There is little or no kinetic data on individual reactions in order to evaluate the mechanisms. Only empirical data are available at present and the results of observation.

(1) Amorphous FeS → Mackinawite. As long ago as 1968 Rickard (76) showed that mackinawite x-ray lines were first observed from the initial amorphous precipitate on simple aging at 25°C. The rate of the transformation is slow: the first lines indicating long-range ordering were observed after 24 hours, but the complete mackinawite pattern might take up to 2 years to develop. Independently both Vaughan et al. (74) and Parise et al. (75) confirmed this observation using EXAFS. After 24 hours the initial FeS precipitate developed a local structure similar to mackinawite.

(2) Mackinawite → Greigite. Horiuchi et al. (83) showed that mackinawite converts to greigite on the addition of elemental sulfur. Schoonen and Barnes (81) thought that this reaction was unlikely in solution and proposed aqueous sulfur species as intermediates. However, it is still unclear if this transformation involves a dissolved stage or whether it is a solid state conversion. The problem noted by Taylor et al (84) that exposure of mackinawite to air for only a few seconds will predispose the precipitate to transform to greigite has still not been addressed.

(3) Greigite → Pyrite. Sweeney and Kaplan (85) initially proposed that pyrite could be formed by the successive oxidation of FeS through Fe_3S_4 to pyrite. Their reaction, which resulted in the formation of perfect pyrite framboids, has not yet been repeated and little is known about this reaction. Schoonen and Barnes (81) noted that this transformation required a considerable structural reorganization and the transformation to marcasite was stereochemically more compatible. Marcasite will change to the stable pyrite through a solid-state transformation.

(c) The H_2S Pathway. Taylor et al. (84) initially demonstrated that Fe(II) monosulfides could react with H_2S to form pyrite at temperatures as low as 100°C

in a few days. One product is hydrogen gas: $FeS + H_2S \rightarrow FeS_2 + H_2$
    The reaction was repeated by Drobner et al. *(86)* and Rickard *(63)*. Drobner et al. *(86)* observed the reaction at 100°C and Rickard *(63)* used Eh poises to obtain pyrite and measured H_2 gas from the reaction at 25°C. Initial results from the kinetic study by Rickard *(63)* suggest that the rate varies with pH, FeS surface area, temperature and H_2S concentration and that the reaction stoichiometry closely agrees with that predicted. The reaction is interesting for a variety of reasons: (i) it is a competitive reaction with (a) and (b) above and may lead to a re-evaluation of the experimental observations in those systems; (ii) the production of H_2 may be microbially significant in natural sedimentary systems; (iii) the large volumes of H_2 gas associated with deep sea hydrothermal vents, which have previously been assumed to result from magma degassing may need to be re-evaluated and (iv) it may be involved in the origin of life. However, there is a caveat. The reaction has not been observed in a number of carefully controlled experiment. Further kinetic work is required to establish the mechanism and evaluate the role of this process in the natural system.

Marcasite Formation. Experimental work by Murowchick and Barnes *(82)* shows that below pH 5, marcasite becomes the dominant disulfide. This work, consistent with much earlier experimental work by Allen et al. *(79)* is supported by several more recent experimental studies *(81,87)*. However, these studies were synthetic rather than kinetic and there is still lack of knowledge regarding the mechanism. Murowchick and Barnes *(82)* proposed that the speciation of disulfide species in solution controls the relative rate of pyrite growth versus marcasite growth. Their argument centers on the notion that aqueous disulfide species have to be incorporated into a FeS_2 surface. The aqueous speciation of the disulfide ion is pH dependent, with H_2S_2 the dominant species below pH 5, HS_2^- the dominant species between pH 5 and 10, and the S_2^{2-} species the dominant species above pH 10 *(81)*. Murowchick and Barnes proposed that the protonated ends of H_2S_2 and HS_2^- are repelled from pyrite growth sites but not from marcasite growth sites; whereas, the charged ends of the HS_2^- and S_2^{2-} species are strongly attracted to the pyrite growth sites. The presence of protons on the H_2S_2 species and the protonated end of the HS_2^- species makes these species isoelectronic with AsS^{2-} and As_2^{2-}. As pointed out by Tossel *(88)* the interaction between AsS^{2-} and As_2^{2-} species and Fe^{2+} produces the loellingite structure, which is closely related to that of marcasite. Hence, based on the work by Tossel, Murowchick and Barnes argued that in acidic solutions where H_2S_2 dominates the disulfide speciation marcasite grows much faster than pyrite. Recently, Rakovan et al *(89)* described marcasite overgrowths on euhedral pyrite and explained this phenomena as a reconstruction of the pyrite surface to a marcasite structure induced by the presence of H_2S_2 on the surface.

FeS_2 Reaction Chemistry. Pyrite reactivity is directly related to reduction or oxidation of the S_2^{2-} ligand *(90)* because the low spin Fe(II) electron configuration is kinetically inert to dissociation in non-oxidizing acids such as HCl. Pyrite reacts with oxidizing acids *(91)* and other oxidants including Fe(III) and molecular halogens *(92)*. In these redox reactions, one of the π^* orbitals of S_2^{2-} in pyrite binds to the oxidant in a σ-bond at the surface. For halogens and NO_x species in nitric acid, complete electron transfer occurs along the σ-bond axis and results in the formation of halide ions and the oxidation of the S_2^{2-} in pyrite. For Fe(III) complexes, σ bond attachment of the Fe(III) occurs to the pyrite surface via one of the π^* orbitals of S_2^{2-} and one of the Fe(III) σ orbitals. The other π^* orbital, which is perpendicular to the one binding to Fe(III) in the σ bond, transfers an electron to one of the t_{2g} orbitals in Fe(III). Moses and Herman (93) have shown that Fe(II) can also bond to the surface of the pyrite and that the addition of O_2 accelerates the oxidation of pyrite by first oxidizing Fe(II) to Fe(III). O_2 is not an effective oxidant for pyrite because it has partially filled π^* orbitals which cannot accept two electrons from the S_2^{2-} in pyrite and thus attach to the surface of pyrite (90). Oxygen stable isotope analyses indicate that O atoms from O_2 do not become a significant part of the sulfate produced (94). The O atoms come from the water.

Acid extraction is the most commonly used method to determine the amount of iron monosulfides present in a sediment. Several different protocols relying on the difference in dissolution rate of amorphous FeS, mackinawite, greigite and pyrrhotite versus that of FeS_2 in non-oxidizing acids have been developed (95, 96,97,98,99,100). Hence, with these acid extraction techniques the amount of (amorphous FeS + mackinawite + greigite + pyrrhotite) is determined, and this is often referred to as the amount of Acid Volatile Sulfides (AVS). The extraction methods are, however, not without problems (101) and application to ancient rocks is less than straightforward (100). Sequential determination of FeS and FeS_2 should not generally be performed since acidifying a natural sample results in dissolving amorphous Fe(III) minerals as well as the FeS. The Fe(III) released to solution can then oxidize pyrite with high reaction rates (102,93,103,104). Either a reductant such as Sn(II) needs to be added to reduce Fe(III) to Fe(II) (96) or a separate analysis on another subsample needs to be performed for accurate pyrite determination. However, Sn(II) will reduce pyrite to acid soluble sulfides when heated. This reaction is grain size dependent so the reduction can occur at low temperatures. However, for trace element determination in pyrite and other fractions by wet chemical methods there is no other alternative to sequential determination since high purity acids are required for leaching (91) and Sn(II) salts are a further possible source of contamination. The analysis of AVS in natural systems might further be constrained by particle size. Solutions are commonly filtered through 0.45 m filters which are larger than the maximum critical nucleus size for precipitated Fe(II)S. Rapid crystal growth and/or flocculation in solutions with high ionic strength like seawater, might act to reduce the analytical uncertainties. However, even the commonly observed crystal size of sedimentary pyrite, whether or not in the framboidal organization, is 0.1 m or substantially less than these filters. Landing and Westerlund (105) noted a slight difference in dissolved Fe(II) in solutions filtered through 0.2 and 0.45 m filters.

The molecular halogens are isoelectric with S_2^{2-} (Fig. 1) and the addition of electrons to the σ_p^* orbital results in molecular halogen reduction with halogen-halogen bond breakage and halide formation. Pyrite can be reduced in the same fashion by Cr(II) in acid (96,90). The electronic configuration for Cr(II) is $t_{2g}^3 e_g^1$. Electron transfer from the e_g^* orbital of σ symmetry from Cr(II) to the σ_p^* orbital of S_2^{2-} results in reduction of pyrite and formation of H_2S. In addition, Cr(II) reduces the Fe(III) released in the acid medium. The success of the Cr(II) reduction method in the determination of pyrite by trapping and analysis of the H_2S formed has been one of the important analytical feats in understanding pyrite formation and turnover in the environment *(106)*.

The Geochemistry of Iron Monosulfides

The Occurrence of Iron Monosulfides in Sediments. Because the presence of iron monosulfides in modern and ancient sediments may provide constraints on the depositional environment and diagenetic history *(107,108,109)* a large number of studies have examined the presence of iron monosulfides in modern and ancient sediments using several different techniques. While acid extraction methods determine AVS, several studies have been able to demonstrate the presence of greigite and pyrrhotite in modern and ancient sediments using methods based on the magnetic properties of iron sulfides *(110,111,112,113)* and Mössbauer Spectroscopy *(114,15,115)*. A study by van Velzen *(116)* showed the presence of Fe-Ni in sulfides in a Pleistocene marine marl in Calabria, Italy, indicating that it is possible that the chemical composition of sedimentary magnetic iron monosulfide phases is far from ideal. Reynolds and co-workers *(111)* found early diagenetic greigite as well as epigenetic greigite in Cretaceous strata in Alaska and showed that the diagenetic and epigenetic greigite dominates the magnetic properties of these rocks.

The number of direct observations using optical microscopy, Scanning Electron Microscopy or Transmission Electron Microscopy of iron monosulfides in sediments are limited. In the last ten years, greigite has been found in Recent[1] marine sediments (117,118,98), Holocene lake sediments *(119,120)*; Pliocene/Pleistocene marine marls *(121)*, Middle Miocene brown coal *(122,123,124)*, and Cretaceous marine sediments *(111)*. Greigite has been observed inside foraminifera *(112)* and in plant tissue *(22)*. An interesting occurrence of intracellular greigite (and pyrite) in magnetotactic bacteria has been documented *(125,126)*. These bacteria use ferrimagnetic greigite to navigate and may account for some of the iron sulfides found in euxinic environments *(127)*. Mackinawite is seldom observed in Recent or ancient sediments. While significant amounts of mackinawite may be present as demonstrated by AVS determinations, it may be difficult to observe it due to very small particle size, its reactivity (conversion to greigite or pyrite is fast), and rapid oxidation which make recovery difficult. In a study of Recent marine sediments Lein and coworkers *(128)* demonstrated using x-ray diffractometry that mackinawite was present in a magnetic separate. Presumably, the sample contained mackinawite adhering to ferrimagnetic greigite. The presence of mackinawite was also demonstrated in Black Sea sediments *(129)*. Pyrrhotite has not been found in Recent sediments but it has been found in ancient sediments. It is common as an epigenetic iron sulfide mineral along with greigite and pyrite in oil-bearing strata *(130,111,131,132)*.

---

[1]Recent: An epoch of geologic time following the Pleistocene.

The Geochemistry of Iron Disulfides

Crystal Growth and Morphology of Sedimentary Iron Disulfides. Crystal growth of sedimentary iron disulfides becomes an important process once FeS_2 nuclei are formed. Numerous studies have shown that porewaters in reduced sediments are often very close to saturation with respect to iron monosulfide phases, which are about nine orders of magnitudes more soluble than iron disulfides *(81)*. Hence, porewaters in reduced sediments and in euxinic waters are typically vastly supersaturated with respect to both pyrite and marcasite. This supersaturation is the driving force for FeS_2 nuclei to grow. The kinetic barrier for crystal growth is considerably less than for nucleation from solution because the growing surface of a pyrite or marcasite is likely to continuously provide growth sites, avoiding the need for an additional nucleation step. However, the fact that marine porewaters remain near saturation with respect to iron monosulfides even if most of the iron sulfides are present as pyrite, indicates that the rate of pyrite growth is considerably slower than the dissolution rate of any iron monosulfide that may be left.   Overall growth of iron disulfides from solution may be the most important process in terms of the mass of disulfides formed. Because the morphology of a crystal is related to the conditions during growth *(133)*, it is important to understand what may determine the morphology of iron disulfides. Finally, it is important to understand what controls whether a system will predominantly form pyrite versus marcasite. Both a thorough understanding of the controls on crystal morphology and pyrite-marcasite abundance may provide powerful indicators of conditions during the early diagenesis of a sediment.
      Murowchick and Barnes *(134)* showed that under hydrothermal conditions the striation and morphology of pyrite (i.e. cubic, octahedral or pyritohedral habitus) is controlled by the temperature and supersaturation with respect to pyrite, but it is not clear whether supersaturation is also the controlling factor at low temperatures. For example, preferential sorption of aqueous species on one growth surface versus another may control the morphology of the crystal *(135, 136, 137)*

Pyrite versus Marcasite Abundance. Pyrite vs marcasite abundance in sediments is a function of pH. Marcasite is never formed during early diagenesis in marine environments where the pH is not acidic, but it does occur in coals and acidic swamps *(138,139)*.
      Crucial to all the models for the crystal growth of iron disulfides is the notion that a disulfide ion must be incorporated into the growing crystal lattice. Because the total concentration of polysulfides appears to be not higher than 10^{-5} M in most natural waters (work by Visscher et al. *(140)* on biomats is a notable exception) and polysulfides are unstable in acidic solution, it is not clear whether incorporation of aqueous disulfide species is a viable mechanism under all conditions. The direct formation of pyrite through the reaction between precursor Fe(II) sulfide species and H_2S, reported above, thus becomes an attractive candidate for the major sedimentary pyrite-forming pathway. Rickard *(63)* called the process the "sulfide pump" since it simply required that sulfide was continuously pumped into an iron source for iron (II) sulfides and, ultimately, pyrite to form. The kinetics of iron (II) monosulfide formation may also suggest that the precipitation of Fe(II)S could be a by-product of

the process since dissolved iron (II) sulfide species or iron (II) bisulfide clusters might react faster than solid mackinawite. Schoonen and Barnes *(81)* have argued that formation of the disulfide moiety on the surface through reactions involving other, often more abundant S-species including organosulfur compounds *(22)* might be a more plausible mechanism. Recently, Xu and Schoonen *(141)* have shown that thiosulfate forms tetrathionate at the pyrite surface, which indicates that S-S bonds may be formed on the surface.

Finally, it is important to understand the mechanism by which elements other than Fe and S are incorporated into a growing pyrite or marcasite surface. Although it is well documented that sedimentary pyrite is an important sink for siderophilic, chalcophilic and some lithophilic elements *(91,142,143,144)*, the mechanisms of incorporation are poorly understood. Several studies have examined the interactions between aqueous species and the pyrite surface *(62,77,145)*, but no studies have been conducted with growing surfaces.

Morphology of Sedimentary Pyrite. Iron(II) monosulfides may act as precursors for pyrite formation; however, once FeS_2 nucleii are formed they will tend to grow from solution. Because the morphology of the crystal is determined by conditions during the crystal growth stage, the shape of the crystal provides no constraints on the conditions during the nucleation step, including the saturation state of the porewaters with respect to iron monosulfides.

Authigenic pyrite formed in low-temperature marine environments occurs either as micron-size crystals aggregated in a raspberry-like morphology (framboidal pyrite) or as micron-sized euhedral crystals. In euxinic environments, pyrite forms both in the water column as well as in the sediments *(146,147,148,149, 150)*. By contrast, pyrite formation is strictly diagenetic in environments with oxygenated bottom waters. Pyrite formed in the water column is mostly framboidal *(151,152)*, while diagenetic pyrite is both present as framboids and euhedral crystals. The difference in pyrite morphology has sometimes been interpreted as a difference in formation mechanism due to changes in the saturation state of the porewaters *(153, 154,155,156)*. For example, it has been argued that euhedral pyrite are indicative of porewaters undersaturated with respect to iron monosulfides, because it is assumed that euhedral crystals are strictly formed via nucleation and growth from solution. Framboidal pyrite, on the other hand, is assumed to form via replacement of precursor monosulfides, which requires porewaters to be saturated with respect to iron monosulfides. Both models are probably oversimplifications which do not take into account the role of crystal growth.

It is probably an oversimplification to assume that the individual crystals forming a framboid are formed exclusively by replacement of an iron monosulfide precursor. Replacement of ferrimagnetic greigite (Fe_3S_4) as a step in framboid formation, as first suggested by Sweeney and Kaplan *(85)*, explains the aggregated morphology by the magnetic forces among precursor greigite particles *(157)* but it does not explain the morphology of the pyrite crystallites. Recent attempts to reproduce the Sweeney and Kaplan *(85)* experiments have failed *(158)*.

Rickard and Butler *(159)* described natural framboids in which the individual octahedral microcrysts were hollow. Close examination with Scanning Electron Microscopy suggested orthogonal spindles of pyrite within these hollow microcrysts.

They suggested that the microcrysts grew rapidly by successive nucleation from high energy sites on the apices of pre-existing microcrysts. The process is thought to be initiated by nucleation of one or more microcrysts at sites within a volume that is saturated with respect to pyrite *(158)*. Subsequent growth from these sites gives the familiar framboidal microarchitectures *(160,161)*.

The only low temperature syntheses of pyrite framboids reported since 1987 have been by Rickard's group[2] (op cit.). Interestingly these framboids were produced via the reaction involving H_2S oxidation of precursor iron (II) monosulfide. These framboids are small and poorly developed and are described as protoframboids. This may suggest that solid iron(II) monosulfides as the iron reactant inhibits framboid formation. None of the many experiments performed with iron (II) sulfides at low temperatures have yet produced framboids with the same apparently ordered microarchitectures as those found in nature.

The Occurrence of Iron Disulfides in Sediments. As shown by Canfield et al. *(96)*, the reactivity of iron minerals toward dissolved sulfidic sulfur may be an important control in the morphology and distribution of authigenic pyrite in marine sediments. Studying sediment cores collected at the FOAM site in Long Island Sound, Connecticut, USA, Canfield and co-workers demonstrated that small framboidal pyrite is the dominant morphology during the earliest stage of shallow diagenesis, while deeper in the sediment most of the pyrite is found as small crystals forming on Fe-bearing sheet silicates (e.g., biotite, chlorite) and magnetite *(162)*. Canfield and co-workers note that the switch from predominantly framboidal pyrite to pyrite adhered to iron-containing minerals coincides with a build up of sulfidic sulfur (H_2S + HS^-) in the porewater. They explain the absence of sulfidic sulfur in the top of the sediment core (0-10 cm) by rapid consumption via reaction with the most reactive source of iron in the sediments, iron oxyhydroxides. In essence, as soon as sulfidic sulfur is formed via sulfate reduction, the sulfidic sulfur reacts with iron oxyhydroxides in the sediment. The nearly exclusive formation of framboids during the initial stages may reflect the presence of micro-environments with a high rate of sulfate reduction coupled with an immediate consumption of sulfidic sulfur through reaction with iron oxyhydroxides or any other source of reactive iron *(163)*. However, a difference on the sulfur isotopic signature between the framboidal and single crystal pyrite morphologies, as predicted by the model, may not always be present *(164)*.

As the pool of reactive iron oxyhydroxides becomes exhausted, the sulfidic sulfur concentration in the porewaters can increase because the rate of sulfidic sulfur production is faster than that of consumption via reaction with less reactive iron-containing minerals *(165)*. The sulfidation of less reactive iron-containing minerals may lead to the formation of small framboidal pyrite and small euhedral pyrite crystals at the surface of iron-bearing minerals *(162,166)*. Ultimately, the production of sulfidic sulfur is limited by the availability of sulfate in the porewaters. Bioturbation and physical reworking of surface sediments contributes extra organic matter and sulfate, which sustains sulfate reduction; however, it also advects oxygen,

---

[2] Wilkin, R.T. and Barnes, H.L. (personal communication 1995) have reported framboid synthesis at low temperatures.

juxtaposes oxides to sulfides *(167)*, promotes diffusive loss of H_2S through burrows, and mixes pyrite formed at different depths *(168)* walls of burrows and tubes providing evidence for significant mass transfer across these interfaces *(169)*.

Although there is a net production of pyrite in marine sediments, these systems are clearly very dynamic with much of the pyrite and iron sulfides formed during early diagenesis later destroyed by oxidative dissolution *(167)*. Hence, the amount of pyrite found in ancient sediments may reflect only a small portion of the total amount of pyrite formed in the sediment during early diagenesis. Because of extensive bioturbation and physical reworking during early diagenesis, pyrite found in ancient sediments may include pyrites formed at different stages during diagenesis and in euxinic environments some pyrite may also be syngenetic *(170,149,150)*. In addition, each type of pyrite (diagenetic, euhedral or framboidal; syngenetic, framboidal) may have its own characteristic trace element content *(143,101)* and S-isotopic signature *(164)*, which complicates the reconstruction of the depositional environment based on the isotope geochemistry or trace-element geochemistry of pyrites *(171,152)*.

Conclusions

A first attempt to model iron sulfide formation in sediments, using "Rickard kinetics", was made by Gardner and Lerche *(172)*. Since that time, significant progress has been made in understanding the different competitive routes to pyrite formation in sediments and to develop rate laws for these routes. First order models of the sedimentary process, based on Gardner and Lerche's initial attempt, should soon be more refined. Subsequently, second order models which take into account recycling of sulfides in the sediments and deep bacterial sulfide production should provide a key link in quantifying the global sulfur cycle. Establishment of realistic, dynamic models of this cycle will input into global systems allowing both prediction of future earth environments and evaluation of past environments.

At present, comparison of the rates of the three pathways proposed to be involved in sedimentary pyrite formation suggest that the reaction between H_2S and iron monosulfides is much faster than that between polysulfides and iron monosulfides. In the natural system the polysulfide pathway may be inhibited by the apparent low standing concentrations of polysulfide in solution. However, analyses of polysulfide concentrations in sediments are not sufficiently reliable at present to determine natural sedimentary polysulfide concentrations with confidence. Even so, these two pathways appear to be much faster in the laboratory than the solid state, sequential, oxidation of iron monosulfide through greigite. The kinetic comparisons are difficult since the oxidation of FeS by H_2S was not excluded from the designs of experiments involving the solid state oxidation pathway nor the polysulphide pathway. It will be important to reconsider previous work in the light of H_2S oxidation kinetics when these are determined in detail.

A major outstanding problem is the lack of any experimental evidence for direct precipitation of pyrite from solution *(81)*. Circumstantial evidence at present includes only fast pyrite formation in the laboratory *(e.g.63)*, some sediments *(e.g.153)* and in the water column *(e.g.148)*. Theoretical interpretations suggesting a direct precipitation mechanism *(e.g.70,80,52)* have been supported by the discovery of strong iron (II) bisulfide complexes *(32,55,52,54)*. However, there is no evidence at present for direct precipitation of pyrite from solution at low temperatures in the absence of precursor iron sulphides.

Chemical studies of the iron sulfides are beginning to contribute to the general understanding of how solid phases form from solution. Particular interest is focused on the cluster state, between the pure solute and the true solid, and its role in the processes of precipitation and crystal growth. The iron sulfides were some of the earliest natural solids to be described using molecular orbital models. These models, based initially on Mössbauer studies, have now been refined with EXAFS. The introduction of widespread semi-empirical molecular modelling techniques combined with the frontier molecular orbital approach suggests that the understanding of iron sulfide reactions on a molecular level will provide a valuable insight into these processes.

The interrelationships between the various iron sulfide solids are not understood. No kinetic studies have been made of marcasite formation, and its formation mechanism remains a mystery. Its elucidation may contribute to the understanding of dimorph formation in general. While it is widely recognized that metastable iron sulfide phases play an important role in iron sulfide geochemistry at low temperatures in aqueous systems, the kinetics of the conversions between metastable Fe-S phases, and from metastable states to pyrite, are poorly understood.

The surface properties of iron (II) monosulfides have been implicated in the mechanism of pyrite nucleation *(81)*. Research on the surface chemistry of these minerals appears to have particular potential for determining the processes controlling iron sulfide formation and dissolution, as well as contributing to understanding the role of surface properties in the incorporation of other metals during growth.

The suggestion that framboids themselves may be a novel form of crystal may have considerable implications for the materials industry, especially in the electronics and energy applications. Despite all the attempts, well-organized framboids have not yet been synthesized under conditions analogous to those in sedimentary environments. Yet they form readily and rapidly in natural environments. The identification of what framboids actually represent, may catalyze this quest.

In the last decade, here has been an effort to better characterize the nature of iron sulfides in sediments. Acid digestion methods continue to be refined, recalibrated and tested in both ancient and modern sediments. There appears to be a great potential in using voltammetric methods not only to analyse ultralow concentrations of iron and sulfur species in sediments but also to determine speciation.

The oxidation of iron (II) sulfides with H_2S produces H_2 gas. It is thus of considerable interest in the natural environment since the H_2 is a valuable energy source for micro-organisms. Even if the process is not, finally, implicated in the origin of life it is still of major potential importance in sedimentary environments. The next stage appears to be attempts to detect H_2 gas in anoxic sediments and to relate this quantitatively to sedimentary pyrite formation. It is possible that H isotopic studies may contribute to the quantification of this process in ancient and modern natural systems.

Acknowledgements

We thank Bill Davison for comments on an early version of the manuscript and two anonymous ACS reviewers. The work was funded by NERC grants GR9/603 and GR3/7476. Dr M.A.A. Schoonen's contribution to this paper was supported by NSF-EAR 9407099.

Literature cited

1. Berner R.A. *Amer. Jour. Sci.* **1987**; *287*, 177-198.
2. Kump, J.A. *Am. J. Sci.* **1989**; *289*, 390-410
3. Berner, R.A., Canfield, D.E. *Am. J. Sci.* **1989**, *289*, 333-361
4. Dean, W.E.; Arthur, M.A. *Am. J. Sci.* **1989**; *289*, 708-743.
5. Jorgenson, B.B. *Nature* **1982**; *296*, 643-645
6. Ohmoto, J., Kakegawa, T., Lowe, D.R. *Science* **1993**; *262*, 555-557
7. Corliss, J.B.; Dymond, J.; Gordon, L.I.; Edmond, J.M.; von Herzen, R.P.; Ballard, R.D.; Green, K.; Williams, D.; Bainbridge, A.; Crane, K.; van Andel,T.H. *Science* **1979** *203*, 1073-1083.
8. German, C. and HEAT Scientific Team, *BRIDGE Newsletter* **1994**; No 7, 3-5.
9. Cairns-Smith A.G.; Hall A.J; Russel M.J. *Origins of Life and Evolution of the Biosphere* **1992**; *22*, 161-180.
10. Wächtershäuser, G. *Syst. Appl. Microbiol.* **1988**; *10*, 207-210
11. Wächtershäuser, G. *Microbiol. Rev.* **1988**; *52*, 452-484
12. Wächtershäuser, G. *Proc. Natn. Acad. Sci. USA*, **1990**; *87*, 200-204
13. Parkes, R.J., Cragg, B.A., Bale, S.J., Getliff, J.M., Goodman, K.J., Rochelle, P.A., Fry, J.C., Weightman, A.J., Harrey, S.N., *Nature*, **1994**, *371*, 410.
14. Deming, J.W., Baross, J.A. *Geochim. Cosmochim. Acta.* **1993**; *57*, 3219-3230
15. Manning, P.G.; Ash, L.A. *Can. Mineral.* **1979**; *17*, 111-115.
16. Davison, W.; Grime, G.W.; Woof, C. *Limnol. Oceanogr.* **1993**; *37*, 1770-1777.
17. Wicks, C.M. *Early diagenesis of iron and sulfur in sediments of lakes that receive acid mine drainage,* Masters Thesis, **1989**; University of Virginia, 152pp.
18. Tiercelin, J.J.; Pflumio, C.; Castrec, M.; Boulegue, J.; Gente, P.; Rolet, J.; Coussement, C.; Stetter, K.O.; Huber, R.; Buku, S.; Mifundu, W *Geology* **1993**; *21*, 499-502.
19. Tuttle, M.L; Goldhaber, M.B. *Geochem. Cosmochim. Acta.* **1993**; *57*, 3023-3029.
20. Tuttle, M.L.; Goldhaber, M.B.; Williamson, D.L. *Talanta* **1990**; *33*, 953-961.

21.   Jakobsen, B.H. *J.Soil Sci.* **1988**; *39*, 447-455.

22.   Altschuler, Z.S.; Schnepfe, M.N.; Silber, C.C.; Simon, F.O. *Science* **1983**; *221*, 221-227.

23.   Jakobsen, R. *208th ACS Meeting Washington, DC,* **1994**; Abstract.

24.   Morse, J.W.; Millero, F.J.; Cornwell, J.C.; Rickard, D. *Earth-Sci. Rev* **1987**; *24*,1-42.

25.   Liu, C.; Bard, A.J *J.Phys. Chem.* **1989**; *93*, 7047-7049.

26.   Tsay, T.; Huang, Y.S.; Chen, Y-F. *J. Applied Phys.* **1993**; *74*, 2786-2789.

27.   Fan, F-R.; Bard, A.J. *J. Phys. Chem.* **1991**; *95*, 1969-1976.

28.   Ferrer, I.J.; de las Heras, C.; Menendez, N.; Tornero, J.; Sanchez, C. *J. Material Sci.* **1993**; *28*, 389-93.

29.   Lalvani, S.B.; DeNeve, B.A.; Weston A. *Corrosion* **1991**; *47*, 55-61.

30.   McMeil, M.B.; Little, B.J. *Corrosion* **1990**; *46*, 599-600.

31.   Walker, M.L.; Dill, W.R.; Besler, M.R.; McFatridge, D.G. *J. Petroleum Technology* **1991**; *43*, 603-607.

32.   Luther,G.W.III; Ferdelman, T. *Env. Sci.Tech.* **1993**; *27*, 1154-1163.

33.   Millero, F.J; Plese,T; Fernandez, M. *Limnol.Oceanogr.* **1988**; *33*, 269-274.

34.   Schoonen, M.A.A.; Barnes, H.L *Geochim. Cosmochim. Acta.* **1988**; *52*, 649-654.

35.   Meyer, B.; Ward, K.; Koshlap, W.K.; Peter, L. *Inorg. Chem.* **1983**; *22*, 2345-2346.

36.   Giggenbach, W., *Inorg. Chem.***1971**; *10*, 1333-1338.

37.   Davison, W. *Aquatic Sciences* **1991**; *53*, 309-329.

38.   Dyrrsen, D. *Mar. Chem.* **1988**; *24*,143-160.

39.   Elliott, S. *Mar. Chem.* **1989**; *24*, 203-213

40.   Raiswell, R.; Whaler, K.; Dean, S.; Coleman, M.L.; Briggs, D.E.G. *Mar. Chem.* **1993**; *113*, 89-100.

41.   Kenrick, P.; Edwards, D. *Botanical J. Linnean Soc.* **1988**; *97*, 95-123

42.   Cutter, G.A., Krahforst, C.F. *Jour. Geophys. Letters* **1988**; *15*, 1393-1396

43.   Luther, G.W.III.; Tsamakis, E. *Mar. Chem.***1989**; *27*, 165-177.

44.   Elliott, S., Lu, E., Rowland, F.S. *Geophys. Res. Letters* **1987**; *14*, 131-134

45.   Walsh, R.S., Cutter, G.A., Dunstan, W.M., Radford-Knoery, J. & Elder, J.T., *Limnol. Oceanogr.* **1994**, *39;* 941-948.

46.   Millero, F.J.; Hubinger, S.; Fernandez, M.; Garnett, S. *Env. Sci.Tech.* **1987**; *21*, 439-443.

47.   Jorgensen B.B., Fossing, H. Wirsen, C.O.; Jannasch, H.W. *Deep-Sea Res.* **1991**; *38*, Suppl. Issue No 2a, S1083-S1103.

48.   Luther, G.W.III; Ferdelman, T; Tsamakis, E. *Estuaries.* **1988**; *11*, 281-285.

49.   Millero, F.J. *Estuarine Coastal Shelf Sci.* **1991**; *33,* 521-527.
50.   Millero, F.J. *Limnol.Oceanogr.* **1991**; *26,*1007-1014.
51.   Davison, W.; Buffle J.; de Vitre, R.R. *Pure Appl. Chem.* **1988**; *60,* 1535-1548.
52.   Rickard, D. *Chem. Geol.* **1989**; *78,* 315-324.
53.   Rickard, D. *Min. Mag.* **1989**; *53,* 527-530.
54.   Zhang, J-Z; Millero, F.J. *Anal. Chim. Acta,* **1994**; *284,*497-504.
55.   Luther, G.W.III.; Rickard, D.; Oldroyd, A.O.; Theberge, S. *208th American Chemical Society Meeting* **1994**; Abstract No 93.
56.   Luther, G.W.III.; Rickard, D.; Oldroyd, A.O.; Theberge, S. *208th American Chemical Society Meeting* **1994**; Abstract No 48
57.   Tossel, J.A.; Vaughan, D.J. *Theoretical Geochemistry: Applications of quantum mechanics in the earth and mineral sciences.* Oxford University Press, Oxford **1992**; 514pp.
58.   Tossel, J.A.; Vaughan, D.J. *Geochim. Cosmochim. Acta* **1993**; 57, 1935-1945.
59.   Goodenough, J.B. *Mat. Res. Bull.* **1978**; *13,* 1305-1314.
60.   Arakaki, T., Morse, J.W. *Geochim. Cosmochim. Acta* **1993**; *57,* 9-14
61.   Morse, J.W.; Arakaki T. *Geochim. Cosmochim. Acta* **1993**; *57,* 3635-3640.
62.   Dekkers, M.J.; Schoonen, M.A.A *Geochim. Cosmochim. Acta* **1994**; 58, 4147-4153.
63.   Rickard, D. *Min. Mag.* **1994**; *58A* p772-773.
64.   Rickard, D. *208th American Chemical Society Meeting* **1994**; Abstract No 86.
65.   Buffle J.; de Vitre, R.R.; Perret, D.; Leppard, G.G. In *Metal Speciation :Theory, Analysis and Application* (Editors, J.R. Kramer; H.E. Allen) Lewis Publishers Inc. Chelsea, MI. **1988**.
66.   Herron, N; Wang,Y; Eckert, H. J.*Amer. Chem. Soc.* **1990;** *1112,*1322-1326.
67.   Kortan, A.R.; Hull, R.; Opilla, R.L.; Bawendi, M.G.; Steigerwald, M.L.; Caroll, P.J.; Brus, L.E. *J. Amer. Chem. Soc.* **1990**; *112,* 1327-1332.
68.   Nedeljkovic, J.M.; Patel, R.C.; Kaufman, P.; Joyce-Pruden, C.; O'Leary, N.J. Chem.Ed. 1993; *70,* 342-345.
69.   Shea, D; Helz, G.R. *Geochim. Cosmochim. Acta* **1988**; 52, 1815-1825.
70.   Luther, G.W.III. *Geochim. Cosmochim. Acta* **1991**; *55,* 2839-2849.
71.   Adamson, A. *Physical Chemistry of Surfaces* Wiley New York, **1990;** 775p
72.   Kouvo O.,Vuorelainen, Y. and Long, J.V.P. *American Mineralogist* **1963;** *48,* 511-554.
73.   Takeno S., Zoka, H. and Niihara, T. *American Mineralogist* **1970**; *55,* 1639-1649.

74. Vaughan, D.J.; Lennie, A.R.; Charnock, J.; Pattrick, R; Garner D. *First International Iron Sulfide Workshop.*University of Wales Cardiff, **1991**; p6.

75. Parise J.B., Schoonen M.A.A. and Lamble G. *Geol. Soc. Am. Prog. Abstract* 22 **1994**; A293.

76. Rickard, D. *Stockholm Contr. in Geology* **1968**; *20*, 67-95.

77. Kornicker, W.A., Morse, J.W. *Geochim. Cosmochim. Acta* **1991**; *55*, 2159-2171

78. Schoonen, M.A.A.; Barnes, H.L *Geochim. Cosmochim. Acta* **1991**; *55*, 1495-1504.

79. Allen, E.T.; Crenshaw, J.L.; Merwin, H.E. *Am. J. Sci.* **1914**; *38*, 169-236.

80. Rickard, D. *Amer. J. Sci.* **1975**; *275,* p.636-652.

81. Schoonen, M.A.A.; Barnes, H.L. *Geochim. Cosmochim. Acta* **1991**; *55,* 1505-1514

82. Murowchick, J.B.; Barnes, H.L. *Geochim Cosmochim. Acta* **1986**;*50,* 2615-2630.

83. Horiuchi, S; Wada, H; Noguchi,T. *Naturwiss.* **1970**; *57*, 670.

84. Taylor, P,; Rummery, T.E.; Owen, D.G. *J.Inorg. Nucl. Chem.* **1979**; *41*, 1026-1030

85. Sweeney, R.E.; Kaplan, I.R. *Econ. Geol.* **1973**; 618-634

86. Drobner E.; Huber, H.; Wächtershäuser, G.; Rose, D.; Stetter, K.O. *Nature* **1990**; *346*, 742-744.

87. Goldhaber, M.B.; Stanton, M.R. *Geol. Soc. Amer. Abstr. Prog.*, **1987**; 19.

88. Tossel, J.A. *Phys. Chem. Minerals* **1983;** *9,* 115-12.

89. Rakovan, J.; Schoonen, M.A.A.; Reeder, R.J.; Tyrna, P.; Nelson, D.O. *Geochim. Cosmochim. Acta* **1994**

90. Luther, G.W.III *Geochim. Cosmochim. Acta* **1987**; *51*, 3193-3199.

91. Huerta-Diaz, M.A.; Morse, J.W. *Mar. Chem.* **1990**; *29,* 119-144.

92. Luther, G.W.III In *Aquatic Chemical Kinetics* (Editor, W.Stumm) John Wiley and Sons, New York. **1990**; pp173-198.

93. Moses, C.O.; Herman, J.S.*Geochim. Cosmochim. Acta* **1991**; *55*, 471-482.

94. Reedy, B.J.; Beattie, J.K. & Lowson, R.T., *Geochim. Cosmochim. Acta* **1991,** *55; 1609-1614.*

95. Davison, W., Lishman, J.P., *J. Geochim. Cosmochim. Acta* **1985**; *49,* 1615-1620.

96. Canfield, D.E.; Raiswell, R.; Westrich J.T.; Reaves, C.M.; Berner, R.A. *Chem. Geol.* **1986**; 54, 149-155.

97. Chanton, J.P; Martens, C.S. *Biogeochemistry* **1985**; *1*, 383-384

98. Morse, J.W.; Cornwell, J.C. *Mar. Chem.* **1987**, *22*, 55-69.

99.   Raiswell, R.; Berner, R.A. *Geochim. Cosmochim. Acta,* **1986**; *50,* 1967-1976.

100. Rice, C.A., Tuttle, M.L., Reynolds, R.L. *Chem. Geol.* **1993;** *107,* 83-95

101. Raiswell, R.; Canfield D.E.; Berner R.A. *Chem. Geol.***1994;** 111, 101-117.

102. Moses, C.O.; Nordstrom, D.K.; Herman, J.S. & Mills, A.L. *Geochim. Cosmochim. Acta* **1987**, *51,* 1561-1571.

103. McKibben, M.; Barnes, H.L. *Geochim. Cosmochim. Acta* **1986**; *50,* 1509-1520.

104. Luther, G.W.III.; Kostka, J.E.; Church, T.M.; Sulzberger, B; Stumm, W. *Mar. Chem.* **1992**; *40,* 81-103.

105. Landing , W.M.; Westerlund, S. *Mar. Chem.* **1988**; 23, 329-343.

106. Fossing, H.; Jorgensen, B.B. *Biogeochem.* **1989**; *8,* 205-222.

107. Berner R.A. *American Jour. Sci.* **1967**; *265,* 773-785.

108. Berner, R.A. *Geochim. Cosmochim. Acta* **1984**; *48,* 605-615

109. Berner, R.A.; Baldwin, T.; Holdren, G.R. *J. Sed. Petr.* **1979**; *49,* 1345-1350.

110. Hilton, J.; Lishman, J.P.; Chapman, J.S. *Chem. Geol.* **1986***56,* 325-333.

111. Reynolds, R.L.; Tuttle, M.L.; Rice, C.; Fishman, N.S.; Karachewski, J.A.; Sherman, D.; *Am. J. Sci.* **1993**; *294,* 485-528.

112. Roberts, A.P.; Turner, G.M. *Earth and Planet. Sci. Lett.* **1993**; *115,* 257-273.

113. Stuttill, R.J.; Turner, P.; Vaughan, D.J. *Geochim. Cosmochim. Acta* **1982**; *46,* 205-217.

114. Hilton, J.; Long, G.J.; Chapman, J.S.; Lishman, J.P. *Geochim. Cosmochim. Acta* **1986**; *50,* 2147-2151.

115. Manning, P.G.; Murphy, T.P.; Mayer, T. *Can. Mineral.* **1988**, *26,* 965-972

116. van Velzen, A.J., Dekkers, M.J.; Zijderveld, J.D.A. *Earth Planet. Sci. Lett.* **1993**; *115,* 43-55.

117. Demitrack, A. In *Magnetite Biomineralization and Magnetoreception in Organisms*; Kirschvink, J.L.; Jones, D.S.; McFadden, B.J.; Plenum, New York, **1985**; 626-645.

118. Pye, K. *Mar. Geol.* **1984**; *56,* 1-12.

119. Snowball, I.F.; Thompson, R. *J. Geophys. Res.* **1990**; *95,* 4471-4479.

120. Snowball, I.F. *Phys. Earth Planet. Inter.* **1991**; *68,* 32-40.

121. Horng, C.S.; Ric, E.; Cadoret, T.; Jehanno, C.; Lai, C.; Lee, T.Q.; Kissel, C. *EOS,* **1989**; *70,* 1066.

122. Hofmann, V. *Phys. Earth Planet. Inter.* **1992**; *70,* 288-301.

123. Krs, M.; Krsova, M.; Pruner, P.; Zeman, A.; Novak, F.; Jansa, *J. Phys. Earth Planet. Inter.* **1990; 63,** 98-112.

124. Krs, M.; Novak, F.; Krsova, M.; Pruner, P.; Koulikova, L.; Jansa, *J. Phys. Earth Planet. Inter.* **1992**; *70,* 273-287.

125. Farina, M.; Esquivel, D.M.S.; Lins de Barras, H.G.P. *Nature,* **1990;** *343,* 256-258.

126. Mann, S.; Sparks, N.H.C.; Frankel, R.B.; Bazylinski, D.A.; Jannash, H.W. *Nature* **1990**; *343,* 258-261.

127. Skinner, H.C.W. *Precambrian Res.* **1993**; *61,* 209-229.

128. Lein, A.Y.; Sidorenko, G.A.; Volkov, I.I.; Shevchenko, A.Ya. *Doklady Akad. Nauk. USSR,* **1980**; *238,* 167-169.

129. Bonev, I.K.; Krischev, K.G.; Neikov, H.N. and Georgiev, V.M. *Compt. Rend. Acad. Bulg. Sci.* **1989**; *42,* 97-100.

130. Goldhaber, M.B.; Reynolds, R.L. *Geophysics* **1991**; *56,* 748-757.

131. Reynolds, R.L.; Fishman, N.S.; Hudson, M.R. *Geophysics,* **1991**; *56,* 606-617

132. McManus, K.M.; Hanor, J.S. *Abstracts Soc. of Economic Paleontologists and Mineralogists, Annual Mid Year Meeting* **1987**; *4,* 56.

133. Hartman, P. *Geologie und Mijnbouw,* **1982**; *61,* 313-320

134. Murowchick, J.B.; Barnes, H.L. *Am. Min.* **1987**, *72,* 1241-1250

135. Bliznakov, G. *Fortschr. Min.* **1958**; *36,* 149-165

136. Franke, W. *Fortschr. Min.* **1987**; *69,* Beihelft 1, 50-51

137. Staudt, W.J., Reeder, R.J., Schoonen, M.A.A., *Geochim. Cosmochim. Acta* **1994**, *58,* 2087-2098

138. Frankie, K.A., Hower, J.C. *Int. J. Coal Geol.* **1987**, *7,* 349-364

139. Wiese, R.G.; Fyfe, W.S. *Int. J. Coal Geol.* **1986**; *6,* 251-276.

140. Visscher, P.T.; van Gemerden, H., *In:"Biogeochemistry of Global Change",* (Editor, Ormeland, C.R.) Chapman Hall, **1993;** p.672-689.

141. Xu, Y., Schoonen, M.A.A. *208th ACS Meeting, Washington, D.C.* **1994**; Abstract.

142. Raiswell, R., Plant, J. Econ. Geol. **1980**; *75,* 684-699

143. Dill, H.; Kemper, E. *Sedimentology,* **1990**; *37,* 427-443

144. Belzile, N., Lebel, J. *Chem. Geol.* **1986**; *54,* 279-281

145. Schoonen, M.A.A.; Fisher, N.S.; Wente, M.A. *Geochim. Cosmochim. Acta* **1992**; *56,* 1801-1814.

146. Beier, J.A.; Hayes, J.M. *Geol. Soc. Amer. Bull.* **1989**; 101, 774-783.

147. Fisher I.S; Hudson J.D. In *Marine Petroleum Source Rocks* (Editors Brooks J.; Fleet A.J.) Geol. Soc. London. Spec. Pub. **1987**; *26,* 121-135.

148. Muramoto, J.A.; Honjo S.; Fry B.; Hay B.J.; Howarth R.W.; Cisne, J.L *Deep Sea Res.* **1991**; 38 Suppl. 2A, 1151-1187.

149.   Raiswell, R., Berner, R.A., Pyrite formation in euxinic and semi-euxinic sediments, *Am. J. Sci.* **1985**, *285*, 710-724

150.   Saelen, G.; Raiswell, R.; Talbot, M.R.; Skei, J.M.; Bottrell, S.H. *Geology* **1993**; *21*, 1091-1094.

151.   Middelburg, J.J., de Lange, G.J., van der Sloot, H.A., Emberg, P.R., Sophian, S., *Mar.Chem* **1988**; *23, 353-364*.

152.   Skei, J.M. *Mar. Chem.* **1988**; *23, 345-352*

153.   Howarth, R.W. *Science* **1979**; *203*, 49-51.

154.   Raiswell, R. *Am. J. Sci.* **1982**; *282*, 1244-1263

155.   Giblin, A.E.; Howartth, R.W. *Limnol. Ocean.* **1984**; *29*, 47-63

156.   Giblin, A.E. *Geomicrobiol. J.*   **1988**; *6*, 77-97

157.   Wilkin, R.T., Barnes, H.L., *Geol. Soc. Am. Abstr. Prog.* **1993**; *25*, A200

158.   Butler, I.B *Framboid Formation.* Unpublished PhD thesis University of Wales, Cardiff **1995**; 300pp.

159.   Rickard, D. and Butler, I.B. *Organic Inorganic Interface Abstracts*; **1991**; Ross Priory April 1991, Glasgow University publication

160   Love, L.G.; Amstutz, G.C. *Fortschr.Miner.* **1966**; *43, 273-309*.

161.   Rickard, D. *Lithos* **1970**; *3*, 269-293

162.   Canfield, D.E.; Berner, R.A. *Geochim. Cosmochim. Acta* **1987**; *51,* 645-659.

163.   Raiswell, R., Plant, J. *Econ. Geol.* **1980**; *75*, 684-699

164.   McKibben, K.A. & Eldridge, C.S. *Am. J. Sci.* **1989**; *289*, 661-707

165.   Raiswell, R. *Chem. Geol.* **1993**; *107*, 467-469

166.   Canfield, D.E., Raiswell, R., Bottrel, S. *Am. J. Sci.* **1992**, *292,* 659-683

167.   Aller, R.C.; Rude, P. D. *Geochim. Cosmochim. Acta* **1988**; *52*, 751-765.

168.   Berner, R.A., Westrich, J.T. *Am. J. Sci.* **1985**; *285*, 193-206

169.   Thomsen, E.; Vorren, T.O. *Sedimentology* **1984**; *31*, 481-492

170.   Henneke, E. *Early diagenetic processes and sulphur speciation in pore waters and sediments of the hypersaline Tyro and Bannock Basins, eastern Mediterranean.* PhD thesis. University of Utrecht. **1993**; pp150.

171.   Fisher, I. *Geochim. Cosmochim. Acta* **1986**; *50*, 517-523

172.   Gardner, L.R.; Lerche, I. *Computers and Geosci.* **1990**; *16*, 441-460.

RECEIVED August 11, 1995

Chapter 10

Reactions Forming Pyrite from Precipitated Amorphous Ferrous Sulfide

Yoko Furukawa[1] and H. L. Barnes

Ore Deposits Research Section, Pennsylvania State University, University Park, PA 16802

The replacement reactions that convert precipitated ferrous sulfide to pyrite were evaluated as possible kinetic paths by comparing solid molar volume changes of the reactions ($\Delta \overline{V}_{solids}$). They may proceed by iron loss rather than by sulfur addition, the conventional conversion process:

$$FeS(s) + S(s) \rightarrow FeS_2(py).$$

Alternative conventional reactions include the addition of sulfur from aqueous species, such as polysulfides or thiosulfate. However, these sulfur-addition reactions result in positive $\Delta \overline{V}_{solids}$ that slows the reaction to ineffective rates by causing armoring of precursor mineral surfaces. Instead, reactions that combine the loss of Fe^{2+} and oxidation of precursor phase such as

$$2FeS(s) + 1/2O_2 + 2H^+ \rightarrow FeS_2(py) + Fe^{2+} + H_2O$$

cause a decrease in $\Delta \overline{V}_{solids}$ with resulting shrinkage cracks that promote solute mobility and speed further replacement reactions.

Authigenic pyrite is common in both modern and ancient marine sediments. Its remarkable post-depositional persistence in earth surface environments makes pyrite an ideal geochemical indicator. In many natural aquatic systems, such as water columns in anoxic basins and pore waters in marine sediments, authigenic pyrite formation correlates with specific combinations of oxidation state, sulfate concentration, and iron concentration of the systems (*e.g., 1*). Thus understanding the

[1]Current address: Naval Research Laboratory, Code 7431, Stennis Space Center, MS 39529

mechanisms of authigenic pyrite formation would allow pyrite to be used as an indicator of geochemical variables such as oxidation state, sulfate concentration and influx of detrital iron minerals.

Authigenic iron sulfide formation is also critical to sedimentary paleomagnetic studies. Among the intermediate phases leading to sedimentary pyrite, greigite is ferrimagnetic and therefore its diagenetic formation in post-depositional environments has a strong effect in the interpretation of paleomagnetic data (*e.g., 2*). One of the major minerals to carry remanent magnetism in sediments, magnetite, is a major iron source for diagenetic iron sulfide formation and therefore its conversion by this process is a major cause of magnetic instability in marine sediments (*3*). Thus an understanding of the mechanisms of authigenic pyrite formation enables us to better understand the magnetization processes in sediments and to better interpret paleomagnetic data.

Berner (*1, 4*) divided the authigenic pyrite formation process into three stages. In the first stage, aqueous sulfate (SO_4^{2-}) is reduced to sulfide (H_2S or HS^-) by bacterial reduction that accompanies the biologic oxidation of organic matter. In the second stage aqueous sulfide, formed during the first stage, reacts with Fe^{2+} provided by the bacterial dissolution-reduction of detrital iron minerals in the sediments, or suspended in the water column, to form an amorphous ferrous sulfide precipitate. This amorphous ferrous sulfide precipitate subsequently reacts to increasingly ordered mackinawite (Fe_9S_8) (*5*). The third stage is a series of replacement reactions that produce pyrite progressively through a series of sulfur-rich phases from mackinawite as the initial ordered phase. Experimental studies have shown that direct precipitation of pyrite nuclei from low temperature aqueous solutions is kinetically prohibited, although growth of pyrite nuclei is possible (*6, 7*). Schoonen and Barnes (*6, 7*) experimentally showed that direct pyrite precipitation does not occur in solutions undersaturated with respect to precursor phases such as amorphous ferrous sulfide unless pyrite nuclei are present.

Although this three-stage process has been the topic of many previous studies (*e.g., 1, 4 - 11*), our knowledge of the reaction mechanisms of pyrite formation at low temperatures remains incomplete. Sulfate reduction is reviewed by Berner (*1*). Rickard (*8*) used T-tube experiments to study amorphous ferrous sulfide precipitation from Fe^{2+} and aqueous sulfide, and subsequent conversion to mackinawite was inferred by an EXAFS study (*5*). However, our understanding of the third stage, Fe-S mineral replacement processes, is far from complete. Experimental studies have indicated that elemental sulfur ($S°$), or aqueous sulfur species with intermediate oxidation states such as polysulfides (S_n^{2-}), are essential to form pyrite through replacement reactions (*1, 4, 7*) as seen in the reactions,

$$FeS(s) + S°(s) \rightarrow FeS_2(s) \qquad (1)$$

amorph.   pyrite

$$FeS(s) + S_n^{2-} \rightarrow FeS_2(s) + S_{n-1}^{2-} \qquad (2)$$
amorph.   pyrite

However, sufficient elemental sulfur and adequate intermediate sulfur species concentrations are not found in many pyrite-forming environments.

This paper will review previous experimental studies on the third stage, the replacement sequence from mackinawite to pyrite, in low temperature aqueous environments, and discuss those experiments in terms of possible alternative reaction mechanisms that do not require elemental sulfur or polysulfides as reactants. It is important to understand the mechanism of the replacement sequences because pyrite growth depends on the presence of initial pyrite nuclei, that must have formed by replacement because direct precipitation is kinetically prohibited. All sedimentary pyrite, both framboidal and euhedral, should originate as pyrite nuclei that are products of replacement reactions. What is presented here is a new interpretation of previous experimental studies pertinent to this process.

Previous Studies

Pyrite was synthesized by Berner (4) at 65°C by the replacement of freshly precipitated amorphous ferrous sulfide (described as FeS in the study) in aqueous solution with neutral pH. A suspension of solid elemental sulfur ($S°$) and its immediate products upon hydrolysis, hydrogen sulfide, aqueous polysulfides, and thiosulfate were present in the solution. Pyrite formation depended on the presence of excess elemental sulfur on the surface of which pyrite crystallized. No pyrite was recovered from runs in which elemental sulfur was completely dissolved to yield aqueous sulfur species. Due to the absence of pyrite in runs without elemental sulfur, the following reaction was suggested as dominant:

$$FeS(s) + S°(s) \rightarrow FeS_2(s). \qquad (3)$$
amorph.   pyrite

This reaction can be modified to accommodate the more recent knowledge that mackinawite occurs as the first ordered phase in the replacement reaction following its recrystallization from the amorphous ferrous sulfide precipitate:

$$Fe_9S_8(s) + 10S°(s) \rightarrow 9FeS_2(s). \qquad (4)$$
mackinawite   pyrite

On the other hand, Schoonen and Barnes (7) proposed the overall reaction;

$$FeS(s) + S_n^{2-} \rightarrow FeS_2(s) + S_{n-1}^{2-}, \qquad (5)$$
amorph.   pyrite

which also can be better written as:

$$Fe_9S_8(s) + 10S_n^{2-} \rightarrow 9FeS_2(s) + 10S_{n-1}^{2-}. \qquad (6)$$

mackinawite   pyrite

In the study by Schoonen and Barnes (7), pyrite resulted when amorphous ferrous sulfide (described as FeS in the study) was aged either in a polysulfide solution (S_4^{2-} = 0.066 mol/liter) at 65°C with no elemental sulfur in suspension, or in a solution containing both aqueous intermediate sulfur species and suspended solid elemental sulfur. Mackinawite was the first ordered phase formed immediately after the amorphous ferrous sulfide in the chain of replacement reactions. Next, mackinawite was occasionally replaced by either greigite or marcasite, and always by pyrite as the final product. Greigite was favored as an intermediate product where the solution was slightly oxidized, and marcasite was favored if the solution was acidic (pH < 5).

Aqueous sulfur species of intermediate oxidation states, such as polysulfides or thiosulfate, were essential in order to form pyrite in the experiments of Schoonen and Barnes (7). The intermediate aqueous sulfur species were considered to be the sulfur source for sulfidation of mackinawite in a reaction like (6). Reaction (6) (or (5)) was favored over reaction (4) (or (3)) because elemental sulfur is virtually inert in low temperature aqueous solutions. The reason why Berner (4) failed to obtain pyrite in experimental runs that contained aqueous intermediate sulfur species but no solid elemental sulfur is unknown. However, the surfaces of elemental sulfur tend to be saturated with polysulfides (12) and, therefore, can be considered as the most likely site for pyrite crystallization, as was described by Berner (4). Schoonen and Barnes (7) concluded that pyrite forms through sulfidation of mackinawite rather than by iron loss from mackinawite. Sulfidation reactions such as (6) were preferred over iron-loss reactions such as

$$2FeS(s) + 2H^+ \rightarrow FeS_2(s) + Fe^{2+} + H_2(g) \qquad (7)$$

amorph.   pyrite

because no pyrite was obtained from experimental runs that contained no sulfur species with intermediate oxidation states. Reaction (7) is an iron-loss reaction in which iron diffuses from the original, solid structure and dissolves into the aqueous solution in order to stoichiometrically balance the solid-phase conversion.

Other experimental studies agree with the studies by Berner (4) and by Schoonen and Barnes (7) that sulfur species with intermediate oxidation states are necessary for converting mackinawite to pyrite. Freshly precipitated mackinawite was converted to greigite and then pyrite in 5 to 11 days at 60~85°C in the presence of elemental sulfur in the experimental study by Sweeney and Kaplan (9), in which isotopic measurements indicated addition of sulfur in the mass balanced conversion of precursor phases to pyrite. Rickard (13) quantified the rate of pyrite formation in reaction (3). Luther (11) investigated pyrite formation by mixing Fe^{2+} and Na_2S_n (n=2, 4, 5) solutions at 25°C and neutral pH. The mixing resulted in the immediate black coloration of the solution which was described as due to the mixture of soluble Fe^{2+} - (HS^-) - S_n^{2-} complexes such as $[Fe(SH)(S_n)]^-$ and an amorphous solid such as

$Fe(HS)_2$. The initial black mixture was aged to form pyrite after several hours to months in the presence of dissolved polysulfides.

In summary, the previous experimental studies support the conclusion that replacement of amorphous ferrous sulfide or mackinawite proceeds when aqueous sulfur species of intermediate oxidation states, such as polysulfides or thiosulfate, are available. The replacement reaction is expressed as sulfidation of mackinawite as in reaction (6). The above authors presumed that pyrite did not form by iron loss reactions such as (7).

However, the proposed sulfidation reactions have positive changes in the overall molar volume of the solids ($+\Delta \overline{V}_{solids}$) as iron atoms are preserved in solids and additional sulfur is introduced to the solid structures. Generally, replacement reactions are favored by $-\Delta \overline{V}_{solids}$ in order for efficient transport of atoms among participating solid phases and an aqueous solution, and replacement reactions among iron sulfide minerals should follow this principle. In addition, the scarcity of intermediate sulfur species in at least some pyrite-forming environments makes such sulfidation reactions unrealistic in geological environments especially at low temperature. The next sections discuss the $\Delta \overline{V}_{solids}$ of replacement reactions, and consequently, alternative replacement mechanisms among iron sulfides will be proposed that lead to $-\Delta \overline{V}_{solids}$.

Theory of Replacement

In a mineral replacement process, a pre-existing mineral reacts to become a successor mineral, as a result of changing chemical potential gradients (*14 - 16*). In a hydrothermal or low temperature aqueous process, the cause of replacement process is an aqueous solution which has concentrations in disequilibrium with the original solid phase, and the replacement process begins by surface reaction between the original phase and the aqueous solution. Replacement reactions require a physical or chemical link between the original phase and the replacing phase. The original phase and the replacing phase may share common elements (*e.g.*, replacement of hydrothermal pyrrhotite by pyrite, calcite by dolomite) or share a similar crystal structure as well (*e.g.*, replacement of calcite by magnesite).

Replacement reactions are more likely to proceed efficiently when there is a net decrease of total solid volume (i.e., $-\Delta \overline{V}_{solids}$). The resulting void space in the solid promotes reactant transport by diffusion between the original solid phase and aqueous solution, and thus promotes further replacement reaction. This process is demonstrated in Figure 1(a), in which the replacing phase, shown in solid black, occupies a smaller volume than the original phase (shaded). Because of the net decrease in solid volume, there is an increase in porosity that is represented in the figure by cracks. The porosity promotes more rapid transport between the original phase and aqueous solution, and the replacement reaction may proceed to completion.

If a replacement reaction leads to a net increase in solid volume ($+\Delta \overline{V}_{solids}$), the secondary phase armors the initial phase with a reaction rim and limits mass transfer between the solution and the original phase. Figure 1(b) illustrates the later scenario. The replacing phase (solid black) has negligible surface porosity and permeability due to the net increase in solid volume and armors the original phase to prevent further direct interaction between the original phase (shaded) and aqueous solution. Once the armoring occurs, the exchange of chemical components between the original phase and solution necessary for the replacement reaction can proceed only by solid state diffusion through the rim of the replacement phase. This will decrease the efficiency of replacement reaction significantly because solid state diffusion is a much slower process than aqueous transport of elements especially at low temperature.

Murrowchick (*17*) showed that the replacement of pyrrhotite by pyrite proceeds by conserving sulfur as in:

$$Fe_7S_8(s) + 6H^+ + 3/2O_2(aq) \rightarrow 4FeS_2(s) + 3Fe^{2+} + 3H_2O. \qquad (8)$$
$$\text{pyrrhotite} \qquad\qquad\qquad \text{pyrite}$$

When 4 moles of pyrite replace 1 mole of monoclinic pyrrhotite, the solid volume change is negative (-32%) and the reaction proceeds by iron loss rather than sulfur addition. Sulfur addition would have resulted in a net increase of solid volume. A $-\Delta \overline{V}_{\text{solids}}$ was apparent in photomicrographs of samples from Kutna Hora, Czechoslovakia where replacement of pyrrhotite by pyrite produced increased pore space. "Bird's eye" texture of pyrite and marcasite (*14*) is another example of $-\Delta \overline{V}_{\text{solids}}$ during mineral replacement.

In summary, replacement reactions proceed much more rapidly in the direction of a net decrease in the volume of solids. Replacement reactions with $-\Delta \overline{V}_{\text{solids}}$ favors interaction between the original phase and aqueous solution, as well as among two or more solid phases including the original phase and replacing phases.

$\Delta \overline{V}_{\text{solids}}$ during Fe-S System Replacement Reactions

Previous studies show that pyrite formation at low temperature is initiated by replacement of precursor phases because direct nucleation of pyrite is kinetically prohibited. Evidence for the replacement origin of pyrite in nature is the residual magnetic character from greigite cores within some pyrite framboids where the replacement has not gone to completion (*2*). The question here is which reactions provide a $-\Delta \overline{V}_{\text{solids}}$ which would permit fast replacement at low temperatures.

Published experimental studies of authigenic pyrite formation support sulfidation rather than iron loss as the mechanism of the replacement reactions that convert mackinawite to pyrite (*e.g.*, *1, 4, 7 - 11*). Experimental conversion of iron monosulfides to pyrite always required the presence of a sulfur source, such as elemental sulfur, thiosulfate, or polysulfides, and Sweeney and Kaplan (*9*) showed that isotopic compositions of their experimental products implies the sulfur addition mechanism. Sulfidation reactions would proceed by the preservation of iron and addition of sulfur to solids, whereas iron-loss reactions would proceed by the preservation of sulfur in solids and loss of iron to an aqueous solution. Aqueous sulfur species of intermediate oxidation states, such as polysulfides, must be the sulfur source. For example, the replacement of mackinawite by greigite would be written as;

$$Fe_9S_8(s) + 4S_n^{2-} \rightarrow 3Fe_3S_4(s) + 4S_{n-1}^{2-}, \tag{9}$$
mackinawite   greigite

and the replacement of mackinawite by pyrite is written as;

$$Fe_9S_8(s) + 10S_n^{2-} \rightarrow 9FeS_2(s) + 10S_{n-1}^{2-}. \tag{6}$$
mackinawite   pyrite

The solid volume change ($\Delta \overline{V}_{solids}$) for reaction (9) is given by

$$\begin{aligned} \Delta \overline{V}_{solids}(\%) &= (3 \times \overline{V}_{gr} - \overline{V}_{mk}) \div \overline{V}_{mk} \times 100 \\ &= (3 \times 72.5 - 184.3) \div 184.3 \times 100 = +18\%, \end{aligned} \tag{10}$$

and is positive. (Molar volumes of iron sulfide minerals are calculated using density data in (18)). Similarly in the case of reaction (6), mackinawite to pyrite,

$$\begin{aligned} \Delta \overline{V}_{solids}(\%) &= (9 \times \overline{V}_{py} - \overline{V}_{mk}) \div \overline{V}_{mk} \times 100 \\ &= (9 \times 23.9 - 184.3) \div 184.3 \times 100 = +17\% \end{aligned} \tag{11}$$

and is again positive. In both examples, the $+\Delta \overline{V}_{solids}$ should make these reactions prohibitively slow at low temperatures where solid state diffusion rates are negligible.

Figure 2 summarizes such $\Delta \overline{V}_{solids}$ for all of the Fe-S minerals in replacement reactions that have been suggested to occur based on experimental evidence. All but one reaction would lead to a $+\Delta \overline{V}_{solids}$ when the replacement reactions proceed by sulfidation. The greigite to pyrite step is the only sulfidation reaction that has a $-\Delta \overline{V}_{solids}$.

An alternative mechanism to sulfidation in the conversion of mackinawite to pyrite is replacement by iron loss to the solution. In such reactions, sulfur is conserved in the solid structure whereas iron is removed from the original solid structures and released into an aqueous solution. Such reactions result in a $-\Delta \overline{V}_{solids}$, as shown by the example of pyrite replacing mackinawite below. A possible reaction for this process is the oxidation of mackinawite by combined generation and loss of hydrogen, and can be written as;

$$Fe_9S_8(s) + 10H^+ \rightarrow 4FeS_2(s) + 5Fe^{2+} + 5H_2(g) \tag{12}$$
mackinawite   pyrite

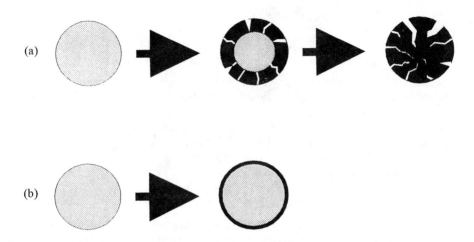

Figure 1. Schematic diagrams of replacement reactions. (a): When solid volume change is negative; (b): When solid volume change is positive.

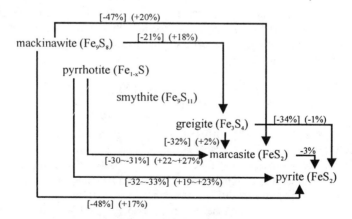

Figure 2. Solid volume change in Fe-S mineral replacement reactions. [%] represents solid volume change by Fe loss; (%) represents solid volume change by sulfidation.

The solid volume change for reaction (12) is given by

$$\Delta \overline{V}_{solids}(\%) = (4 \times \overline{V}_{py} - \overline{V}_{mk}) \div \overline{V}_{mk} \times 100$$
$$= (4 \times 23.9 - 184.3) \div 184.3 \times 100 = -48\%, \qquad (13)$$

which is negative (-48%). Figure 2 summarizes values for $\Delta \overline{V}_{solids}$ for replacement reactions with iron loss in addition to the sulfidation reactions. Contrary to the case of sulfidation, all replacement reactions have negative $\Delta \overline{V}_{solids}$ when the reactions proceed by iron loss. However, it should be noted that no direct evidence has been found for such iron loss reactions as (13) (7). Previous experiments on mackinawite conversion to pyrite always required the presence of aqueous sulfur species of intermediate oxidation states, species not involved in reaction (13).

Possible Reaction Mechanisms

Replacement reactions in aqueous solutions are likely to proceed in the direction of a $-\Delta \overline{V}_{solids}$. Of the reaction steps from mackinawite to pyrite with $-\Delta \overline{V}_{solids}$, all but one replacement reaction must proceed by iron loss. The replacement of greigite by pyrite has a negative $\Delta \overline{V}_{solids}$ by either iron loss or sulfidation. On the other hand, the described experimental studies on replacement processes support sulfidation, because aqueous sulfur species with intermediate oxidation states were always necessary.

This apparent discrepancy can be resolved if the intermediate sulfur species act as oxidizing agents rather than simply as sources of sulfur. To illustrate, take as an example the overall mackinawite → pyrite replacement reaction with polysulfides as the aqueous intermediate sulfur species:

$$Fe_9S_8(s) + 5(n-1)S_n^{2-} \rightarrow 4FeS_2(s) + 5Fe^{2+} + 5nS_{n-1}^{2-}, \qquad (14)$$
$$\text{mackinawite} \qquad\qquad\qquad \text{pyrite}$$

Here, replacement takes place by diffusion of iron from the mineral lattice into the aqueous solution as Fe^{2+}. The polysulfides are not merely sources of sulfur for sulfidation, but are oxidizing agents. Fe^{2+} and S_{n-1}^{2-}, polysulfides with shorter mean chain length may be products of the reaction and will then continue to react, as found by Luther (11), to precipitate additional amorphous iron sulfide, which then transforms to pyrite by repeating reaction (14). Also Fe^{2+} and polysulfides can then react to precipitate pyrite directly because the pyrite formed by earlier replacement can now act as pyrite nuclei. The later pyrite will form an overgrowth. Hence, although the previous studies implicate reactions requiring intermediate sulfur species for the steps between mackinawite and pyrite, the process could have proceeded by oxidation rather than apparent sulfidation. The elementary steps of the initial pyrite formation may have actually been a series of iron-loss reactions with associated reprecipitation of amorphous ferrous sulfide, and further conversion to mackinawite and to pyrite by iron loss.

Such iron-loss replacement means that the availability of intermediate sulfur species is not a limiting factor in the conversion of amorphous ferrous sulfide to pyrite. Instead, the availability of all appropriate oxidizing agents directly controls the extent of authigenic pyrite formation. In general, the conversion of mackinawite to pyrite can be written using $[O_2]$ schematically to indicate any oxidizing agent and not necessarily only dissolved oxygen,

$$Fe_9S_8(s) + 10H^+ + 5/2[O_2] \rightarrow 4FeS_2(s) + Fe^{2+} + 5H_2O \qquad (15)$$
$$\text{mackinawite} \qquad\quad \text{oxidizer} \quad\ \text{pyrite}$$

implying that diagenetic pyrite formation is possible even in sulfur-poor, fresh water environments. As long as there is an appropriate oxidizing agent present, mackinawite must convert to pyrite.

Based on these considerations, the formation scheme of sedimentary pyrite by Berner (*1, 4*) can be modified as shown in Figure 3. The initial two stages remain unchanged, whereas the third stage involves oxidizing agents in general instead of only intermediate sulfur species. Intermediate sulfur species are effective oxidizing agents as the previous experimental studies all successfully converted mackinawite to pyrite in the presence of polysulfides, thiosulfate, or elemental sulfur. However, they are not the only possible oxidizing agents.

Although Sweeney and Kaplan (*9*) implied S-addition by their isotopic study, the isotopic analysis was for an overall reaction. Fe^{2+} released by the Fe-loss reactions must react with the intermediate sulfur species to form overgrowths around the original pyrite thereby acquiring the new isotopic signature. The sulfur isotopic study is inconclusive in explaining the pyrite formation mechanisms because it cannot separate the replacement step from overgrowth step. So the resulting bulk pyrite shows the isotope characteristics expected for S-addition.

The fact that there is often observed euhedral overgrowth of pyrite reveals that once nuclei are available, then direct pyrite precipitation should be expected. However, in order to begin pyrite formation, Fe-loss replacement of precursor phases is necessary and then this pyrite may act as nuclei for direct precipitation of pyrite overgrowth.

Summary

The replacement sequence that transforms amorphous iron sulfide to pyrite proceeds by loss of iron accompanied by a decrease in the total solid molar volume, $\Delta \overline{V}_{solids}$. A $-\Delta \overline{V}_{solids}$ is necessary for the efficient exchange of atoms between solid phases and the aqueous solution. This mechanism has not been considered in earlier work. Previous experimental work suggests that the replacement occurs via a sulfidation mechanism, because sulfur species with intermediate oxidation states were always necessary to convert amorphous iron sulfide to pyrite in laboratory studies. However,

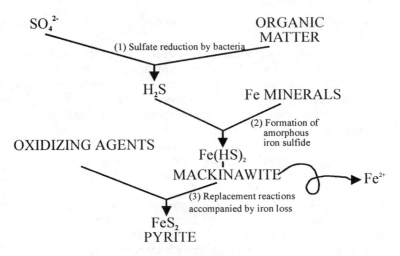

Figure 3. Revised pyrite formation mechanism, based on Berner (*1*).

intermediate sulfur species are not always present in natural pyrite-forming environments. This discrepancy can be resolved if such intermediate sulfur species act as oxidizing agents rather than as a sulfur source in the experiments. In such a case, intermediate sulfur species oxidize the precursor iron sulfide so that iron is removed from the precursor to from pyrite by reactions such as the replacement of mackinawite by pyrite. Such replacement reactions proceed by iron loss and do not require intermediate sulfur species, but they do need oxidizing agents. Although intermediate sulfur species are efficient oxidizing agents as the previous experimental studies demonstrated, there may be other oxidizing agents present to transform amorphous iron sulfide to pyrite. Oxidizing agents other than intermediate sulfur species may be especially important in sulfur-poor, fresh water environments. Further work is necessary to test possible oxidizing agents such as oxide minerals, H_2O_2 or O_2(aq) and mechanisms for the oxidation of amorphous iron sulfide to form pyrite.

Acknowledgments. This research was supported by Ore Deposits Research Section of The Pennsylvania State University and by NSF Grant EAR-8903750 to HLB.

Literature Cited

(*1*) Berner, R. A. *Geochim. Cosmochim. Acta* **1984**, *48*, 605-615.
(*2*) Roberts, A. P.; Turner, G. M. *Earth Planet. Sci. Lett.* **1993**, *115*, 257-273.
(*3*) Karlin, R. *J. Geophys. Res.* **1990**, *95*, 4405-4419.
(*4*) Berner, R. A. *Amer. J. Sci.* **1970**, *268*, 1-23.
(*5*) Parise, J.; Schoonen M. A. A.; Lamble G. GSA abstract **1990**.
(*6*) Schoonen, M. A. A.; Barnes, H. L. *Geochim. Cosmochim. Acta* **1991**, *55*, 1495-1504.

(*7*) Schoonen, M. A. A.; Barnes, H. L. *Geochim. Cosmochim. Acta* **1991**, *55*, 1505-1514

(*8*) Rickard, D. T. *Chem. Geol.* **1989**, *78*, 315-324.

(*9*) Sweeney, R. E.; Kaplan, I. R. *Econ. Geol.* **1973**, *68*, 618-634.

(*10*) Drobner, E.; Huber, H.; Wachtershauser, G.; Rose, D.; Stetter, K. O. *Nature* **1990**, *346*, 742-744.

(*11*) Luther, G. W. III *Geochim. Cosmochim. Acta* **1991**, *55*, 2839-2849.

(*12*) Graham, U. M. *Relationships between formational mechanisms and variations in properties of pyrite crystals synthesized between temperatures of 150 and 350 degree Celsius*; Ph.D. Dissertation; The Pennsylvania State University: University Park, PA, 1991.

(*13*) Rickard, D. T. *Amer. J. Sci.* **1975**, *275*, 636-652.

(*14*) Ramdohr, P. *The ore minerals and their intergrowths, 2nd ed.*; International Series in Earth Sciences; Pergamon: New York, NY, 1980; Vol. 35.

(*15*) Gould, W. W. *Hydrothermal replacement of calcite by sphalerite in a temperature gradient*; Ph.D. Dissertation; The Pennsylvania State University: University Park, PA, 1989.

(*16*) Craig, J. R. In *Advanced Microscopic Studies of Ore Minerals*; Jambor, J. L.; Vaughan, D. J., Eds.; Mineralogical Association of Canada Short Course Handbook; Mineralogical Association of Canada: Nepean, Ontario, Canada, 1990, Vol. 17; pp213-262.

(*17*) Murrowchick, J. B. *Econ. Geol.* **1992**, *87*, 1141-1152.

(*18*) Nickel, H. E.; Nichols, M. C. *Mineral Reference Manual*; Van Nostrand Reinhold: New York, NY, 1991.

RECEIVED April 17, 1995

Chapter 11

Laboratory Simulation of Pyrite Formation in Anoxic Sediments

Qiwei Wang and John W. Morse[1]

Department of Oceanography, Texas A & M University,
College Station, TX 77843

The formation of pyrite was investigated using the silica gel method under conditions similar to those occurring in natural anoxic sediments. Three different combinations of reactants, FeS(mackinawite) + S^0(s) + S(-II)(aq), Fe_3S_4(greigite) + S(-II)(aq), and FeOOH(goethite) + S(-II)(aq), were studied. Solid reactants were dispersed in silica gel that was overlain by sodium sulfide solutions. These solutions covered a wide range of pH and ΣS(-II) values. Results indicate that mass transport probably plays an important role during sedimentary pyrite formation. Solution pH and sulfide concentration have major influences on the rate of pyritization of mackinawite, greigite and goethite. The rate of pyrite formation generally increases with decreasing pH. The effect of sulfide concentration is not the same in different reaction systems.

The morphology of pyrite formed at room temperature is apparently controlled by the degree of supersaturation with respect to pyrite in the solution. When the solution is supersaturated with respect to amorphous-FeS, mackinawite and greigite, iron monosulfides will be initially produced. Then pyrite crystals that subsequently form, via pyritization of the precursors, may be euhedral, spherulitic, or in clusters of small grains, depending on the availability of dissolved Fe(II) and polysulfides. When the solution is undersaturated with respect to iron monosulfides, pyrite crystals appear to form via direct nucleation from solution. The crystals have cubic, cube-octahedral, and spherulitic morphologies.

Iron sulfide minerals are among the most studied authigenic minerals in marine sediments. They are important sinks for iron, sulfur and trace metals, and play an important role in their global geochemical cycles (1). The relations of iron sulfide minerals to other sediment components, such as organic matter, have proven useful as paleoenviromental indicators (2). While much is known about these minerals and the diagenetic conditions necessary for their formation (3), fundamental questions still remain to be resolved. Many of these remaining questions are associated with the reaction pathways by which pyrite forms and how they may relate to the morphology of sedimentary pyrite.

---

[1]Corresponding author

0097–6156/95/0612–0206$12.00/0

Experimental studies have demonstrated that pyrite can form through different mechanisms (*3*). Formation of sedimentary pyrite may involve precursor iron sulfide phases, such as amorphous-FeS, mackinawite or greigite. These phases can react with dissolved sulfide or elemental sulfur to form pyrite (*4-6*). Berner (*4*) suggested that overall reaction for pyrite formation as: $FeS + S^0 \rightarrow FeS_2$, and the presence of excess solid elemental sulfur was necessary. Sweeney and Kaplan (*8*) proposed that greigite was an important intermediate during transformation of mackinawite to pyrite. In this reaction sequence, mackinawite combines with elemental sulfur to produce greigite and then greigite produces pyrite either through dissociation, $Fe_3S_4 \rightarrow 2FeS + FeS_2$, or by further sulfidation, $Fe_3S_4 + 2S^0 \rightarrow 3FeS_2$.

Rickard (*7*) has suggested that reaction mechanisms may change with temperature. At room temperature, both mackinawite and elemental sulfur must dissolve prior to pyrite formation. Polysulfides and ferrous ions are respectively produced through dissolution of elemental sulfur and ferrous sulfide under reducing conditions. Pyrite may then precipitate directly through the reaction between aqueous ferrous and polysulfide ions. This reaction mechanism has been supported by polarographic observations (*9*). Luther (*9*) demonstrated that the initially formed "FeS" consisted of solid FeS and a soluble complex of $Fe(SH)^+$. The $Fe(SH)^+$ complex can react with excess polysulfide resulting in a complex of the form $[FeSH(S_x)]^-$. A reduction of the polysulfide by sulfide producing S_2^{2-} and change of Fe^{2+} from high spin to low spin can subsequently result in pyrite formation.

Drobner et al. (*10*) have reported that pyrite could be formed through reaction of amorphous-FeS with hydrogen sulfide at 100°C. The overall reactions were: $FeS + H_2S \rightarrow Fe(SH)_2$, and $Fe(SH)_2 \rightarrow FeS_2 + H_2$. The proton was used as an oxidant to produce zero-valence sulfur. However, Berner (*4*) and Schoonen and Barnes (*6*) found that this reaction did not occur at lower temperatures.

In marine and lacustrine sediments, pyrite typically occurs as euhedral crystals ranging 1 to 10 μm in size and as framboids. Euhedral pyrite crystals with cubic, octahedral, pyritohedral and spherical shapes have been found in marine sediments (*8, 11*). Pyrite framboids are spherically or spheroidally microscopic aggregates of discrete, equant, and morphologically uniform crystallites. Most framboids fall in the 5 to 20 μm size range. The ratio of crystallite size to framboid diameter is usually 1:15 to 1:10. Individual crystallites may be packed randomly or with some geometric regularity (*12, 13*). Framboidal pyrite may account for about 80% of sedimentary pyrite (*7*). Because of its importance and unique structure, numerous investigations of framboidal pyrite formation have been carried out using several approaches. However, the processes leading to the formation of pyrite with a framboidal morphology still remain largely an enigma.

It is well known that the morphology of a crystal is largely determined by the environmental conditions during its growth (*14*). The textures of sedimentary pyrite probably reflect the *in situ* sediment and porewater chemistry during its formation. Consequently, pyrite morphology may be useful as a paleo-geochemical indicator, if the links between the crystal shape and growth conditions can be established.

Methods

General Considerations. The major difference in this study from the previous experimental investigations of pyrite formation is that all experiments were conducted under conditions reasonably similar to those found in natural anoxic sediments. In particular, reactions were performed at room temperature (~ 23°C) and slow mixing of reactants was obtained by their diffusion in silica gel. Three different combinations of reactants, which have been previously proven being capable to produce pyrite within a reasonable time period ($7, 8, 15$), were employed to produce pyrite crystals of different morphology. These were: FeS(mackinawite) + S^0(s) + S(-II)(aq), Fe_3S_4(greigite) + S(-II)(aq) and FeOOH(goethite) + S(-II)(aq).

Materials and Characterization. Working solutions were made with deoxygenated water, which was prepared by deaeration of Milli-Q water with N_2 for at least 2 hours. Experiments were done in a glove bag with flowing N_2 to avoid oxidation of reactants.

All minerals used in this study, except elemental sulfur, were freshly prepared. Mackinawite was obtained by quickly adding $FeCl_2 \cdot 4H_2O$ powder to a $Na_2S \cdot 9H_2O$ solution and aging the precipitate for 4 days at room temperature. The synthesis of greigite was started with precipitation of amorphous-FeS by mixing $Na_2S \cdot 6H_2O$ and $Fe(NH_4)_2(SO_4)_2 \cdot 6H_2O$ solutions. This amorphous-FeS slurry was then boiled overnight with a polysulfide solution to form greigite (16). Goethite (α-FeOOH) was prepared by aging ferrihydrite in a strongly alkaline solution at 70°C for 3 days, according to Schwertmann and Cornell (17).

The synthetic minerals were analyzed using X-ray powder diffraction with Cu Ka to verify phase purity. Goethite and greigite exhibited sharp diffraction peaks, whereas mackinawite displayed broad, low intensity peaks, implying poor crystallinity and/or fine grain size. The phase purity of mackinawite and greigite was further examined by SEM/EDS. No other phases were observed.

Examination of goethite by SEM revealed uniform lath-shaped crystals of typically 2 μm length and 0.15 μm width. The specific surface area estimated from size and density is about 12.5 m^2g^{-1}. Greigite and mackinawite appeared as irregular coagulates of fine grains which were formed during the freeze-drying process (18). Elemental sulfur (α-S_8) consisted of close to uniform ~10 μm rhombs with rough surfaces, and usually two or three grains aggregated together.

Experimental Procedure. Silica gel was prepared by the slow addition of 0.75 M sodium metasilicate solution to 1.5 M acetic acid with constant stirring. Solid reactants, goethite, greigite, mackinawite and elemental sulfur, were introduced after the reacting solution reached a pH = 5. Both mackinawite and greigite were used as wet precipitates to avoid any oxidation. After the pH was adjusted to the desired value, 20 ml of this mixture solution was poured into 70 ml test tubes, where it polymerized to form a solid-like gel. During the gelation period, solid minerals tend to sink toward the bottom of the test tubes. Most of the mineral particles fell close to the bottom in pH 5 and pH 6 gels due to their long gelling time. But in the case of pH 7 gels, most particles are trapped in gel column because a continuous network forms quickly.

The supernatant sulfide solutions were made using $Na_2S \cdot 9H_2O$ and their pH was adjusted with degassed HCl acid. The solution with the desired concentration and pH values was added to the test tubes slowly. The test tubes were then tightly capped and sealed with electric tape. For each given experimental condition, 10 replicates were set up. Individual replicates were terminated after various periods of reaction time.

At the end of an experiment, the pH and sulfide concentration of overlying solution were measured. Then the test tubes were frozen. This was done because silica gel collapses with freezing and thawing. After the solid phases, including silica sheets, unreacted minerals and products, sank to the bottom of test tube, the supernant solution was saved for determination of dissolved sulfide and iron concentrations in the gel. Finally the silica debris was dissolved in 1 N NaOH solution and minerals were recovered on 0.45 μm filters.

For all three reaction systems, the dissolved sulfide in overlying solutions reached a steady state with the sulfide in gel column after six months of reaction, as shown in Figure 1. The lower concentration in the gel was possibly contributed to the loss of sulfide during sample freezing and thawing processes. Whereas the pH values of overlying solution and gel were very close. The difference was less than 0.04 for experiments at pH around 5 and less than 0.07 for experiments of pH 7-9.

Analytical Methods. Total dissolved sulfide concentrations were determined by potentiometric titration with lead perchlorate. Iron concentrations were measured on a Perkin-Elmer 2380 atomic absorption spectrophotometer. pH values of the sulfide solution were measured using a combination electrode.

The recovered solid samples were divided into two portions. The first portion was used for X-ray diffraction analysis, and then dissolved with 1 N HCl for study of the morphology and surface structure of pyrite. The second portion of the sample was analyzed for crystal morphology and crystal distribution using the SEM/EDS methods described in Morse and Cornwell (*19*).

Results

FeS(s) + S⁰(s) + S(-II)(aq) Experiments. Results of experiments are summarized in Table I, over a wide range of reaction conditions, where it can be observed that the rate of pyrite formation is surprisingly slow. Except for experiment #A6, pyrite was not detected by X-ray diffraction analysis even after 18 months of reaction. The presence of pyrite was observed only by SEM/EDS analysis. SEM observations demonstrated that the pyritization of mackinawite was very heterogeneous, and that the degree of pyritization varied considerably from mackinawite cluster to cluster.

The effect of sulfide concentration on the rate of pyrite formation changes at different pH values. Under acidic conditions, the formation of pyrite was enhanced with increasing sulfide concentration; whereas at neutral and alkaline pH conditions, the formation of pyrite was retarded by increasing sulfide concentration.

Pyrite crystals were usually formed inside porous clusters of fine-grained mackinawite. No pyrite was found on the surface of elemental sulfur, as observed in experiments at elevated temperatures (*4*). Pyrite crystals primarily appeared as micron-size euhedra, as shown in Figure 2. In experiment #A6, relatively large-sized (7-10 μm) spherical pyrite was also observed. With increasing reaction time, both the size and number of pyrite grains were found to increase. However, the growth rate of pyrite is very slow. Based on the observations of polished sections, the linear growth rate of pyrite was less than 1 μm yr⁻¹ for euhedral grains and ~10 μm yr⁻¹ for large spherulites under our experimental conditions.

Fe₃S₄(s) + S(-II)(aq) Experiments. The formation of pyrite in this system was observed to be substantially faster than in FeS(s) + S⁰(s) + S(-II)(aq) system, which was consistent with the results obtained in solution reaction at 65°C (*6*). All samples had distinct XRD diffraction peaks for pyrite after only 3 months of reaction. The changes in the degree of pyritization of greigite, semi-quantitatively expressed as the intensity ratio of strongest XRD peaks of pyrite to greigite, with sulfide concentration is plotted in Figure 3. It can be observed that the transformation rate of greigite to pyrite was strongly affected by sulfide concentration and pH. The formation of pyrite was promoted with increased sulfide concentration and decreased pH. The effect of sulfide concentration is more significant at low pH and ΣS(-II) values. It should be noted, that

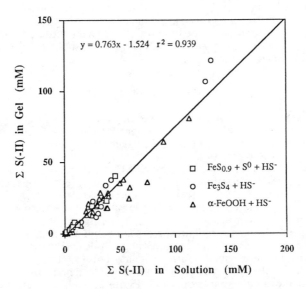

Figure 1. A linear relationship between the dissolved sulfide concentrations in overlying solution and gel column after 6 months of reaction for three different reactant combinations.

Table I. Summary of experimental conditions and results of $FeS_{(S)} + S^0_{(S)} + S(-II)_{(aq)}$ system

Run #	Initial solution		Final solution		Duration	Pyrite description
	pH	$\Sigma S(-II)(mM)$	pH	$\Sigma S(-II)(mM)$	(month)	
A1	7.21	45.7	7.51	32.0	6	no pyrite found
A2*	7.24	49.5	7.24	40.5	18	several clusters of ~1um grains
A3	7.24	9.6	7.82	6.2	6	few ~0.8um grains
A4*	7.24	9.8	7.68	8.7	18	several clusters of ~1um grains
A5	7.19	5.0	8.24	2.1	6	some ~1um grains
A6	7.19	5.0	8.76	1.1	12	lot of ~1um grains and 7-10um spherulites
A7	5.21	42.5	5.40	21.5	6	few ~1um grains
A8	5.21	42.5	5.35	12.0	12	some 2-3um grains
A9*	5.20	9.5	5.53	7.2	18	few 1.5um grains
A10*	5.20	4.2	5.44	1.7	18	no pyrite found

* Overlying solutions were changed after 12 months reaction.

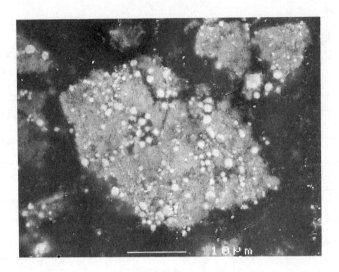

Figure 2. A cluster of micro-sized pyrite grains formed in FeS(mackinawite) + $S^o_{(s)}$ + S(-II)(aq) experiments.

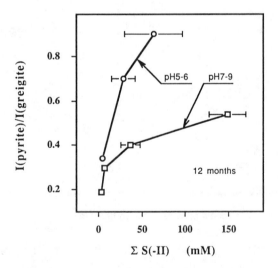

Figure 3. A plot showing the effects of sulfide concentration and pH value on the rate of the pyritization of greigite. The degree of pyritization is expressed as the intensity (I) ratio of strongest XRD peaks which locate at (2θ) 33.23 degrees for pyrite and 30.23 degrees for greigite. Error bars represent the differences of initial and final sulfide concentrations in overlying solutions.

due to the differences in crystal size, crystallinity, intrinsic structure and chemical composition, pyrite exhibits much stronger diffraction peaks than greigite. Consequently, the content of pyrite, as observed under SEM, was less than that determined by the I(pyrite)/I(greigite) ratios.

In all experiments, pyrite was the only new phase detected by both XDR and SEM/EDS analysis. Pyrite crystals were more abundant than in $FeS(s) + S^o(s) + S(-II)(aq)$ experiments, and they formed both inside and on the surface of greigite clusters. Except for a few ~10 μm spherulites, pyrite crystals were ~1 μm polyhedrons and spherulites.

FeOOH(s) + S(-II)(aq) Experiments. Compared to the previous two systems, the formation of pyrite via sulfidation of goethite was much faster, except for experiments at high pH and $\Sigma S(-II)$ values, which produced elemental sulfur.

The solid products of the FeOOH(s) + S(-II)(aq) reaction depended on the chemistry of the overlying solutions ($\Sigma S(-II)$, pH) and reaction duration. The minerals produced under various conditions after 6 months of reaction are presented in Figure 4. Amorphous FeS was the first mineral formed, and quickly converted to mackinawite in low pH experiments. However amorphous-FeS persisted for months in high pH experiments. Greigite was only found in some experiments at low pH values (< 5.2) and usually in trace amounts. Marcasite was never observed. The transformation of mackinawite and greigite to pyrite was inhibited by high $\Sigma S(-II)$ at pH values less than 6. This is demonstrated by the observation that large amounts of pyrite were formed within 3 months when $\Sigma S(-II)$ was less than 30 mM, whereas only a few pyrite crystals were found under SEM when $\Sigma S(-II)$ was greater than 50 mM. Pyrite was the only phase observed by XRD analysis after 18 months of reaction. After this time, greigite was totally pyritized and only a trace amount of mackinawite was left.

Pyrite morphology was also closely related to experimental conditions. When the pH was greater than 7, clusters of small euhedral grains were the dominant form of pyrite (Figure 5). When the pH was less than 6, large-sized spherulites (~7 - 15 μm) were the primary form of pyrite grown in the silica gel column. Clusters of small-sized spheroids (~1 μm) were the main form of pyrite produced close to the bottom of the test tube where goethite had settled. Four spherulitic pyrite crystals with different surface structures are shown in Figure 6.

Several different types of euhedral and fibrous-radial pyrite crystals were found in the low pH (~5) and high $\Sigma S(-II)$ experiments after 6 months reaction. Two of them, a smooth-faced cube and a combination of cube and octahedron, are shown in Figure 7.

Discussion

Formation of Pyrite at Room Temperature. The occurrence of pyrite inside of mackinawite clusters and the dependence of the rate of pyrite precipitation on solution chemistry suggest the precipitation of pyrite from $FeS(s) + S^o(s) + S(-II)(aq)$ system at room temperature is a dissolution - precipitation process. This is consistent with the mechanism proposed by Rickard (7). However, the rate of pyrite formation in this study was much slower than that determined by Rickard, implying the significant effect of mass transport during the formation of sedimentary pyrite.

Based on the results shown in Figure 3, it also can be concluded that the pyritization of greigite is a dissolution - precipitation process. Greigite probably

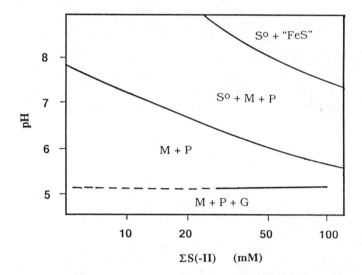

Figure 4. A schematic diagram showing the solid reaction products, determined by XRD and SEM/EDS analysis, in FeOOH(goethite) + S(-II)(aq) experiments after 6 months of reaction. The dashed line is used to indicate that only trace amounts of greigite were found in some experiments ("FeS" = amorphous-FeS; M = mackinawite; G = greigite; P = pyrite).

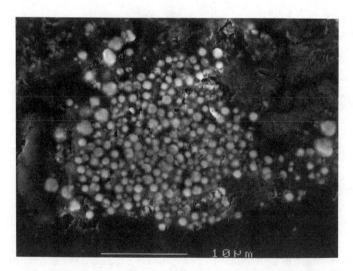

Figure 5. A cluster of small euhedral pyrite crystals formed via sulfidation of goethite after 6 months.

dissolved under attack from H^+ to produce ferrous iron and zero-valence sulfur $S(0)$: $Fe_3S_4 + 3H^+ \rightarrow 3Fe^{2+} + S(0) + 3HS^-$. $S(0)$ further reacted with HS^- to form polysulfides which is essential for pyrite formation (9): $(n-1)S(0) + HS^- \rightarrow HS_n^-$. Formation of polysulfides species are favored at high $\Sigma S(-II)$ concentrations. When the solution becomes supersaturated with respect to pyrite, pyrite nuclei will be formed after an induction period: $Fe^{2+} + HS_n^- \rightarrow FeS_2 + H^+ + (n-2)S(0)$. The overall reaction can be expressed as: $Fe_3S_4 + 2H^+ \rightarrow FeS_2 + 2HS^- + 2Fe^{2+}$.

With the formation of each mole of pyrite, 2 moles of dissolved Fe^{2+} are produced. Consequently, Fe^{2+} can build up in the solution as the reaction proceeds. This process was confirmed by direct measurements of dissolved iron concentrations in silica gel (Figure 8).

In the $FeOOH_{(s)} + HS^-_{(aq)}$ system, hydrogen sulfide is oxidized by ferric iron on the surface of goethite. The oxidation products can be elemental sulfur, polysulfides (S_4^{2-} and S_5^{2-}), sulfate, thiosulfate, and sulfites (20, 21). Based on the rate of pyrite formation and the results of SEM analysis, we conclude that the oxidation products were dominantly determined by the sulfide solution chemistry. At pH values of less than 6, dissolved zero-valence sulfur is the main products. When pH values are greater than 7, elemental sulfur is the dominant product at high $\Sigma S(-II)$, and sulfate and dissolved zero-valence sulfur are the main products at low $\Sigma S(-II)$.

Except for $FeOOH_{(s)} + S(-II)_{(aq)}$ experiments at pH greater than 8, in which the reductive dissolution of goethite was inhibited by insufficient protons, and $FeS(s) + S^o(s) + S(-II)(aq)$ experiments of high $\Sigma S(-II)$, in which the dissolution of mackinawite was inhibited by $\Sigma S(-II)$, the slow formation of pyrite in this study is probably related to limitation by the availability of dissolved zero-valence sulfur. This is indicated by the presence of substantial concentrations of dissolved iron in the silica gel. In $FeS(s) + S^o(s) + S(-II)(aq)$ experiments, polysulfides were formed through dissolution of α-S_8 in the sulfide solution. Experimental studies have shown that polysulfides can be rapidly produced at room temperature (22), but the equilibrium concentrations of polysulfides are very low (23). Under the experimental conditions of this study, the equilibrium state in the $HS^- - S_8 - H_2O$ system can be expected to be achieved rapidly at the surface of elemental sulfur. Further dissolution of α-S_8 would depend on the diffusion of polysulfide species away from the surface of elemental sulfur. Also, polysulfide ions could be absorbed on the surface of silica during the diffusion process. Only a limited amount of polysulfides may reach the mackinawite cluster where pyrite precipitation takes place. Consequently, formation of pyrite is probably controlled by the availability of polysulfides because dissolved iron can be rapidly produced *in situ* through dissolution of mackinawite. This will be most efficient at low pH and $\Sigma S(-II)$ values.

In anoxic marine sediments, elemental sulfur formed via biological activity and chemical reactions may be different in size, crystallinity and surface properties from that used in this study. It is also possible that different sulfur allotropes or different chemical forms other than α-S_8 are formed. In fact, elemental sulfur can exist at least 12 different forms at room temperature (24). The thermodynamic properties of these allotropes are very different, α-S_8 being the most and S_6 and S_7 the least stable (25). Therefore, the concentration of polysulfides produced through dissolution of elemental sulfur in natural sediments could be significantly higher than in this study, and could possibly lead to faster pyrite formation. Direct measurements (26) have shown much higher polysulfide concentrations in porewaters than values calculated

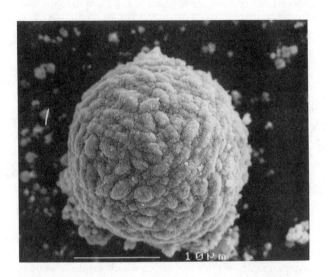

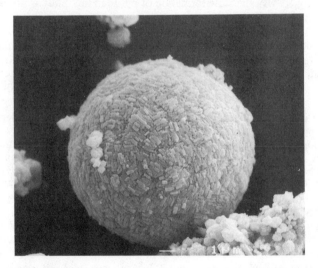

Figure 6. Four different forms of spherulitic pyrite crystals.

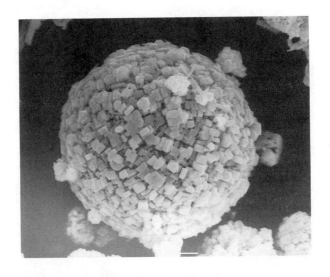

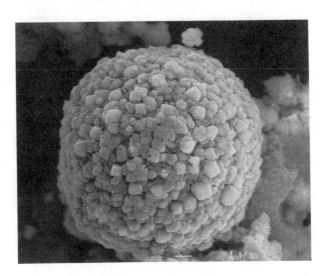

Figure 6. Continued.

Figure 7. Euhedral pyrite crystals: a). a perfect smooth-faced cube and b). a crystal with rough {100} faces and smooth {111} faces.

based on equilibrium with α-S_8. In some cases, the measured values are about three orders of magnitude higher than the calculated ones.

Morphology of Pyrite. The morphology of a crystal is determined by the supersaturation state (chemical potential difference between dissolved and solid phases) and structure (in atomic scale) of the aqueous solution - crystal interface (*14*). For an atomically rough interface, adhesive or continuous growth occur. In this case, the relationship between growth rate and supersaturation degree is linear. For an atomically smooth interface, two-dimensional nucleation or screw-dislocation growth mechanisms often operate. In this case, the relationship between growth rate and supersaturation degree usually follow a power law in which the power order is determined by the growth mechanism.

At low values of supersaturation degree, crystal growth often occurs by the screw-dislocation mechanism. Growth steps are generated by screw dislocations that emerge on the crystal surface, and advance outward with the continuous incorporation of atoms, molecules, or ions (*27*). This results in a euhedral crystal with smooth faces, although microscopically the profile is a stepped hill with the summit at the center (*28*). With an increase in supersaturation degree, the surface-nucleation mechanism overwhelms the screw-dislocation mechanism because the growth rate by the surface-nucleation mechanism increases much faster with supersaturation degree than the screw-dislocation mechanism does. When supersaturation degree is relatively low, the rate of two-dimensional nucleation on the surface is low, and the rate of crystal growth is limited by the formation of new nuclei. When supersaturation degree is high, the rate of nuclei formation is great enough that young layers may overtake older layers. The crystal surface then becomes macroscopically rough. If a crystal is of a fairly large size (~100 μm), the nucleation rate will be higher at corners and edges than at the center of faces. The growth layer will then spread inward due to the Berg phenomenon (*29*), with a high probability of forming a hopper crystal (*30*). With a further increase in supersaturation degree, the solution-crystal interface can become rough on the atomic scale. The morphology of the crystal will be spherulitic, or dentritic with curved surfaces and branches (*14*).

The forms of pyrite crystals observed in this study can be possibly understood in terms of the mechanisms of crystal growth discussed above. When the activity product of ferrous iron and sulfide, $[Fe^{2+}][S^{2-}]$, exceeds the solubility of iron monosulfides, metastable phases are formed first and dissolve later. Under these conditions, the growth rate and morphology of pyrite crystals depend on the rates of dissolution of iron monosulfide minerals and generation of dissolved zero-valence sulfur. According to Nielsen (*31*), the number of pyrite nuclei formed in the clusters of fine-grained metastable phases increases with an increase in supersaturation with respect to pyrite. Therefore, only a few pyrite nuclei are formed in the cluster of metastable phases, if supersaturation is low. In this case, large-sized crystals are formed. If initial supersaturation with respect to pyrite is high, many pyrite crystals will be produced. Pyrite formed under such conditions could be a cluster of euhedral or spherical grains, depending on the supply rates of Fe^{2+} and dissolved zero-valence sulfur during growth process.

When supersaturation is low, the morphology of a crystal is more likely determined by thermodynamics than kinetics. Crystals will then be bounded by the faces with minimum interfacial free energy, i.e., {100} faces for pyrite (*32*). Under these conditions, screw-dislocation growth is usually assumed to be the growth mechanism. This results in smooth faces, as shown in Figure 7a. With an increase in supersaturation, surface-nucleation becomes the dominant growth mechanism. If the rate of step initiation (two-dimensional nucleation) is greater than that of step spreading, steps will pile up, resulting in a rough surface. Increase in rates of

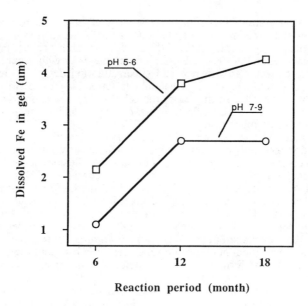

Figure 8. Changes of dissolved iron in silica gel with reaction period in $Fe_3S_{4(greigite)} + S(-II)_{(aq)}$ experiment.

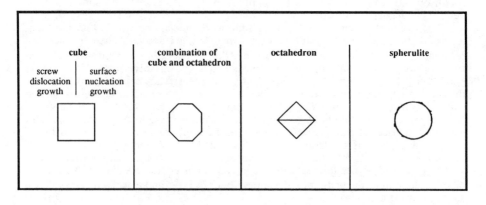

Increasing pyrite supersaturation   ⟶

Figure 9. A schematic drawing showing the change of pyrite morphology with degree of supersaturation.

nucleation and step spreading also promote octahedral habit (*33*). Consequently, pyrite crystals with {111} faces (as in Figure 7b) are formed at medium supersaturation state. With a further increase in supersaturation degree, both {100} and {111} faces become atomically rough, and rates of crystal growth are controlled by volume diffusion. In a uniform diffusion field, spherical morphology is produced (Figure 6a and 6b). Therefore, the changes of pyrite morphology depend on solution supersaturation as can be summarized as in Figure 9.

In natural anoxic sediments, pyrite crystals may be formed via replacement of iron monosulfides or direct nucleation from solution and subsequent growth (*3, 4, 34, 35*), depending on the saturation state of porewaters with respect to iron monosulfides. However, no direct evidence currently up to date suggests the morphology of individual pyrite crystals is not determined by the pathway of pyrite formation. The occurrence of different forms of pyrite crystals appears to be primarily determined by the concentrations of dissolved ferrous iron and polysulfides. Certainly, the presence of other dissolved organic and inorganic species may modify crystal morphology (*36*).

In many instances, porewaters are supersaturated with respect to amorphous-FeS, mackinawite, or greigite (*37*). Consequently iron monosulfides are frequently first precipitated in sediments. This is followed by the dissolution of these metastable phases, and both euhedral and framboidal pyrite can be formed. If the concentrations of dissolved zero-valence sulfur and Fe^{2+} are low, euhedral crystals will be produced. However, if the concentrations of dissolved zero-valence sulfur and Fe^{2+} are relatively high, poly-nucleation will take place and many pyrite crystals will be formed in clusters of iron monosulfides. The growth of pyrite crystals depends on further dissolution of iron monosulfides. This process may lead to the formation of a cluster of euhedral grains. We believe this is a possible pathway for formation of framboidal pyrite. The morphology of individual pyrite crystals in cluster is also determined by the solution supersaturation with respect to pyrite during growth. Framboids with cubic crystals are formed under lower supersaturation conditions than the framboids with octahedral crystals.

Conclusions

The formation of pyrite can be significantly affected by mass transport. This is demonstrated by the much slower rate of pyrite formation in this study than in previous investigations. The rate of pyrite formation is strongly dependent on the solution sulfide concentration and pH. Pyrite slowly forms in FeS(mackinawite) + S^0(s) + S(-II)(aq) experiments. Under acidic conditions, a high sulfide concentration enhances the transformation of mackinawite to pyrite. In the neutral to alkaline pH range, a high sulfide concentration inhibits pyrite formation. The formation of pyrite from greigite is substantially faster than from mackinawite. The rate increases with increasing solution sulfide concentration and decreasing pH. The pyritization rate of goethite is also significantly affected by the solution sulfide concentration and pH value. When pH is less than 6, pyrite is quickly formed due to the probable elevated concentration of dissolved zero-valence sulfur and more rapid dissolution of goethite. When pH is greater than 7, formation of pyrite is much slower probably because of limited availability of dissolved zero-valence sulfur.

The morphology of pyrite is controlled by mechanisms of crystal growth that are determined primarily by the supersaturation state of the solution. Screw-dislocation and surface-nucleation growth mechanisms produce euhedral pyrite crystals. Adhesive or continuous growth mechanisms produce spherical crystals. At increased supersaturations pyrite morphology changes in following order: smooth-faced cube → rough-faced cube → octahedron → spherulite. Due to the slow dissolution of iron monosulfides and iron (hydr)oxides in natural sediments, porewaters generally remain at lower degrees of supersaturation with respect to pyrite than in this study. Consequently, pyrite crystals often appear in cubic and octahedral forms.

Acknowledgments

This research was supported by a grant from the Chemical Oceanography Program of the National Science Foundation.

Literature Cited

1. Garrels, R. M.; Lerman, A. *Am. J. Sci.* **1984**, *284*, 989-1007.
2. Berner, R. A. *Geochim. Cosmochim. Acta* **1984**, *48*, 605-615.
3. Morse, J. W.; Millero, F. J.; Cornwell, J. C.; Rickard, D. *Earth Sci. Rev.* **1987**, *24*, 1-42.
4. Berner, R. A. *Am. J. Sci.* **1970**, *268*, 1-23.
5. Schoonen, M. A. A.; Barnes, H. L. *Geochim. Cosmochim. Acta* **1991**, *55*, 1495-1504.
6. Schoonen, M. A. A.; Barnes, H. L. *Geochim. Cosmochim. Acta* **1991**, *55*, 1505-1514.
7. Rickard, D. T. *Am. J. Sci.* **1975**, *275*, 636-652.
8. Sweeney, R. E.; Kaplan, I. R. *Econ. Geol.* **1973**, *68*, 618-634.
9. Luther III, G. W. *Geochim. Cosmochim. Acta* **1991**, *55*, 2839-2849.
10. Drobner, E.; Huber, H.; Wachtershauser, G.; Rose, D.; Steller, K. O. *Nature* **1990**, *346*, 742-744.
11. Hein, J. R.; Griggs, G. B. *Deep-Sea Res.* **1972**, *19*, 133-138.
12. Love, L. G.; Al-Kaisy, A. T. H.; Brockley, H. *J. Sed. Petrol.* **1984**, *54*, 869-876.
13. Love, L. G.; Amstutz, G. C. *Fortschr. Miner.* **1966**, *43*, 273-309.
14. Sunagawa, I. *Estud. Geol.* **1982**, *38*, 127-134.
15. Stanton, M. R.; Goldhaber, M. B. In: *Geochemical, Biogeochemical, and Sedimnetological Studies of the Green River Formation, Wyoming, Utah, and Colorado*; Tuttle, M. L., Ed.; USGS Bull., 1991; pp E1-20.
16. Wada, H. Bull. *Chem. Soc. Japan* **1971**, *50*, 2615-2617.
17. Schwertmann, U.; Cornell, R. M. *Iron Oxides in the Laboratory*; VCH: Weinheim. 1991; pp 137.
18. Arakaki, T.; Morse, J. W. *Geochim. Cosmochim. Acta* **1993**, *57*, 9-14.
19. Morse, J. W.; Cornwell, J. C. *Mar. Chem.* **1987**, *22*, 55-69.
20. Pyzik, A. J.; Sommer, J. E. *Geochim. Cosmochim. Acta* **1981**, *45*, 687-698.
21. Santos Afonso, M.; Stumm, W. *Langmuir* **1992**, *8*, 1671-1675.
22. Boulegue, J.; Lord III, C. J.; Church, T. M. *Geochim. Cosmochim. Acta* **1982**, *46*, 453-464.
23. Boulegue, J.; Michard, G. J. *J. Fr. Hydrol.* **1978**, *25*, 37-34.
24. Meyer, B. *Sulfur, Energy, and Enviromental*; Elservier: Amsterdam, 1977; pp448.
25. Meyer, B. *Chem. Rev.* **1976**, *76*, 367-388.
26. Luther III, G. W.; Giblin, A.; Varsolona, R. *Limnol. Oceanogr.* **1985**, *30*, 727-736.
27. Burton, W. K.; Cabrera, N.; Frank, F. C. *Phil. Trans.* **1951**, *A243*, 299-358.
28. Sunagawa, I.; Bennema, P. In: *Preparation and properties of Solid State Materials*; Wilcox, W. R. Ed.; Marcel Dekker, Inc.: New York, 1982; pp 1-129.
29. Berg, W. F. *Proc. Roy. Soc. (London)* **1938**, *A164*, 79-95.

30.   Kurada, T.; Irisawa, T.; Ookawa, A. *J. Cryst. Growth* **1977,** *42*, 41-46.
31.   Nielsen, A. E. *Krist. Tech.* **1969,** *4,* 17-38.
32.   Babic, D. D. *Neues Jah. fur Mineral. Monat.* **1981,** *5*, 225-227.
33.   Murowchick, J. B.; Barnes, H. L. *Amer. Mineral.* **1987,** *72*, 1241-1250.
34.   Howarth, R. W. *Science* **1979,** *203*, 49-51.
35.   Raiswell, R. *Am. J. Sci.* **1982,** *282*, 1244-1263.
36.   Davey, R. J.; Polywka, L. A.; Maginn, S. J. In: *Advances in Industrial Crystallization*; Davey; R. J.; Garside, J. J. Ed.; Butterworth-Heinemann Ltd.: Oxford, 1991; pp 150-165.
37.   Swider, K. T.; Mackin, J. E. *Geochim. Cosmochim. Acta* **1989,** *53*, 2311-2323.

RECEIVED April 17, 1995

Chapter 12

Environmental Controls on Iron Sulfide Mineral Formation in a Coastal Plain Estuary

Jeffrey C. Cornwell and Peter A. Sampou

Horn Point Environmental Laboratory, University of Maryland Center for Environmental and Estuarine Studies, Cambridge, MD 21613–0775

Salinity and organic matter loading gradients in estuaries can be used to examine iron sulfide mineral formation. Chesapeake Bay has low salinity, terrestrially-dominated sediments in the northern bay, seasonally anoxic mesohaline environments in the mid-bay region, and coarser-grained, bioturbated sediments in the more saline southern bay. High rates of sediment sulfur reoxidation and low inputs of labile organic matter limit the burial of authigenic sulfide minerals in northern bay sediments, while the activity of bioturbating organisms in south bay sediments enhances the reoxidation of sulfur. High iron sulfide $\delta^{34}S$ values in the northern bay are consistent with low sulfate concentrations. A strong bay-wide relationship between organic carbon and iron sulfide concentrations was evident when refractory terrestrial organic matter was considered.

Microbial sulfate reduction and the production of iron sulfide minerals are ubiquitous processes in estuarine and coastal marine sediments. There is considerable interest in the sedimentary sulfur cycle in coastal systems because a) sulfate reduction is often the dominant pathway for sediment metabolism and nutrient regeneration (*1-3*), b) sulfide production can maintain anoxia in the overlying water column (*4*), c) the conversion of iron oxides to sulfides can have a significant impact on phosphorus cycling (*5,6*), d) sulfur cycling influences the solution chemistry and solid phase speciation of trace metals (*7-9*), e) iron monosulfides are useful indicators for potential metal toxicity (*10*) and f) the form and quantity of iron sulfide minerals may be used as indicators of environments of deposition (*11-13*).

A number of factors may limit sulfate reduction and the formation of iron sulfide minerals (Figure 1). In low sulfate waters, sulfate supply may limit the production of hydrogen sulfide, though in most marine and estuarine systems, sulfate is not limiting (*14-16*). In general, the limit to sulfate reduction in estuarine systems is organic matter supply. Hydrogen sulfide produced via sulfate reduction has three

0097–6156/95/0612–0224$12.00/0

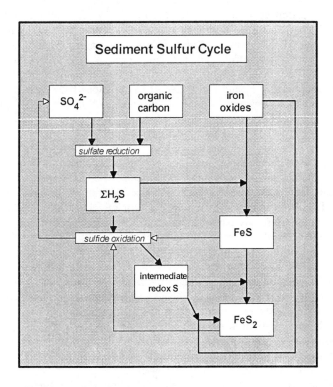

Figure 1. Simplified diagram of the sediment sulfur cycle as it applies to the formation of iron sulfide minerals.

main fates: reoxidation to sulfate or intermediate sulfur redox species such as thiosulfate, sulfite, elemental sulfur, polysulfides or polythionates (17-19); incorporation into iron monosulfides or pyrite (20-22); or incorporation into organic matter (23, 24). The first two pathways are quantitatively more important in estuarine sediments, with reoxidation dominating (2). In oxidizing or reducing horizons, sulfide oxidation can occur directly with oxygen or by interaction with metal oxides respectively (25, 26). The formation of pyrite requires a source of reduced sulfur which has been partially oxidized. Iron monosulfides or acid volatile sulfides (AVS) have often been considered a necessary precursor of pyrite formation (21), but pyrite synthesis from polysulfides and iron have demonstrated the feasibility of direct precipitation (27). The availability of "labile" iron mineral sources is a major limit to the formation of iron sulfide minerals (28) when sulfate reduction is a major metabolic process. The presence of pore water sulfide indicates that reactive iron phases are no longer available for sulfide mineral formation. Pyrite is the dominant iron sulfide mineral in estuarine sediments (22) and the degree of pyritization (DOP) may be calculated from pyrite and HCl-extractable iron concentrations (20):

$$DOP = Pyrite-Fe/(pyrite-Fe + HCl-Fe)$$

The use of DOP to indicate iron limitation of iron sulfide mineral formation may be valid only under moderate to high sulfate reduction rates. Canfield et al. (28) have cautioned that DOP can be an unreliable indicator of iron mineral reactivity.

The formation of iron sulfide minerals on relatively short time scales (days to months) does not necessarily lead to their permanent burial. The preservation of iron sulfide minerals depends on several different physical and biogeochemical processes, including degree of overlying water oxygenation, intensity of physical and biological reworking of sediment, and the presence of oxidants within the sediment. Iron sulfide mineral preservation is high in permanently anoxic systems where reoxidation is minimized. Physical reworking of sediments in bioturbated sediments mixes sulfide minerals into more oxidized environments and transports oxidants such as oxygen, Mn(IV) oxide and Fe(III) into reducing zones (26).

Sedimentary C/S ratios have been used to discriminate between freshwater and marine sediments (11). Freshwater sediments have high C/S ratios due to lower rates of sulfate reduction and high inputs of refactory organic carbon. Sediments deposited under oxygen-limited conditions exhibit considerably higher DOP (12).

The sulfur isotopic composition of iron sulfide minerals reflects the depositional environment. The microbial reduction of sulfate to hydrogen sulfide enriches the product in ^{32}S, leaving the remaining sulfate enriched in ^{34}S (29). Under high sulfate conditions, the sulfate pool is minimally enriched with heavier ^{34}S because of exchange with overlying water sulfate. The resultant hydrogen sulfide is relatively light. Under sulfate limited conditions, virtually the whole sulfate pool is reduced and the hydrogen sulfide has an isotopic composition similar to overlying water sulfate (30). Consequently, iron sulfide minerals in fresh water sediments generally have a heavier δ^{34}S than that found in marine sediments.

In this paper, we examine both the factors which control the distribution of iron sulfide minerals in Chesapeake Bay and the utility of C/S ratios, DOP and sulfur isotopic composition as indicators of depositional environments. Chesapeake Bay is an excellent system for the study of sulfur biogeochemistry because of gradients in salinity, organic production and bottom-water oxygenation. Considerable information on the biology, chemistry, physics and geology of Chesapeake Bay has been developed in the past two decades because of the need to effectively manage this heavily impacted system; this information also provides useful background on the biogeochemical processes which influence the sedimentary sulfur cycle.

Chesapeake Bay Biogeochemical Setting

Chesapeake Bay, a >300 km long coastal plain estuary, is characterized by large land-derived inputs of nutrients, high rates of plankton production, seasonal bottom water anoxia in the mesohaline region, and high rates of sediment metabolism. The mainstem Chesapeake Bay can be divided into three major regions: 1) the oligohaline upper bay, 2) the mesohaline mid-bay region characterized by high rates of organic production and seasonal anoxia, and 3) the shallow lower bay. High turbidity in the upper bay limits organic productivity and Susquehanna River-derived nutrients fuel primary production in the mid-bay region (*31*).

Chesapeake Bay sediments are generally fine-grained in the upper bay and at greater depths in the mid-bay region, while shallow water mid-bay and most southern bay sediments are sandy (*32, 33*). Sediment accretion rates are high in the northern bay because of high inputs of fluvial particulates. Sand from the continental shelf is a major source of sediment for the southern bay while the lower rates of sedimentation in the mid-bay region are largely derived from shoreline erosion (*34*).

Upper bay sediments have the highest concentrations of organic matter, most of which is fluvially-derived (*32, 33*). Deep-water organic carbon concentrations in the mid-bay are generally in the 2-4% range, with low concentrations in the lower bay. Sediment metabolism is low in the northern bay, high in the mid-bay region, and moderate in the southern bay (*35*). Rates of sulfate reduction in mid-bay sediments are high (15-20 mol m^{-2} yr^{-1}) (*36*) and summer fluxes of hydrogen sulfide contribute to bottom water anoxia (*4*).

Methods

In June 1988 and August 1989 cores were collected at 18 sites to obtain a system-wide description of solid phase and pore water chemistry (Figure 2). Seasonal measurements of sulfate reduction were made at many of these sites in 1988 and 1989. In 1989, spring, summer and fall profiles of sulfur species were measured at 5 sites on an east-west transect in the mesohaline region of the bay. Two cores for geochronological studies were collected in 1991 (*37*).

Sediments were collected using box corers. Cores were extruded in N_2-filled glove bags and pore water was obtained by centrifugation. Water samples were filtered (0.4 μm) and hydrogen sulfide (ΣH_2S) and iron were analyzed colorimetrically (*38, 39*). Sulfate analysis was via colorimetry (*40*) or ion chromatography.

Sediment for acid volatile sulfide (AVS) and total reduced sulfur (TRS) analysis was stored frozen and analyzed using cold 6 N HCl for AVS extraction (*41*) and acidic Cr(II) for TRS (*42*). The TRS analysis was not sequential and includes AVS, elemental S and pyrite-S. Hydrogen sulfide was analyzed via Pb titration (*41*).

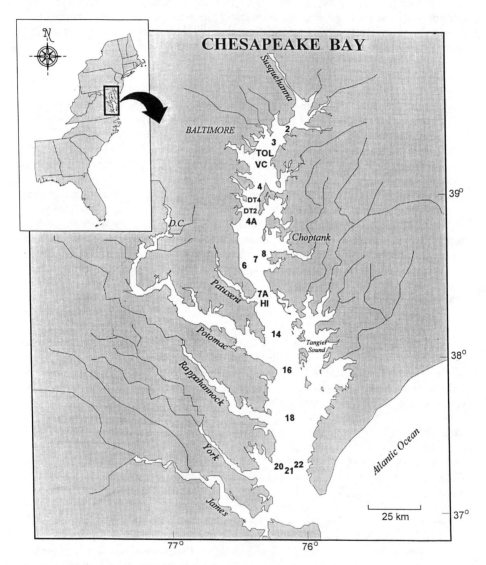

Figure 2. Map of Chesapeake Bay showing locations of sample sites. In the vicinity of site 7, five additional sites on a west to east transect (M1, M2, M3, M4, M6) were used for pore water and solid phase chemistry.

Oxalate-Fe was extracted using ammonium oxalate/oxalic acid (43) on frozen samples and HCl-Fe was extracted from dried sediments using 6 N HCl for 1 hr (22). Organic carbon was calculated as the difference between total C (CHN analyzer) and carbonate C analyses (44). The isotopic composition of TRS was analyzed via mass spectrometry at Coastal Science Laboratories (Austin, Texas) on AgS precipitates (45). Sulfate reduction was estimated by sediment incubation (46) of samples from 0-2, 5-7 and 12-14 cm core intervals. Homogenized sediment was packed into 50 mL centrifuge tubes and the sulfate decrease was determined after 1-2 weeks.

Results and Discussion

Pore Water and Solid Phase Profiles. Pore water sulfur and iron chemistry exhibited high spatial and temporal variability in Chesapeake Bay sediments (Figure 3). Low sulfate concentrations, rapid sulfate depletion, undetectable ΣH_2S concentrations, and large vertical dissolved iron gradients were observed at the north bay site. In the seasonally anoxic mid-bay region sulfate was depleted below 10 cm during summer months, similar to other coastal sediments (47). High rates of sulfate reduction are found in these mesohaline sediments (36) leading to high concentrations of ΣH_2S. During summer, low concentrations of ΣH_2S (<10 µmol L^{-1}) are commonly observed below the pycnocline due to fluxes across the sediment/water interface (4). Dissolved Fe concentrations were high in surficial sediments during spring months and low during the period of peak sulfate reduction (late spring to summer). Southern Chesapeake Bay sediments had abundant macrofauna, a minimal depletion of pore water sulfate, no detectable ΣH_2S, and high pore water Fe concentrations. Despite low organic carbon concentrations, south bay sulfate reduction rates are quite high.

Consistent with data from other estuarine and coastal sediments (48-50), iron sulfide mineral formation is often complete in the top 5 cm of sediment (Figure 4). At several stations, AVS concentrations approached 200 µmol g^{-1}, though concentrations were generally less than 20 µmol g^{-1}. TRS concentrations below 5 cm ranged from ~ 30 to 500 µmol g^{-1}. Pyrite-sulfur, represented as the difference between TRS and AVS, is the dominant form of sulfur in Chesapeake Bay sediments; elemental S is also included in the pyrite pool, but elemental S concentrations are generally small and limited to near-surface sediments (51, 52).

Oxalate- and HCl-extractable Fe concentrations were used to estimate the poorly crystalline (43) and total iron oxide concentrations (20, 22). Both extractions recover Fe associated with the AVS pool (41, 28). About half of the HCl-Fe was extracted with oxalate. Vertical profiles showed an Fe enrichment in surficial sediments from several cores, with down-core extractable Fe decreases resulting from pyrite formation. The profiles of DOP (Figure 3) show that the maximum DOP is generally reached in the top 5 cm and considerable oxalate-Fe and HCl-Fe is buried.

Latitudinal Concentration Trends. Gradients in carbon, iron, AVS, and TRS were evident along the axis of Chesapeake Bay (Figure 5), reflecting changes in salinity and the sources and input rates of organic matter. Upper bay sediments had high organic carbon and HCl-Fe concentrations while mid-bay sediments had high TRS concentrations; concentrations of AVS were low at most sites.

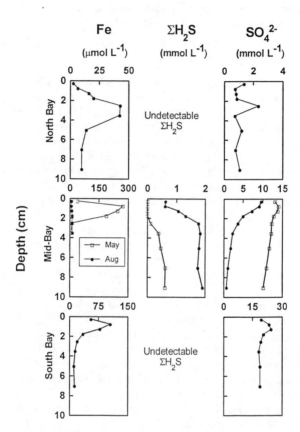

Figure 3. Vertical profiles of pore water sulfate, sulfide and iron in the northern, middle and southern Chesapeake Bay. Pore water chemistry at the mid-bay site M3 (near station 7) is shown for spring and summer conditions. The north bay site is at station 2 and the south bay site is at station 21.

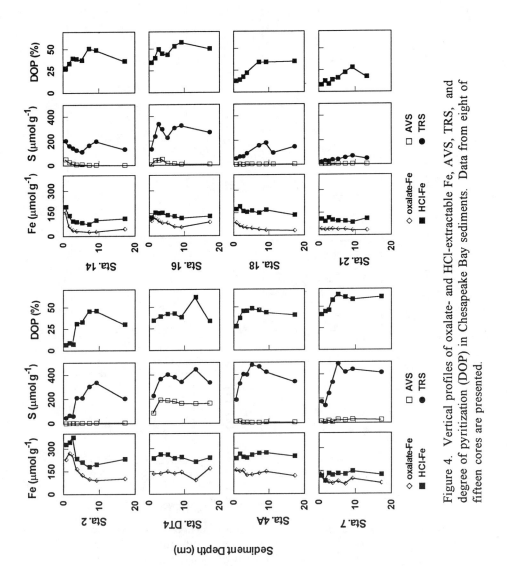

Figure 4. Vertical profiles of oxalate- and HCl-extractable Fe, AVS, TRS, and degree of pyritization (DOP) in Chesapeake Bay sediments. Data from eight of fifteen cores are presented.

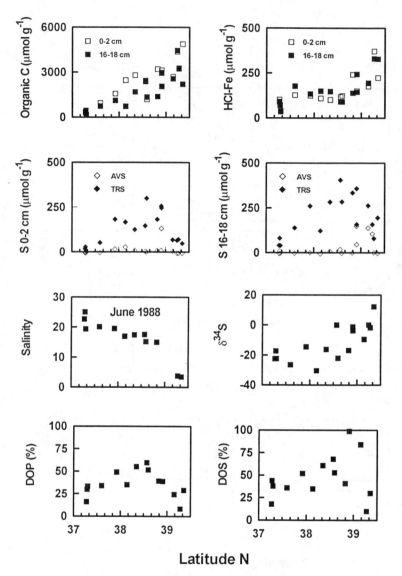

Latitude N

Figure 5. Latitudinal trends in Chesapeake Bay sediment chemistry. Surficial (0-2 cm) and deep (16-18 cm) sediment samples were analyzed for organic carbon, HCl-Fe, acid volatile sulfide (AVS) and total reduced sulfur (TRS). Salinity from June 1988 was estimated from the pore water chloride content of surficial sediments. The sulfur isotopic content of TRS ($\delta^{34}S$), DOP, and the degree of sulfidization (DOS) are plotted for 16-18 cm depth intervals.

Fluvial inputs from the Susquehanna River are the source of high concentrations of both organic carbon and iron. Analysis of $\delta^{13}C$ showed that upper bay sediments have organic carbon primarily from terrestrial sources, while that in the mid-bay and southern bay is derived from estuarine phytoplankton (*53*; Cornwell unpublished data). Iron concentration trends were similar to carbon with higher concentrations close to the Susquehanna River.

The downbay distribution of TRS is related to the reactivity of organic matter, availability of sulfate and the availability of iron. Berner et al. (*54*) determined AVS and pyrite-S concentrations in northern and mid-bay sediments and observed concentrations of TRS similar to those from this study; the use of a hot stannous chloride extraction resulted in an overestimation of the AVS pool (*55*).

Indicators of Sedimentary Environments.

Degree of Pyritization. The degree of pyritization is a measure of the efficiency of oxide-Fe to pyrite-Fe conversion. From the viewpoint of the overall sulfur cycle and particularly in systems with high AVS concentrations, a measure of the degree of sulfidization (DOS) is perhaps a better indicator (*49*):

$$DOS = (AVS\text{-}Fe + pyrite\text{-}Fe)/(HCl\text{-}Fe + pyrite\text{-}Fe)$$

AVS-Fe is estimated from the AVS concentration. The DOP and DOS values followed TRS concentration trends (Figure 5). The highest DOP values (> 0.5) are found in sulfidic sediments in the mid-bay area, with two high DOS values at two AVS-rich sites. Chesapeake Bay DOP values (10-55%) were similar to those found in other estuarine and coastal sediments (*9, 28 50*). The range of DOS values in the Chesapeake Bay (11-100%) is higher than that observed in anoxic sediments of the Baltic (20-50%) and other marine sediments (*49*). Fundamental questions exist concerning the utility of using DOP as a measure of iron reactivity (*28*). While DOP may have limited utility for intersystem comparisons, it may help explain site-to-site differences within a single system.

The high DOP values from Chesapeake Bay sites with seasonally anoxic bottom waters are consistent with the observations of Raiswell et al. (*12*) that DOP can be an indicator of bottom-water oxygenation; DOP's higher than 45% indicate low bottom water oxygen concentrations. In the Chesapeake mid-bay region, Cooper and Brush (*56*) have observed a transition of DOP values from < 40% in sediments greater than 100 yr old to values over 45% in modern sediments, suggesting that increased bottom water anoxia may result from cultural eutrophication.

Carbon-Sulfur Relationships. Organic carbon to sulfur ratios have been used by Berner and Raiswell (*11*) and others to distinguish between freshwater and marine environments; the used several sediment samples from the Chesapeake Bay to develop this index. Freshwater sediments are generally characterized by high molar C:S ratios (> 20) while marine sediments have a characteristic molar C:S ratio of 7.5 ± 4.0. Mid-bay and south bay sediments generally conform to the marine C:S ratio (Figure 6), with two of the northern bay sites exhibiting high C:S ratios indicative of low salinity environments. Other oligohaline sites do not show such a characteristically

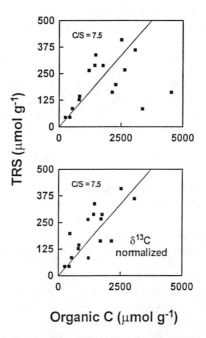

Figure 6. Plots of total reduced sulfur versus organic carbon concentration for Chesapeake Bay sediments. The "corrected" organic carbon concentrations are adjusted for the proportion of estuarine (versus terrestrial) organic matter; this correction is based on the measured $\delta^{13}C$ content of organic matter. All data are from 16-18 cm core depths.

high ratio. We can calculate the proportion of "estuarine" carbon in oligohaline sediments by using an estuarine endmember $\delta^{13}C$ value of 21.8 per mil (Cornwell, unpublished data) and a fluvial input $\delta^{13}C$ value of 26 per mil. By normalizing total organic carbon concentrations to their estuarine carbon concentrations, the C:S ratios follow the predicted marine C:S ratios much more closely.

While the differences between C:S ratios in freshwater and marine sediments may reflect sulfate limitation of sulfate reduction, the correlation between carbon and sulfur in these estuarine sediments may arise from different mechanisms. If iron limits the burial of sulfur in marine sediments, the basis of C:S correlations is most likely indirect. A strong correlation between grain size and the concentrations of both organic carbon and sulfur exists in mid and south-bay sediments (*32, 33*). Extractable Fe has been shown to be useful as a grain size surrogate in estuarine sediments (*57*). Furthermore, Mayer (*58*) has shown a strong relationship between mineral surface area and organic carbon concentrations. In iron-limited sediments, the C:S ratio in mesohaline and marine sediments may be relatively constant because surface area may control both organic carbon and iron concentrations. In sulfate-limited freshwater and oligohaline sediments, low rates of sulfate reduction and higher concentrations of refractory organic carbon from terrestrial sources may be the cause of high C:S ratios.

Sulfur Stable Isotopes. Sulfur isotopes appear to be an excellent indicator of salinity in this system, with higher values in the upper bay (Figure 5). The six most northern sites have a mean $\delta^{34}S$ of 0.0 ± 7.1, while the remainder (excluding one value of +0.6 per mil) have a mean $\delta^{34}S$ of -20.2 ± 5.2. Higher $\delta^{34}S$ values in the upper bay are indicative of lower isotopic fractionation of reduced sulfur from it's parent sulfate. Our data from the southern and mid-bay sites are similar to those observed in other coastal marine sediments (*28, 59, 60*). Under low sulfate conditions, near-total utilization of sulfate results in a minimal fractionation. In the south and mid-bay regions, a more plentiful sulfate supply allows a greater degree of fractionation from water column sulfate which has a $\delta^{34}S$ of about 20 (*29*).

The range of $\delta^{34}S$ values observed in southern and mid-bay sediments is relatively high and may be related to the mode of transport of pore water solutes. In an oxidizing system with mixing by processes such as bioirrigation, pore water $^{34}SO_4^{2-}$ enrichment is minimal because of exchange with overlying water. Diffusive systems should have a lower $\delta^{34}S$ fractionation of TRS relative to sulfate. Sulfate reduction rates and depth distributions may also determine TRS $\delta^{34}S$ values, though no significant correlations were observed between sulfate reduction rates and $\delta^{34}S$ values.

Temporal Trends in Sulfur Distribution. The concentrations and mineral forms of iron sulfide minerals may have distinct seasonal and long-term trends. In a dynamic system such as Chesapeake Bay, seasonal changes in organic matter loading, temperature, sediment metabolism and overlying water chemistry can have a profound impact on the formation and preservation of iron sulfide minerals. Many coastal ecosystems have been perturbed by a number of man's activities and sediments can contain useful indicators of environmental change (*61*).

Seasonal Changes in Mid-Bay Sediments. Changes in surficial sediment sulfur chemistry were monitored seasonally in the mid-bay region of Chesapeake Bay

(Figure 7). Pore water ΣH_2S was evident at all stations, although only very low concentrations were observed in surficial sediments during May and November at stations M1 and M2. During August, all sites except M1 had high concentrations of ΣH_2S. At the seasonally anoxic sites, AVS concentrations increased dramatically in the summer, decreasing to low levels at most sites in late fall. There were no clear seasonal patterns evident in the TRS data, possibly because of proportionately small seasonal changes and the effects of spatial heterogeneity at the sites. The conversion of iron oxides to sulfides results in the release of iron-bound phosphate (62); AVS oxidation contributes to high rates of autumn sediment oxygen demand.

Long-Term Trends. Cooper and Brush (56) have shown that over the last 2000 years there have been changes in the burial of sulfur in Chesapeake Bay mid-bay sediments, caused primarily by changing sediment deposition rates. To examine recent (< 100 yr) changes in the burial of AVS and TRS in the Chesapeake Bay, we used two dated cores with ~ 0.5 cm yr^{-1} sedimentation rates (Figure 8). As in most other Chesapeake Bay sediments, pyrite is the dominant iron sulfide mineral and AVS makes up only a small proportion of the TRS. Higher AVS concentrations in surficial sediments may not result from temporal changes in estuarine conditions; indeed, such enrichments are common in 0-5 cm sediment horizons where high rates of sulfate reduction are observed (19). No discernable TRS concentration trends were observed. Increasing organic matter concentrations are related both to increases in the carbon loading and steady-state organic matter diagenesis (63). These data do not support the idea that recent changes in the ecosystem leave an interpretable signature in the forms and concentrations of iron sulfide minerals. No recent changes in sedimentation were evident. Observational data from other studies have shown that at some locations, anoxic sediments overlay sediments which supported macrofauna (64).

Controls on Burial of Iron Sulfide Minerals. The production of ΣH_2S via sulfate reduction does not necessarily result in high rates of iron sulfide burial. The proportion of sulfide that is reoxidized or buried may be estimated by comparing annual rates of sulfate reduction and estimates of iron sulfide burial rates (Figure 9). Most reduced sulfur is reoxidized Chesapeake Bay sediments; such a high proportion of reoxidation is typical of most estuarine environments (2, 36, 65). Additionally, sulfide reoxidation may occur in the water column after diffusion of sulfide across the sediment-water interface (4). Despite high rates of sulfate reduction in south bay sediments, burial fluxes of iron sulfide minerals were low. Low pore water sulfate gradients in south bay sediments may arise both from bioirrigation or by rapid sulfide reoxidation. In north bay sediments, Roden and Tuttle (66) noted very rapid sulfide reoxidation during $^{35}SO_4^{2-}$ incubation experiments. The presence of ΣH_2S in mid-bay sediments shows that the production of reduced sulfur does not limit the burial of iron sulfide minerals; if ΣH_2S titrates the available iron, there is a large oversupply of titrant. In Chesapeake Bay, as in most productive estuarine and coastal systems, iron sulfide burial is controlled by the supply and efficient utilization of reactive iron.

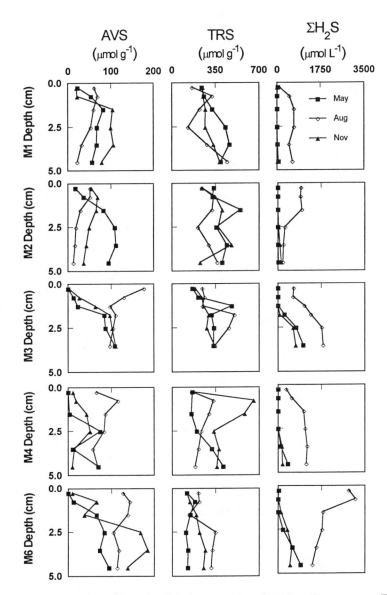

Figure 7. Seasonal profiles of solid phase AVS and TRS and pore water ΣH_2S in an east-west transect of the mesohaline Chesapeake Bay. Sites M3, M4 and M6 experience seasonal anoxia while M1 and M2 generally do not.

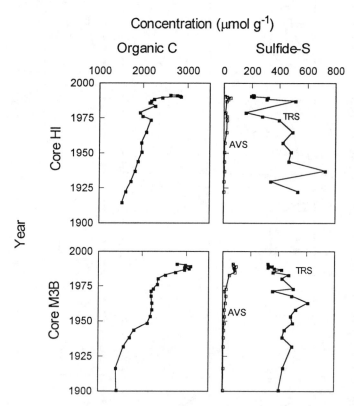

Figure 8. Vertical profiles of organic carbon, AVS and TRS in two mesohaline cores. Sediment ages were obtained by ^{210}Pb dating (*37*). Site M3B is near station 7.

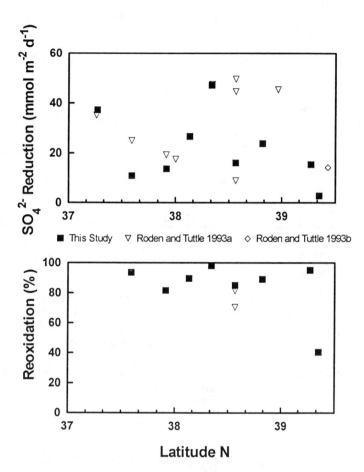

Figure 9. Latitudinal trends in sulfate reduction rate (July/August data) and annual proportion of sulfide reoxidation in Chesapeake Bay sediments. The sulfate reduction rates from this study were obtained from sulfate disappearance from packed centrifuge tubes; Roden and Tuttle (*36, 66*) used ^{35}S sulfate reduction methodologies. The efficiency of reoxidation was calculated from the annual sulfate reduction rate and the rate of burial of TRS estimated from published sedimentation rates for different regions of the bay (*67*); such sedimentation rate data was not available for all sites.

Conclusions

Gradients in organic loading, salinity, and degree of bottom water oxygenation make the Chesapeake Bay an excellent system for examining the process of iron sulfide mineral formation and preservation. Specific conclusions include:

1. The highest concentrations of iron sulfide minerals occur in the seasonally-anoxic mid-bay region.
2. As in other estuarine and coastal systems, pyrite is the predominant iron sulfide mineral.
3. High rates of sediment sulfate reduction are found in much of the Chesapeake; organic matter, not sulfate, is generally the limiting factor.
4. Both C:S ratios and $\delta^{34}S$ are useful indicators of the salinity regime.
5. High DOP values in the mid-bay region are consistent with the use of DOP as an indicator of restricted bottom water conditions.
6. The long-term fate of 80-100% of the sulfide produced via sulfate reduction is reoxidation to sulfate.
7. The key limiting factors for the burial of iron sulfide minerals appears to be 1) the rate of input of "labile" iron oxide minerals and 2) the efficient utilization of iron oxides. With DOP's generally < 0.6, considerable extractable iron must be unavailable for iron sulfide mineral formation.

Future research directions in the study of iron sulfide mineral formation in estuaries include: 1) determination of the reactivity of iron oxide minerals; 2) understanding the processes which control the transformation of iron oxides in sediments; 3) examining the controls of sulfur isotope abundance in sediment, specifically the importance of sulfate oxidation and reduction rates and 4) determining the sulfur burial budget for the entire system. Observations of environmental distributions provide a convenient starting point for the more detailed study of the processes which control such distributions.

Acknowledgments

Funding for this research was provided by the National Science Foundation-sponsored Land Margin Ecosystem Research Program (BSR-8814272), the United States Environmental Protection Agency and the Maryland Department of the Environment. We thank Mike Owens for field and laboratory assistance and the crews of the R/V Cape Henlopen and R/V Aquarius for assistance with coring. The manuscript was improved by comments from two anonymous reviewers.

References

(1) Jorgensen, B. B. *Nature* **1982**, *296*, 643-645.
(2) Berner, R. A.; Westrich, J. T. *Am. J. Sci.* **1985**, *285*, 193-206.
(3) Canfield, D. E. *Deep-Sea Res.* **1989**, *36*, 121-138.
(4) Roden, E. E.; Tuttle, J. H. *Limnol. Oceanogr.* **1992**, *37*, 725-738.
(5) Caraco, N.; Cole, J.; Likens, G. E. *Biogeochemistry* **1990**, *9*, 277-290.

(6)   Cornwell, J. C.; Boynton, W. R.; Owens, M.; Cowan, J. Phosphorus cycling in Chesapeake Bay sediments: control by iron and sulfur diagenesis; Estuarine Research Federation Conference, San Francisco, CA.

(7)   Boulegue, J.; Lord, C. J., III; Church, T. M. *Geochim. Cosmochim. Acta* **1982**, *46*, 453-464.

(8)   DeLaune, R. D.; Smith, C. J. *J. Environ. Qual.* **1985**, *14*, 164-168.

(9)   Huerta-Diaz, M. A.; Morse, J. W. *Geochim. Cosmochim. Acta* **1992**, *56*, 2681-2702.

(10) DiToro, D. M.; Mahony, J. D.; Hansen, D. J.; Scott, K. J.; Carlson, A. R.; Ankley, G. T. *Environ. Sci. Technol.* **1992**, *26*, 96-101.

(11) Berner, R. A.; Raiswell, R. *Geology* **1984**, *12*, 365-368.

(12) Raiswell, R.; Buckley, F.; Berner, R. A.; Anderson, T. F. *J. Sed. Petrol.* **1988**, *58*, 812-819.

(13) Raiswell, R.; Berner, R. A. *Am. J. Sci.* **1985**, *285*, 710-724.

(14) Boudreau, B. P.; Westrich, J. T. *Geochim. Cosmochim. Acta* **1984**, *48*, 2503-2516.

(15) Capone, D. G.; Kiene, R. P. *Limnol. Oceanogr.* **1988**, *33*, 725-749.

(16) Lin, S.; Morse, J. W. *Am. J. Sci.* **1991**, *291*, 55-89.

(17) Luther, G. W., III; Ferdelman, T. G.; Kostka, J. E.; Tsamakis, E. J.; Church, T. M. *Biogeochemistry* **1991**, *14*, 57-88.

(18) Luther, G. W., III; Giblin, A.; Howarth, R. W.; Ryans, R. A. *Geochim. Cosmochim. Acta* **1982**, *46*, 2665-2669.

(19) Morse, J. W.; Millero, F. J.; Cornwell, J. C.; Rickard, D. *Earth Sci. Rev.* **1987**, *24*, 1-42.

(20) Berner, R. A. *Am. J. Sci.* **1970**, *268*, 1-23.

(21) Berner, R. A. *Geochim. Cosmochim. Acta* **1984**, *48*, 605-615.

(22) Morse, J. W.; Cornwell, J. C. *Mar. Chem.* **1987**, *22*, 55-69.

(23) Kohnen, M. E. L.; Sinninghe Damste, J. S.; ten Haven, H. L.; de Leew, J. W. *Nature* **1989**, *341*, 640-641.

(24) Ferdelman, T. G.; Church, T. M.; Luther, G. W., III. *Geochim. Cosmochim. Acta* **1991**, *55*, 979-988.

(25) Fossing, H.; Jorgensen, B.B. *Geochim. Cosmochim. Acta* **1990**, *54*, 2731-2742.

(26) Aller, R. C.; Rude, P. D. *Geochim. Cosmochim. Acta* **1988**, *52*, 751-765.

(27) Luther, G. W., III. *Geochim. Cosmochim. Acta* **1991**, *55*, 2839-2849.

(28) Canfield, D. E.; Raiswell, R.; Bottrell, S. *Am. J. Sci.* **1992**, *292*, 659-683.

(29) Goldhaber, M. B.; Kaplan, I. R. In *The Sea*; Goldberg, E. A., Ed.; Wiley-Interscience: New York, 1974; Vol. 5, pp. 569-655.

(30) Jorgensen, B. B. *Geochim. Cosmochim. Acta* **1979**, *43*, 363-374.

(31) Malone, T. C.; Crocker, L. H.; Pike, S. E.; Wendler, B. W. *Mar. Ecol. Prog. Ser.* **1988**, *48*, 235-249.

(32) Hobbs, C. H., III. *J. Sed. Petrol.* **1983**, *53*, 383-393.

(33) Hennessee, E. L.; Blakeslee, P. J.; Hill, J. M. *J. Sed. Petrol.* **1986**, *56*, 674-683.

(34) Hobbs, C. H., III; Halka, J. P.; Kerhin, R. T.; Carron, M. J. *J. Coast. Res.* **1992**, *8*(2), 292-300.

(35) Boynton, W. R.; Kemp, W. M. *Mar. Ecol. Prog. Ser.* **1985**, *23*, 45-55.

(36)   Roden, E. E.; Tuttle, J. H. *Mar. Ecol. Prog. Ser.* **1993**, *93*, 101-118.
(37)   Owens, M.; Cornwell, J. C. *Ambio* **1994**. (*accepted*)
(38)   Cline, J. D. *Limnol. Oceanogr.* **1969**, *14*, 454-458.
(39)   Gibbs, M. M. *Wat. Res.* **1979**, *13*, 295-297.
(40)   Tabatabai, M. A. *Environ. Letters* **1974**, *7*, 237-243.
(41)   Cornwell, J. C.; Morse, J. W. *Mar. Chem.* **1987**, *22*, 193-206.
(42)   Canfield, D. E.; Raiswell, R. R.; Westrich, J. T.; Reaves, C. M.; Berner, R. A. *Chem. Geol.* **1986**, *54*, 149-155.
(43)   Williams, J. D. H.; Syers, J. K.; Shukla, S. S.; Harris, R. F.; Armstrong, D. E. *Environ. Sci. Technol.* **1971**, *5*, 1113-1120.
(44)   Stainton, M. P. *J. Fish. Res. Board Can.* **1973**, *30*, 1441-1445.
(45)   Hall, G. E. M.; Pelchat, J.-C.; Loop, J. *Chem. Geol.* **1988**, *67*, 35-45.
(46)   Goldhaber, M. B.; Aller, R. C.; Cochran, J. K.; Rosenfeld, J. K.; Martens, C. S.; Berner, R. A. *Am. J. Sci.* **1977**, *277*, 193-237.
(47)   Berner, R. A. *Geochim. Cosmochim. Acta* **1964**, *28*, 1497-1503.
(48)   Aller, R. C. *Adv. Geophys.* **1980**, *22*, 237-350.
(49)   Boesen, C.; Postma, D. *Am. J. Sci.* **1988**, *288*, 575-603.
(50)   Morse, J. W.; Cornwell, J. C.; Arakaki, T.; Lin, S.; Huerta-Diaz, M. *J. Sed. Petrol.* **1992**, *62*, 671-680.
(51)   Troelsen, H; Jorgensen, B.B. *Est. Coast. Shelf Sci.* **1982**, *15*, 255-266.
(52)   Thode-Andersen, S.; Jorgensen, B.B. , *Limnol. Oceanogr.* **1989**, *34*, 793-806.
(53)   Kerhin, R. T.; Halka, J. P.; Wells, D. V.; Hennessee, E. L.; Blakeslee, P. J.; Zoltan, N.; Cuthbertsen, R. H. *Surficial Sediments of Chesapeake Bay, Maryland: Physical Characteristics and Sediment Budget*; Report of Investigations No. 48; Maryland Geological Survey: Baltimore, MD, 1988.
(54)   Berner, R. A.; Baldwin, T.; Holdren, G. R. *J. Sed. Petrol.* **1979**, *49*, 1345-1350.
(55)   Chanton, J. P.; Martens, C. S. *Biogeochemistry* **1985**, *1*, 375-383.
(56)   Cooper, S. R.; Brush, G. S. *Science* **1991**, *254*, 992-996.
(57)   Trefry, J. H.; Presley, B. J. *Environ. Geol.* **1976**, *1*, 283-294.
(58)   Mayer, L. M. *Geochim. Cosmochim. Acta* **1994**, *58*, 1271-1284.
(59)   Kaplan, I. R.; Emery, K. O.; Rittenberg. *Geochim. Cosmochim. Acta* **1963**, *27*, 297-331.
(60)   Goldhaber, M. B.; Kaplan, I. R. *Mar. Chem.* **1980**, *9*, 95-143.
(61)   Valette-Silver, N. J. *Estuaries* **1993**, *16*, 577-588.
(62)   Krom, M. D.; Berner, R. A. *Geochim. Cosmochim. Acta* **1981**, *45*, 207-216.
(63)   Cornwell, J. C.; Stevenson, J. C.; Conley, D. J.; Owens, M. *Estuaries* **1995**, *(in press)*.
(64)   Reinharz, E.; Nilsen, K. J.; Boesch, D. F.; Bertelsen, R.; O'Connell., A. E. *A Radiographic Examination of Physical and Biogenic Structures in the Chesapeake Bay;* Report of Investigations No. 36; Maryland Geological Survey: Baltimore, MD, 1982.
(65)   Jorgensen, B. B. *Limnol. Oceanogr.* **1977**, *22*, 814-832.
(66)   Roden, E. E.; Tuttle, J. H. Biogeochemistry **1993**, *22*, 81-105.
(67)   Officer, C.B.; Lynch, D.R.; Setlock, G.H.; Helz, G.R. *The Estuary as a Filter;* cademic Press: New York, NY, 1984; pp 131-157.

RECEIVED August 17, 1995

Chapter 13

Digestion Procedures for Determining Reduced Sulfur Species in Bacterial Cultures and in Ancient and Recent Sediments

R. E. Allen and R. J. Parkes

Department of Geology, University of Bristol, Wills Memorial Building, Queens Road, Bristol BS8 1RJ, England

A detailed study was made to determine the recovery of different sedimentary metal sulfide minerals, polysulfides and elemental sulfur by digestion procedures. This was a part of an investigation into the influence of bacterial sulfate reduction on the formation of authigenic pyrite and other sulfide minerals in the geological record. The recovery of different varieties of metal sulfide and forms of reduced sulfur from both artificial compounds, sediments and mineral specimens, indicates that hot AVS digestions, and cold followed by hot chromous chloride digestions produce the best recoveries combined with the greatest separation of sulfur pools for both ancient and Recent samples. The reaction time required to recover pyrite (up to 4 hours) is much greater than that previously reported (2 hours), especially when applied to Recent sediments and minerals. Recoveries of different sulfide minerals and elemental sulfur indicate that there may be an over-estimation of components used in degree of pyritization and sulfidation ratios. The application of the standardized methods to cultures of sulfate-reducing bacteria demonstrates that pyrite can be formed within three days. This method is also capable of producing accurate divisions of sulfur pools in sediments and sedimentary rocks.

Role of Sulfate Reduction in the Authigenic Production of Sulfide Minerals. Dissimilatory bacterial sulfate reduction is one of the dominant decay processes in marine sediments, accounting for up to half of the breakdown of organic matter in coastal environments (1, 2). Sulfate reduction occurs mainly in anoxic sediments, although it can also occur in oxic sediments within anoxic micro-environments such as decaying fragments of organic matter and animal burrow linings (3, 4, 5). Generally, sulfate reducing environments occur where bacterial processes have exhausted the supplies of oxygen, nitrate, manganese and iron used as electron acceptors in organic matter degradation (3,6). During sulfate reduction, hydrogen sulfide is produced, a small fraction of which reacts with metals and residual organic matter within the sediment and porewaters to produce various varieties of authigenic sulfide minerals, such as pyrite and organo-sulfur compounds (7). The remainder is oxidised by reaction with minerals, or diffuses

0097–6156/95/0612–0243$12.00/0

into more oxic regions of the sediment, where it is both microbially and chemically oxidised. This oxidation will produce various forms of sulfur, such as sulfur, thiosulfate and sulfate.

Low temperature authigenic sulfide mineralisation can therefore be considered a biological process requiring metabolisable organic matter, sulfate and metal ions either dissolved in porewaters or from reactive minerals. It produces both economic metalliferous deposits and some of the commonest examples of fossilisation (8). As sulfide formation in sediments is intimately linked to the metabolic activities of sulfate-reducing bacteria, it is appropriate to use bacterial cultures as an approach to the study of early sulfide mineralisation. This approach is especially relevant as the production of sulfide minerals in bacterial cultures has long been used as a method of detecting bacterial growth (9) (pyrite formation in cultures of sulfate reducers was first reported in 1912 by Issatchenko (10)). As sedimentary sulfate-reducing bacteria are found in environments where pyritization occurs, they provide a means of reproducing pyritization at low temperature and under realistic chemical and biological conditions. However, the solid phases produced in microbiological cultures can be amorphous and, as such, cannot be identified unequivocally, either microscopically or using x-ray diffraction, although the elemental abundances of particles can be determined using an electron microprobe. Therefore it is essential to separate the different forms of sulfur in these systems by chemical methods . It is also necessary to be able to relate the forms of sulfur formed in bacterial cultures to those formed in sedimentary environments and geological samples, in order to understand the role of bacteria in the formation of sedimentary sulfide minerals.

The concentration of pyrite in an ancient sedimentary rock can be used as a palaeoenvironmental sediment redox indicator, by means of the degree of pyritization ratio (DOP, 11). This ratio relates the concentration of pyritized iron in the sediment to the concentration of reactive iron. Pyritic iron concentrations are usually determined from the concentration of sulfide or iron released during hot chromous chloride digestions.

$$DOP= (Sulfur\ in\ pyrite\ *0.891)\ /\ (Sulfur\ in\ pyrite\ *\ 0.891) + reactive\ iron$$

In sediments where the solid sulfide pool is dominated by monosufides the degree of sulfidation (DOS, 12) ratio is used as an indicator of the redox of the environment. This ratio relates the concentration of all the iron sulfides to the concentration of reactive iron in the sediment. Unlike the DOP ratio, the DOS ratio includes both the acid volitile sulfide and pyritic sulfide concentrations, as a means of expressing the amount of sulfide which has reacted with iron in a particular sediment.

$$DOS= Total\ Reduced\ Inorganic\ Sulfur\ /Total\ Reduced\ Inorganic\ Sulfur + reactive\ iron$$

If these ratios are to be applied to the formation of pyrite and other sulfide minerals in all ages of sedimentary environments and bacterial cultures, then the techniques for the separation and quantitation of each sulfur pool must be standardized for all conditions.

Methods of Determining the Proportion of Different Types of Reduced Sulfur Compounds in Sediments. Although, methods of determining both the Acid Volatile Sulfide (AVS) and the Total Reduced Inorganic Sulfide (TRIS) (*13, 14*) have been previously checked with various sulfur compounds the details of the procedures have often varied considerably between research groups. Futhermore, a comprehensive range of reduced sulfur species and minerals have not been subjected to a standard procedure incorporating both chromous chloride (TRIS) and acid volatile digestions. Cornwell and Morse (*15*) studied the greatest range of extraction conditions, however their study was limited to iron sulfides. TRIS digestions are a common method for determining the presence and quantity of bacterially reduced sulfur in a sediment and use reduced chromic chloride and concentrated HCl (*16*). The analysis is conducted in an oxygen-free environment, and the hydrogen sulfide released is then actively transported into a chemical solution (eg. zinc acetate) where it is trapped. The concentration of the trapped sulfide can then be determined spectrophotometrically (*17*), gravimetrically (*18*), by ICP-MS (*19*), polarographically (*20*) or by titration (*14*). Nearly all the published methods for determining TRIS use 1M chromous chloride in an acidified solution, concentrated HCl, and ethanol (*14, 21*). Heating of the solution and sediment is required to release any elemental sulfur in the sample although Fossing & Jørgensen (*14*) observed the reduction of a small amount of elemental sulfur even under cold digestion conditions.

The AVS procedure is a less vigorous reaction, which uses acid to digest the more reactive sulfides, including iron monosulfides, such as mackinawite, and intermediate iron sulfides, such as greigite. Cornwell and Morse (*15*) compared different methods for extracting AVS using synthesized mackinawite, greigite and pyrite, with varying concentrations of hydrochloric acid, and with or without heat and stannous chloride. In their comparison of 6M HCl digestions, hot 6M HCl, cold 6M HCl and cold 6M HCl in conjunction with $SnCl_2$ produced similar high recoveries of monosulfides and intermediate sulfides, although the latter method could include some sulfide from cryptocrystalline pyrite. Stannous chloride is added in order to negate the oxidation of the hydrogen sulfide produced in the digestion by ferric iron (*22*). Rice *et al.* (*23*) suggested that stannous chloride should be included in AVS digestions for ancient sediments as it significantly improved the recovery of monosulfides from their samples, however the inclusion of stannous chloride in a cold AVS digestion followed by a cold chromic chloride digestion would lead to an over-estimation of AVS and an under-estimation of pyrite in Recent sediments and bacterial cultures. These errors are particularly significant in bacterial cultures and $^{35}SO_4$ radiotracer measurements of sulfate reduction activity (*1, 2, 14, 24*) as the initially formed amorphous and cryptocrystalline pyrite produced would be susceptible to breakdown (*15*). Chanton and Martens (*25*) indicated that the use of stannous chloride in low concentrations at room temperatures will not extract pyrite. The euhedral 98% hydrothermal pyrite, which they used as a standard, however, is not representative of the whole of the pyrite pool found in Recent sediment and bacterial cultures, due to the presence of a cryptocrystalline pyrite. Heating with excess 6M HCl appears to produce the greatest recovery of AVS and the lowest recovery of artificial pyrite (*15*). The recovery of the pyrite and elemental sulfur pools is minimal under hot 6M HCL digestion conditions, due to the sulfur-sulfur covalent bonds present in these reduced species. Elemental sulfur can be present in a sample, either as a primary component or as a by-product of the breakdown under acid conditions of more intermediate reduced sulfur compounds, such as greigite and polysulfide (*26*). The AVS digestion method for determining the concentrations of monosulfides and intermediate sulfides is usually applied to sediments, but it can also be used for sedimentary rocks (*19, 23*).

As a small proportion of elemental sulfur is digested in the cold chromous chloride digestion (14), these digestions may be preceded by an extraction for zero-valent sulfur, so that pyrite concentrations may be estimated more accurately. The removal of elemental sulfur is apparently not necessary in most sediments, as it occurs in very low concentrations (14). However, it is necessary in laboratory experiments, as it may be produced by oxidation within the incubation/reaction vessel, as well as during acid digestion. If sulfur is extracted before and after acid digestion, then the combined concentrations of polysulfide and greigite can be estimated, independently of elemental sulfur as these two forms of sulfide breakdown into elemental sulfur and hydrogen sulfide under hot acid conditions. There are three main methods of determining zero-valent sulfur concentrations in sediments. The first is the sodium cyanide method of Troelsen and Jørgensen, (26), which involves dissolving sulfur in carbon disulfide and reacting this with sodium cyanide in order to produce a compound which can then be measured spectrophotometrically. The second uses dichloromethane as a sulfur solvent, followed by a hot chromous chloride digestion. This was used by Mossman *et al.* (17) who refluxed samples in a dichloromethane - methanol mixture, and Schimmelmann and Kastner. (27), who used two twelve hour extractions with 20 ml of dichloromethane. Other nonpolar solvents, such as cyclohexane, may also be used to extract elemental sulfur from samples. The third method is a continuous extraction of sulfur in Soxhlet with acetone. The sulfur-bearing acetone then reacts with copper precipitating copper sulfide, and the sulfide content of the trap is then determined using an HCl digestion. Acetone digestions (13, 28) were not performed as acetone does not partition with water and as such it is not suitable for bacterial cultures or high porosity sediments.

Methods

Sample Preparation and Selection. In order to produce a consistent method of determining the distribution of different forms of sulfur within ancient, modern and bacterial systems, samples of each type were selected to validate previously used reaction schemes. Sedimentary sulfides, such as chalcopyrite, marcasite and galena were included in this study as previous work had been limited in the main to iron sulfide minerals. Sphalerite was used in Canfield *et al.* (21), but only under chromous chloride digestion conditions. Pyrrhotite, a monosulfide, was not included in this study as it is rare in normal sedimentary environments (29). Pyritic sediments from the Liassic of Devon, UK, and the Tamar Estuary, Plymouth, UK were included in this study in order to provide a comparison between modern and ancient sedimentary sulfides. Sulfate-reducing bacterial cultures provide samples of the earliest formed iron sulfides and are analogous to those formed during $^{35}SO_4$ radiotracer incubations and early diagenesis. In addition artificial samples of the monosulfides, mackinawite, and the intermediate sulfide, greigite, were prepared along with elemental sulfur and polysulfide solutions, to check the separation of solid sulfide pools .

The rock and mineral samples listed in Table I were crushed in order to remove the matrix and increase the available reactive surface area. Rock samples were initially crushed in a jaw-crusher. Oxidised surfaces were removed by picking fragments from the interior of the samples and these fragments were then powdered to a 60µm mesh size. The composition of these samples were confirmed using XRD analysis. Digestions with unpowdered rock and mineral samples gave extremely poor recoveries compared to those which had been powdered. Sediment samples from the Tamar Estuary, Plymouth, UK were used wet as the recovery of elemental sulfur pool is adversely effected by drying (14). Sediment samples were stored frozen in 20% zinc acetate in order to minimize

Table I. Mineral samples analysed

Mineral/ Sample	Characteristics	Preparation/Storage
Pyrite (euhedral) FeS_2	Brassy yellow striated cubes from a hydrothermal vein. Unknown age	Crushed to a 60μm powder after removing rock matrix
Pyrite (rock nodule)	Cryptocrystalline nodule from black Liassic marine shales	As for euhedral pyrite
Pyrite (artificial)	Subhedral pyrite	Made following the method of Cornwell & Morse (*15*).Washed in deoxygenated water, 6M HCl and dichloromethane Dried anaerobically
Pyrite (vein)	Cryptocrystalline vein from a black slate. Llandovery in age	As for euhedral pyrite
Sphalerite (ZnS)	Brown, resinous material showing subhedral crystal edges. With galena in a hydrothermal vein.	Crushed to 60 μm powder after removing galena from around the crystals
Chalcopyrite ($CuFeS_2$)	Subhedral yellow mineral. Softer than pyrite. In a hydrothermal vein.	As for euhedral pyrite
Galena (PbS)	Silver, heavy metallic sulfide, Hydrothermal origin, age unknown	Crushed after picking from fragments of the sphalerite sample
Marcasite FeS_2	Orthorhombic dimorph of pyrite. Nodule from Liassic marine shales	Crushed as for euhedral pyrite.
Iron Monosulfide FeS	Amorphous black mass of monosulfide	Made following the method of Cornwell & Morse (*14*).Washed with deoxygenated water and stored wet in the anaerobic cabinet
Greigite Fe_3S_4	Intermediate sulfide in sediments	Made following the method of Berner (*22*) Washed with deoxygenated water and acetone Dried in the anaerobic cabinet and stored in a sealed nitrogen filled vial
Polysulfide Solution $S_n\text{-}S^{2-}$	Colourless solution	Made following the method of Wada (*29*). Stored in a sealed vial filled with nitrogen.
Sulfur suspension $S°$	Yellow suspension in acetone.	Made following the method of Fossing & Jorgensen (*14*) Stored as a suspension in a sealed vial with nitrogen
Flowers of Sulfur $S°$	Yellow, crystals of sulfur	Precipitated by evaporating sulfur-bearing dichloromethane Dried and stored anaerobically

oxidation of the reduced sulfur. The artificial sulfides were prepared using the methods shown in Table I. These were washed with deoxygenated water and dichloromethane in order to remove residual reagents. Pyritic samples were also washed in 1M HCl in order to remove the monosulfides produced during the reaction. After drying in an anaerobic atmosphere at room temperature, the composition of each sample was confirmed using XRD. The artificial samples were then stored in sealed deoxygenated vials prior to digestions. Samples were not stored for more than 4 days. Subsamples were then taken in an anaerobic cabinet (Forma Scientific, Marietta, Ohio, USA). Elemental sulfur and polysulfide solutions were prepared according to the methods shown in Table I and stored in a similar manner to the artificial sulfides. The sulfur content for each sample was calculated theoretically from the purity of the phase in the XRD. Although this is not as accurate as a total sulfur analysis, in a comparison of digestion methods it was considered sufficient.

Analysis of Samples. All samples were treated in the same way. Three 15ml aliquots were removed from each liquid and sediment sample, while three 1.5g aliquots were removed from each solid sample. The first two aliquots were mixed with either cyclohexane or dichloromethane in order to determine the concentration of elemental sulfur. The solvent to sample ratio used was 0.1g of sample to 20ml of solvent for solids and 1ml to 20 ml of solvent for bacterial cultures. The mixtures were shaken for 24 hours prior to the separation of the organic phase, and it's subsequent removal using a pasteur pipette. This was digested with hot chromous chloride to determine the concentration of elemental sulfur. Each extraction was performed in triplicate. The third aliquot was stored in 15ml of 20% zinc acetate and then subsequently digested. Four types of digestions were performed; 6M HCl acid digestions with and without heating and chromous chloride digestions with and without heating (*13, 14, 21, 23, 25*). Each digestion was performed in triplicate and the remainder of the aliquot was then used to run sequential digestions

Additional samples from the bacterial cultures and sediment porewaters were filtered and stored in 10% zinc acetate so that the contribution from the free sulfide pool could be determined. The concentration of sulfide was determined using a colorimetric method (*17*).

The Digestions. From the literature it was clear that a protocol involving a hot 6M HCl digestion would separate the monosulfide from the pyritic sulfide pool (*15*). Despite the recommendation of Rice *et al.* (*23*) stannous chloride was not used in the 6M HCl digestions as it has been shown to digest synthetic pyrite (*15*), and would probably remove the large amount of amorphous pyrite produced in bacterial cultures. Cold AVS digestions were only applied to the varieties of minerals with a metal to sulfur ratio of 1:1 to determine if there was any difference in reactivity between them. In order to separate the pyrite and elemental sulfur pools for all the groups of samples a cold chromous chloride digestion was tried as this only removes a minimal amount of the elemental sulfur and removes the majority of the pyritic pool. To avoid delays when changing from hot to cold digestions, hot and cold digestions were normally run in parallel on replicate samples. However, some sequential runs were conducted to assess subsample variability.

In the AVS digestions, 2ml of bacterial sample stored in zinc acetate, or 1ml of sediment or 0.1g of finely powdered rock or mineral sample was placed in a distillation flask. To each flask 5ml deoxygenated water, 5ml of 20% zinc acetate and a magnetic stirrer were added. The flasks were then attached to the distillation apparatus (see Figure 1) where they were flushed with OFN (commercial oxygen-free nitrogen passed through an oxygen scrubber) and stirred. The system was checked for leaks. An all glass and PTFE apparatus was used for the distillations (2). The apparatus was left to flush for 10 minutes, before 2ml of, or excess, 6M HCl was injected into each flask. The amount of HCl required for the digestion was dependant upon the concentration of carbonate in the sediment. The volume required can be determined prior to the digestions by adding acid to a subsample of sediment until effervesence has stopped. In heated digestions, the temperature was then brought up to 80°C and the flasks left to react . The trap tubes filled with zinc acetate were changed every 15 minutes for artificial samples and every half an hour for all other types of sample, and the reaction continued until there was no noticeable change in colour of the trap tubes for two subsequent tubes. 1ml from the zinc acetate trap was then used to determine the AVS content of the sample using a colormetric method.(*17*)

The same amount of sample as in the AVS digestions was placed in a distillation flask for chromous chloride digestions. To each flask 5ml of 95% ethanol, 5ml of 20% zinc acetate and a flea were added to each flask The flasks were then attached to the distillation apparatus as previously described. The apparatus was left to flush for 10 minutes, before 25ml of 1M chromous chloride in 0.5N HCl (reduced in a Jones Reductor column) and 10ml of concentrated hydrochloric acid were injected into each flask. In heated digestions the temperature was then brought up to 80°C. Distillation times and sulfide concentrations were determined as in the AVS procedure

Additional digestions were performed on mixtures of samples in order to determine whether they could be separated by using sequential and parallel digestions. In these cases mixtures of two or three samples were treated in a similar manner to the single digestions with a cold AVS being followed by a hot AVS digestion. The AVS digestions were then followed by a cold chromous chloride digestion prior to heating again. In sequential digestions such as these 20 minutes was allowed before the introduction of further reagents in order to allow the temperature changes to occur. Replicate samples were digested in parallel, with cold and hot AVS extractions being performed on one sample while cold and hot chromous chloride digestions were performed on a replicate.

Results and Discussion

All four pyrite samples behaved similarly under different reaction conditions (Table II). Hot and cold chromous chloride digestions produced similar recoveries from all types of pyrite, except artificial which showed a increase from 75% to 93%. Only the length of digestion time required to obtain complete recovery varies. In previous studies (*2, 19* and unpublished results) it was found that between 40 minutes and 1 hour was sufficient to reduce all the sulfide compounds in a bacterial culture or $^{35}SO_4$ radiotracer incubation sample, while two hours was recommended for ancient shales (*21*). This variation in digestion times was considered by Canfield *et al.* (*21*) to be due to the lithification of the pyrite crystals, while Chanton and Martens (*25*) and Rice *et al.* (*23*) have shown that grainsize is a contributing factor to digestion time. However, Rice *et al.* (*23*) also commented that the age of pyrite used will effect its recovery in AVS digestions containing stannous chloride. In this study, which is concentrated on defining a method of determining pyrite, elemental sulfur and AVS concentrations in a range of samples aged from 1 day to millions of years, maturity is likely to be the primary controlling factor, while an associated increase in crystallinity and grainsize are important only in the division between sediments and rocks. The difference in grainsize between Recent and ancient pyrite can be removed by crushing the rock

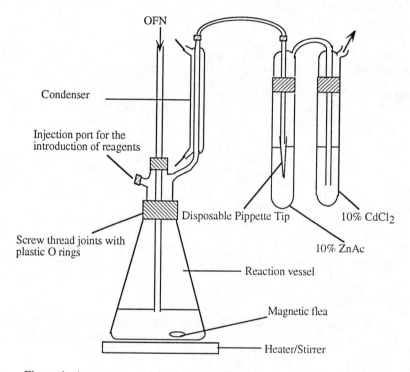

OFN

Condenser

Injection port for the
introduction of reagents

Disposable Pippette Tip

10% CdCl$_2$

Screw thread joints with
plastic O rings

10% ZnAc

Reaction vessel

Magnetic flea

Heater/Stirrer

Figure 1.-Apparatus used for Solid sulfide digestions. (1 of a bank of 6)

samples. A 40 minute chromous chloride digestion of a sediment would therefore remove the least mature pyrite, for example, that which contains $^{35}S^{2-}$ in radiotracer experiments, but continued digestion for up to 3 hours is required to remove the remaining pyrite fraction in sediments (Figure 2). This older pyrite fraction represented up to 35% of the total pyrite concentration of Tamar Estuary sediments (Figure 2), and is considered to be framboidal or crystalline, while the newly formed material is amorphous. The elemental sulfur content is low in these sediments and there is, therefore, a small increase in the sulfide recoveries after heating with chromous chloride (shown by higher TRIS concentrations in Figure 2). The euhedral pyrite digestion times were similar to those of the Tamar sediments, with the majority of sulfide recovered within 3 hours (c. 85% +/- 20%), indicating that the sediment contains pyrite of a similar crystallinity. This behaviour means that the concentration of sulfide produced during $^{35}SO_4$ radiotracer incubations for sulfate reduction activity measurements and the concentration of iron sulfide minerals in a sediment (its degree of sulfidation (*12*)) are not synonymous, if the digestion times used are short. In order to compare ancient and Recent sulfidic sediments the whole of the pyrite pool must be analysed and the recovery of pyrite from sediments after a 40 minute cold chromous chloride digestion is not representative of this.

Results for marcasite were similar to those for pyrite. It is recovered under hot and cold chromous chloride digestions. However, as the sample is a nodule from a sedimentary succession it also contains clays. This means it has a lower concentration of sulfide than the theoretical value and hence lower recoveries (Table II). Chalcopyrite, unlike pyrite does not contain sulfur-sulfur covalent bonds, and hence, should be recovered with AVS digestions. However, only 37% was recovered under these conditions while the remainder required chromous chloride digestion. Since chalcopyrite is usually found intergrown with pyrite this behaviour is likely to be due to the contamination of the sample with pyrite. This was later confirmed by XRD. The other mineral samples which contain a metal to sulfur ratio of 1:1, reacted as expected and in a similar manner to that of iron monosulfide minerals (Table II). Galena was digested completely in hot and cold acid digestions, while sphalerite required heating for complete recovery (Table II). The recovery of the non-iron monosulfide minerals (Table I and II) during AVS digestions is not unexpected, but it does indicate that the DOS ratio may be distorted. Hence, for a DOS ratio to accurately represent the sulfidation of the sediment, the concentration of the HCl-soluble metal fraction should be analsyed for other metals in addition to iron, as their sulfides contribute to the sulfide pool. This may be significant in DOS ratios in deposits such as the Kupferschiefer (copper shales) of Germany where large concentrations of copper are precipitated as sulfides (*30*), or in sediments where anthropogenic influences are high (*31*). The DOP ratio will also be distorted as small amounts of iron are also included in chalcopyrite and other sulfide minerals, for example sphalerite can contain up to 15% iron (*30*).   Thus DOP ratios need to take the proportion of other minerals into account because they contribute to the available iron pool but their sulfides are not included in the pyrite pool.

The artificial greigite showed recoveries of 34% in cold AVS and 88% in hot AVS digestions. In contrast at least 81% of iron monosulfide was recovered in each digestion (Table II). This clearly demonstrates that these two compounds cannot be separated on the basis of a cold and hot AVS digestion. Mixtures of different pyrite samples with greigite or FeS showed good separation when treated in hot AVS and cold chromous chloride digestions. During hot acid digestions, 92% FeS (+/- 6%) and 87% (+/- 21%) greigite was recovered while pyrite was only recovered during cold chromous chloride digestions (90% (+/- 20%)).

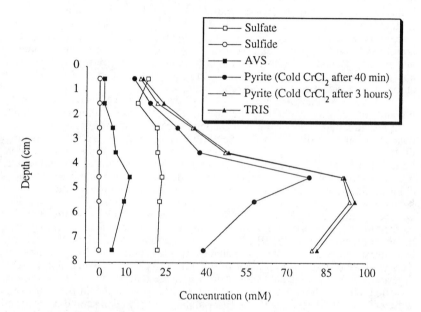

Figure. 2.- Depth distribution of sulfur pools in a short sediment core from the Tamar Estuary, Uk, demonstrating the variation in pyritic sulfur concentrations after digestions for 40 minutes and 3 hours. (TRIS- Total Reduced Inorganic Sulfide)

Table II- Recoveries of Various Mineral Samples under Different Digestion Conditions

Sample	Method	Digestion Time	Percentage recovery	Standard deviation
Pyrite (euhedral) FeS_2	Hot CrCl2	4 hours	94	24
	Cold CrCl2	4 hours	92	29
	Hot AVS	2 hours	0	0
Pyrite (rock nodule)	Hot CrCl2	4 hours	106	23
	Cold CrCl2	4 hours	91	25
	Hot AVS	2 hours	0	0
Pyrite (artificial)	Hot CrCl2	2 hours	93	9
	Cold CrCl2	2 hours	75	25
	Hot AVS	2 hours	0	0
Pyrite (vein)	Hot CrCl2	4 hours	105	19
	Cold CrCl2	4 hours	85	12
	Hot AVS	2 hours	0	0
Sphalerite (ZnS)	Hot CrCl2	4 hours	93	12
	Cold CrCl2	4 hours	92	13
	Hot AVS	2 hours	111	14
	Cold Avs	2 hours	68	15
Chalcopyrite (CuFeS2)	Hot CrCl2	4 hours	108	23
	Cold CrCl2	4 hours	125	26
	Hot AVS	2 hours	12	7
	Cold Avs	2 hours	37	13
Galena (PbS)	Hot CrCl2	4 hours	106	44
	Cold CrCl2	4 hours	80	13
	Hot AVS	2 hours	119	8
Marcasite FeS_2	Hot CrCl2	4 hours	72	13
	Cold CrCl2	4 hours	61	3
	Hot AVS	2 hours	1	1
Iron Monosulfide FeS	Hot CrCl2	1hour	87	13
	Cold CrCl2	1 hour	87	7
	Hot AVS	1 hour	81	3
	Cold Avs	1 hour	104	14
Greigite Fe_3S_4	Hot CrCl2	1 hour	78	21
	Cold CrCl2	1 hour	91	24
	Hot AVS	1 hour	88	4
	Cold Avs	1 hour	34	12
Polysulfide Solution $S_n - S^{2-}$	Hot CrCl2	1 hour	93	4
	Cold CrCl2	1 hour	50	1
	Hot AVS	1 hour	11	2
Sulfur suspension $S°$	Hot CrCl2	1 hour	85	7
	Cold CrCl2	1 hour	0	0
	Hot AVS	1 hour	0	0
Flowers of Sulfur $S°$	Hot CrCl2	1 hour	105	28
	Cold CrCl2	1 hour	1	2
	Hot AVS	1 hour	0	0

Samples of other minerals and various pyrites showed similar behaviour when treated in this way.

The sulfur suspension in acetone and the flowers of sulfur were recovered only in hot chromous chloride digestions (Table II). Extractions for elemental sulfur using cyclohexane and dichloromethane were not particularly successful due to a poor partition between sulfur and the solvents, especially in aqueous solutions. Separation of the phases is also difficult. After extracting sulfur for 24 hours the maximum sulfur recovered, when compared to the recovery of the whole sample under hot chromous chloride conditions, was 72% using dichloromethane. Dichloromethane consistently gave higher recoveries than cyclohexane, although both solvents had similar problems with phase separation. Recoveries were especially poor when the extractions were applied to cultures of sulfate-reducing bacteria. Dichloromethane only recovered 20% of the elemental sulfur in this sample, while cyclohexane did not recover any. When applied to Recent and ancient sedimentary samples the problem of separating the solvent from the sample was increased as there was often a suspension of sample at the interface. However, hot chromous chloride digestions recover the majority of elemental sulfur (between 85% and 105%, Table II) and as most of the pyrite can be removed without heating (75% to 92%, Table II) a cold followed by a hot chromous chloride digestion would provide a reasonable estimate of the concentration of elemental sulfur in a sample. This was tested with a mixture of elemental sulfur and pyrite and although the recovery of pyritic sulfide increased slightly (possibly due to the recovery of some sulfur during cold chromous chloride digestions) this was within experimental error.

Polysulfide breaks down into hydrogen sulfide and elemental sulfur upon acidification (20). This means that the sulfide evolved during chromous chloride digestions is separated equally between cold and the additional sulfide being recovered during hot digestions (Table II). AVS digestions produce less sulfide than cold chromous chloride digestions. The increased yield in cold chromous chloride digestions is due to the increased reducing capacity of chromous chloride. Heating in chromous chloride digestions breaks the sulfur-sulfur bonds in the polysulfide ion. Extractions on a polysulfide-elemental sulfur mixture confirmed that dichloromethane is better at extracting sulfur than cyclohexane.

When the hot AVS, and cold followed by hot chromous chloride digestions are applied to cultures of sulfate-reducing bacteria modelling sedimentary pyrite formation (Figure 3) iron sulfide minerals including pyrite, elemental sulfur and hydrogen sulfide can be observed coexisting within 3 days. This digestion scheme (Figure 4) allows the chemical changes in the sulfides, and the ratio of various reduced sulfur species to pyrite to be monitored, despite the amorphous nature of the solid phases, and the presence of organic films draping the minerals. The characterisation of solid phases was subsequently confirmed by electron microprobing precipitates from the samples. Recent sediments when treated in the same way also produce reliable results and efficiently separated different sulfide pools, enabling the formation of authigenic sulfide minerals to be studied (Figure 2).

Conclusions

In conclusion, a hot acid digestion in addition to a cold/hot chromous chloride digestion appears to be applicable to all ages of sulfide mineralisation, allowing the separation of reduced sulfide pool, into AVS, pyritic sulfide and elemental sulfur. There will, however, be a considerable difference between the digestion times required to recover mineral sulfides during early bacteria sulfide formation (ie. in bacterial cultures and $^{35}SO_4$ radiotracer incubations) and those for mature sulfide minerals (ie. Recent sedimentary framboids, geological samples and fossils). The

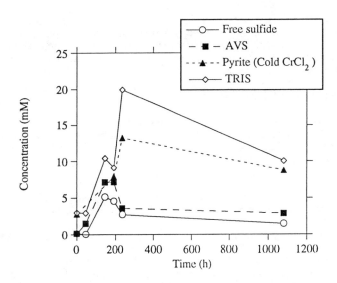

Figure 3.- Analysis of a culture of sulfate-reducing bacteria, showing the changing concentrations of reduced sulfide pools over 10 days (TRIS- Total Reduced Inorganic Sulfide)

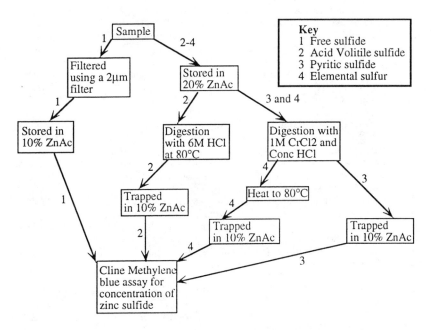

Figure 4.- Schematic representation of the protocol suggested for ancient, Recent and bacterially produced sulfides.

hot chromous chloride digestion, also, appears to be a more accurate way of determining the concentration of elemental sulfur in the sample, than the use of solvent extraction of sulfur. As each environment or experimental sample varies it is recommended that precise digestion times are established prior to the determination of sulfur species.

Acknowledgments.

This work is funded by NERC studentship GT4/92/120. We would like to thank Prof. D.E.G. Briggs and Dr Simon Bale, Cindy Rice and Martin Schoonen for their critical comments, Dr Peter Wellsbury and Mr Toni Woodward for the sediment samples.

Literature Cited

1.   Thode-Anderson, S.; Jorgensen, B.B., *Limnol. Oceanogr.*, **1989**, *34*, pp. 793-806.
2.   Parkes, R.J.; Buckingham, W.J., *Proc. Int. Sym.Bact.Ecol.*, **1986**, *4* , pp. 617-624.
3.   Froelich, P.N., G.P. Klinkhammer, M.L. Bender, Luedtke, N.A., G.R. Heath, Doug Cullen, P. Dauphin, D. Hammond, B. Hartman, and V.Maynard., *Geochim. Cosmochim. Acta*, **1979**, *43*, pp. 1075-1090.,
4.   Jørgensen, B.B., *Mar. Biol.,* **1977**, *41*, pp. 7-17.
5.   Aller, R.C., In *Animal-sediment relations. The biogenic alteration of sediments;* McCall, P.L.; Tevesz, M.J.S., Ed.;   Plenum Press:Topics in Geobiology, Vol. 2; **1982**, pp. 96-102.
6.   Nedwell, D.B., *Adv. Bact. Ecol.* **1984**, *7*, pp. 93-131.
7.   Kohen, M.E., *Nature*, **1989**, *341*, pp. 640-641.
8.   Canfield, D.E., and R. Raiswell, In: *Taphonomy: Releasing the data*
locked   *in the fossil reoord*, P.A.Allison and D.E.G.Briggs, Ed., Plenum
Press,   New York, **1991**, pp. 337-387
9.   Postgate, J.R.,*The sulfate-reducing bacteria.*; Cambridge University Press: **1979**,pp.208
10.   Issatchenko, B.L., The deposits of iron sulfide in bacteria.; *Bull. Jord. Inmp. Bot.*, Vol. 12; : St Petersburg,**1912**, 12.
11.   Raiswell, R.; Buckley, F.; Berner, R.A.; Anderson, T.F., *Jour. Sed. Pet.*, **1988**, *58*, pp. 812-819.
12.   Boesen, C.; Postma, D., *Am. J . Sci*, **1988**, *288*, pp. 575-603.
13.   Zhabina, N.N.; I.I. Volkov, In *Environmental biogeochemistry and geomicrobiology; 3. Methods, metals and assessment* , W.E. Krumbein, Ed.; Ann Arbor Science: **1978**, pp. 735-746.
14.   Fossing, H; Jørgensen, B.B., *Biogeochem.*, **1989**, *8*, pp. 205-222
15.   Cornwell, J.C.; J.W. Morse, *Mar.Chem.,* **1987**, *22*, pp. 193-206
16.   Howarth, R.W.; Jørgensen, B.B., *Geochim. Cosmochim. Acta*, **1984**, *48*, pp. 1807-1818
17.   Cline, J.D., *Limnol. Oceanogr.*, **1969**, *14*, pp. 454-458.
18.   Mossman, J.-R., Aplin, A.C., Curits, C.D.; Coleman, M.L., *Geochim. Cosmochim. Acta*, **1991**, *55*, pp. 3581-3592.
19.   Krairaponond, N., DeLaune, R.D.; Patrick, W.H., *Org. Geochem.*, **1992**, *18*, pp. 489-500.
20.   Luther, G.W., A.E. Giblin, and R. Varsolona, *Limnol. Oceanogr.*, **1985**, *30*, pp. 727-736.
21.   Canfield, D.E.; Raiswell, R.; Westrich, J.T.; Reaves, C.M.; Berner,
R.A.,   *Chem.Geol.*, **1986**, *54*, pp. 149-155.

22. Pruden, G., And C. Bloomfield, *Analyst*, **1968**, *93*, pp. 532-534
23. Rice, C.A., M.L. Tuttle, and R.L. Reynolds, *Chem.Geol.*, **1993**, *107*, pp. 83-95.
24. Jorgensen, B.B., *Nature*, **1982**, *296*, pp. 643-645.
25. Chanton, J.P.; C.S. Martens, *Biogeochem.*, **1985**, *1*, pp. 375-383.
26. Troelsen, H.; Jørgensen, B.B., *Estuar., Coast. Shelf Sci.*, **1982**, *15*, pp. 255-266.
27. Schimmelmann, A.; Kastner, M., *Geochim., Cosmochim. Acta*, **1993**, *57*, pp. 67-78.
28. Berner, R.A., *Mar. Geol.*, **1964**, *1*, pp. 117-140.
29. Wada, H., *Bull. Chem. Soc. Japan*, **1977**, *50*, pp. 2615-2617
30. Deer, W.A.; R.A. Howie; J. Zussman, *An introduction to the rock-forming minerals.*; Longman: London,**1966**,pp. 445-462
31. Bryan, G.W. and W.J. Langton, *Environ. Poll.*, **1992**, *76*, pp. 89-131.

RECEIVED May 11, 1995

OXIDATIVE TRANSFORMATION OF HYDROGEN SULFIDE

Chapter 14

Oxidation of Hydrogen Sulfide by Mn(IV) and Fe(III) (Hydr)Oxides in Seawater

Wensheng Yao and Frank H. Millero

Rosenstiel School of Marine and Atmospheric Science,
University of Miami, Miami, FL 33149

We review here recent results on the rates of oxidation of hydrogen sulfide by Mn and Fe (hydr)oxides as a function of environmental variables, including pH (3.0 - 9.0), temperature (5 - 45 °C), and ionic strength (0 - 4 m) in seawater media. The discussion includes the effects of major ions (Ca^{2+}, Mg^{2+}, SO_4^{2-}, $B(OH)_4^-$ and HCO_3^-), minor solutes (PO_4^{3-}, $Si(OH)_4$, NH_4^+, Fe^{2+} and Mn^{2+}) and some organic ligands (EDTA, TRIS, oxalate, humic acid and fulvic acid) on the rates of oxidation. The mineral phases used included δMnO_2, $\alpha FeOOH$, $\beta FeOOH$, αFe_2O_3, and freshly precipitated $Fe(OH)_3(s)$. The reactions are surface-controlled processes, and first order, with respect to both total hydrogen sulfide, and metal oxides. The strong pH dependence on the reaction rates was attributed to the formation of surface complexes. The mode of formation of $Fe(OH)_3(s)$ seemed to control its reactivity. The major product from H_2S oxidation by Mn(IV) and Fe(III) (hydr)oxides was elemental sulfur, except at high MnO_2/H_2S ratios when $S_2O_3^{2-}$ and SO_4^{2-} became important.

Hydrogen sulfide forms in a number of areas in natural waters. In anoxic marine environment, the production of H_2S is primarily due to bacterial anaerobic respiration using sulfate as an electron acceptor,

$$SO_4^{2-} + 2CH_2O \rightarrow H_2S + 2HCO_3^- . \qquad (1)$$

Hydrogen sulfide is generated in hydrothermal systems by geochemical processes, and reacts with metal ions depositing metal sulfide minerals, mainly as pyrite,

$$2SO_4^{2-} + 4H^+ + 11Fe_2SiO_4 \rightarrow FeS_2 + 2H_2O + 7Fe_3O_4 + 11SiO_2 . \qquad (2)$$

The formation of H_2S and its removal through oxidation and precipitation can affect

0097–6156/95/0612–0260$12.00/0

the cycles of other elements (trace metals, carbon, etc.) in the ocean and the global sulfur budget. Although a number of kinetic studies on the oxidation of H_2S by dissolved oxygen have been made both in the laboratory and in the field (*1-15*), relatively few have dealt with oxidation by Mn and Fe oxides (*16-24*). At some oxic/anoxic interfaces, the downward flux of oxygen is insufficient to oxidize the upward flux of hydrogen sulfide (*25,26*). The oxides of Mn(IV) and Fe(III) may be alternative oxidants. Solid-phase manganese and iron-rich particles are often found at high concentrations above the oxic/anoxic (O_2/H_2S) interface in marine and freshwater systems (*27-31*). Below the redox boundary, high concentrations of dissolved Mn^{2+} and Fe^{2+} are produced from the reductive dissolution of Mn and Fe oxides by H_2S. When the dissolved Mn^{2+} and Fe^{2+} diffuse above the interface, they are oxidized back to oxides. These redox cycles of Mn and Fe can have an important impact on the recycling of other minor elements, such as PO_4^{3-} and trace metals, in water column (*32,33*), sediments (*16,34*) and hydrothermal systems (*35,36*). For example, the unique PO_4^{3-} distribution at the oxic/anoxic interface in the Black Sea has been suggested to be the result of the redox reactions of Fe and Mn (*33*). The reaction between H_2S and Fe oxides is the starting point of a complex sequence of processes resulting in the formation of sedimentary pyrite (*24*).

Burdige and Nealson (*18*) made the first chemical and microbiological study of sulfide-mediated manganese reduction. Studies by Aller and Rude (*16*) suggested that solid phase sulfides can be completely oxidized by manganese and bacteria in anoxic marine sediments. Earlier studies on the kinetics and mechanism of the sulfidation of goethite (αFeOOH) were made by Rickard (*24*) and Pyzik and Sommer (*23*). More recently, Canfield and coworkers (*19,20*) studied the reactivity of the solid-phases of iron with sulfide in marine sediments. Dos Santos Afonso and Stumm (*21*) investigated the reductive dissolution of hematite (αFe$_2$O$_3$) by H_2S and interpreted the rate law by assuming a surface-controlled reaction. Pelffer *et al.* (*22*) studied the initial reaction between hydrogen sulfide and the surface of lepidocrocite (γFeOOH), and applied a surface speciation model (*21*) to explain the pH maximum of the reaction rate. Biber *et al.* (*17*) examined the effects of some inorganic and organic ligands on the reductive dissolution of αFe$_2$O$_3$, and αFeOOH by H_2S. Recently, we have expanded these previous studies by examining the effect of pH, temperature, ionic strength and media composition on the rate of oxidation of hydrogen sulfide (*37-39*). In this review paper we summarize our recent studies (*37-39*), together with the work of others (*16-24*), on the chemical oxidation of H_2S by Mn(IV) and Fe(III) (hydr)oxides in seawater.

Experimental Methods

We summarize here the methods used in our studies since some of our results have not been published. The setup of the oxidation experiments was given in Yao and Millero (*37*). The rates of oxidation of H_2S were followed by monitoring the decrease of total sulfide concentration with the methylene blue method of Cline (*40*). δMnO_2 was prepared according to Murray (*41*) and had a specific surface area (BET) = 206 and 341 m^2 g^{-1} for aged and freshly precipitated solids, respectively. αFeOOH,

βFeOOH and αFe$_2$O$_3$ were prepared according to Schwertmann and Cornell (42) and had a specific surface area (BET) = 48.8, 114.2 and 22.7 m^2 g^{-1}, respectively. The hydrous Fe(III) oxide, Fe(OH)$_3$(s), discussed in this paper was freshly derived from the hydrolysis and oxidation of Fe(II) in the solutions and the specific surface area was assumed to be 120 m^2 g^{-1} (43). The elemental sulfur (S^o) formed from the oxidation of H$_2$S was determined by high-performance liquid chromatography (HPLC) based on derivatization with triphenyphosphine (TPP) to form triphenylphosine sulfide (TPPS) (44). The concentrations of sulfite and thiosulfate were determined by HPLC after derivatization with 2,2-dithiobis (5-nitropyridine) (45). The concentration of SO$_4^{2-}$ was determined using ion chromatography.

Overall Rate of Oxidation of H$_2$S

The overall rate equation can be expressed by

$$-d[H_2S]_T/dt = k \, [H_2S]_T^a \, [MO_x]^b \tag{3}$$

where [H$_2$S]$_T$ is the total concentration of hydrogen sulfide, k is the overall rate constant, a and b are the orders with respect to H$_2$S and metal oxides MO$_x$, respectively. When the experiments are performed with an excess of metal oxides, the rate equation can be reduced to

$$-d[H_2S]_T/dt = k'[H_2S]_T^a \tag{4}$$

The order with respect to hydrogen sulfide was determined by fitting the data to various rate equations with different values of a. Plots of ln [H$_2$S]$_T$ versus time produced straight lines with the equivalent slope for different initial [H$_2$S]$_T$ concentrations, indicating that the reactions are first order with respect to [H$_2$S]$_T$ (37-39).

The value of the pseudo-first-order rate constant k' is related to k by

$$k' = k \, [MO_x]^b \tag{5}$$

The value of b was determined from measurements of k' at different concentrations of MO$_x$. The plots of log k' versus log [MO$_x$] for reactions with MnO$_2$ and Fe (hydr)oxides gave slopes of 1.0 $\pm$ 0.1 and demonstrate that the reactions are first order with respect to the concentration of the metal oxides (37-39). These results indicate that the reactions are surface dependent, since the oxide surface area is proportional to the amount of metal oxides added.

The effect of temperature on the oxidation of H$_2$S by Mn(IV) and Fe(III) (hydr)oxides in seawater was examined from 5 to 45 ^{o}C. The apparent activation energy of the reaction with MnO$_2$ was found to be 12 $\pm$ 2 kJ mol^{-1} (37). It is much lower than the value of 66 $\pm$ 5 kJ mol^{-1} for the oxidation of H$_2$S by dissolved oxygen in seawater (7). The apparent activation energy of the reaction between H$_2$S and Fe(OH)$_3$(s) was found to be 4.5 $\pm$ 0.2 kJ mol^{-1} (38). The value for less reactive αFeOOH was found to be 10 times higher (40 $\pm$ 2 kJ mol^{-1}) (39).

The reactivity of MnO_2 has been found to change upon aging (*37,46*), probably due to the gradual recrystallization of the oxides and/or a decrease in the surface area by the coalescence of particles. Manganese dioxide suspensions used in our previous study (*37*) had been aged for years. The rate constant for H_2S oxidation in seawater (pH = 8.17) at 25 °C was found (*37*) to be 436 M^{-1} min^{-1}, or 0.0244 m^{-2} l min^{-1} when [MnO_2] is expressed in surface area (m^2/l). Later studies using freshly made MnO_2 gave a rate constant in seawater (pH = 8.0) at 25 °C of 4.15 x 10^3 M^{-1} min^{-1} or 0.140 m^{-2} l min^{-1}. The oxidation rate of hydrogen sulfide by δMnO_2 was found to be 5 times faster than that by γ-MnOOH (based on per unit surface area). Both microbial manganese reduction coupled to the oxidation of organic matter and chemical reduction of manganese oxides by Fe^{2+} have been found to be strongly dependent upon the manganese oxide mineralogy and crystal structure (*47*).

The rate constant for H_2S oxidation by freshly precipitated $Fe(OH)_3(s)$ in seawater (pH = 8.0) at 25 °C was found to be 1.48 x 10^2 M^{-1} min^{-1} or 0.0138 m^{-2} l min^{-1}, which is one order of magnitude lower than that by δMnO_2. The magnitude of the rate constants (based on per unit surface area) of H_2S oxidation by different forms of Fe(III) (hydr)oxides was in the order $Fe(OH)_3(s) > \beta FeOOH > \alpha FeOOH > \alpha Fe_2O_3$ (*39*). The rates were found to increase with decreasing free energy of the half redox reactions of $Fe(OH)_3(s)$, FeOOH(s) and $Fe_2O_3(s)$. The free energy values were taken from the literature (*48,49*). These results are similar to other studies (*17-24*). It is difficult, however, to make meaningful comparison of the absolute values of rate constants due to the differences in experimental conditions and treatment of the data.

Effect of Ionic Strength and Major Sea Salts

The effect of ionic strength on the rates of oxidation in seawater and NaCl solutions at 25 °C is shown in Figure 1. The rate of H_2S oxidation by MnO_2 was not strongly dependent on ionic strength (Figure 1a). The reaction rate with $Fe(OH)_3(s)$ decreased with increasing ionic strength in seawater (Figure 1b). The effect of ionic strength on the rate of reductive dissolution of manganese oxides by some organics was also very small (*46*). On the contrary, log k for the oxidation of H_2S by dissolved O_2 has been found to increase in a linear manner with the square root of ionic strength (*5,7*).

At the same ionic strength the oxidation rates in NaCl solution for both Mn and Fe oxides were faster than those in seawater (Figure 1). To elucidate these results, the effect of the major ionic components of seawater on the rates of H_2S oxidation was examined. A series of measurements were made in 0.57 M NaCl solution (the Cl^- concentration in seawater) with the addition of the major sea salts at their oceanic levels (Ca^{2+}: 0.011 M, Mg^{2+}: 0.055 M, SO_4^{2-}: 0.029 M). These results are shown in Figure 2.

For the reaction between H_2S and MnO_2, the addition of SO_4^{2-} caused the rate to decrease slightly; while the addition of Ca^{2+} and Mg^{2+} to the NaCl solution decreased the reaction rate by nearly 40% (Figure 2a) (*37*). The SO_4^{2-} ion was found not to be adsorbed on MnO_2 in the pH range of natural waters (*50*). The decreases in the rates caused by Mg^{2+} and Ca^{2+} were related to their adsorption on the surface of MnO_2 and the resultant decrease of the surface sites. However, only 8.4% of the surface sites of MnO_2 in major ion seawater at pH 8 were found to be complexed by

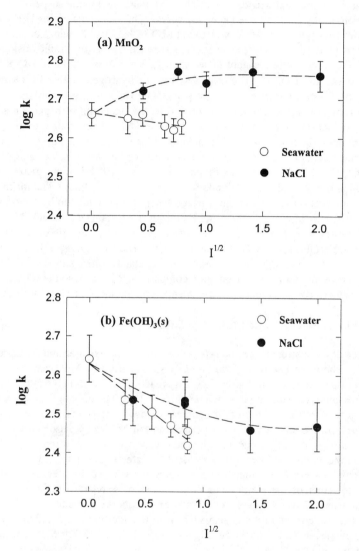

Figure 1. The effect of ionic strength on the rate of H_2S oxidation (k, M^{-1} min^{-}1) in seawater and NaCl solution at 25 °C: (a) MnO_2 (at pH = 8.17); (b) $Fe(OH)_3$(s) (at pH = 7.50) (*37, 38*).

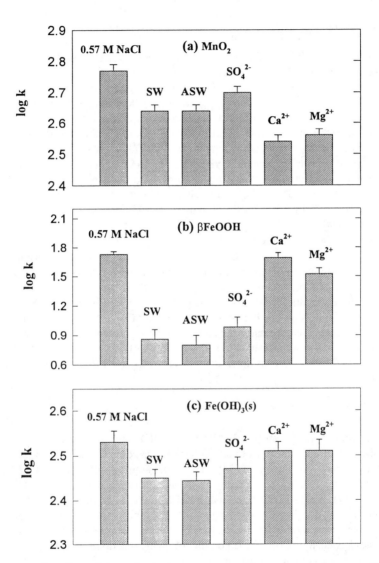

Figure 2. The effect of the major ionic components of seawater on the rates of H_2S oxidation (k, M^{-1} min^{-1}): (a) MnO_2 (at pH = 8.17); (b) $\beta FeOOH$ (at pH = 7.50); (c) $Fe(OH)_3(s)$ (at pH = 7.50) (*37-39*).

Mg^{2+}, and 4.6% by Ca^{2+} (50). On a molar basis, the effect of Ca^{2+} was 5 times larger than Mg^{2+}. This is consistent with the fact that Ca^{2+} binds nearly 5 times more strongly with MnO_2 ($p*K_{Ca}^{int} = 3.3$) than does Mg^{2+} ($p*K_{Mg}^{int} = 3.9$) (50). The effect of the individual Mg^{2+} and Ca^{2+} ions was less than expected in seawater, possibly due to the interaction of Mg^{2+} and Ca^{2+} with SO_4^{2-}. The rate of the reaction in artificial seawater (Na^+, Mg^{2+}, Ca^{2+}, Cl^-, SO_4^{2-}) was in good agreement with the rate in real seawater (Figure 2a). Two other major anions in seawater, HCO_3^- and $B(OH)_4^-$, were found to have no effect on the oxidation rate of H_2S by MnO_2 in NaCl solutions.

For the reactions between H_2S with Fe(III) (hydr)oxides, the effect of Mg^{2+} and Ca^{2+} was small relative to NaCl; while SO_4^{2-} was the major factor causing the rates to decrease in seawater (Figure 2b,c) (38,39). By taking the intrinsic equilibrium constants from Balistrieri and Murray (51) and Van Geen et al. (52), the distribution of the surface species of αFeOOH (500 μM) at pH 8 in seawater (S = 35) was calculated by using the solution equilibrium computer program HYDRAQL (53). Results indicate that 18% of the surface sites are bound by hydrogen, 31% by Mg^{2+}, 32% by SO_4^{2-}, 12% by Ca^{2+}, 4% by Cl^- and 1% by CO_3^{2-} (39). The effect of SO_4^{2-} on the reaction with αFeOOH was smaller than that with βFeOOH (39). No studies have been made on the surface chemistry of βFeOOH with the major ions of seawater. It is likely that βFeOOH has more affinity for SO_4^{2-} than does αFeOOH. SO_4^{2-} was also found to greatly reduce the rate of reductive dissolution of αFe_2O_3 by H_2S at pH = 4 (17). By taking the intrinsic equilibrium constants from Zachara et al (54), the model results for $Fe(OH)_3(s)$ at pH = 7.5 in seawater (S = 35) show that 35% of the surface sites are bound by Mg^{2+}, 6% by Ca^{2+}, 49% by SO_4^{2-} and 3% by CO_3^{2-}. The rates for artificial seawater were in agreement with the rates in real seawater within the experimental error (Figure 2b,c). Major anions $B(OH)_4^-$ and HCO_3^- showed little effect on the oxidation rates by Fe(III) (hydr)oxides in seawater (38,39).

Effect of pH and the Reaction Mechanism

The rates of H_2S oxidation by Mn(IV) and Fe(III) (hydr)oxides were found to be highly dependent upon the pH of the solution . A maximum oxidation rate for MnO_2 was at a pH near 5.0 (Figure 3) (37). The maximum rate for hydrous Fe(III) oxide was found at pH = 6.5 (38). The maximum value for αFeOOH and βFeOOH was at pH = 7.0 (39), which is the same as γFeOOH (22). The influence of pH on the reaction rates is due in part to changes in the chemical speciation of H_2S and the surface sites on the oxides. A mechanism for the reactions between hydrogen sulfide and oxides has been proposed based on the following reaction sequence (21,22):

(i) surface complex formation

$$>MOH + HS^- \underset{k_{-1}^0}{\overset{k_1^0}{\Leftrightarrow}} >MS^- + H_2O \tag{6}$$

(ii) electron transfer

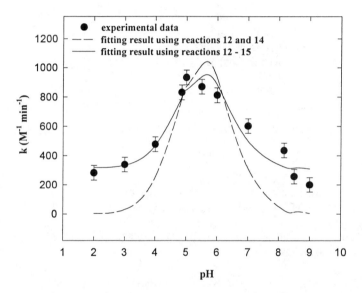

Figure 3. The effect of pH on the rate constants (k, M^{-1} min^{-1}) for the oxidation of H_2S in seawater by MnO_2 at 25 °C (*37*).

$$>MS^- \underset{k_{-et}}{\overset{k_{et}}{\Leftrightarrow}} >M^{II}S \tag{7}$$

(iii) release of the oxidized product and detachment of M(II)

$$>M^{II}S + H_2O \underset{k_{-2}^0}{\overset{k_2^0}{\Leftrightarrow}} >M^{II}OH_2^+ + S^{\cdot\cdot} \tag{8}$$

$$>M^{II}OH_2^+ \overset{k_3^0}{\longrightarrow} \text{ new surface site} + M^{2+} \tag{9}$$

According to this reaction scheme, the consumption rate of HS$^-$ in terms of the surface species >MS$^-$ can be derived as

$$R = -\frac{d[HS^-]}{dt} = \frac{k_{et}\,k_2 k_3}{k_2\,k_3 + k_{-et}(k_3 + k_{-2}[S^{\cdot\cdot}])}\,\{>MS^-\}$$

$$= k_1\,\{>MS^-\} \tag{10}$$

assuming a steady-state concentration for intermediate S$^{\cdot\cdot}$. In step (i), other surface sites (such as >MOH$_2^+$ and >MO$^-$) can also react with HS$^-$ as well as H$_2$S to form surface complexes such as >MS$^-$, >MHS and >MOHS^{2-}. An expression similar to equation 10 can be derived for surface species >MHS and >MOHS^{2-}. The overall rate equation can be written as

$$-d[H_2S]_T/dt = k_1\{>MS^-\} + k_2\,\{>MHS\} + k_3\{>MOHS^{2-}\} \tag{11}$$

The mechanism discussed above can be used to examine the observed pH dependence of the rate of oxidation of H$_2$S with Fe and Mn oxides (here we take MnO$_2$ as an example). Four possible reactions for the formation of surface complexes, >MnS$^-$, >MnHS and >MnOHS^{2-}, followed by electron transfer can be expressed as:

$$>MnOH + HS^- \overset{K_1}{\Leftrightarrow} >MnS^- + H_2O \tag{12}$$

$$>MnOH + H_2S \overset{K_2}{\Leftrightarrow} >MnHS + H_2O \tag{13}$$

$$>MnO^- + H_2S \overset{K_3}{\Leftrightarrow} >MnS^- + H_2O \tag{14}$$

$$K_4$$
$$>MnO^- + HS^- \Leftrightarrow >MnOHS^{2-} \tag{15}$$

where K_1, K_2, K_3 and K_4 are the formation constants of surface species (e.g. $K_1 = k_1^0/k_{-1}^0$, in equation 6). Two more reactions between $>MnO^-$ and HS^- to form $>MnHS$ and $>MnS^-$ are omitted since they can be related to equation 13 and 14 using ionization constants for H_2S and the surface of MnO_2 (K_1^* and $*K_{a2}^S$, see below). The substitution of the concentration of the surface complexes (equation 12-15) into equation 11 gives

$$-d[H_2S]_T/dt = k_1K_1\{>MnOH\}[HS^-] + k_1K_3\{>MnO^-\}[H_2S] + k_2K_2\{>MnOH\}[H_2S]$$
$$+ k_3K_4\{>MnO^-\}[HS^-] \tag{16}$$

The pH dependence of the H_2S speciation is related to the following ionization equilibrium:

$$K_1^*$$
$$H_2S \Leftrightarrow HS^- + H^+ \tag{17}$$

The fractions of H_2S and HS^- can be calculated

$$\alpha_{H_2S} = 1/(1+K_1^*/[H^+]) \tag{18}$$
$$\alpha_{HS^-} = 1/(1+[H^+]/K_1^*) \tag{19}$$

where K_1^* is the stoichiometric dissociation constant for the ionization of H_2S (*55*). Similarly, the pH dependence of the surface speciation of MnO_2 is related to the following ionization equilibrium:

$$*K_{a1}^S$$
$$>MnOH_2^+ \Leftrightarrow >MnOH + H^+ \tag{20}$$

$$*K_{a2}^S$$
$$>MnOH \Leftrightarrow >MnO^- + H^+ \tag{21}$$

where $*K_{a1}^S$ and $*K_{a2}^S$ are the conditional acidity constants of the positively and negatively charged surface groups. It is well known (*50*) that the surface of MnO_2 is very acidic compared with other oxide surfaces found in the marine environment. The $>MnOH_2^+$ species only exists in very acidic solution. The fractions of $>MnOH$ and $>MnO^-$ are given by:

$$\alpha_{>MnOH} = 1/(1 + *K_{a2}^S/[H^+]) \tag{22}$$
$$\alpha_{>MnO^-} = 1/(1 + [H^+]/*K_{a2}^S) \tag{23}$$

where $p*K_{a2}^S = 4.6$ (*37*). We assume that the concentration of H^+ at the interface is equal to the bulk value and make no electrostatic corrections.

Substituting equations 18,19, 22 and 23 into equation 16 gives

$$-d[H_2S]_T/dt = [k_1' \; \alpha_{>MnOH} \, \alpha_{HS^-} + k_2' \; \alpha_{>MnO^-} \, \alpha_{H_2S} + k_3' \; \alpha_{>MnOH} \, \alpha_{H_2S}$$
$$+ k_4' \; \alpha_{>MnO^-} \, \alpha_{HS^-}] \, [H_2S]_T \, S_T$$
$$= k \, [H_2S]_T \, [MnO_2] \qquad (24)$$

where $k_i' = k_i K_i$ (equation 16) and S_T is the total concentration of the reactive surface sites. Using the site density $N_S = 18$ sites nm^{-2} (56), the concentration of MnO_2 can be related to S_T by 1 M $MnO_2 = 0.76$ M S_T. Note that at a given pH equation 24 is equivalent to equation 3 with a and b equal to 1. Thus, the overall rate constant, k, can be represented as

$$k = k_1' \; \alpha_{>MnOH} \, \alpha_{HS^-} + k_2' \; \alpha_{>MnO^-} \, \alpha_{H_2S} + k_3' \; \alpha_{>MnOH} \, \alpha_{H_2S} + k_4' \; \alpha_{>MnO^-} \, \alpha_{HS^-} \qquad (25)$$

The fractional products ($\alpha_i \, \alpha_j$) in equation 25 are a function of pH. The variations in the products $\alpha_{>MnOH} \, \alpha_{HS^-}$ and $\alpha_{>MnO^-} \, \alpha_{H_2S}$ with pH pass through a maximum at pH $\approx$ 5.5 which is close to the experimental result (37). The fractional products of $\alpha_{>MnOH} \, \alpha_{H_2S}$ and $\alpha_{>MnO^-} \, \alpha_{HS^-}$ decrease and increase with pH, respectively, with no maximum (37). The values of k_1', k_2', k_3' and k_4' in equation 25 can be determined by fitting the fraction products ($\alpha_i \, \alpha_j$) to the experimental values of k. The overall pattern of the pH dependence on the reaction rate can be represented by using one rate $k_1' = 7.98 \times 10^4$ M^{-1} min^{-1} or $k_2' = 9.58 \times 10^2$ M^{-1} min^{-1}, or the combination of two rates $k_1' = 3.97 \times 10^4$ M^{-1} min^{-1} and $k_2' = 4.79 \times 10^2$ M^{-1} min^{-1}, all of which give a maximum rate at pH = 5.5 and the same fitting result (dashed line in Figure 3). These results indicate that the formation of surface complex >MnOH + HS$^-$ and/or >MnO$^-$ + H_2S (which has opposite dependence on pH) play more important roles in the overall sulfide oxidation than the formation of >MnOH + H_2S and >MnO$^-$ + HS$^-$. The value of k_1' (for >MnOH + HS$^-$) is 100 times greater than k_2' (for >MnO$^-$ + H_2S) as the fraction product of {>MnO$^-$} [H_2S] is two orders of magnitude higher than that of {>MnOH} [HS$^-$] over the entire pH range (37). It is not possible to confirm which pair of these two reacting species is more important based on the kinetic results. In order to get a better fit of the experimental data at lower and higher pH the reactions between >MnOH with H_2S (k_3') and >MnO$^-$ with HS$^-$ (k_4') have to be included with either k_1' or k_2' (solid line in Figure 3), which gives values of $k_1' = 6.89 \times 10^4$ or $k_2' = 8.28 \times 10^2$ with the same values of $k_3' = 2.40 \times 10^2$ and $k_4' = 2.33 \times 10^2$ M^{-1} min^{-1} .

In summary, the formation of surface complexes discussed above can explain both the order and pH dependence of the reaction of metal oxides with H_2S. The earlier work (24) on the sulfidation of goethite proposed that the reaction mechanism involves the dissolution of goethite and the subsequent reaction in solution with dissolved sulfide species. This mechanism would lead to an increase in the reaction rate with a decrease in pH and no maximum value in the reaction rate. Another mechanism (23) was suggested based on the formation of an outer-sphere binuclear complex between HS$^-$ and surface ferric ion and subsequent reductive dissolution of surface Fe. An outer-sphere electron transfer is expected to be slower than an inner-sphere process (57). In addition, the formation of a binuclear complex requires more energy to remove simultaneously two central atoms from the crystalline lattice (58).

Effect of PO_4^{3-} and Minor Solutes

Some minor solutes, such as PO_4^{3-}, NH_4^+, $Si(OH)_4$, Mn^{2+}, and Fe^{2+}, usually have relatively high concentrations in anoxic environments. Measurements were made in seawater with the addition of these species to study their effects on the rate of H_2S oxidation (*37-39*).

Phosphate was found to inhibit the reactions due to the adsorption on the oxide surfaces (Figure 4). The addition of 10 μM phosphate in seawater (initial $[MnO_2]$ = 25 μM) caused the rate of H_2S oxidation by MnO_2 to decrease about 50% (Figure 4a). However, the adsorbed PO_4^{3-} occupied only about 1% of the total surface sites of MnO_2, reflecting the heterogeneity in the reactivity of surface sites. The addition of 25 μM PO_4^{3-} had the same effect as 10 μM. The addition of an excess of PO_4^{3-} showed no further effect probably due to the saturation of high energy surface sites. Phosphate was also found to inhibit the reaction between manganese oxides and hydroquinone and 2, 5-DiOH (*46*).

The addition of phosphate also caused the rate of H_2S oxidation by αFeOOH and βFeOOH to decrease (*39*). Biber *et al.* (*17*) found that PO_4^{3-} significantly reduced the rates of reductive dissolution of αFeOOH (at pH = 5) and αFe$_2$O$_3$ (at pH = 6) by H_2S. For hydrous Fe(III) oxide, the effect of phosphate depended on the condition used to form the $Fe(OH)_3(s)$ (Figure 4b). When PO_4^{3-} was added to the solution before the formation of $Fe(OH)_3(s)$ (500 μM), no effect on the reaction rate was found. This might be due to the fact that the added PO_4^{3-} (10 μM and 20 μM) was incorporated into the structure of hydrous oxide instead of being adsorbed on the surface. Formation of $FePO_4 \cdot 2H_2O(s)$ (log K_{sp} = -26.4) was unlikely based on thermodynamic calculations. As expected, the addition of PO_4^{3-} after the formation of hydrous Fe(III) oxide caused the rate of H_2S oxidation to decrease significantly.

A number of other minor solutes (50 μM NH_4^+, 50 μM $Si(OH)_4$, 25 μM Mn^{2+}, 2 μM Fe^{2+}) had no effect on the rate of H_2S oxidation by MnO_2. The reaction rate of H_2S with $Fe(OH)_3(s)$ showed a decrease with the addition of $Si(OH)_4$ probably due to its adsorption on the surface (*38*).

Effect of Organic Ligands

Organic material is usually found to be abundant in anoxic environments. Some organics, both naturally derived organics (10 mg/l humic acid and 10 mg/l fulvic acid) and model ligands (500 μM EDTA, 500 μM oxalate, 25 mM TRIS) were used to study the potential role of organic compounds on the H_2S oxidation with Mn(IV) and Fe(III) (hydr)oxides. The results for Fe(III) are shown in Figure 5. All the ligands showed small and no effect on the rate of H_2S oxidation by MnO_2 in seawater. This may be due to the fact that these ligands (as anions) have a small affinity for the surface of MnO_2 at the pH of seawater.

The organic ligands caused the reaction rates with βFeOOH in NaCl solution to decrease significantly due to their interactions with the oxide surface (Figure 5a) (*39*). However, their effects in seawater were found to be much smaller. The interactions between these organic ligands with the surface of βFeOOH in seawater might be inhibited by the presence of SO_4^{2-}. Biber *et al.* (*17*) have found that various

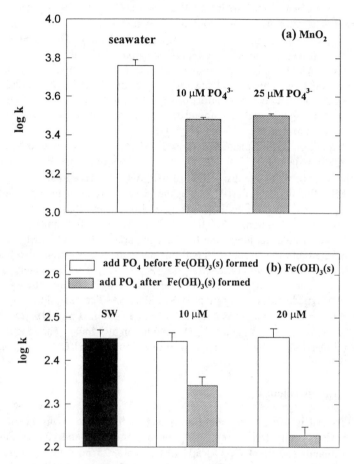

Figure 4. The effect of phosphate on the rate of H_2S oxidation (k, M^{-1} min^{-1}) in seawater at 25 °C: (a) MnO_2 (at pH = 7.30); (b) $Fe(OH)_3(s)$ (at pH = 7.50) (*38*).

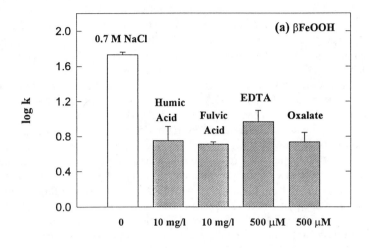

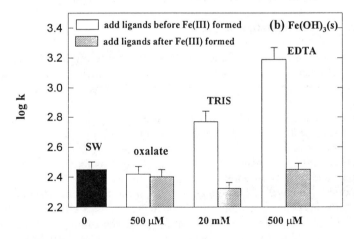

Figure 5. The effect of some organic ligands on the rate of H$_2$S oxidation (k, M^{-1} min^{-1}) in seawater at pH = 7.50 and 25 °C: (a) βFeOOH; (b) Fe(OH)$_3$(s) (*38,39*).

organic ligands, especially EDTA and oxalate, significantly inhibit the rate of dissolution of αFeOOH (at pH = 5) in the presence of H_2S by occupying the surface sites and shifting the major dissolution pathway from reductive to nonreductive.

For $Fe(OH)_3(s)$, oxalate showed no effect on the reaction rate while the effect of TRIS and EDTA depended on the formation conditions of $Fe(OH)_3(s)$ (Figure 5b) (*38*). When TRIS and EDTA were added before the addition of Fe(II) into the solution, the rate of sulfide oxidation was dramatically increased (about 2 times and 10 times, respectively). In the presence of TRIS, the extensive polymerization was prevented by forming outer-sphere complexes between TRIS and Fe(III) on the colloid surface and Fe(III) formed more reactive colloids with larger surface areas (*59,60*). EDTA strongly complexed Fe(III) formed from Fe(II) oxidation and kept Fe(III) in solution as an Fe(III)-EDTA complex. Not surprisingly, this complex had much higher reactivity than the solid phases of Fe(III). It is also known that some organic ligands increase the rate of Fe(II) oxidation by O_2, especially at low concentration of dissolved oxygen (*61*). These kinds of organically mediated processes could have important implications in enhancing the Fe redox cycle and thus the cycling of electrons in the environment. Luther *et al.* (*62*) have shown the importance of organic ligand complexes with Fe(II) and Fe(III) in the dissolution of Fe(III) minerals and pyrite in the salt-marsh sedimentary environment.

The addition of TRIS after the formation of $Fe(OH)_3(s)$ decreased the oxidation rate probably due to its adsorption on the surface (Figure 5b). The addition of EDTA after the formation of $Fe(OH)_3(s)$ had no effect (Figure 5b).

Products from the Oxidation of H_2S

Products formed from the oxidation of H_2S are of interest in understanding the reaction mechanism and the formation of metal sulfide minerals. The distributions of products were measured in seawater (pH = 7.5) at 25 °C and the results are shown in Figure 6.

The reaction between H_2S and MnO_2 was found to produce elemental sulfur (S°), thiosulfate and sulfate. When the ratio of $[MnO_2]/[H_2S]_T = 1$, S° was the dominant product (about 90%). The fractions of $S_2O_3^{2-}$ and SO_4^{2-} increased when the $[MnO_2]/[H_2S]_T$ ratio was increased (Figure 6a). SO_3^{2-} was found at the beginning of the reaction and quickly disappeared. This suggested that SO_3^{2-} might also be an initial product. Removal of SO_3^{2-} might be due to the further oxidation of SO_3^{2-} by MnO_2 to form SO_4^{2-} and reaction with S° (or S_x^{2-}) to form $S_2O_3^{2-}$. Perry *et al.* (*63*) showed that SO_3^{2-} reacted with HS^- to form $S_2O_3^{2-}$ in the cultures of *S. putrefaciens*, but no $S_2O_3^{2-}$ was found within 2 hours when we reacted HS^- with SO_3^{2-} in O_2 free seawater. Quantum mechanical considerations (*64*) have shown that Mn oxides, but not Fe(III) oxides, can be reduced and thus dissolved by SO_3^{2-}. A constant concentration of $S_2O_3^{2-}$ was reached after removal of SO_3^{2-}, which suggested that $S_2O_3^{2-}$ was not the initial product, but was formed from the reaction between S° and SO_3^{2-}. Burdige and Nealson (*18*) reported that only 47-56% of the removed sulfide could be accounted for as elemental sulfur collected on a 0.2 μm filter. This lower recovery of S° was unlikely due to the formation of polysulfides since the reaction was completed leaving no sulfide to form S_x^{2-}. Our measurements showed that the

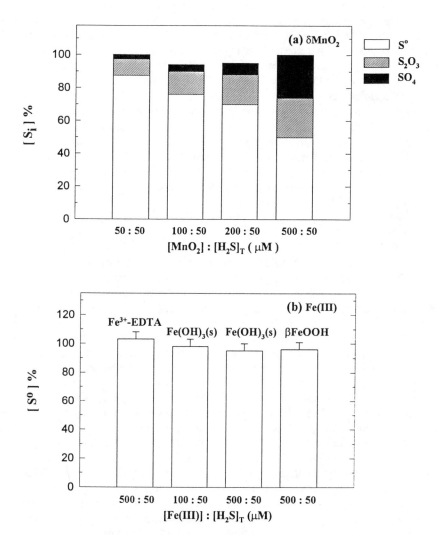

Figure 6. The distribution of products from H₂S oxidation in seawater at pH = 7.50 and 25 °C: (a) MnO₂; (b) Fe(III)-EDTA complex and Fe(III) oxides (*38*).

recovery of S^o from direct extraction by organic solvent ($CHCl_3$) was higher than from filtration, indicating that colloidal S^o might be too small to be retained by a 0.2 µm filter. Aller and Rude (16) showed that solid phase sulfides could be completely oxidized to SO_4^{2-} by manganese oxides in anoxic marine sediments.

Elemental sulfur (S^o) was found to be the dominant product (95 - 100%) from the oxidation of H_2S by Fe(III) (hydr)oxides independent of the ratio of reactants and the mineral phases (Figure 6b). Similar results have been reported by other studies (22-24). The formation of elemental sulfur from this redox reaction could be an important pathway leading to the formation of pyrite (24).

The differences in the formation of oxidation products of hydrogen sulfide by Mn(IV) and Fe(III) oxides have been examined by Luther (56) using the frontier-molecular-orbital theory approach. Based on the model, $\pi(p_y)$ to $\pi(d_{yz})$ electron transfer occurs from sulfide to Fe(III) resulting in the formation of S^o as the dominant product. For reaction between sulfide and Mn(IV), it is a σ to σ type of electron transfer resulting in the formation of a zero-valent sulfur atom. Further oxidation on the MnO_2 surface leads to elemental sulfur (S_8) and sulfate. These predicted products are in agreement with our measurements.

Summary

The oxidation of hydrogen sulfide by Mn(IV) and Fe(III) (hydr)oxides was found to be a surface-controlled process and first order with respect to both H_2S and oxides. The abiotic oxidation of H_2S with these metal oxides may play an important role in sediment diagenesis. Among the major ions in seawater, Ca^{2+} and Mg^{2+} decreased the rate of H_2S oxidation by MnO_2; while SO_4^{2-} caused the rate by Fe(III) (hydr)oxides to decrease. PO_4^{3-} was found to inhibit the reactions for both Mn and Fe oxides. These inhibiting effects were due to the interaction between these ions with the oxide surfaces. The oxidation rates were strongly pH dependent with maximum values near a pH of 5.0 for MnO_2 and 6.5 - 7.0 for Fe(III) (hydr)oxides. This pH dependence can be explained by a reaction mechanism based on the formation of surface complexes, which are sensitive to pH, followed by electron transfer. The reactivity of hydrous Fe(III) oxide was highly dependent on the conditions of its formation. For example, some organic ligands could enhance the Fe(II)-Fe(III) redox cycle and thus the cycling of electrons in the marine environment. Products from sulfide oxidation by MnO_2 included S^o, $S_2O_3^{2-}$ and SO_4^{2-} with S^o as the major product. More H_2S was completely oxidized to SO_4^{2-} when the ratio of $[MnO_2]/[H_2S]_T$ was increased. Oxidation of hydrogen sulfide by Fe(III) (hydr)oxides yielded elemental sulfur as the dominant product independent of the ratio of reactants and the mineral phases, which has an important implication for the formation of pyrite (FeS_2).

However, the rates of these redox processes in the field could be different from the laboratory measurements because redox cycling of Mn and Fe are microbiologically-mediated (16,18,65) in some natural waters. The rates of oxidation of H_2S by Mn and Fe oxides in the waters near the O_2/H_2S interface in the Framvaren Fjord were found to be slightly faster than the observed values in the laboratory, probably due to the presence of bacteria (26). Furthermore, the reactivity of Mn(IV) and Fe(III) oxides is dependent on the composition of the media and the conditions of their formation as discussed above. Further surface spectroscopic measurements are needed to better understand the reaction mechanism and the inhibiting effects of the ionic components in the medium.

Acknowledgments

The authors wish to acknowledge the support of the Office of Naval Research (N00014-89-J-1632) and the Oceanographic Section of the National Science Foundation (OCE88-00411) for this study.

Literature Cited

1. Almgren, T.; Hagström, I. *Water Res.* **1974**, *8*, 395-400.
2. Avrahami, M.; Golding, R. M. *J. Chem. Soc.* **1969**, (*A*), 647-651.
3. Chen, K. Y.; Morris, J. C. *Environ. Sci. Technol.* **1972**, *6*, 529-537.
4. Cline, J. D.; Richards, F. A. *Environ. Sci. Technol.* **1969**, *3*, 838-843.
5. Hoffmann, M. R.; Lim, B. C. *Environ. Sci. Technol.* **1979**, *13*, 1408-1414.
6. Millero, F. J. *Mar. Chem.* **1986**, *18*, 121-147.
7. Millero, F. J.; Hubinger, S.; Fernandez, M.; Garnett, S. *Environ. Sci. Technol.* **1987**, *21*, 439-443.
8. Millero, F. J. *Limnol. Oceanogr.* **1991**, *36*, 1007-1014.
9. Millero, F. J. *Estuar. Coastal Shelf Sci.* **1991**, *33*, 21-527.
10. Millero, F. J. *Deep-Sea Res.* **1991**, *38*, S1139-S1150.
11. O'Brien, D. J.; Birkner, F. G. *Environ. Sci. Technol.* **1977**, *11*, 1114-1120.
12. Vazquez, G. F.; Zhang, J-Z.; Millero, F. J. *Geophys. Res. Lett.* **1989**, *16*, 1363-1366.
13. Zhang, J-Z.; Millero, F. J. *Deep-Sea Res.* **1993**, *40*, 1023-1041.
14. Zhang, J-Z.; Millero, F. J. *Geochim. Cosmochim. Acta* **1993**, *57*, 1705-1718.
15. Zhang, J-Z.; Millero, F. J. In *Environmental Geochemistry of Sulfide Oxidation*; Alpers, C. N.; Blowes, D., Ed.; Symp. Ser. 550, ACS Press: Washington, D. C., **1994**, pp 393-409.
16. Aller, R. C.; Rude, P. D. *Geochim. Cosmochim. Acta* **1988**, *52*, 751-765.
17. Biber, M. V.; Dos Santos Afonso, M.; Stumm, W. *Geochim. Cosmochim. Acta* **1994**, *58*, 1999-2010.
18. Burdige, D. J.; Nealson, K. H. *Geomicrobiol. J.* **1986**, *4*, 361-387.
19. Canfield, D. E. *Geochim. Cosmochim. Acta* **1989**, *53*, 619-632.
20. Canfield, D. E.; Raiswell, R.; Bottrell, S. *Amer. J. Sci.* **1992**, *292*, 659-683.
21. Dos Santos Afonso, M.; Stumm, W. *Langmuir*, **1992**, *8*, 1671-1675.
22. Pelffer, S.; Dos Santos Afonso, M.; Wehrll, B.; Gächter, R. *Environ. Sci. Technol.* **1992**, *26*, 2408-2413.
23. Pyzik, A. J.; Sommer, S. E. *Geochim. Cosmochim. Acta* **1981**, *45*, 687-698.
24. Rickard, D. T. *Amer. J. Sci.* **1974**, *274*, 941-952.
25. Murray, J. W.; Codispoti, L. A.; Friederich, G. E. In *Aquatic Chemistry*; Huang, C. P.; O'Melia, C. R.; Morgan, J. J., Ed.; ACS Press, **1995**, in press.
26. Yao, W.; Millero, F. J. *Aquatic Geochemistry* **1995**, in press.
27. Buffle, J.; De Vitre, R. R.; Petrret, D.; Leppart, G. G. *Geochim. Cosmochim. Acta* **1989**, *53*, 399-408.

28. Burdige, D. J.; Gieskes, J. M. *Amer. J. Sci.* **1983,** *283*, 29-47.
29. Emerson, S. A.; Kalhorn, S.; Jacobs, L.; Tebo, B. M.; Nealson, K. H.; Rosson, R. A. *Geochim. Cosmochim. Acta* **1982,** *46*, 1073-1079.
30. Tebo, B. M.; Nealson, K. H.; Emerson, S.; Jacobs, L. *Limnol. Oceanogr.* **1984,** *29*, 1247-1258.
31. Wells, M. L.; Goldberg, E. D. *Nature* **1991,** *353*, 342-344.
32. Jacobs, L.; Emerson, S.; Skei, J. *Geochim. Cosmochim. Acta* **1985,** *49*, 1433-1444.
33. Shaffer, G. *Nature* **1986,** *321*, 515-517.
34. Canfield, D. E.; Thamdrup, B.; Hansen, J. W. *Geochim. Cosmochim. Acta* **1993,** *57*, 3867-3883.
35. German, C. R.; Klinkhammer, G. P.; Edmond, J. M. *Nature* **1990,** *345*, 516-518.
36. Feely, R. A.; Trefry, J. H.; Massoth, G. J.; Metz, S. *Deep-Sea Res.* **1991,** *38*, 617-623.
37. Yao, W.; Millero, F. J. *Geochim. Cosmochim. Acta* **1993,** *57*, 3359-3365.
38. Yao, W.; Millero, F. J. *Mar. Chem.* **1995,** Submitted.
39. Yao, W.; Millero, F. J. *Environ. Sci. Technol.* **1995,** Submitted.
40. Cline, J. D. *Limnol. Oceanogr.* **1969,** *14*, 454-458.
41. Murray, J. W. *J. Colloid Interface Sci.* **1974,** *46*, 357-371.
42. Schwertmann, U.; Cornell, R. M. *Iron Oxides in the Laboratory*, Weinhein, New York, **1991.**
43. Crosby, S. A.; Glasson, D. R.; Cuttler, A. H.; Butler, I.; Turner, D. R.; Whitfield, M.; Millward, S. E. *Environ. Sci. Technol.* **1983,** *17*, 709-713.
44. Taylor, B, F.; Hood, T, A.; Pope, L. A. *J. Microbio. Methods,* **1989,** *9*, 221-231.
45. Vairavamurthy, A.; Mopper, K. *Environ. Sci. Technol.* **1990,** *24*, 333-337.
46. Stone, A. T.; Morgan, J. J. *Environ. Sci. Technol.* **1984,** *18*, 450-456.
47. Burdige, D. J.; Dhakar, S.; Nealson, K. H. *Geomicrobiol. J.* **1992,** 10, 27-48.
48. Fisher, W. R. In *Iron in Soils and Clay Minerals*; Stucki, J. W.; Goodwan, B. A.; Schwertmann, U., Ed.; Dordrecht, **1988.**
49. Stumm, W.; Morgan, J. J. *Aquatic Chemistry*; Wiley, New York, **1981.**
50. Balistrieri, L. S.; Murray, J. W. *Geochim. Cosmochim. Acta* **1982,** *46*, 1041-1052.
51. Balistrieri, L. S.; Murray, J. W. *Amer. J. Sci.* **1981,** *281*, 788-806.
52. Van Geen, A.; Robertson, A. P.; Leckie, J. O. *Geochim. Cosmochim. Acta* **1994,** *58*, 2073-2086.
53. Papelis, C.; Hayes, K. F.; Leckie, J. O. *Tech. Rept.* No. 306. Dept. of Civil Engineering, Stanford Univ., **1988.**
54. Zachara, J. M.; Girvin, D. C.; Schmidt, R. L.; Resch, C. T. *Environ. Sci. Technol.* **1987,** *21*, 589-594.
55. Millero, F. J.; Plese, T.; Fernandez, M. *Limnol. Oceanogr.* **1988,** *33*, 269-274.
56. Catts, J. G.; Langmuir, D. *Appl. Geochem.* **1986,** *1*, 255-264.
57. Luther, G. W. III, In *Aquatic Chemical Kinetics*; Stumm, W., Ed., Wiley, New York, N.Y., **1990.**
58. Bondietti, G.; Sinniger, J.; Stumm, W. *Colloids Surfaces A: Physicochem. Eng. Aspects,* **1993,** *79*, 157-167.

59. Deng, Y.; Stumm, W. *Appl. Geochem.* **1994,** *9*, 23-36.
60. Von Gunten, U; Schneider, W. *J. Colloid Interface Sci.* **1991,** *145*, 127-139.
61. Liang, L.; McNabb, J. A.; Paulk, J. M.; Gu, B.; McCarthy, J. F. *Environ. Sci. Technol.* **1993,** *27*, 1864-1870.
62. Luther, G. W. III; Kostka, J. E.; Church, T. M.; Sulzberger, B.; Stumm, W. *Mar. Chem.* **1992,** *40*, 81-103.
63. Perry, K. A.; Kostka, J. E.; Luther, G. W. III.; Nealson, K. H. *Science,* **1993,** *259*, 801-803.
64. Petrie, L. M. *Geochem. Newsletter*, ACS, **1994,** Abstract 66.
65. Lovley, D. R. *Microbiol. Rev.* **1991,** *55*, 259-287.

RECEIVED April 17, 1995

Chapter 15

Characterization of a Transient +2 Sulfur Oxidation State Intermediate from the Oxidation of Aqueous Sulfide

Murthy A. Vairavamurthy and Weiqing Zhou

Department of Applied Science, Geochemistry Program, Applied Physical Sciences Division, Brookhaven National Laboratory, Upton, NY 11973

The oxidation to sulfate of H_2S, a common reduced sulfur compound in sedimentary systems, involves a net transfer of eight electrons (a change of sulfur oxidation state from -2 to +6), and occurs through the formation of several partially oxidized intermediates. The known intermediates include elemental sulfur (oxidation state: 0), polysulfides (outer sulfur: -1, inner sulfur: 0), sulfite (+4) and thiosulfate (outer sulfur: -1, inner sulfur: +5). A noticeable gap in this series of intermediates is that of a +2 sulfur oxidation state oxoacid/oxoanion species, which was never detected experimentally. We present here evidence for the transient existence of a +2 oxidation state intermediate in the Ni(II)-catalyzed oxidation of aqueous sulfide. X-ray absorption near-edge structure (XANES) spectroscopy and Fourier-transform-infrared (FT-IR) spectroscopy were used to characterize this species; they suggest that it has a sulfoxylate ion (SO_2^{2-}) structure.

The biogeochemical sulfur cycle is extremely complex because this element can exist in several oxidation states between -2 and +6, and it forms a large variety of inorganic and organic compounds. Hydrogen sulfide and the sulfate ion, the commonest inorganic sulfur species which play important roles in the atmospheric, aquatic, and sedimentary transformations of sulfur, contain sulfur atoms in the highly reduced (-2) and the highly oxidized (+6) oxidation states, respectively. In anoxic marine environments, hydrogen sulfide is formed mainly by the bacterial reduction of sulfate (1). The formation of hydrogen sulfide in hydrothermal systems probably occurs by a thermochemical reduction mechanism (2), although high-temperature bacterial reduction also is possible (3). Oxidation converts H_2S back to sulfate, but also forms several partially oxidized intermediates such as polysulfides, elemental sulfur, sulfite, and thiosulfate (4-10). Several of these intermediates (for example, polysulfides, thiosulfate) play major roles in a number of important biogeochemical processes, such as pyrite formation (11), sulfur incorporation into

0097–6156/95/0612–0280$12.00/0

organic matter (*12,13*), cycling of metal elements (*14*), and bacterial energetics (*15*). In natural waters, dissolved oxygen usually is the principal oxidant although redox transformations of some metals and nonmetals are frequently coupled to the oxidation of sulfide (*16*). In addition to chemical oxidation, bacteria also play an important role in sulfide oxidation which is carried out in conjunction with their energy metabolism (*17*).

The oxidation of H_2S to SO_4^{2-} involves an overall transfer of eight electrons. The formation of several intermediates with oxidation states in between those of the end members (elemental sulfur: oxidation state 0; polysulfides: outer sulfur, -1, inner sulfur, 0; sulfite: +4; thiosulfate: outer sulfur, -1, inner sulfur, +5 (*15*)) suggests that oxidation proceeds through a series of step-wise electron-transfer reactions. A noticeable gap is that of a +2 sulfur oxidation state oxoacid/oxoanion species, which was never detected experimentally. In fact, there has been no clear experimental evidence for its existence under any circumstance, although the occurrence of species with stoichiometry H_2SO_2 (*18-20*) and HSO_2^- (7) was proposed. According to Schmidt (1972), the oxoacid H_2SO_2 (and its corresponding oxoanion), "...at best may be regarded as a postulated intermediate in some complicated reactions of sulfur chemistry" (*21*). In this paper, we present evidence, primarily from x-ray absorption near-edge structure (XANES) spectroscopy, for the transient occurrence of a +2 oxidation state intermediate in the oxidation of Ni(II)-catalyzed oxidation of aqueous sulfide. This technique (i) provides simultaneous qualitative and quantitative information on all the forms of sulfur present, (ii) does not require cumbersome preparation of the sample so greatly minimizing the problems of artifact formation and alteration in the samples' composition, and (iii) allows non-destructive analysis. We used Fourier-transform-infrared spectroscopy (FT-IR) to further characterize the intermediate. These spectroscopic measurements suggest that the +2 sulfur oxidation state intermediate has a sulfoxylate ion (SO_2^{2-}) structure.

Methods

We conducted time-series experiments on sulfide oxidation using air-saturated solutions of 100 mM anhydrous NaHS (Johnson Matthey Co., Ward Hill, MA) in deionized water at room temperature (25 ± 1 °C). Ni(II) (NiCl$_2$ at a concentration of 100 μM) was added to the solutions as this transition metal ion was shown to be an effective catalyst of sulfide oxidation (8). The pH ranged from 11.5-12.0. We did not adjust the pH with buffers because the IR absorption peaks of the compounds commonly used (for example, sodium tetraborate) might interfere with the peaks of sulfide oxidation intermediates. The sulfide solutions were prepared with nitrogen-sparged Milli-Q water, and were maintained in a nitrogen atmosphere. Samples for the oxidation-time series were prepared by passively exposing small volumes of the stock sulfide solution (5 - 50 mL) in air in relatively large volume flasks (50 - 500 mL). Our previous observations suggest that with such high ratios of exposed area to volume, oxygen concentrations rapidly reach equilibrium with air through diffusion.

Sulfide and its oxidation products were determined using XANES spectroscopy at the National Synchrotron Light Source (NSLS) X-19A beam line, and

FT-IR spectroscopy, based on the cylindrical internal reflection technique. For x-ray analysis, ca 1-2 mL of the samples were packaged in thin Mylar film bags (Chemplex Industries, Tuckahoe, NY). Sulfur standards were analyzed in the same way. The XANES data were collected as fluorescence excitation spectra, using a fluorescence detector placed at 90 degrees to the x-ray beam. Samples were run in a helium atmosphere to minimize the attenuation of the x-ray beam by air. The spectra were recorded so that the scanning procedure yielded sufficient pre-edge and post-edge data for precise determination of the background. The x-ray energy was calibrated using XANES spectra of elemental sulfur measured between sample runs, assigning 2472.7 eV to the "white-line" maximum of elemental sulfur spectrum. A non-linear least-squares fitting procedure, which uses linear combinations of normalized spectra of model compounds, gave quantitative information on the different forms of sulfur (10,22). Further details of the XANES methodology were described elsewhere (23).

The FT-IR spectra were recorded with a Nicolet 205 spectrometer fitted with a CIRCLE accessory (Spectra-Tech, Stamford, Connecticut) based on the cylindrical internal reflection technique for analyzing aqueous samples. The CIRCLE device was equipped with a ZnSe crystal with a frequency range of 20,000 cm^{-1} - 650 cm^{-1}.

Results and Discussion

A major reason for using XANES spectroscopy for sulfur speciation is that it provides an experimental approach to infer the charge density, and hence, the oxidation state of sulfur. As usually practiced, measuring the oxidation state is an empirical counting scheme; nevertheless, it is a simple, useful approach for expressing the relative electronic charge of an atom in a molecule without considering its electronic structure in detail. By definition, the oxidation state is a number which represents the charge that an atom would have if the electrons in a compound were assigned to the atoms in a certain way (15). Simply stated, this formal number reflects charges on atomic components assigned by regarding all compounds as totally ionic. In contrast to the traditional counting approach, calibration with edge energies from XANES spectroscopy is a direct experimental approach to determining the atomic charge density which can be correlated with the formal oxidation state (24).

Figure 1 shows the XANES spectra of the terminal oxidation state species S^{2-} and SO_4^{2-}, various intermediate oxidation state species (polysulfides, elemental sulfur, sulfite and thiosulfate), and an organic sulfoxide (dibenzyl sulfoxide). In the x-ray absorption spectrum, evidence for a change in atomic charge density is a shift in the position of the absorption edge, which corresponds to electronic transitions from 1s to empty outer p orbitals for sulfur. Earlier, Kunzl suggested that there is a linear relationship between an edge shift and oxidation state based on his critical study on the shift of the K-absorption discontinuities of the oxides of several elements (25). As Figure 2 shows, a plot of edge energy vs. sulfur oxidation state gives a nearly linear correlation for most inorganic anions of sulfur. Thus, Kunzl's law provides a valuable way to determine the oxidation state of sulfur by XANES Spectroscopy. Ideally, Kunzl's law is valid for a monoatomic ionic species where the changes in its atomic charge density, represented by edge shifts, can be equated unambiguously to changes in its formal oxidation state (25). For atoms in polyatomic ions, and in

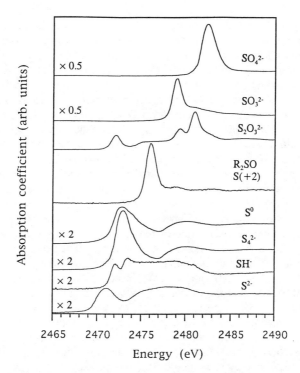

Figure 1. Normalized XANES spectra of the terminal oxidation state species S^{2-} and SO_4^{2-}, known H_2S oxidation intermediates, polysulfides, elemental sulfur, sulfite and thiosulfate, and dibenzyl sulfoxide with a charge density of $+2$ on sulfur.

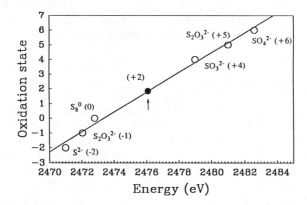

Figure 2. Plot of XANES peak energy vs. sulfur oxidation state for S^{2-} (iron sulfide), S_8^0, SO_3^{2-}, $S_2O_3^{2-}$, SO_4^{2-} fitted with a linear least-squares line (correlation coefficient, r =0.99). The energy position (2476.1) of the unknown intermediate is shown by the filled circle on the regression line, and corresponds to $+2$ sulfur oxidation state.

molecules which are not fully ionic, deviations may occur because of significant covalent bonding, although other factors also may be involved. Thus, for organic sulfur compounds dominated by covalent bonding, XANES peak-energy correlation directly estimates the relative charge density of sulfur, although this may not correlate with the formal oxidation state derived empirically. However, as demonstrated in Figure 2, for oxoanions of sulfur (such as sulfate), there is a nearly linear relationship between edge energy and (formal) sulfur oxidation state, mainly because of the strong electron withdrawing capability of the number of oxygen atoms attached to the sulfur atom, which gives it a large positive charge. Thus, to a first-order approximation, the Kunzl's law can be applied for oxoanions of sulfur. In a recent study, this method was the basis for assigning new oxidation states for the two different sulfur atoms in thiosulfate (-1 and +5 for the terminal and inner sulfur atoms, respectively) (15).

Figure 3 shows the XANES spectra of 0.1 M sulfide in deionized water, and in solution containing Ni(II), after different periods of exposure to air. In the spectra of the latter series, a striking feature is the presence of a peak at 2476.1 eV during the initial period of oxidation (up to ca. 70 hours); the peak had completely disappeared after about 100 hours. We deconvoluted these spectra to obtain quantitative information on the oxidation products, using a non-linear least- squares fitting procedure. The spectra of samples were fitted with various linear combinations of the spectra of different sulfur standards, and the results of the best fit were taken to indicate the actual sulfur composition. For the deionized water series, and the Ni(II) series after about 100 hours of exposure to air (those spectra with no peak at 2476.1 eV), we could deconvolute the spectra fully with selected combinations from the spectra of hydrogen sulfide, sulfate, polysulfide, sulfite, and thiosulfate. However, spectra from the initial period of oxidation could not be fitted entirely with any combination of these standards, mainly due to this 2476.1 eV peak. Therefore, we infer that this peak corresponds to a novel intermediate because none of the known intermediates have peaks at this position. In the plot of peak energy vs. oxidation state for various sulfur anions (Figure 2), this position corresponds to an oxidation state of +2, indicating that the novel species contains a +2 oxidation state sulfur.

A +2 charge density for sulfur also is characteristic of organic sulfoxides, as shown experimentally with XANES edge-energy correlations (Figure 1). Here, the charge density of sulfur reflects mainly the effect of oxygen, as the carbon atoms attached to sulfur do not affect its charge density because both have similar electron-withdrawing capabilities. Thus, we could take advantage of the existence of organic sulfoxides with +2 charge density to quantitatively estimate the unknown intermediate. Figure 4 shows a typical analysis of an XANES spectrum, with benzyl sulfoxide as a surrogate for this unknown species, and Figure 5 gives an example of the time-dependence of the fraction of sulfur in various oxidation states. Thus, with 0.1 M initial sulfide and 100 μM NiCl$_2$, the +2 species reached a steady-state concentration of about 5 mM (5 % of the initial H$_2$S concentration) between 20 and 70 h, but decreased thereafter as HS$^-$ was depleted.

Further information about this intermediate was obtained using attenuated total reflectance FT-IR spectroscopy. For aqueous solutions, this technique is usable in

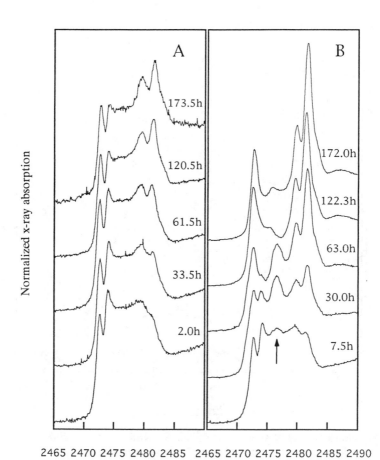

Figure 3. Sulfur K-edge XANES spectra of oxidation-time series of 0.1 M NaHS in deionized water at 25 ±1 °C: (A) pure deionized water, (B) with added NiCl₂ (100 μM). The arrow indicates the XANES peak at 2476.1 eV corresponding to the unknown species.

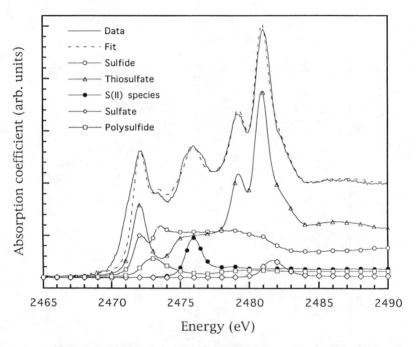

Figure 4.   Non-linear least-squares fit of the XANES spectrum of 0.1 M aqueous sulfide in deionized water with added $NiCl_2$ (100 μM) after 63 hours of oxidation in air.   Benzyl sulfoxide was used as a surrogate to quantify the unknown sulfide oxidation intermediate with a peak at 2476.1 eV.

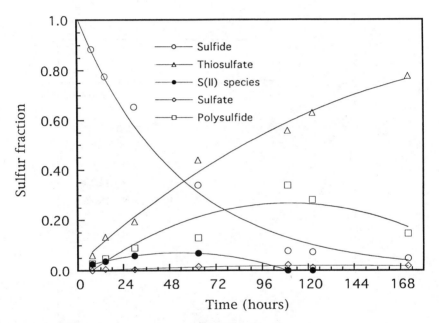

Figure 5. Distribution of sulfur species during the oxidation of 0.1 M NaHS in water containing NiCl$_2$ (100 μM).

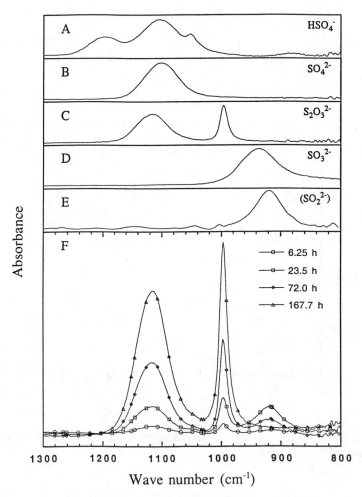

Figure 6. A-D, Fourier-transform infrared spectra of aqueous sulfur standards; F, spectra of oxidation-time series of aqueous NaHS (0.2 M) with added NiCl$_2$ (100 μM) at 25 $\pm$1 ^{0}C. The spectrum of the sulfoxylate ion (E) was obtained by subtracting the thiosulfate spectrum from that of the sample corresponding to 23.5 h oxidation.

the 750 - 1500, and 1775 -3700 cm^{-1} regions of the spectrum. Initially, we obtained FT-IR spectra of aqueous solutions of the major oxoanions of sulfur to examine their IR absorption peaks due to sulfur - oxygen stretching modes. As shown in Figure 6 (A-D), their sulfur - oxygen stretching vibrations have large infrared extinction coefficients in the 800 -1300 cm^{-1} range. The FT-IR spectra of aqueous sulfate and sulfite each show single symmetrical peaks with maxima at 1101.4 cm^{-1} and 935.5 cm^{-1}, respectively. However, for the protonated oxoanions, HSO$_4^-$ and HSO$_3^-$, the spectra assume a split, asymmetrical shape, reflecting structural changes from symmetric to asymmetric (Figures 6A and 6B). Thiosulfate exhibits two absorption peaks, a broad peak at 1118.8 cm^{-1} and a narrower one at 997.3 cm^{-1} (Figure 6C). Aqueous polysulfides have no peaks in this region. Under conditions similar to those of the XANES experiments, the IR spectra of the oxidation-time series with added Ni(II) display three absorption peaks (1116.9 cm^{-1}, 997.3 cm^{-1}, 918.2 cm^{-1}) in the 800 - 1300 cm^{-1} frequency region (6F). The peaks at 1116.9 cm^{-1} and 997.3 cm^{-1} clearly belong to thiosulfate, but that at 918.2 cm^{-1} cannot be attributed to either sulfite or sulfate. This peak exhibited a similar time behavior to that of the +2 XANES feature (Figure 7), strongly supporting our belief that this intermediate was the same as that identified by XANES spectroscopy. This unknown peak in the spectral region where the sulfur - oxygen stretching modes of aqueous-phase sulfur oxoanions, such as SO$_4^{2-}$ and SO$_3^{2-}$, also are present suggests that the new intermediate is an oxygenated sulfur species, either an oxoanion or an oxoacid. An oxoanion structure was more likely because of the basic pH (range 11.5 - 12.0) of the sulfide solutions.

Based on relative charge density of +2 for sulfur calculated from the XANES results, we propose three structures for the oxoanion intermediate: sulfoxylate ion (SO$_2^{2-}$) (I), bisulfoxylate ion (HOSO$^-$) (II), and sulfinate ion (HS(O)O$^-$) (III).

(I)   (II)   (III)

Although we cannot be as certain of its specific structure from XANES, the IR spectrum advantageously distinguishes among them. We obtained the IR spectrum characteristic of the intermediate by subtracting the thiosulfate spectrum from that of the sample (Figure 6E). The symmetrical shape of the spectrum suggests that the intermediate has a symmetrical structure, in analogy with our observations for aqueous-phase sulfate and sulfite (6B,6D). Thus, the structure is likely to be SO$_2^{2-}$, and not HOSO$^-$ or HS(O)O$^-$. Furthermore, if the structure was HOSO$^-$, it would have produced a split, asymmetrical IR spectrum, analogous to those for aqueous-phase HSO$_4^-$ and HSO$_3^-$ ions (6A). The HS(O)O$^-$ structure also would produce a asymmetrical IR spectrum because of its asymmetry.

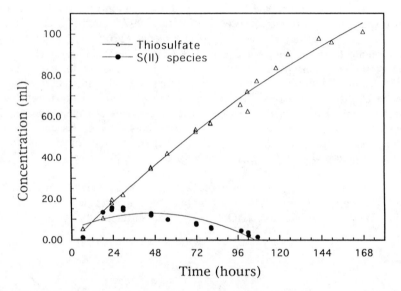

Figure 7. Distribution of thiosulfate and sulfoxylate from the IR data of the oxidation-time series shown in Figure 6F. Thiosulfate was calibrated using the broad peak in the spectrum. Sulfoxylate ion concentrations were estimated using the same thiosulfate calibration.

Both XANES and FT-IR spectroscopies suggest that the sulfoxylate ion is the initial intermediate formed in the Ni(II) catalyzed oxidation of aqueous sulfide. This is the lowest oxidation end member in the homologous series of sulfur oxoanions which include SO_3^{2-} and SO_4^{2-}. Although the mechanisms which generate and stabilize the sulfoxylate intermediate are not well understood, our experiments suggest that strongly alkaline pH (11-12) is important in stabilizing it in aqueous solutions, and seems to facilitate its accumulation to significant levels during the initial period of oxidation. Upon acidification to 8.5, the IR absorption peak at 918.2 cm^{-1}, corresponding to the sulfoxylate ion, disappeared immediately and completely; there was no indication of its conversion into the bisulfoxylate form (HOSO⁻), analogous to the formation of bisulfate from sulfate (Figure 6A and 6B). However, at such a low pH, bisulfoxylate may be converted to other species as soon as it is formed.

In summary, our study provides direct experimental evidence for the existence of a +2 sulfur oxidation state oxoanion, the sulfoxylate ion, which was observed as a transient intermediate in the Ni(II) catalyzed oxidation of sulfide with molecular oxygen. In pure deionized water only, sulfoxylate may be formed at a slower rate, so that its concentration is below the detection limit of the XANES technique (about 0.5 mM for most sulfur compounds at the X19A beamline). Although Hoffmann and Lim proposed that a similar species with HSO_2^- stoichiometry was formed as the initial step in the oxidation of aqueous sulfide catalyzed by metal-phthalocyanine complexes (7), such a species had not been detected. The role of this intermediate in forming other oxidation products is not understood; however, its observation is an important step forward in our understanding of the mechanism of sulfide oxidation.

Acknowledgments

This research was performed under the auspices of the U.S. Department of Energy Division of Engineering and Geosciences of the Office of the Basic Energy Sciences under Contract No. DE-AC02-76CH00016 (KC-04). We thank Mark Sweet for the use of Nicolet 205 FT-IR spectrometer. We acknowledge the valuable comments of Avril Woodhead and Stephen Schwartz (Brookhaven National Laboratory).

Literature Cited

1.   Jørgensen, B.B. *Nature* **1982**, *296*, 643-645.
2.   Orr, W. L. *AAPG Bull.* **1974**, *58*, 263-276.
3.   Jørgensen, B. B.; Isaksen, M. F.; Jannasch, H. W. *Science* **1992**, *258*, 1756-1757.
4.   Chen, K. Y.; Morris, J. C. *Environ. Sci. Technol.* **1972**, *6*, 529-537.
5.   O'Brien, D. J.; Birkner, F. G. *Environ. Sci. Technol.* **1977**, *11*, 1114-1120.
6.   Hoffmann, M. R. *Environ. Sci. Technol.* **1977**, *11*, 61-66.
7.   Hoffmann, M. R.; Lim, B. C. *Environ. Sci. Technol.* **1979**, *13*, 1406-1414.
8.   Weres, O.; Tsao, L.; Chatre, R. M. *Corrosion-NACE* **1985**, *41*, 307-316.
9.   Zhang, J-Z.; Millero, F. *Geochim. Cosmochim. Acta* **1993**, *57*, 1705-1718.

10. Vairavamurthy, A.; Manowitz, B.; Zhou, W.; Jeon, Y. In *The Environmental Geochemistry of Sulfide Oxidation*; Alpers, C. and Blowes, D. eds.; ACS Symposium Series 550; American Chemical Society: Washington D.C., 1994, pp 412-430.

11. Luther, G. W. III *Geochim. Cosmochim. Acta* **1991**, *55*, 2839-2849.,

12. Kohnen, M. E. L.; Sinninghe Damste, J. S.; ten Haven, H. L.; de Leeuw, J. W. *Nature* **1989**, *341*, 640-641.

13. Vairavamurthy, A.; Mopper, K. In *Biogenic sulfur in the environment*; Saltzman, E. S. and Cooper, W. J. eds.; ACS Symposium Series No. 393; ACS, Washington DC, 1989, pp 231-242.

14. Luther, G.W. III; Ferdelman, T.G. *Environ. Sci. Technol.* **1993**, *27*, 1154-1163.

15. Vairavamurthy, A; Manowitz, B; Luther, G. W. III; Jeon, Y. J. *Geochim. Cosmochim. Acta*, **1993**, *57*, 1619-1623.

16. Jacobs, L.; Emerson, S. *Earth Planet. Sci. Lett.* **1982**, *60*, 237-252.

17. Jørgensen, B. B.; Fossing, H.; Wirsen, C. O.; Jannasch, H. W. *Deep-Sea Research* **1991**, *38*, Suppl. 2, S1083-1103.

18. Plummer, P.L.; Chen, T.S.; Law, K. Y. *Atm. Environ.* **1984**, *18*, 2769-2774.

19. Laakso, D.; Marshall, P. *J. Phys. Chem.*, **1992**, *96*, 2471-2474.

20. Steiger, T.; Steudel, R. *J. Mol. Str.*, **1992**, *257*, 313-323.

21. Schmidt, M. In *Sulfur in Organic and Inorganic Chemistry*, Vol. 2; Senning A., Ed.; Marcel Dekker, Inc.: New York, 1972, pp 71-112.

22. Waldo, G.S.; Carlson, R.M.K.; Moldowan, J.M.; Peters, K.E.; Penner-Hahn, J.E. *Geochim. Cosmochim. Acta* **1991**, *55*, 801-814.

23. Vairavamurthy, A.; Zhou, W.; Eglinton, T.; Manowitz, B. *Geochim. Cosmochim. Acta* **1994**, *21*, 4681-4687.

24. Wong, J.; Lytle, F. W.; Messmer, R. P.; Maylotte, D. H. *Physical Review* **1984**, *B30*, 5596-5607.

25. Kunzl, V. *Collect. Czech. Commun.* **1932**, *4*, 213-224.

RECEIVED May 11, 1995

STUDIES OF SULFUR SPECIATION IN SEDIMENTARY SYSTEMS

Chapter 16

Temporal Relationship of Thiols to Inorganic Sulfur Compounds in Anoxic Chesapeake Bay Sediment Porewater

William MacCrehan and Damian Shea[1]

Analytical Chemistry Division, National Institute of Standards and Technology, Gaithersburg, MD 20899

Seasonal variations in the sediment porewater concentration profiles of organic and inorganic sulfur species at a Chesapeake Bay channel site provide new insight into an esturarine sulfur cycle. Oxidation of sulfide, as evidenced by the presence of zero-valent sulfur, occurs in this sulfidic sediment despite bottom water column and sediment anoxia. Thiosulfate shows a bimodal occurrence profile with a maximum in the late spring and fall, and is absent in midsummer. Two thiols, cysteine and glutathione, show maximal concentrations at 4 cm in mid-summer, which overlaps the highest gradient of sulfide and zero-valent sulfur occurrence patterns, suggesting a role in elemental sulfur redox or transport processes. Mercaptoacetate and mercaptopyruvate, were found in much lower concentrations and appear to be linked to the diagenesis of cysteine. Two mercaptopropanoic acids show temporal profiles increasing with sediment depth and may be linked to deeper sediment methanogenic bacterial activities.

Intense seasonal periods of suboxic and anoxic conditions in the Chesapeake Bay water column have been observed at mid-bay sites (1). One factor contributing to water-column oxygen-demand is the oxidation of benthically-generated sulfide (2,3). When oxygen and nitrate are depleted, sulfate becomes the primary oxidant used for the bacterial diagenesis of organic detritus in shallow marine sediments (3). A fraction of the resulting hydrogen sulfide diffuses upward and is reoxidized (4). The impact of this benthic oxygen demand on the Bay ecosystem remains to be determined.

Arguably, recent progress in determining the role of the individual sulfur species in this sulfur cycle has been a direct result of improvements in the analytical methods for their selective determination. Sediment sulfide, oxygen and pH have been measured in situ using microelectrodes (5,6). New methods for the determination of solid-phase zero-valent sulfur (7-10) have been developed. Early measurements of organic thiols in marine waters were limited to total thiols using non-selective techniques that were prone

[1]Current address: Department of Toxicology, North Carolina State University, 3709 Hillsborough Street, Raleigh, NC 27607

to interferences (7,10-12). More selective analytical methods employing a liquid chromatographic (LC) separation have recently been developed to measure individual thiols in the marine environment. Fluorescence detection has been applied using pre-injection derivatizations of the thiol moiety with monobromobimane (13) and orthodiphthaldehyde (14). Nanomolar concentrations of glutathione have been measured in suspended marine particles (15) and sub-μM concentrations of eight thiols have been determined in coastal marine sediments using these approaches (16). We used an amperometric LC detection approach (17,18) for the measurements presented here because it may be used directly on the underivatised thiols, eliminating the possible errors of cross-reactivity with other compounds such as biogenic amines found with the orthodiphthaldehyde derivatization. This technique was recently used to determine μM concentrations of four thiols in the Black Sea water column (19).

At this early stage in the understanding of the sulfur cycle in anoxic sediments, an important first step is to identify the inorganic and organic sulfur species present and to determine their variation with the season and sediment depth. In this study, we have measured porewater concentrations of 3 inorganic sulfur species (sulfide HS$^-$, total soluble zero-valent sulfur S^o and thiosulfate S$_2$O$_3^-$), as well as 6 organic thiols (cysteine CSH, glutathione GSH, mercaptopyruvate MPV, mercaptoacetate MAC, 2- and 3-mercaptopropanoate 2MPA and 3MPA) over the period of June to December 1986 at a single, mid-bay channel site. Hypotheses are drawn on the processes and relationships that give rise to the observed temporal data set.

Methods

Sampling - Samples were collected at CBI station 848C (N38°48'30", W76°23'30"), using a 10-cm diameter gravity corer by a procedure designed to minimize oxidation (17) on June 13, July 15, August 11, October 5, and December 5, 1986. This location is in the mid-channel of the Bay, approximately 25 km SE of Annapolis, MD. The overlying water column is seasonally anoxic, and sulfide levels in the sediment porewater are known to reach millimolar concentrations during the summer (2,17,20). Water overlying the sediment surface was collected by a Niskin bottle within 0.5 m of the bottom.

Sediment cores were collected using a 10-cm diameter butyrate-lined gravity corer. The core and liner were maintained in an upright position on ship-board and immediately placed in a polyethylene tube while nitrogen gas (99.5%) was blown through the tube. The tube was tightly closed, placed on dry ice in a closed cooler, and returned to the laboratory the same day. Immediately upon return to the laboratory, the semi-frozen cores were transferred to a nitrogen-filled glove box, extruded and sectioned from 0 to 2 cm, 3 to 5 cm and then at 4 cm intervals to the bottom of the core. Porewater was expressed from each core section using a pneumatic (20 psi nitrogen) nylon porewater press inside the glove box. The porewater was filtered sequentially through 5-μm glass fiber filter and 0.4-μm Nuclepore membranes. The filtrate was collected and sealed in septum-enclosure vials under nitrogen and stored in a nitrogen-filled desiccator. The sealed desiccator was removed from the glove box and stored at -20 °C until analysis, which was usually within a few days.

The horizontal homogeneity of these samples was evaluated by taking duplicate and triplicate cores on two sampling dates from the same ship position. Agreement between porewater concentrations of the sulfur species was within a reasonable experimental error ($\pm$ 20%) for the duplicate and for two of the triplicate cores. However, the third sample showed levels of the sulfur species significantly different (>50%) from the other two cores. This indicates some horizontal heterogeneity of samples taken from this same sample site. The horizontal homogeneity within a single 10 cm core was determined using 1 cm width subsamples. Excellent agreement was found for the analytical results (17).

Analytical methods - Sulfide was measured using the polarographic method of Luther (10,11). Thiosulfate and zero-valent sulfur were also measured polarographically (11). Thiols were measured using ion-pair, reverse-phase LC with electrochemical detection at a gold/mercury-film electrode (17). Qualitative identifications were based on the chromatographic retention times of authentic standards and the inherent selectivity of the amperometric detector.

Results

Inorganic Sulfur Species

The presence of reduced inorganic sulfur compounds in marine sediments have been the focus of a number of studies, which are summarized in Table 1. Additionally, references are cited regarding the chemical reactions that produce the compounds and their role in the sediment sulfur cycle.

The occurrence of biotic sulfate reduction in the Chesapeake Bay sediment is recognized by the rise in porewater concentration of hydrogen sulfide measured during the summer, shown in Figure 1. In these depth/time/concentration (d/t/c) profiles, sediment depth is represented on the x-axis, time of sampling on the y-axis and the concentration is presented on the z-axis. The larger, left plot provides a perspective on the data set that focuses on shallower sediment depths and an earlier time in the year. The smaller, right plot provides a view rotated $180°$ from the first plot, focusing on the data at greater depths and at a later time in the year. The lowest isopleth drawn represents the detection limit of the analytical method used for the measurement. The concentration of sulfide in the bottom water was found to be in the 10's of μM range and is arbitrarily plotted at the "-2 cm" depth. An abrupt increase in sulfide concentration was observed for the surface porewater sample. The sulfide concentration then increases with depth to 4 cm, where the concentration becomes approximately constant for depths from 4 to 18 cm on a given sampling date. The decrease in the sulfide concentration profile in the upper ≈ 4 cm of sediment follows the general sigmoidal shape often observed for transport-limited processes.

The occurrence of zero-valent sulfur in the Bay samples is shown in Figure 2. The maximum concentration was found on the July 15 date in the 0 to 2 cm depth sample. For any given date, the maximum concentration occurred near the sediment surface and then declined with increasing depth. The $S°$ concentration profile shows a maximum at depths where sulfide concentration declines rapidly toward the oxic zone. The relative profiles of HS^- and $S°$ found in this Bay sediment are in reasonable agreement with those found for an anoxic coastal fjord sediment (9) and in the Black Sea water column (19).

Given the concurrent occurrence of HS^- and $S°$ in many of these samples, we calculated the concentration of the polysulfide species expected, via the measured $[HS^-]$, $[S°]$ and pH and the known equilibrium distribution of polysulfides (21). Based on these results, most of the zero-valent sulfur should be bound as inorganic polysulfides.

The temporal variations of thiosulfate concentration are presented in Figure 3. Since the limit of detection is plotted as the lowest isopleth, this Figure appears inset. However, the maximum concentration is found in the surface sediments, with no detectable levels found in the overlying water column. The thiosulfate pattern shows some correlation with the d/t/c pattern for $S°$ in the spring and fall. However, most all of the samples showed $[S_2O_3^=]$ below the detection limit for the July and August sampling dates. The highest concentrations of thiosulfate were found in June.

Table 1: Reduced Inorganic Sulfur Species in Marine Sediments

Species	Source/role	Sediment Type	Concentration
sulfide HS⁻	product dissimilatory $SO_4^=$ reduction (3,30,40,44); energy resource (3); pyrite deposition (29)	salt marsh (12)	2×10^{-4} - 5.5 mM
		fjord (6)	0 - 1 mM
		coastal (51)	0 - 1.4 mM
		coastal (10)	0.4 - 2.3 mM
		salt marsh (11)	0 - 3.3 mM
		salt marsh (58)	3×10^{-6} - 1.5 mM
		estuary (*)	0.5 µM - 5.6 mM
zero-valent sulfur S°	intermediate HS⁻ oxidation (12,32,39,40,42); product $S_2O_3^=$ reduction (39); respiration (39,41)	salt marsh (12)	saturated
		coastal (10)	0.1 - 0.2 mM
		salt marsh (11)	0 - 0.3 mM
		fjord (8)	< 1 µM - 20 µM
		estuary (*)	< 0.1 - 67 µM
thiosulfate $S_2O_3^=$	intermediate HS⁻ oxidation (8); product S° + $SO_3^=$ (8,10,32)	salt marsh (12)	0.2 - 0.5 mM
		coastal (10)	0.01 - 0.03 mM
		salt marsh (7,10)	0 - 0.01 mM
		salt marsh (58)	0.1 - 0.6 mM
		estuary (*)	0.1 - 7.1 µM
sulfite SO_3^-	intermediate $SO_4^=$ reduction (40); intermediate HS⁻ oxidation via $S_2O_3^=$ (32)	coastal (16)	0.07 mM
		salt marsh (12)	$< 10^{-4}$ - 0.2 mM
		salt marsh (7,10)	
tetrathionate $S_4O_6^=$	oxidation of $S_2O_3^=$ (32); intermediate $SO_4^=$ reduction (40)	salt marsh (11)	0.17 - 0.31 µM
polysulfides $S_nS^=$ (n = 2-5)	reaction of HS⁻ and S°; mobilization S° and transition metal ions (10,12)	salt marsh (12)	10^{-3} - 0.43 mmolal
		salt marsh (10,11)	< 0.33 mM
		estuary (21)	0.02 - 0.2 mM

* this study

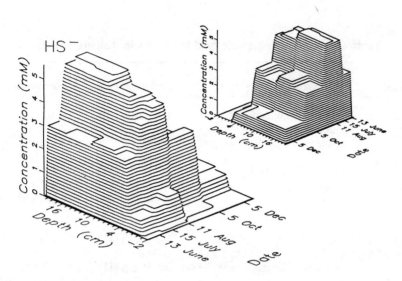

Figure 1 Temporal Variation of Sulfide Concentration with Depth and Time in a Chesapeake Bay Channel Sediment. Inset view rotated 180°

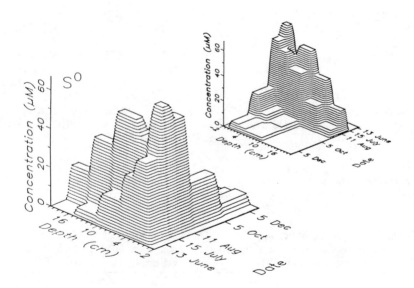

Figure 2 Temporal Variation of Zero-Valent Sulfur Concentration with Depth and Time in a Chesapeake Bay Channel Sediment. Inset view rotated 180°

Organic Sulfur Species

The role of volatile organic sulfur species on the marine sulfur cycle has received much recent attention (23). Less studied, but no less crucial to our understanding of this cycle, are the identities and concentrations of the hydrophilic organosulfur species. The results of recent studies of these compounds are given in Table 2. Note that the sediment thiols identified to date are generally found in nM to low µM concentrations, whereas the inorganic sulfur species in the same porewaters were generally measured to be in 10's of µM to low mM concentrations.

The measured concentrations of the organic thiols in the Chesapeake Bay sediments throughout the year are shown in Figures 4 through 8. In no case do the d/t/c profiles of the organic thiols directly trace those of sulfide, in agreement with observations at other sites (3,11). The d/t/c patterns of two of the thiols are particularly closely linked - cysteine and glutathione (see Figures 4 and 5), showing a fairly sharp concentration maximum at a 3 to 5 cm depth in August. The d/t/c patterns for CSH and GSH nearly identically overlap, with [CSH] present at approximately twice [GSH]. Some correlation is also observed for the pattern of occurrence of cysteine and mercaptopyruvate, particularly in the June to August sampling interval as shown in Figures 4 and 6. Mercaptoacetate was also determined and followed the general pattern for [CSH] with a concentration maximum of 0.6 µM in August.

The concentration maxima in the d/t/c profiles of these CSH-related thiols [GSH, CSH, MPV], correlate well with the sediment depth where sulfide depletion occurs, and overlaps the zone where zero-valent sulfur is found. Luther et al (7) also observed that the maximum concentration of a sediment thiol, tentatively identified as glutathione, was in the 6 to 9 cm depth in a sample taken in Great Marsh, DE in the summer. The sulfide concentration was depleted in this shallow zone but increased at 12 to 30 cm depths from ≈ 1 to 3 mM. A peak shaped concentration profile was found for S°, with maximum at the 12 to 15 cm sample depth, which was exactly between the regions where [HS⁻] and [thiol] were depleted. It was concluded that glutathione was a possible link between the inorganic and organic sulfur pools in this marsh sediment.

Two additional organic thiols, 2MPA and 3MPA, measured in the Bay sediment samples (Figures 7 and 8), show much different d/t/c profile from the CSH-related thiols. The concentration of these thiols increase with depth to ≈ 12 cm and then becomes relatively constant with depth. The pattern is somewhat similar to that found for the sulfide d/t/c profile particularly for the July and August sampling dates. However, the 2MPA and 3MPA concentrations are proportionately much lower in the data set on dates in the spring and fall than is sulfide. 3MPA is found at concentrations as high as 12 µM in these samples, which is higher than the maximum concentration found by Kiene and Taylor (24), but similar to that found by Vairavamurthy and Mopper (25) in anoxic marine sediments. 2MPA, a thiol not previously measured temporally in anoxic sediment, shows a maximum concentration of ≈ 4 µM in the Bay samples. It has been suggested that sediment 2MPA results from the addition of sulfide to diagenetically-produced acrylic acid via a Markovnikov addition mechanism (16).

Thiomalic acid, when detected in the Bay sediments, showed a d/t/c profile with broad maximum concentration of 0.08 µM at the 3 to 5 cm sampling depth in the July through October sampling dates. Monothioglycerol was also detected in some of the samples but was not quantitated because of its low concentration (≤ 0.08 µM).

Table 2: Hydrophilic Organosulfur Compounds in Marine Sediments

Species	Source/role	Sample type	Concentration
dimethylsulfonio-propanoate DMSP	plant osmolyte (24,52)	salt marsh (57)	100 - 200 µM
3-mercaptopropanoate 3MPA	DMSP demethylation product (51.57); abiotic addition of HS⁻ to acrylate (25,51)	salt marsh (16) coastal (51) estuary (*)	0.5 - 20 µM 20 - 230 nM < 0.04 - 11.8 µM
2-mercaptopropanoate 2MPA	abiotic acrylate addition of HS⁻ (16)	salt marsh (16) estuary (*)	< 0.5 µM < 0.04 - 3.8 µM
cysteine CSH	assimilitory $SO_4^=$ reduction (40,47,49); microbial protein degradation (50)	estuary (*)	< 0.04 - 12.4 µM
glutathione GSH	redox cofactor (47,48), cysteine repository (49,15); S° metabolism (40,42)	salt marsh (11) estuary (*)	10 - 2400 µM < 0.04 - 5.4 µM
mercaptopyruvate MPV	cysteine metabolite (47); sulfur donor (42)	coastal (16) estuary (*)	0.5 - 20 µM < 0.04 - 1.4 µM
mercaptoacetate MAC	mercaptopyruvate metabolite (49,59)	coastal (16) estuary (*)	< 0.5 µM < 0.04 - 0.60 µM
monothioglycerol		coastal (16) estuary (17)	< 0.05 - 20 µM 1.6 - 4.3 µM
N-acetylcysteine		coastal (16) estuary (17)	< 0.5 µM < 0.03 - 0.3 µM
thiomalate	mercaptopyruvate metabolite (49)	estuary (*)	< 0.04 - 0.08 µM

Also detected: Coenzyme M (16), 2-mercaptoethanol (16)

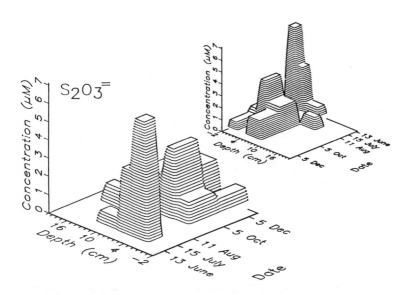

Figure 3 Temporal Variation of Thiosulfate Concentration with Depth and Time in a Chesapeake Bay Channel Sediment. Inset view rotated 180°

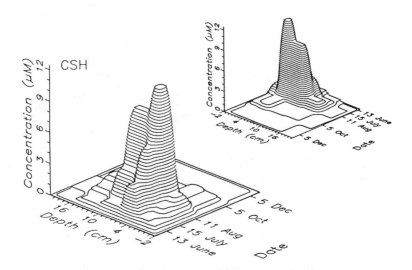

Figure 4 Temporal Variation of Cysteine Concentration with Depth and Time in a Chesapeake Bay Channel Sediment. Inset view rotated 180°

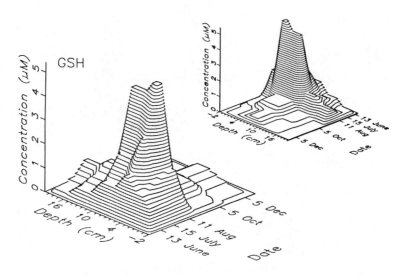

Figure 5 Temporal Variation of Glutathione Concentration with Depth and Time in a Chesapeake Bay Channel Sediment. Inset view rotated 180°

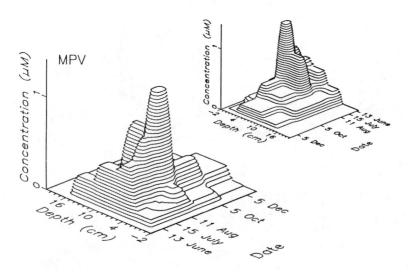

Figure 6 Temporal Variation of Mercaptopyruvate Concentration with Depth and Time in a Chesapeake Bay Channel Sediment. Inset view rotated 180°

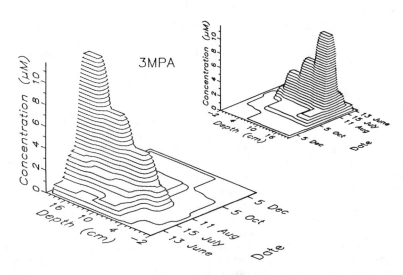

Figure 7 Temporal Variation of 3-Mercaptopropanoate Concentration with Depth and Time in a Chesapeake Bay Channel Sediment. Inset view rotated 180°

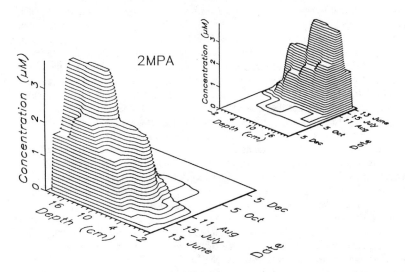

Figure 8 Temporal Variation of 2-Mercaptopropanoate Concentration with Depth and Time in a Chesapeake Bay Channel Sediment. Inset view rotated 180°

Discussion

Sulfide

The temporal occurrence pattern of sulfide in the Chesapeake Bay sediment may be the result of several factors. The concentration of HS⁻ is relatively low in the uppermost 4 cm of this sediment and increases with depth to a relatively constant value between 4 and 20 cm. Because sulfide is being produced in the uppermost sediments and overlying water column in midsummer by sulfate reduction, the sigmoidal concentration profile of sulfide in the shallow sediments must reflect the concurrent removal of sulfide in the uppermost layer of sediment and in the overlying water column.

Three processes could be candidates to control the concentration profile found for sulfide in the top ≈ 4 cm of this sediment. First, HS⁻ and its conjugate acid H₂S, are diffusing into the overlying water column from sulfate reduction in the deeper sediment, ultimately to be oxidized to SO₄⁼. This process should occur most rapidly in the lower density surface sediment which interfaces to the water column reservoir. However, during the midsummer anoxia, upper surface sediment and overlying water column sulfate reduction will only add to the observed porewater [HS⁻] in the uppermost sediments. Sulfate reduction to sulfide is known to occur during midsummer in the Bay's surface sediments with rates as high as 55 mmol HS⁻/m²/d (2). Sulfate reduction also occurs in the channel water column in mid-bay sites (20), with rates as high as 12-20 mmol/m²/d, during summer anoxia. Thus other mechanisms must be invoked to account for the lack of accumulation of HS⁻ in the uppermost sediment layer.

The second possible HS⁻ removal process is mineral deposition. Cornwell et al (29) have recently measured the temporal variation of iron sulfide minerals in mid-bay sediments. They found pyrite to be the major depositional species, with iron monosulfide formation confined to the topmost 0 to 0.5 cm sediment layer under conditions of high midsummer [HS⁻]. They concluded that for much of the year, titration of iron oxides by HS⁻ controls diffusion into the water column. However, during the midsummer months when SO₄⁼ reduction was highest, the iron binding capacity was exceeded. In an anoxic marine fjord, Jorgensen (4) estimated that free porewater sulfide exceeded the binding capacity of the iron present at 0 to 2 cm depths, with solid iron phases accounting for only ≈ 5% of total sulfide. Therefore, mineral deposition probably only partly accounts for low sulfide concentration at the sediment surface at this Bay site.

The third HS⁻ removal process is biotic and abiotic oxidation of sulfide in the surface sediment. Since oxygen is completely depleted in immediate overlying water column and sediments, other oxidants must play a role. HS⁻ oxidation can either result in complete regeneration of sulfate, or result in the formation of sulfur species of intermediate oxidation state including S°/polysulfides, which may be removed via deposition of pyrite in sediments (26). Jorgensen et al (27) studied the oxidation of HS⁻ in water column samples from the Black Sea, simultaneously monitoring the disappearance of sulfide and the appearance of the oxidation products SO₄⁼ and S₂O₃⁼. Decline of [HS⁻] occured in collected samples despite careful attempts to maintain oxygen-free conditions (19). Curiously, Jorgensen found the rate of HS⁻ removal by oxidation exceeded the expected stoichiometric production of the SO₄⁼ + S₂O₃⁼. This observation held true through a number of test conditions including unfiltered samples (with living bacteria), samples filtered to remove living bacteria, and sterilized samples. They speculated that formation of sulfur species of intermediate oxidation state, particularly S°, were being formed by both biotic and abiotic processes. A clear determination of the source of the oxidant in these completely anaerobic samples remains elusive, although chemoautotrophic assimilation of CO₂ and other forms of oxidized organic carbon coupled to HS⁻ oxidation were demonstrated (27,28). Luther hypothesised that organic Mn (II/III) complexes may provide a direct pathway for the catalysis of abiotic oxidation of sulfide that was observed in these Black Sea water samples (19).

For sediment depths greater than 4 cm, a relatively constant [HS⁻] was found with concentrations of 3 mM in June, rising to 5 mM for July and August, and then falling back to 3 mM by October and 1 mM by December. Two phenomena could account for the constancy of sulfide concentration at 4 to 18 cm depths: [1] a combination of sulfate depletion below a 4 to 6 cm sediment horizon (2) with an absence of sulfide oxidation in the 4 to 18 cm zone, limited by lack of oxidants and [2] sulfide is biotically regulated in such a way that formation and consumption of HS⁻ are balanced in the 4 to 18 cm zone, possibly limited by toxicity to sediment microbes (30) or by inhibition of the reduction of oxidized forms of sulfur (31). The overall sulfide increase in the warmer summer months is a result of the combination of the seasonal increase in organic carbon loading from the overlying water column in the more productive months, and the increased rate of $SO_4^=$ reduction by sulfate reducing bacteria at the higher temperatures. A factor of three increase in sediment sulfate reduction rate for a 10°C temperature increase has been previously measured in sediments at the outflow of the Choptank river into the Chesapeake (2).

Zero-valent sulfur
Abiotic Formation
Since no known reductive processes produce zero-valent sulfur from sulfate, its occurrence in anoxic sediments is thought to be a result of the oxidation of sulfide (9,27,32) and possibly the disproportionation of thiosulfate (3). In order for sediment sulfide oxidation to S° to take place under anoxic conditions, oxidants such as hydrous Fe(III) and Mn(III/IV) oxides may play an important role. The oxidation of sulfide occurs rapidly by an abiotic route on freshly prepared goethite, FeOOH (33) and vernadite, δ-MnOOH (34); however, reductive dissolution rates of metal oxides are much slower after a surface layer of insoluble metal sulfide product forms (33) and when adsorption of indifferent ions and organic compounds occurs (35). Burdige and Nealson (34,36) have shown that sulfide, produced by sulfate reducing bacteria, may be important for manganic oxide reduction in sediments and a model has been developed for manganese redox cycling between the oxic and anoxic zones (37).
Biotic Formation
However, using the same source of oxidants, biotic processes may also compete for the oxidation of HS⁻ to S°, since the activity of sulfuretum-type bacteria is frequently quite high in sulfidic sediments (9). Many organisms are adapted to life in the sediment transition zone between oxygen and sulfate-dominated respiration (3,5). Anaerobes such as Thiobacillus denitrificans, T. concretivoris and Thiomicrospira denitrificans as well as aerobic microorganisms such as Beggiatoa and Thiovulum (5,38), and Pseudomonas (27) take advantage of the chemical energy derived from the oxidation of biotically-produced sulfide. Even under anoxic conditions, biota can take advantage of such available oxidants as NO₃⁻ and carbohydrates. Organisms such as Beggiatoa leptomitiformis can oxidize sulfide with the concurrent formation of internal globules of zero-valent sulfur. This S° can represent as much as 20% of their cell weight, providing a store sufficient to support respiration for several days under anaerobic conditions via S° reduction (39). External S° may also be formed by this and other organisms (40), and may be present as either solid amorphous or orthorhombic S_8 (39,40). An experiment on living coastal sulfidic sediment found that the pattern of occurrence of solvent-extractable S° corresponded exactly to the observed distribution of Beggiatoa filaments (9). This relationship held true even when

the bacteria were encouraged to migrate under light and aeration stimulus. S^o may also be produced by the disproportionation of $S_2O_3^=$, mediated by a common marine anaerobe *Shewanella putreficans* (31).

Jorgensen et al (5) have argued that S^o may represent an important redox intermediate for respiration of thiophilic bacteria not located at the optimum of the redoxcline. In bacteria mats, under conditions with μM O_2 at the sediment surface, S^o may be favorably converted all the way to $SO_4^=$; however, below the redoxcline the reduction of S^o to HS^- may be thermodynamically favored. In this way S^o could provide an energy resource for these redoxcline-adapted microbial communities.

In the presence of inorganic sulfide, as found in these Bay samples, solid/colloidal zero-valent sulfur is solubilized by the formation of polysulfides:

$$nS^o \ + \ HS^- \ \rightleftarrows \ S_nS^= \ + \ H^+ \text{ where } n = 2 \text{ to } 5.$$

Equilibrium formation of $S_2S^=$ through $S_5S^=$ may impact the complexation of trace metals such as Cu(I), Pb(II) and Hg(II) (22,41). Sediment polysulfides also influence the deposition of solid iron phases. Pyrite formation, kinetically slow in the absence of polysulfides, becomes favored over formation of iron (II) sulfide when polysulfides are present (12,30). The laboratory reaction of polysulfides with Fe (II) to form pyrite has recently been studied in detail (26).

Organic polysulfides, with C bonded to one or both of the terminal sulfurs, are also likely to be present in sulfidic sediments. For example, hydropolysulfides, with a terminal -SH, may be formed directly by the abiotic reaction of zero-valent sulfur with thiols such as cysteine and glutathione (41): $\quad S^o \ + \ GS^- \ \rightarrow \ GSS_n^-$

$$O_2, H_2O$$

or by biotic oxidation (42): $S_8 \ + \ GSH \ \rightarrow \ GSS_6H \ + \ S_2O_3^= \ + \ 2\,H^+$

Based on the overlapping d/t/c occurrence pattern of S^o and CSH/GSH (Figures 4 and 5), zero-valent sulfur should be solubilized by these addition reactions in the Bay sediments. The biotic availability and reactivity of the resulting organic polysulfides may impact the sulfur cycle (43). Suzuki speculated that the glutathione hydropolysulfides are the true substrates for the biotic S^o oxidation system (41). There is evidence that organic polysulfides are also the product of the reaction of inorganic polysulfides with particulate organic carbon (45). The resulting organic polysulfides may make an important contribution to the incorporation of sulfur into the organic matter during the formation of petroleum.

Thiosulfate

The measured thiosulfate concentration goes through a seasonal maxima in June and October. Since sulfide is present in the sediment throughout this time, the formation of this sulfur compound of higher oxidation state may reflect an increased availability of oxidants in the spring and fall relative to midsummer. The lack of formation of this oxidized form of sulfur during the midsummer sulfidic maximum is consistent with the expected increase in the reducing nature of the sediments during July and August.

Sediment thiosulfate may also result from the abiotic reaction of elemental sulfur and sulfite: $\quad SO_3^= \ + \ S^o \ \rightarrow \ S_2O_3^=.$

This reaction can be driven in the reverse direction biotically (37). $S_2O_3^=$ is also produced by the oxidation of pyrite (FeS_2) under oxic conditions (11,43,46) and this route is likely to be of importance to the Bay sediments in the colder months.

Many anaerobic sediment sulfur bacteria, that lack the ability to reduce sulfate directly, i.e., <u>Desulfotomaculum</u> and <u>Desulfomonas,</u> can instead use sulfur compounds of lower oxidation state, such as $SO_3^=$ and $S_2O_3^=$, as electron acceptors (38) and thus may flourish in the upper sediment zone when these substrates are present. A common non-$SO_4^=$ - reducing anaerobe, *Shewanella putreficans*, isolated from the Black Sea, was recently shown to use $SO_3^=$, $S_2O_3^=$ and S^o as electron acceptors producing HS^- (31).

Glutathione and cysteine

Detecting the presence of porewater glutathione (GSH = τ-glutamylcysteinylglycine) in these sediments may be of great significance to understanding the sediment sulfur cycle. Free intercellular GSH is an important cofactor for protein interactions involving disulfide bonds (47,48) and may act as a intercellular reservoir for CSH (49), since intercellular [GSH]>>[CSH]. Membrane-bound GSH has been shown to be an important microbial cell surface component (48), possibly involved in the transport of zero-valent sulfur and other sulfur nutrients through the cell wall via formation of organohydropolysulfides (40,42). The occurrence of free GSH in the porewater may be linked to microbial solubilization of solid S^o via an abiotic reaction or as a cofactor required for exocellular S^o enzyme activity.

Cysteine, present in these porewater samples at up to 12 μM, has not generally been measured in sediment porewater, because of its poor reaction in the LC method (14). As an essential amino acid to protein synthesis, its presence in the porewater may be a result of assimilitory $SO_4^=$ reduction (40) or of the diagenesis of proteins (50). Cysteine also reacts with zero-valent sulfur, in a reaction analogous to that for GSH (42), and thus may be involved in microbial transport and redox assimilation of S^o. We also found that the relative concentration ratio of porewater CSH and GSH in these sediments is remarkably constant, with [CSH]≈ 2 x [GSH]. This may indicate biological control of the porewater concentration of these two thiols, optimized for the transport of inorganic sulfur. Such bioregulation could reflect an important role for CSH and GSH in the sulfur redox chemistry of these recently deposited sediments.

Biotic release of thiols to the porewater may also impact the bioavailability of trace transition metals in sulfidic systems. Difficulty in acquiring essential trace elements such as copper and zinc, strongly bound as sulfide minerals, may limit microbial growth. Evidence for the importance of thiols to metal speciation, in these same Bay sediments, comes from a measurement-supported thermodynamic model, where the ligand cysteine was found to significantly affect soluble Cu(I) in equilibrium with a solid CuS phase (22). This raises the speculative possibility that sediment sulfur microbes release the thiols as metal-binding ligands into the porewater for the express purpose of solublizing essential trace metal nutrients from solid phases, just as siderophores must do to acquire iron in oxic environments.

Mercaptopyruvate and mercaptoacetate

Little is known about the occurrence of mercaptopyruvate and mercaptoacetate in marine sediments. Both of these thiols show the same general temporal d/t/c profiles as cysteine. However, MPV and MAC are found at much lower concentrations than cysteine, and thus may occur in the exocellular porewater from release upon death of sediment microbes of these intercellular cysteine intermediates. MPV can be formed as a cysteine metabolite through a single reversible reaction step with α-oxoglutarate by a transaminase enzyme (47).

MPV has been shown to be an important cellular compound for biotic sulfur transfer reactions such as the reduction of sulfite via the enzyme - mercaptopyruvate sulfurtransferase (49). The presence of MPV in the sediment porewater may thus be an indicator of biotic sulfur redox chemistry.

Mercaptoacetic acid was found to be present at about 1/10 [MPV] and may occur as a metabolite of MPV, transformed by an oxidative decarboxylase enzyme in a single step (49).

3-Mercaptopropanoate

Both biotic and abiotic routes have been discovered for the formation of 3-mercapto-propanoate in marine sediments. Kiene and Taylor (24,51), studying the biotic formation pathways of 3MPA in active marine sediment cultures in vitro, showed that amendment with either homocysteine, methionine or dimethylsulfoniopropanoate (DMSP), increases the production of 3MPA. DMSP is a potentially important precursor to the formation of 3MPA in recently deposited sediments since it is used by marine algae as an osmolyte at 100 mM concentrations (52).

An abiotic route to the formation of 3MPA has also been described by Vairavamurthy and Mopper (25,53), resulting from the addition of sulfide across the vinyl bond of acrylic acid via the Michael reaction mechanism. However, in the single coastal sediment studied in vitro, the abiotic reaction rates were found to be slow compared to the biotic conversion of DMSP (51).

The maximum 3MPA concentration occurs at the 16 to 20 cm sample in this sediment, which is a greater depth than found for the CSH-related thiols. If sediment 3MPA results from the demethylation of DMSP, the bacteria capable of the methyl transfer reactions required for DMSP diagenesis must not be common in the upper 16 cm of the sediment, but must reside at greater sediment depths. In an estuarine fjord, Sorensen (54) found a very similar d/t/c profile for measurements of methanethiol (MeSH) to our observations of 3MPA, with a maximum concentration of 1 μM at a 20 cm depth in summer. Methanogenic bacteria can catalyze a methyl- transfer reaction between methionine and sulfide to produce MeSH (55,56). Such bacteria may also be active at this sediment depth in the Bay samples, and may use DMSP as a substrate for similar methyl-transfer reactions, resulting in the production of 3MPA (24,57). If instead, the abiotic route to 3MPA production predominates, then the maximum occurrence of 3MPA at depths of 16 to 20 cm should be correlated to the concentration of the HS$^-$ and acrylic acid reactants as well as porewater pH (53).

Conclusions

It is well accepted that the production of sulfide in recently deposited marine sediments is a result of the biotic activity of sulfate-reducing bacteria. However, the processes controlling the reoxidation of HS$^-$ to SO$_4^=$ are not as clear. Given the free energy available in the reoxidation of sulfide (3), it seems likely that sediment microbes would be involved in this process during the warmer months if oxidants are available.

Although diagenesis of biotic material containing organic sulfur compounds and abiotic addition reactions of reduced inorganic sulfide to organic compounds are likely to contribute to the profiles of the organic sulfur compounds found in these sediment porewater samples, we contend that the profiles of the biologically significant thiols (particularly CSH and GSH) primarily reflect sulfur cycle microbial activities.

The relationship of the peak-shaped temporal d/t/c profiles of CSH, GSH and S^o to the sigmoidal profile found for HS$^-$ is particularly significant. The presence of a midsummer maximum for the concentration of elemental sulfur provides evidence of sulfide oxidation, even during periods when this sediment is known to be anoxic. The observation that the maximum in metabolically-significant thiol concentrations are found in the zone where sulfide is being depleted and elemental sulfur is formed argues for a major microbial role for thiols in the sulfur cycle of this recently deposited, mid-bay Chesapeake sediment.

Literature Cited

1. Magnien, R. E.; In *Dissolved Oxygen in the Chesapeake Bay*, Mackierman, G. B., Ed.; Maryland Sea Grant Pub; MD, 1988; pp 19-21.
2. Tuttle, J. H.; Roden, E. E.; Divan, C. L.; In *Dissolved Oxygen in the Chesapeake Bay*, Mackierman, G. B., Ed.; Maryland Sea Grant Pub; MD, 1988; pp 100-102.
3. Jorgensen, B. B.; In *The Major Biogeochemical Cycles and Interactions*; Bolin, B. et al., Eds.; Wiley, 1983; pp 477-509.
4. Jorgensen, B. B.; *Nature* 1982, 296, pp. 643-645.
5. Jorgensen, B. P.; Revsbech, N. P.; *Appl. Environ. Microbiol.*, **1983**, 45, pp 1261-1270.
6. Revsbech, N. P.; Jorgensen, B. B.; In *Advances in Marine Ecology*, Marshall, K. C., Ed.; Plenum 1986; pp 293-352.
7. Luther, G. W.; Church,T. M.; Giblin, A. E.; Howarth, R. W.; In *Organic Marine Geochemistry*, Sohn, M., Ed.; ACS Syp. Ser. 305, 1986; pp 340-355.
8. Cutter, G. A.; Oatts, T. J.; *Anal. Chem.* **1987**, 59, pp 717-721.
9. Troelsen, H.; Jorgensen, B. B.; *Estuar., Coast. and Sh. Sci.* **1982**, 15, pp 255-266.
10. Luther, G. W.; Giblin, A. E.; Varsolina, R.; *Limn. Oceanog.* **1985**, 30, pp 727-736.
11. Luther, G. W.; Church, T. M.; Scudlark, J. R.; Cosman, M.; *Science* **1986**, 232 pp 746 -749.
12. Boulegue J.; Lord, C. J.; Church, T. M.; *Geochim. Cosmochim. Acta* **1982**, 46 pp 453-464.
13. Fahey, R. C.; Newton, G. L.; *Meth. Enz.* **1987**, 143, pp 85-95.
14. Mopper, K.; Delmas, D.; *Anal. Chem.* **1984**, 56, pp 2557-2560.
15. Matrai, P. A.; Vetter, R. D.; *Limn. Oceanog.* **1988**, 33, pp 624-631.
16. Mopper, K.; Taylor, B.F.; In *Organic Marine Geochemistry*, Sohn, M., Ed.; ACS Syp. Ser. 305, 1986; pp 324-339.
17. Shea, D.; MacCrehan, W. A; *Anal. Chem.* **1988**, 60, pp 449-454.
18. MacCrehan, W. A. ; Shea, D.; *J. Chromatog.* **1988**, 457, pp 111-125.
19. Luther, G. W.; Church, T. M.; Powell, D.; *Deep Sea Research* **1991**, 38, pp S1121-S1137.
20. San Diego-McGlone, M. L. C.; Cutter, G. A.; *EOS* **1988**, 69, pp 375.
21. Shea, D; Helz, G. R.; *Geochim. Cosmochim. Acta* **1988**, 52, pp 1815-25.
22. Shea, D.; MacCrehan, W. A.; *Sci. Tot. Environ.* **1988**, 73, pp 135-141.
23. Andreae, M. O.; In *The Role of Air-Sea Exchange in Geochemical Cycling*, Buat-Menard, P., Ed.; Reidel 1986; pp 331-362.
24. Kiene, R. P.; Taylor, B. F.; *Nature* **1988**, 332, pp 148-150.
25. Vairavamurthy A.; Mopper, K.; *Nature* **1987**, 329, pp 623-625.
26. Luther, G. W.; *Geochim. Cosmochim. Acta* **1991**, 55, pp 2839-2849.
27. Jorgensen, B. B.; Fossing, H.; Wirsen, C. O.; Jannasch, H. W.; *Deep Sea Research* **1991**, 38, pp S1083-S1103.
28. Jannasch, H. W.; Wirsen, C. O.; Molyneaux, S. J.; *Deep Sea Research* **1991**, 38, pp S1105-S1180.
29. Cornwell, J. C.; Roden, E.; Sampou, P. A.; Capone, D. G.; Abstract 199th ACS Meeting, Boston, April 22-27, 1990.
30. Howarth, R. W.; Teal, J. M.; *Limn. Oceanog.* **1979**, 24, pp 999-1013.

31. Perry, K. A.; Kostka, J. E.; Luther, G. W.; Nealson, K. H.; *Science* **1993**, 259, pp 801-803.
32. Trüper, H. G. in *Sulfur, its Significance for Chemistry, for the Geo-, Bio- and Cosmosphere and Technology*, Müller, A. et al, Eds.; Elsevier, 1984; pp 351-364.
33. Pyzik, A. J.; Sommer, S. E.; *Geochim. Cosmochim. Acta* **1981**, 45, pp 687-698.
34. Burdige, D. J.; Nealson, K. H.; *Appl. Environ. Microbiol.* **1985**, 50, pp 491-497.
35. Stone, A. T.; J. J. Morgan.; In *Aquatic Surface Chemistry*, W. Stumm, Ed.; Wiley, 1987; pp 221-254.
36. Burdige, D. J.; Nealson, K. H.; *Geomicrobiol. J.* **1986**, 4, pp 361-387.
37. Burdige, D. J.; Giekes, J. M.; *Am. J. Sci.* **1983**, 283, pp 29-47.
38. Trüper, H. G. In *Sulfur, its Significance for Chemistry, for the Geo-, Bio- and Cosmosphere and Technology*, Müller, A. et al, Eds.; Elsevier, 1984; pp 367-382.
39. Nelson, D. C.; Castenholz, R. W.; *J. Bacteriol.* **1981**, 147, pp 140-154.
40. Trudinger, P. A.; Loughlin, R. E.; In *Comprehensive Biochemistry*, Neuberger, A. et al Eds.; Elsevier, 1981; pp 165-257.
41. Suzuki, I.; *Biochim. Biophys. Acta* **1965**, 104, pp 359-371.
42. Roy, A. B.; Trudinger, P. A.; *The Biochemistry of Inorganic Compounds of Sulfur*. Cambridge, 1970, pp 207-329.
43. Pickering T. L. Tobolsky, A.V.; In *Sulfur in Organic and Inorganic Chemistry*, Senning, A., Ed.; vol 2, Marcel Dekker 1972, pp 19-38.
44. Howarth, R. W.; Teal, J. M.; *Am. Nat.* **1980**, 116, pp 862-872.
45. Vairavamurthy, A.; Mopper, K.; Taylor, B. F.; *Geophys. Res. Lett,* **1992**, 19, pp 2043-2046.
46. Moses, C. O.; Nordstrom, D. K.; Herman, J. S.; Mills, A. L.; *Geochim. Cosmochim. Acta,* **1987**, 51, pp 1561-1571.
47. Maw, G. A.; In *Sulfur in Organic and Inorganic Chemistry*, Senning, A., Ed.; Marcel Dekker, 1972; pp 113-142.
48. Meister, A.; Anderson, M.; *Ann. Rev. Biochem.* **1983**, 52, pp 711-760.
49. Cooper, A. J. L.; *Ann. Rev. Biochem.* **1983**, 52, pp 187-222.
50. Zinder, S. H.; Brock, T. D.; *Appl. Env. Microbiol.* **1978**, 35, pp 344-352.
51. Kiene, R. P.; Taylor, B. F.; *Appl. Env. Microbiol.* **1978**, 35, pp 2208-2212.
52. Reed, R. H.; *Mar. Biol. Lett.* **1983**, 34, pp 173-181.
53. Vairavamurthy, A.; Mopper, K.; In *Biogenic Sulfur in the Environment*, Saltzman, E. S.; Cooper, W. J. Eds.; ACS Symp. Ser. 393, 1989; pp 231-242.
54. Sorensen, J.; *Biogeochem.* **1988**, 6, pp 201-210.
55. Kiene, R. P.; Visscher, P. T.; *Appl. Environ. Microbiol.* **1987**, 53, pp 2426-2434.
56. Drotar, A.; Burton, G. A.; Travernier, J. E.; Fall, R.; *Appl. Environ. Microbiol.* **1987**, pp 1626-1631.
57. Kiene, R. P.; *FEMS Microbiol. Ecol.* **1988**, 53, pp 71-78.
58. Howarth, R. W.; Giblin, A.; Gale, J.; Peterson, B.J.; Luther, G. W.; *Environ. Biogeochem. Econ. Bull.* **1983**, 35, pp 135-152.
59. Sörbo, B.; *Meth. Enz.* **1987**, 143, pp 178-182.

RECEIVED August 17, 1995

Chapter 17

The Distribution of Free Organic Sulfur Compounds in Sediments from the Nördlinger Ries, Southern Germany

Assem O. Barakat[1] and Jürgen Rullkötter[2]

[1]Department of Chemistry, Faculty of Science, Alexandria University, P.O. Box 426, 21321 Alexandria, Egypt
[2]Institute of Chemistry and Biology of the Marine Environment (ICBM), Carl von Ossietzky University of Oldenburg, P.O. Box 2503, D–26111 Oldenburg, Germany

Black shales in the Miocene crater lake of the Nördlinger Ries (southern Germany) were deposited under largely stagnant conditions in slightly saline, sulfate-rich water in an arid climate. Black shale layers in the sedimentary sequence contain more than 10% organic carbon and up to 5% total sulfur, of which a significant portion is bound to organic matter. The low-molecular-weight organic sulfur compounds in the extractable organic matter consist of thiolanes and thiophenes with a phytane carbon skeleton, structural isomers of C_{33} 2,5-dialkylthiophenes, $C_{33}/C_{37}/C_{38}$ 2,6-dialkylthianes and 2,5-dialkylthiolanes as well as sulfurized steroids. Structures were assigned to these constituents based on gas chromatographic retention times, mass spectrometric fragmentation patterns and analysis of Raney nickel desulfurisation products. There are distinct differences between the distributions of unsulfurized carbon skeletons and carbon skeletons of organosulfur compounds in the extracts, indicating selective interaction of functionalized lipids with inorganic sulfur species during early diagenesis. However, not all of the potentially reactive species are found in the sulfur fraction. For example, ketocarboxylic acids represent up to 50% of some black shale bitumens, and yet no sulfur analogs have been detected.

The Nördlinger Ries is a circular Miocene sedimentary basin 20 km wide located about 100 km NW of Munich in southern Germany (Figure 1). It was formed by meteorite impact into the late Jurassic carbonates of the Schwäbische Alb mountain range about 15 Ma ago (1-2). The extruded material formed a rim around the crater which prevented freshwater supply except by rainfall. Erosion of gypsum deposits at the crater rim supplied sulfate into the lacustrine environment. The climate in southern Germany was semi-arid during the Miocene (3). The basement of the crater

0097–6156/95/0612–0311$12.25/0
© 1995 American Chemical Society

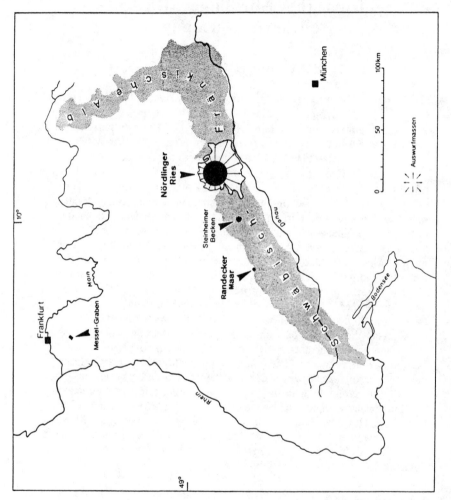

Figure 1. Location map of Nördlinger Ries crater in southern Germany (after 3).

consists of suevite (meteorite impact brecciae) containing high-pressure modifications of quartz. Post-impact sediments were penetrated by research well Nördlingen-1973 near the center of the basin and described in detail by Jankowski (3). They consist of a "basal unit" overlying the suevite, a 140 m thick laminate sequence, which contains two particularly distinct series of highly bituminous black shales, 60 m of marl and a clay layer at the top. Figure 2 shows a stratigraphic correlation of commercial wells NR-10 and NR-30 with the research borehole Nördlingen 1973 (e.g., 1-3). Deposition of the entire laminite series extended over a period of 0.3 to 3 Ma (3).

In a previous study, the organic matter in a variety of black shale samples was characterized using bulk organic geochemical and organic petrographic parameters as well as by the analysis of the aliphatic and aromatic hydrocarbon fractions (4). Several of the sediments - those selected for this study - contained close to or more than 10% organic carbon and were rich in sulfur (Table I). High hydrogen indices were consistent with a predominance of algal and bacterial remains in the sediments. Total bitumen contents were in the range of 80 - 135 mg/g TOC. Microscopically, the organic matter consisted of a mixture of small alginite macerals and homogeneous groundmass (4).

The aromatic hydrocarbon fraction previously was found to actually contain organic sulfur compounds as dominant constituents (4). This study extends the earlier investigation of organic sulfur compounds in the Nördlinger Ries black shales by the assignment of distinct chemical structures to a great number of previously undefined constituents.

Experimental Methods

Sediment samples were supplied by BEB Erdgas und Erdöl GmbH (Hannover), and are from different depths (Figure 2) of commercial wells NR-10 and NR-30 drilled in the Nördlinger Ries in the early 1980s. General information and bulk data of the samples are compiled in Table I.

Sample preparation, extraction, liquid chromatographic separation and gas chromatographic analysis of the aromatic hydrocarbon fractions were described previously (4). The "aromatic hydrocarbon" fractions were desulfurized using Raney nickel in ethanol. Approximately 5 mg of the fraction was dissolved in absolute ethanol (2 ml) and mixed with 0.5 g of a suspension of Raney Ni (Merck, Darmstadt, Germany) in ethanol. The mixture was stirred and refluxed under N_2 for 2 h. The desulfurized products were purified by flash chromatography over silica gel (0.4 x 2 cm) using CH_2Cl_2 as eluent (10 ml), dried, concentrated, and further hydrogenated with PtO_2 in acetic acid at room temperature for 2h.

GC-MS analyses were performed with a Carlo Erba Fractovap 4160 gas chromatograph equipped with a fused silica column (25 m x 0.25 mm i.d.) coated with CP-Sil-5 silicone and coupled to a VG 7070E mass spectrometer operating at 70 eV. Helium was used as carrier gas, and the temperature programmed from 110

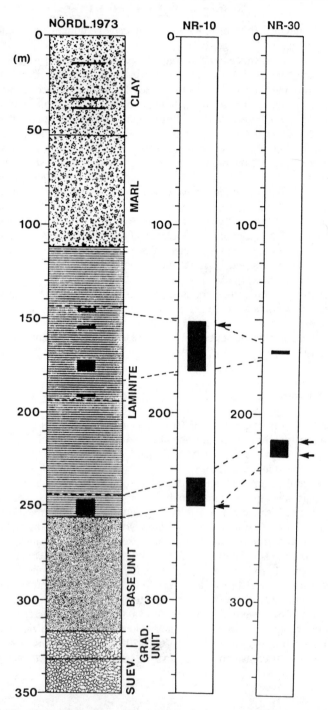

Figure 2. Stratigraphic cross section of boreholes through the Nördlinger Ries sediments showing origin of black shale samples used for this study.

to 310°C at 3°C/min with an initial hold time of 4 min and a final hold time of 20 min. The magnetic field of the mass spectrometer was scanned over a mass range of m/z 45 - 900 at a rate of 2.5 s/scan. Data were acquired, stored and processed using a Kratos DS 90 data system.

Table I. General information and bulk data of black shale samples used in this study

Sample	C_{org} (%)	S_{tot} (%)	Hydrogen Index (mg hc/g C_{org})	Extract yield (mg/g C_{org})	Aromatic fraction (% of extract)
NR-10 (151.5 m)	12.3	1.7	470	132.4	2.2
NR-10 (250.0 m)	14.4	3.0	863	81.5	4.7
NR-30 (215.1 m)	10.3	3.5	626	70.8	7.0
NR-30 (222.9 m)	8.4	4.3	480	61.7	6.5

Results and Discussion

Figure 3 shows gas chromatograms of the "aromatic hydrocarbon" fractions of three black shale extracts. The traces are dominated by a variety of organic sulfur compound classes (A through F). The differences among the samples demonstrate that organic matter accumulation and transformation during early diagenesis was quite variable for the different black shale layers both in time and laterally within the former crater lake. This is not surprising, since fluctuations in the supply of freshwater to this restricted environment may, for example, have caused significant salinity changes and thus ecosystem changes in the lake over relatively short periods.

Type A. Alkylated thiophenes and thiolanes with an acyclic carbon skeleton are the single most abundant components in each fraction (Figure 3). Figure 4 shows the carbon number distributions and absolute concentrations of alkyl thiophenes with unbranched carbon skeletons but different alkyl substitution patterns on the thiophene ring (or different positions of the thiophene ring within the carbon chain) for one sample each from wells NR-10 and NR-30 (both from second black shale event). Differences are obvious both in concentration and distribution, again illustrating lateral organic facies changes. In addition, the concentrations of the C_{20} compounds with a phytane skeleton which are most abundant in both samples and in each isomer class are shown in Figure 4. Isoprenoid thiophene homologs were found as well but at much lower concentrations and, therefore, are not shown.

Table II summarizes the concentration data for C_{20} isoprenoid thiophenes and thiolanes. The two shallower samples in each well are enriched in thiophenes whereas the other two samples show a dominance of thiolanes. Interestingly, the isoprenoid thiophene ratio (ITR) introduced by Sinninghe Damsté et al. (5) and de Leeuw and Sinninghe Damsté (6) to determine paleosalinity, varies with this difference in thiolane and thiophene concentrations. Because thiolanes and thiophenes are thought to be diagenetically related, both compound types should be

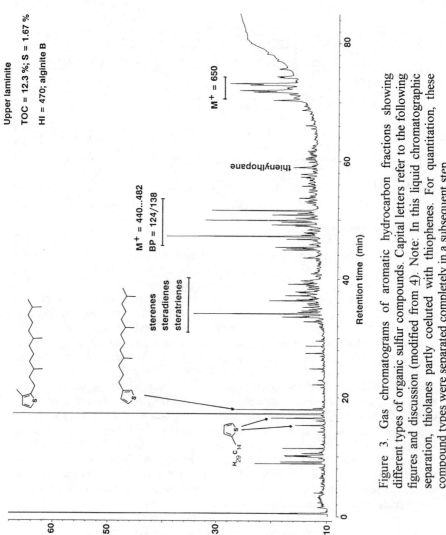

Figure 3. Gas chromatograms of aromatic hydrocarbon fractions showing different types of organic sulfur compounds. Capital letters refer to the following figures and discussion (modified from 4). Note: In this liquid chromatographic separation, thiolanes partly coeluted with thiophenes. For quantitation, these compound types were separated completely in a subsequent step.

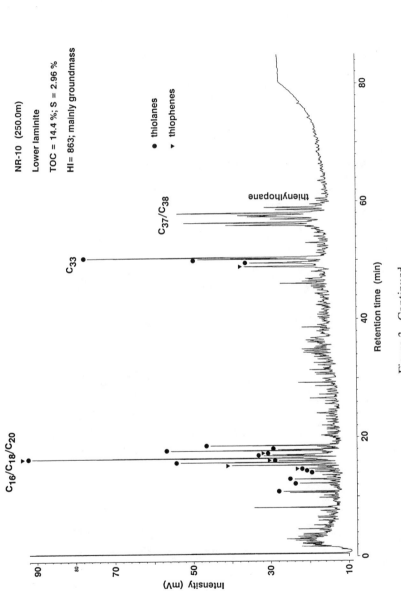

Figure 3. Continued.

Continued on next page

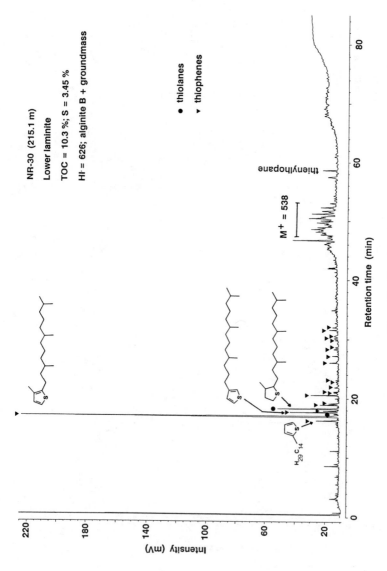

Figure 3. Continued.

included into this ratio in order to avoid diagenetic perturbation effects. Consequently, ITTR (isoprenoid thiophene plus thiolane ratio) values were calculated (Table II). Although there is indeed a change in the absolute values compared to the ITR, the relative relationships among the samples are not changed significantly. This means that despite the observed general variations of the thiophene distributions (Figure 4), the isomer ratio of the isoprenoidal organic sulfur compounds does not change much between thiolanes and thiophenes.

Table II. Concentrations of C_{20} thiophenes and thiolanes and salinitity indicators ITR (6; thiophenes only) and ITTR (including thiolanes)

Sample	C_{20} Isoprenoid Thiophenes				C_{20} Isoprenoid Thiolanes			
	a	b	c	ITR*	a'	b'	c'	ITTR**
	µg/g extract				µg/g extract			
NR-10 (151.5 m)	13	215	71	23	-	-	-	23
NR-10 (250.0 m)	74	133	28	2	372	562	-	1.6
NR-30 (215.9 m)	47	1681	246	41	25	51	-	28
NR-30 (222.9 m)	72	29	17	0.6	738	1255	-	1.6

*ITR = (b + c)/a   **ITTR = (b + c + b' + c')/(a + a')

The diagenetic relationship, on the other hand, does not exhibit a straightforward depth (age, temperature) relationship in terms of conversion of the supposedly less stable thiolanes into the corresponding thiophenes, because the deeper (older) samples from the Nördlinger Ries contain more thiolanes. According to our results, salinity may have varied considerably even during the deposition of a single black shale unit, which may have lasted less than 10,000 years. This conclusion is consistent, however, with the general variability of the organic sulfur compound distributions, if this is considered a facies indicator.

Type B. Steroids with an extended carbon skeleton and a thiophene moiety are the second group of compounds. Figure 5a shows distribution patterns based on mass fragmentograms from GC-MS analysis for several homologous series found in the black shale from 151.5 m depth of well NR-10. Mass spectral data indicate that the steroids contain a carbon chain extension at C-3 ranging from two to six carbon atoms, and the thiophene sulfur atom is attached to C-2. Dominant fragmentation then occurs by benzylic cleavage through ring A leading to key ions at m/z 110+14n (n=0-4) (compare spectrum in Figure 6, top).

Two main series of thiophene steroids were detected, one of them corresponds to the regular 4-desmethyl steroids, while the other one - based on the relative

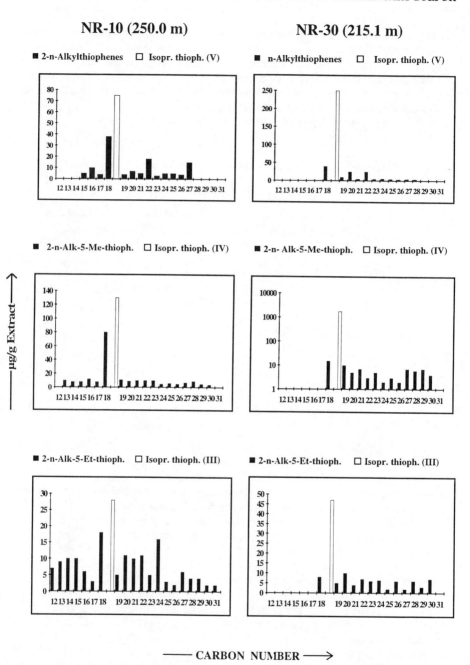

Figure 4. Concentrations and carbon number distributions of *n*-alkyl and isoprenoid thiophenes in Nördlinger Ries black shales. For structural type identification see Table II.

retention time and the fragmentation pattern - carries a 4α-methyl group (hatched peaks in Fig. 5a). Besides the main series of thiophene steroids, three Δ^{22} thiophene steroids, with closely related mass spectra were also found in the sample, although in relatively low concentrations (components marked with asterisks in Fig. 5a). This interpretation is supported by the sterane distribution after Raney Ni desulfurization which contains a series of 3-ethyl- to 3-hexylcholestanes together with some 4-methyl analogs (Figure 5b; Figure 6, bottom; 7-8). Within the series with identical carbon number extension at C-3, pseudohomologous series are due to alkylation at C-24 and isomerization possibly at C-5 as indicated by differences in B-ring fragmentation quite analogous to that observed for regular steranes.

For a given alkylation at C-3, pseudohomologous series are due to alkylation at C-24 and isomerization, possibly at C-5. The fact that 23,24-dimethylsteroids were also found in this group of thiophenes (Figure 5, middle) shows that steroids with extended carbon chains at C-3 are widely distributed among the principal carbon skeletons encountered in biological systems and the geosphere. Since C-3 extended steroids have not been found in natural product surveys, the structural diversity of the steroid thiophenes may indicate that the carbon extension at C-3 is of diagenetic rather than of (planktonic) biogenic origin.

Dahl et al. (7) noted that the length of the carbon chain (≤6) is compatible with the (microbial?) carbon-carbon bond attachment of a sugar unit to sedimentary sterenes, a process which would be analogous to the microbial extension of hopanes. This idea was developed from the observation that reduction of sulfur-bound 3-alkyl steranes with Raney nickel in the presence of deuterium yielded products which were multiply deuteriated in the side-chain. There was also a striking similarity in the distributions of non-alkylated and alkylated steranes in the samples investigated by Dahl et al. (7). The dominance of the 3-pentyl over the 3-butyl and 3-hexyl sterane pseudohomologs was interpreted by them as an indication that compounds with this carbon skeleton may have been particularly useful for the microorganism as membrane constituents, because there is obviously an analogy in the shape and dimensions of these molecules and the carbon skeletons of bacteriohopanetetrols (7).

During diagenesis, the alcohol groups of the sugar moiety may have been a suitable substrate for thiophene formation by reaction with inorganic sulfur species. This was not the only diagenetic reaction pathway of the steroids with carbon-carbon bond extension at C-3, because corresponding 3-alkanoic steroid acids were also formed in the Nördlingeur Ries sediments (9). The fact that small amounts of steranes with a C_7 alkylation at C-3 have also been found by Dahl et al. (7) does not invalidate the sugar moiety hypothesis, because hopanes with side-chain extensions beyond C_5 or C_6 have occasionally been found in sediments, also in very small concentrations (10).

Types C and D. Mid-chain thianes and thiolanes are important constituents in the aromatic bitumen fraction of the black shale from 250.0 m of well NR-10 (11; Figure 3). A partial total ion chromatogram and corresponding molecular ion and key ion traces are shown in Figue 7. Desulfurisation revealed C_{33}, C_{37} and C_{38} n-alkanes as the most abundant products. The mass spectra of selected long-chain organic sulfur compounds indicate that the complex isomer mixture could not fully be resolved by the analytical method used (11).

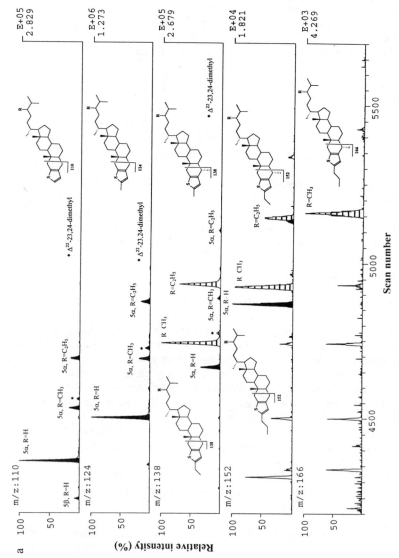

Figure 5. Mass chromatograms of key fragments of sulfur-bearing steroid derivatives in Nördlinger Ries black shales (a) and corresponding saturated hydrocarbons after desulfurization with Raney Ni (b). Black peaks correspond to the 4-desmethyl structures, hatched signatures indicate 4-methyl analogs.

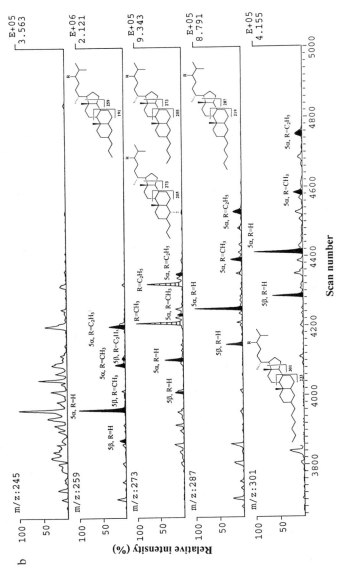

Figure 5. Continued.

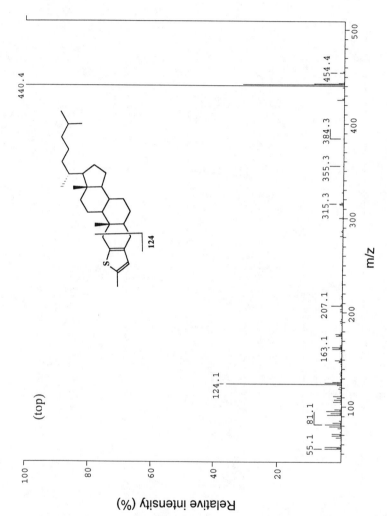

Figure 6. Mass spectra of sulfur-bearing steroid derivatives in Nördlinger Ries black shales (top and middle) and of a corresponding saturated hydrocarbon after desulfurization with Raney nickel (bottom). Structures are based on mass spectral interpretation and com-parison with distribution patterns of 3-alkanoic steroids also found in these samples.

Continued on next page

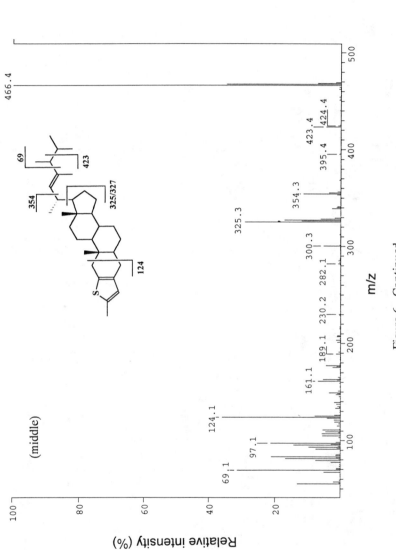

Figure 6. Continued.

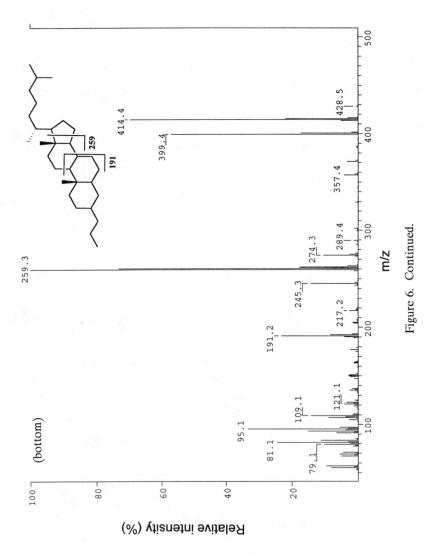

Figure 6. Continued.

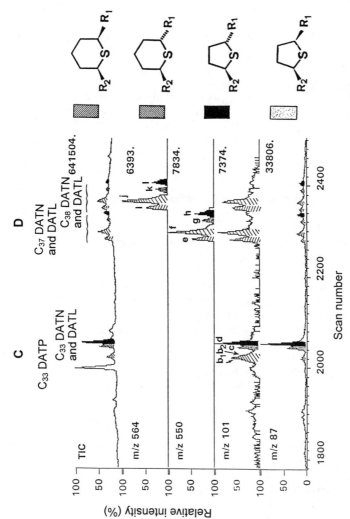

Figure 7. Partial expanded total ion current and mass chromatograms of key fragments and molecular ions of mid-chain C_{33}, C_{37}, and C_{38} dialkylthianes (DATN) and -thiolanes (DATL), and a C_{33} dialkylthiophene (DATP) in the aromatic hydrocarbon fraction of the sample from 250.0 m of well NR-10.

The C_{33} thiolane compounds appear to be mainly comprised of 2-hexyl-5-tricosylthiolane and 2-pentyl-tetracosylthiolane, but other alkylation patterns (position of thiolane ring along the carbon chain) also occur, although slightly less pronounced than with the thianes. In addition, stereoisomers at the alkylation site also occur. The origin of the precursor compound(s) is not known, although a corresponding group of thiophenic organosulfur compounds has been described earlier (e.g., 12).

The C_{37} and C_{38} compounds are also complex isomer mixtures, but no terminal thianes and thiolanes were found. The sulfur compounds are likely to have been formed from long-chain alkenones - or the corresponding long-chain alkadienes and alkatrienes - characteristic of certain Prymnesiophyte algae (cf. 13-14). The absence of terminal sulfur-bearing rings indicates that thiane and thiolane formation by reaction with reduced inorganic sulfur species occurred exclusively via the double bonds and not through the keto group. This does not exclude, however, that the isolated doubled bonds - like the keto group (15) - in a competitive reaction formed intermolecular sulfur bonds as well.

Abundant C_{37} and C_{38} OSC with n-alkane carbon skeletons have been noted before in the upper Cretaceous Jurf ed Darawish oil shale (Jordan; 12,16), a Miocene marl from the Northern Appennines (Italy; 17), the phosphatic unit of the Lower Maastrichtian Timahdit oil shale (Morocco; 18), and the Rozel Point oil (Utah; 19). Of these, only the source rock of the Rozel Point oil is related to a lacustrine depositional environment which obviously was much more saline than the Nördlinger Ries crater lake.

Types E and F. Two groups of organic sulfur compounds are represented by mass spectra in Figure 8, but no structural assessment was possible. Desulfurisation did not yield corresponding high-molecular-weight hydrocarbons. This suggests sulfur compounds consisting of smaller units. The mass spectral isotope pattern of the molecular ions indicate that the compounds contain one or at most two sulfur atoms. The presence of an m/z 280 ion in the mass spectrum of the MW 538 compound may indicate the presence of a C_{20} (phytanyl?) hydrocarbon unit.

Summary and Conclusions

Organic sulfur compound distributions in Nördlinger Ries sediments are highly variable. Drastic differences occur between the two black shale events as well as within a single event. Short-term fluctuations of depositional conditions including the paleoecosystem are likely, considering the unique geological and geographical situation of the Miocene crater lake.

Calculation of ITTR ratios (including thiolanes) from the isomer distributions of isoprenoid organic sulfur compounds yields the same gross trend as the ITR ratios, but different absolute values.

We identified a new series of thiophene steroids based on C-3 alkylated skeletons. The carbon number distribution (alkylation up to C_6) suggests C-C-addition of a hexose unit (probably diagenetic) similar to extension of hopanes in the sidechain. The variation of structural types including that of 23,24-methylsteroids in this series favors the diagenetic formation hypothesis.

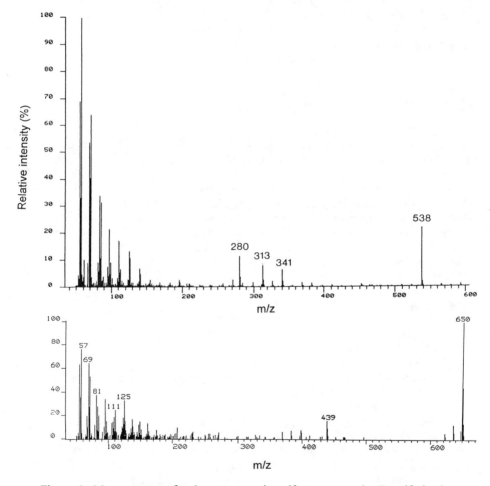

Figure 8. Mass spectra of unknown organic sulfur compounds. Desulfurization does not yield corresponding high-molecular-weight saturated hydrocarbons.

The formation of all sulfur compounds described in this study can be explained by reaction of inorganic sulfur species with double bonds or alcohol groups (potential double bonds). Keto and carboxylic acid groups appear to be less effective substrates for low-molecular-weight thiophene formation. This is concluded from the absence of terminal thiophenes in the C_{37}/C_{38} series and the presence of large amounts of C_{16} and C_{18} 10-oxo-carboxylic acids (20) and of steroid ketones (unpublished) in the Nördlinger Ries black shales. This conclusion is also consistent with the observation of Schouten et al. (15) that saturated ketones tend to form intramolecular sulfur bonds by reaction of the keto group with reduced inorganic sulfur species if no additional double bonds are present.

Acknowledgments

We thank Dr. R. Littke (KFA Jülich) for sample selection and discussions, Drs. M. Radke and R.G. Schaefer (KFA Jülich) and Dr. B. Scholz-Böttcher (ICBM Oldenburg) for supporting extraction/liquid chromatography, gas chromatography and GC-MS analysis, respectively. We appreciate technical support by A. Fischer, U. Disko, R. Harms, J. Hoeltkemeier, B. Kammer, F.-J. Keller, H. Willsch (KFA Jülich) and R. Grundmann (ICBM Oldenburg). Dr. M.A. McCaffrey (Chevron, La Habra) and an anonymous referee thoroughly reviewed the original manuscript and made valuable suggestions for improvement. BEB Erdöl und Erdgas GmbH (Hannover) kindly provided the samples and granted permission to publish. A.O.B. thanks the Alexander von Humboldt Foundation for a grant.

References

1.   Hollerbach A.; Hufnagel H.; Wehner H. Geol. Bavarica 1977, 75, 139-53.
2.   Wolf M. Geol. Bavarica 1977 75, 127-38.
3.   Jankowski B. Bochumer Geol. und Geotechn. Arb., 1981, 6, 315 pp.
4.   Rullkötter J.; Littke R.; Schaefer R.G. In Geochemistry of Sulfur in Fossil Fuels; Orr W.L.; White C.M., Eds.; American Chemical Society: Washington, 1990; pp. 149-69.
5.   Sinninghe Damsté J.S.; Rijpstra W.I.C.; de Leeuw J.W.; Schenck P.A. Geochim. Cosmochim. Acta 1989, 53, 1323-41.
6.   de Leeuw J.W.; Sinninghe Damsté J.S. In Geochemistry of Sulfur in Fossil Fuels; Orr W.L.; White C.M., Eds.; American Chemical Society: Washington, 1990; pp. 417-43.
7.   Dahl J.; Moldowan J.M.; McCaffrey M.A.; Lipton P.A. Nature, 1992, 355, 154-7.
8.   Schaeffer P.; Fache-Dany F.; Trendel M.; Albrecht P. Org. Geochem. 1993, 20, 1227-36.
9.   Barakat A.O.; Rullkötter J. Energy & Fuels 1994, 8, 481-6.
10.   Rullkötter J.; Philp R.P. Nature, 1981, 292, 616-8.
11.   Barakat A.O.; Rullkötter J. Energy & Fuels 1994, 8, 1168-74.

12. Kohnen M.E.L.; Sinninghe Damsté J.S.; Rijpstra W.I.C.; de Leeuw J.W In Geochemistry of Sulfur in Fossil Fuels; Orr W.L.; White C.M., Eds.; American Chemical Society: Washington, 1990; pp. 444-85.
13. Volkman J.K.; Eglinton G.; Corner E.D.S.; Forsberg T.E.V. Phytochemistry, 1980, 19, 2619-22.
14 Marlowe I.T.; Green J.C.; Neal A.C.; Brassell S.C.; Eglinton G.; Course P.A. Br. Phycol. J., 1984, 19, 203-16.
15. Schouten S.; van Driel G.B.; Sinninghe Damsté J.S.; de Leeuw J.W. Geochim. Cosmochim. Acta 1994, 58, 5111-6.
16. Sinninghe Damsté J.S.; Rijpstra W.I.C.; Kock-van-Dalen, A.C.; de Leeuw J.W.; Schenck P.A. Geochim. Cosmochim. Acta, 1989, 53, 1343-1355.
17. Sinninghe Damsté J.S.; Eglinton T.I.; Rijpstra W.I.R.; de Leeuw J.W. In Geochemistry of Sulfur in Fossil Fuels; Orr W.L.; White C.M., Eds.; American Chemical Society: Washington, 1990; pp. 486-528.
18. Meunier-Christmann C. Ph.D. Thesis, 1988, Université de Strasbourg.
19. Sinninghe Damsté J.S.; Rijpstra W.I.C.; de Leeuw J.W.; Schenck P.A. In Advances in Organic Geochemistry 1987; Mattavelli L.; Novelli L., Eds.; Org. Geochem. 1988, 13, 593-606.
20. Barakat A.O.; Peakman T.M.; Rullkötter J. Org. Geochem., 1994, 21, 841-7.

RECEIVED July 28, 1995

Chapter 18

Sulfate Incorporation into Sedimentary Carbonates

Wilfried J. Staudt and Martin A. A. Schoonen

Department of Earth and Space Sciences, State University of New York, Stony Brook, NY 11794–2100

Modern and ancient sedimentary carbonate minerals (calcite, aragonite and dolomite) generally contain sulfate as a trace constituent with concentrations ranging from approximately 200 to 24,000 ppm. Particularly marine skeletal carbonate components, which account for most of the global calcium carbonate accumulation, contain considerable amounts of sulfate. It is estimated that marine calcium carbonate precipitation accounts for the removal of approximately 5% of the pre-anthropogenic annual riverine sulfate input. Carbonate shelves and platforms that account for 33% to 42% of the present-day calcium carbonate accumulation have estimated average sulfate concentrations of 5900 ppm. These environments, which represent only about 3.3% of the total area where calcium carbonate is produced, account for 55% to 77% of the present-day sulfate removal by calcium carbonate accumulation. Low sulfate content of burial calcites and dolomites is consistent with the release of sulfate during recrystallization of sulfate-rich shelf and platform carbonates under burial conditions. Release of carbonate-bound sulfate upon burial recrystallization of sedimentary carbonates may represent a significant sulfur source in sedimentary basins.

Carbonate rocks comprise approximately 15% of the sedimentary rock volume of the continental crust (*1*). Sedimentary carbonates, which have been formed throughout the geologic rock record, are mostly, although not exclusively, of marine origin (*2*) with calcium carbonate secreting organisms accounting for most of the modern carbonate production (*3*). In modern seawater sulfate is the third most abundant ion (after chloride and sodium) and the second most abundant anion. Incorporation of seawater sulfate in marine skeletal carbonates has been reported by Oomori et al. (*4*), Volkov and Rozanov (*5*) and Busenberg and Plummer (*6*). Sedimentary carbonates, which

0097–6156/95/0612–0332$12.00/0

were formed by direct precipitation from seawater or modified seawater, such as dolomites also can contain a considerable amount of sulfate (*7, 8*). How sulfate is present in sedimentary carbonates has received considerable attention. In biologically precipitated carbonates some of the sulfate may be present as an ester sulfate (*9*) or as lattice-bound sulfate. Lattice-bound sulfate in sedimentary carbonates may either substitute for the carbonate ion or it may in principle be present in interstitial sites. Because the sulfate ion is considerably larger than the carbonate ion and it is tetrahedral instead of planar, it has been conventional wisdom that the substitution of sulfate for carbonate is limited. Based on this notion several studies have therefore speculated that the sulfate in biologically precipitated carbonates is predominantly present as sulfate esters (*10*). However, infrared and Raman spectroscopic studies indicate that sulfate in corals, oysters and travertines is dominantly present as inorganic sulfate (*11*). In addition, studies by Busenberg and Plummer (*6*), Takano et al. (*12*), Staudt et al. (*8*) and Staudt et al. (*13*) support the notion that sulfate substitutes for the carbonate ion in synthetic calcite, travertine and natural dolomites. Reeder et al. (*14*) analyzed the Se K-edge XAFS spectra of selenate-bearing synthetic calcite crystals. Their study confirms that Se is present as the tetrahedral selenate complex anion substituting for the carbonate site of calcite. Because sulfate is isostructural with selenate but approximately 10% smaller it is reasonable to assume that sulfate also substitutes for the carbonate ion. Furthermore, based on analysis of the sulfur K-edge x-ray absorption near-edge structure (XANES) of natural carbonates Pingitore et al. (*15*) documented that sulfur is present in the 6+ oxidation state which is consistent with the presence of sulfate.

The objective of this study is to evaluate the importance and geological implications of sulfate incorporation into sedimentary carbonates. To attain this objective more than 250 sedimentary carbonate samples including primary, dominantly biologically precipitated aragonites and calcites, as well as diagenetically formed calcites and dolomites were analyzed for their sulfate content. The results of this study indicate that modern marine calcium carbonate production represents a sink for oceanic sulfate. Upon burial and recrystallization of sulfate-rich sedimentary carbonates most of the initially incorporated sulfate may be released. This release of carbonate-bound sulfate may represent a significant sulfur source for the formation of H_2S, elemental sulfur or metal sulfide deposits in the subsurface of sedimentary basins.

Research Strategy and Methods

To evaluate the importance of sulfate incorporation into sedimentary carbonates we analyzed a large variety of skeletal and non-skeletal carbonate minerals. These samples which represent sedimentary carbonates of various origins from the Ordovician to the Recent generally contain sulfate as a trace constituent. Sulfate incorporation into sedimentary carbonates can be best evaluated using modern marine skeletal components because these have not been altered by diagenetic processes, such as recrystallization. Hence, all marine skeletal carbonates analyzed for sulfate were collected from living organisms. The marine carbonate secreting organisms selected for this study account for the majority of the sediment production in modern carbonate accumulating environments. Non-biogenic carbonates such as calcite cements and dolomites account for only a small fraction of the modern carbonate accumulation. To constrain the amount of sulfate incorporated into calcite cements and dolomites we analyzed samples that were formed by fluids covering a range of sulfate concentrations. The depositional environment and diagenetic history of these dolomites and calcites has been well established based on detailed field and petrographic studies. Using average sulfate concentrations of skeletal and non-skeletal carbonates, their relative abundance in carbonate accumulating environments and an estimate of the global calcium carbonate accumulation we estimated the total removal of oceanic sulfate by calcium carbonate precipitation.

The fate of sulfate in sedimentary carbonates was evaluated by comparing an early-diagenetic dolomite of the Burlington-Keokuk Formation with its later-diagenetic dolomite replacement. Dolomites of the Burlington-Keokuk Formation were chosen because their depositional environment, diagenetic history and relative timing of dolomitization have been well documented.

Origin of Samples. More than 250 sulfate analyses of modern marine skeletal and non-skeletal calcites, aragonites, as well as calcites and dolomites from the geologic record (Ordovician to Recent) form the basis for this study. Carbonate skeletal components (58) were provided from the collections of P. Bretsky and W.J. Meyers. All modern marine skeletal carbonates were collected from living organisms at several locations in the Caribbean Sea and along the east coast of the United States. Sulfate data of modern marine carbonates were supplemented by 20 analyses published by Oomori et al. (4), Volkov and Rozanov (5), and Busenberg and Plummer (6). Previously well characterized dolomites (141) and calcites (43) were also analyzed for their sulfate content. Their location, age and presumed depositional environment are summarized in Table 1.

Sulfate Analysis. A Dionex 2000i Ion Chromatograph (IC) with an AS4A column was used to determine sulfate concentrations in carbonate samples following the method described by Staudt et al. (8). For dolomites a 10 mg portion, and for calcites and aragonites a 2 mg portion of the powdered and rinsed sample material was dissolved in dilute nitric acid and subsequently diluted with deionized water by a factor of 1000 (dolomites) or 2000 (calcites and aragonites). Matrix matching standards for IC analysis were prepared by gravimetric dilution from single stock solutions. Detection limit for sulfate determination in carbonates, determined as three standard deviations of the blank for $6<n<13$ consecutive runs was 35 ppm. The analytical uncertainty, calculated as the standard deviation of the mean was better than $\pm3\%$.

Dissolution tests of mixtures of sulfate-free calcite and pyrite (Ward's: Huanzola, Peru; using the size fraction from 95 - 125 µm) show that some pyrite leaching and subsequent sulfide oxidation occurs, which may result in a too high estimate of the sulfate content of carbonates associated with sedimentary pyrite. Sulfate contributions from samples containing less than 10 wt.% pyrite are below the detection limit, for up to 20 wt.% pyrite 120 ppm, and for 30 to 40 wt.% pyrite between 380 and 600 ppm sulfate. Although pyrite contamination represents a potential source of non-carbonate derived sulfate, this does not significantly affect our data or its interpretation. Firstly, modern marine skeletal carbonates were collected from living organisms and are therefore pyrite-free. Secondly, the petrography and

Table 1.
Location, age and proposed origin of studied diagenetic calcites and dolomites based on field, petrographic and geochemical evidence by referenced authors

	Location	Mineralogy	Age of Formation	Number of samples	Source/Author
Seawater-derived Evaporitic Brines	Mallorca/Spain	Dolomite	Miocene	39	36
	Nijar/Spain	Dolomite	Miocene	5	Packmohr, pers. com.
	Las Negras/Spain	Dolomite	Miocene	6	Packmohr, pers. com.
	Turks & Caicos, West Indies	Dolomite	Holocene	3	37
Seawater	Lower Clear Fork Fm., TX,	Dolomite	Permian	5	Kaufman, pers. com.
Seawater/ Fresh Water Mixtures	NE Yucatan Peninsula, Mexico	Dolomite	Pleistocene	2	38
	Seroe Domi Fm., Curacao, West Indies	Dolomite	Mio-Pleistocene	21	39, 40
	Burlington-Keokuk Fm., Illinois-Missouri, DI	Dolomite	Mississippian	26	28, 29, 30, 31
	Burlington-Keokuk Fm., Illinois-Missouri, DII	Dolomite	Mississippian	14	28, 29, 30, 31
Burial Brines	Swan Hills Fm., Alberta, Canada	Dolomite	Devonian	16	41
	El Paso Group, Franklin Mts., IL	Dolomite	Ordovician/Silurian	4	Oswald, pers. com.
Seawater	Canning Basin, Western Australia	Calcite	Devonian	10	42
	Seroe Domi Fm., Curacao, West Indies	Calcite	Mio-Pleistocene	3	39
Travertines	Turner Falls	Calcite	Holocene	7	Hemming, pers. com.
Cave Cement	Permian Reef, N.M./Canning Basin, Australia	Calcite	Pleistocene/Holocene	3	Ward, pers. com.
Cave Cements	Canning Basin, Western Australia	Calcite	Carboniferous	4	Ward, pers. com.
Burial Brine	Waulsortian Limestone, Ireland	Calcite	Mississippian	8	44
	Burlington-Keokuk Fm., Illinois-Missouri	Calcite	Mississippian	4	43
	Laborcita Fm. Sacramento Mountains, N.M.	Calcite	Permian	4	Ward, pers. com.

geochemistry of diagenetically formed dolomites and calcites used in this study were previously well-characterized (Table 4.1). Although pyrite crystals have been observed in Burlington-Keokuk Dolomite II thin sections (Meyers, pers. commun.), pyrite content in excess of 10 wt.% would have resulted in the rejection of a sample. Even at less than 10 wt.% pyrite content there is a clearly visible insoluble heavy residue in the test sample solutions. Furthermore, a high pyrite content would have been detected by petrographic and XRD analysis.

Results and Discussion

Sulfate in Carbonates. Sulfate concentrations of more than 250 aragonites, calcites and dolomites are summarized in Figs. 1 and 2. Below, the results for modern marine carbonate components and calcites and dolomites from the geologic rock record are discussed in detail.

Modern Marine Carbonate Components. Sulfate concentration ranges and averages of modern marine carbonates are summarized in Fig. 1. Although our data documents that modern marine carbonates generally contain sulfate as a trace constituent the concentrations vary considerably as seen Fig. 1. For example, modern pelagic skeletal components, which account for most of the deep sea carbonate sediments (i.e., coccoliths and planktonic foraminifera), have relatively low sulfate concentrations of approximately 1000 ppm sulfate (6). In contrast, modern shallow-marine carbonates (e.g. corals, codiacean and coralline algae, and benthonic foraminifera) have considerably higher sulfate concentrations between 2300 and 16,000 ppm. Furthermore the sulfate content of different species within the same taxa vary significantly. For example, average sulfate concentrations of aragonitic scleractinian corals, 5750 ppm (e.g. *Acropora cervicornis, Porites porites, Siderastrea, Montastrea annularis*), are significantly lower than in magnesium calcite forming octocorals (e.g. *Gorgonians*) with 10,800 ppm. Even more striking are the differences among foraminifera. The high-magnesium calcite forming benthonic foraminifera *Peneroplids* (7600 ppm) and the encrusting foraminifera *Homotrema rubrum* (12,100 ppm) have a seven to twelve times higher sulfate concentration than the planktonic foraminifera *Globigerina sp.* (1030 ppm) which forms low-magnesium calcite. All marine carbonate-secreting organisms have seawater as the same source of sulfate. Hence, the documented variation in sulfate content is consistent with a dominantly biological effect on the incorporation of sulfate into biologically precipitated calcium carbonate.

Dolomites and Calcites. Sulfate concentrations of calcites and dolomites from the Ordovician to the Pleistocene also vary significantly (see Fig. 2). Unlike biologically precipitated modern marine carbonates, the sulfate content of non-biogenic carbonate minerals appears to correlate with the sulfate concentration, or more precisely, the SO_4/CO_3 ratio of the precipitating fluid. In shallow oxic marine environments sulfate is the second most abundant anion (e.g., seawater and seawater-derived evaporitic brines), whereas it is nearly absent in subsurface fluids due to reduction (*16*). Staudt et al. (*8*) suggested that the sulfate content in dolomites is

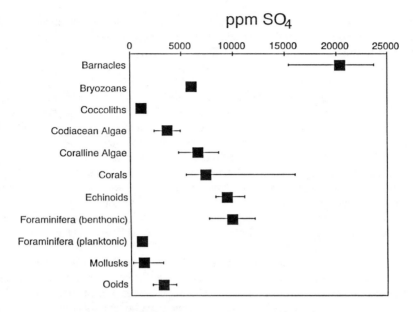

Figure 1.   Sulfate concentration ranges (horizontal bars) and average concentrations (solid squares) of modern marine aragonitic and calcitic carbonate components.

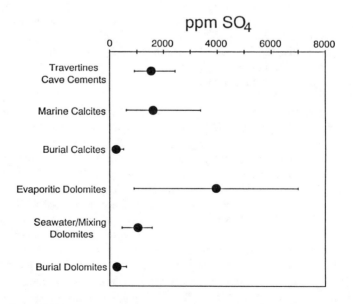

Figure 2.   Sulfate concentration ranges (horizontal bars) and average concentrations (solid circles) of well-characterized calcite and dolomite samples from the geologic rock record (Ordovician to Miocene).

related to the SO_4/CO_3 ratio of the dolomitizing fluid. This is documented by low sulfate concentrations (generally less than 500 ppm) in dolomites that presumably were formed under reducing conditions in subsurface environments, i.e. sulfate-depleted environments. Sulfate concentrations of dolomites formed in the near-surface environment range from 600 to 7000 ppm (8). Dolomites formed from seawater/fresh water mixtures and seawater have sulfate concentrations between 600 and 1800 ppm, those formed by seawater-derived evaporitic brines have up to 7000 ppm. Sulfate concentrations in ancient non-skeletal calcites vary similarly. Like burial dolomites, calcites that are precipitated in reducing subsurface environments have low sulfate concentrations (generally less than 500 ppm SO_4) compared to travertines, cave cements and marine calcites with 600 to 3300 ppm SO_4. It is important to note, however, that dolomites and calcites have significantly lower sulfate concentrations than most modern skeletal carbonates. The lower sulfate concentrations combined with the lack of massive recent dolomite or inorganic calcite formation suggests that these non-skeletal carbonates contribute little to the removal of sulfate from the oceans.

Modern Marine Carbonates: A Sink for Oceanic Sulfur. High sulfate concentrations in carbonates of marine origin indicate that carbonate minerals represent a potentially important sink for oceanic sulfate. To estimate the magnitude of the present-day removal of oceanic sulfate by calcium carbonate precipitation we only considered skeletal carbonate components because modern marine carbonate sediments are almost entirely biologically precipitated (Wilson, 1975). The estimated relative abundance (wt. %) of modern marine carbonate components in major carbonate-accumulating environments are summarized in Table 2. The relative abundance of skeletal and non-skeletal carbonates was estimated based on published data on the average composition of carbonate sediments in such environments by Bathurst (17), Milliman (18, 19) and Wilson (3).

To evaluate the significance of sulfate removal by biological calcium carbonate precipitation, the sulfate data of modern marine carbonate components, summarized in Fig. 1, were combined with a recent detailed compilation of the oceanic calcium carbonate budget by Milliman (19) who calculated the relative rate of calcium carbonate production and accumulation in carbonate-producing environments in the late Holocene oceans. Milliman (19) estimated the global annual calcium carbonate production to be $5.3*10^{13}$ moles per year, of which 60% is preserved ($3.2*10^{13}$ moles). Figure 3 shows the calculated contributions (wt. %) of each carbonate-accumulating environment to the total of $3.2*10^{13}$ moles of calcium carbonate that accumulated in the late Holocene per year (19). Using average concentrations of modern marine carbonates (Fig. 1, Table 2), their relative abundance in carbonate accumulating environments (Table 2), and Milliman's (19) estimate of the calcium carbonate accumulation (Fig. 3), we calculated the contribution of each carbonate-accumulating environment to the total removal of sulfate from the oceans (Fig. 4). The total present-day removal of oceanic sulfate by calcium carbonate precipitation, is estimated to be $1.5*10^{11}$ moles per year. Based on estimates of the natural pre-anthropogenic river flux of $3.2*10^{12}$ moles per year (20, 21, 22), $1.5*10^{11}$ moles account for approximately 5% of the annual riverine sulfate input into the oceans. Removal of oceanic sulfate by modern marine skeletal calcium carbonate precipitation indicates that sedimentary carbonates represent a minor sink for oceanic sulfur.

Table 2.
Sulfate content and relative abundance (wt.%) of modern marine carbonate components in carbonate-accumulating environments

	Average ppm SO_4	Coral Reefs	Banks/ Bays	Clastic Shelves	Carbonate Shelves	Halimeda Bioherms	Slopes	Slopes[b] (imported)	Enclosed Basins	Deep Sea
Barnacles	20300 (3)[a]	[c]	[c]	10%	5%	[c]	10%	2%	[c]	[c]
Bryozoans	6100 (4)	1%	2%	[c]	10%	[c]	20%	8%	[c]	[c]
Coccoliths	1025 (1)	[c]	[c]	[c]	[c]	[c]	5%	[c]	50%	60%
Codiacean Algae	3620 (5)	24%	20%	[c]	10%	100%	[c]	15%	[c]	[c]
Coralline Algae	6630 (7)	26%	20%	[c]	20%	[c]	20%	19%	[c]	[c]
Corals	7290 (21)	26%	15%	[c]	10%	[c]	2%	15%	[c]	[c]
Echinoids	9300 (12)	1%	3%	30%	20%	[c]	18%	10%	[c]	[c]
Forams (benthonic)	9850 (3)	14%	20%	30%	10%	[c]	[c]	13%	[c]	[c]
Forams (planktonic)	1030 (1)	[c]	[c]	[c]	[c]	[c]	10%	[c]	50%	40%
Mollusks	1230 (26)	8%	15%	30%	15%	[c]	10%	17%	[c]	[c]
Ooids	3240 (5)	[c]	5%	[c]	5%	[c]	5%	[c]	[c]	[c]

[a] Number of samples; [b] Sediments imported from coral reefs, banks and bays, and carbonate shelves; [c] Absent or insignificant contribution.

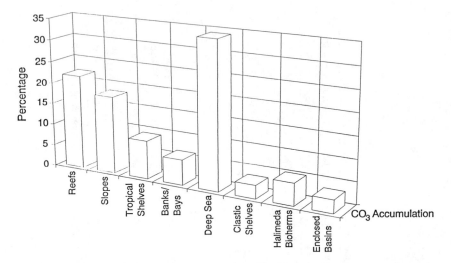

Figure 3. Contribution (wt.%) of each carbonate-accumulating environment to the estimated total global calcium carbonate accumulation of $3.2*10^{13}$ moles per year (after Milliman [19]). The most important carbonate accumulating environment is the deep sea accounting for 34% followed by carbonate reefs (22%) and slopes (18%).

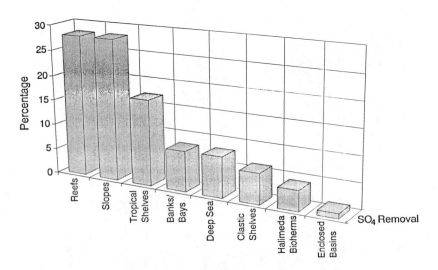

Figure 4. Contribution (wt.%) of each carbonate-accumulating environment (see Fig. 3) to the estimated total removal of oceanic sulfate of $1.5*10^{11}$ moles per year. Note that reefs, slopes and tropical shelves account for 73% of the total annual sulfate removal, whereas the deep sea, with 34% of the total carbonate accumulation, accounts for only 8% of the sulfate removal.

Our estimate for the removal of sulfate by calcium carbonate precipitation is higher than earlier estimates which range from 0% (*20*) over 0.6-1.2% (*23*) to 1.4% (*5*) of the annual riverine sulfate input. Crucial for the estimation of sulfate removal by calcium carbonate precipitation is which of the estimates of the global calcium carbonate precipitation and accumulation is used. We chose Milliman's (*19*) estimate because, to our knowledge, it represents the most recent and detailed estimate that gives areas and rates of calcium carbonate production and accumulation for various environments. It should be mentioned, however, that Holser (*24*) and Morse and Mackenzie (16) calculated significantly different rates of calcium carbonate production. These differences in production rates and their effect on the estimated sulfate removal are summarized in Table 3.

The removal of seawater sulfate by calcium carbonate precipitation represents only a minor sink compared to several major sulfur sinks. In estimates of sulfur sinks, pyrite formation is generally thought to be the largest sedimentary sink (*5, 21, 25, 26*) with up to 66% of the annual natural riverine sulfate input (*24*). The formation of evaporitic sulfate minerals is generally considered to be an important sedimentary sulfate sink with estimates ranging from 20% (*27*) to 50% (*20*), whereas the importance of sulfate removal by hydrothermal circulation through mid-oceanic ridges remains controversial (*28*). Hence, sulfate removal by carbonates is less important than other sinks, but it is not trivial as suggested before. Furthermore, it is important to note that the rate of sulfate removal by biological calcium carbonate precipitation is likely to have varied through geologic time, as it is dependent on factors such as latitude of continents, ocean temperature, sea level, PCO_2, and changes in taxonomic makeup of carbonate precipitating organisms.

Fate of Sedimentary Sulfate upon Burial. In the geologic record there are numerous examples of buried carbonate shelves and platforms. Modern carbonate shelves and platforms are suggested to account for 33% (*16*) to 42% (*19*) of the present-day calcium carbonate accumulation. The estimated average sulfate concentrations of modern carbonate shelves and platforms is 5900 ppm. Importantly, these environments, which represent only about 3.3% (*19*) of the total area where calcium carbonate is produced, account for 55% to 77% of the present-day sulfate removal by calcium carbonate accumulation (depending on which estimate is used). Hence, carbonate shelves and platforms represent relatively confined areas of sulfate enrichment. To evaluate the fate of sulfate-rich buried carbonate buildups we use regional extensive dolomites from the Burlington-Keokuk Formation as an example.

The stratigraphy, petrography, cathodoluminescence, major and trace element geochemistry, O-,C-, Sr isotope geochemistry, and relative time framework of the Mississippian Burlington-Keokuk dolomites have been discussed in great detail (*29, 30, 31, 32*). Burlington-Keokuk dolomites are dominated (> 95%) by two regionally extensive dolomite generations, referred to as Dolomite I and II. The earlier diagenetic Dolomite I was presumably formed by mixing of freshwater with moderately concentrated seawater at the near-surface (*33*). Sulfate concentrations of 26 Dolomite I samples range from 1000 to 3500 ppm. Petrographic and cathodoluminescence evidence indicates replacement of Dolomite I by Dolomite II (*32*). This later diagenetic Dolomite II was presumably formed at a sedimentary burial depth of not more than 500

Table 3.

Comparison of selected calcium carbonate production rate estimates and their effect on the estimated sulfate removal

		Holser et al. (24)	Morse and Mackenzie (16)	Milliman (19)
$CaCO_3$ Production:	Reef-Bank-Shelf	$30 \cdot 10^{12}$ moles C / year	$14 \cdot 10^{12}$ moles C / year	$29 \cdot 10^{12}$ moles C / year
	Open Ocean	$5 \cdot 10^{12}$ moles C / year	$72 \cdot 10^{12}$ moles C / year	$24 \cdot 10^{12}$ moles C / year
$CaCO_3$ Accumulation:	Reef-Bank-Shelf	$15 \cdot 10^{12}$ moles C / year[1]	$6 \cdot 10^{12}$ moles C / year	$14 \cdot 10^{12}$ moles C / year
	Open Ocean	$1.5 \cdot 10^{12}$ moles C / year[2]	$12 \cdot 10^{12}$ moles C / year	$11 \cdot 10^{12}$ moles C / year
Estimated SO_4 Removal	Total Marine	$1 \cdot 10^{11}$ moles S / year[3]	$5.2 \cdot 10^{10}$ moles S / year[3]	$1.5 \cdot 10^{11}$ moles S / year[3]
% of Annual Riverine Input[4]		3 %	1.6 %	5 %

[1] : Assuming preservation of 50% of the $CaCO_3$ shelf production.

[2] : Assuming preservation of 30% of the $CaCO_3$ deep sea production.

[3] : Calculated using average sulfate content and abundance of carbonate sediments in modern marine environments.

[4] : Based on estimates of the pre-anthropogenic annual riverine sulfate input into the oceans (20, 21, 22).

m (*32*) by an extraformational subsurface fluid (30). Sulfate concentrations of 14 Dolomite II samples are generally less than 500 ppm. Hence, during the replacement of the sulfate-rich Dolomite I by a sulfate-depleted Dolomite II a significant amount of sulfate must have been released. Because recrystallization of sedimentary carbonates is a common process in carbonate diagenesis, it is reasonable to assume that significant amounts of sulfate may be released upon burial of carbonate platforms.

Geological Implications

Although the incorporation of sulfate into sedimentary carbonates is not the most important sulfate sink with approximately 5% of the annual river flux, the process has important geological implications. Upon burial, sulfate-rich carbonate shelves and platforms are likely to release sulfate (see above). The generally low sulfate content (< 500 ppm) of carbonates which were formed at depth indicates that initially sulfate-rich sediments (e.g. reef, bank and tropical shelf sediments) release most of their carbonate-bound sulfate upon burial. Sulfate released at depth may subsequently be reduced to H_2S, elemental sulfur, incorporated in organic matter or metal sulfides. This release of carbonate-bound sulfate may be particularly important in sedimentary basins with thick carbonate shelf or reef deposits. For example, burial of the Bahama Platform and subsequent recrystallization under burial conditions would release a significant amount of sulfate. Assuming that Bahama sediments have an average density of 2.83 g/cm^3 and a porosity of 60%, the total sediment accumulation since the last sea level low stand (~ 4000 years) is approximately $7.9*10^{13}$ moles $CaCO_3$. These sediments, containing on the average 5900 ppm sulfate, represent a significant sulfur reservoir of $4.8*10^{11}$ moles. If, after burial recrystallization, the average sulfate content of Bahama sediments would be only 500 ppm, a total of $4.5*10^{11}$ moles would have been released, which is equivalent to approximately 50% of the sulfur present as ZnS in U.S. resources (*34*).

Summary

The determination of the sulfate concentrations of a wide range of skeletal and non-skeletal calcites, aragonites and dolomites show that sedimentary carbonates generally contain sulfate as a trace constituent. Particularly modern marine skeletal carbonates, which account for most of the present-day calcium carbonate production, contain considerable amounts of sulfate (200 to 24,000 ppm). Based on these data we estimate that the rate of present-day oceanic sulfate removal by biological calcium carbonate precipitation accounts for approximately 5% of the natural riverine sulfate input.

Upon sedimentary burial and recrystallization under reducing subsurface conditions, initially sulfate-rich sedimentary carbonates probably release most of their sulfate. The amount of sulfate that can be released by sulfate-rich shelf and reef deposits may represent a significant source of sulfur in sedimentary basins. Given the geological importance of the fate of carbonate-bound sulfate upon burial it is clear that further research is warranted.

Acknowledgments

Acknowledgment is made to the Donors of the Petroleum Research Fund, administered by the American Chemical Society, for providing financial support for this research. We are grateful to P. Bretsky, W.J. Meyers and W.B. Ward for providing samples for this study. We also thank L. Kump, M.A. Arthur (Penn State Univ.), T. Rasbury (SUNY Stony Brook), and an anonymous reviewer for helpful suggestions and discussions.

References

1. Taylor, S. R.; McLennan, S. M. *The Continental Crust: Its Composition and Evolution*; Blackwells: Oxford, 1985, pp 315.
2. Veizer, J. *Chemical Diagenesis of carbonates: Theory and Application of Trace Element Technique*; Soc. Econ. Paleont. Mineral. Short Course 10; Ottowa-Carleton Centre for Geoscience Studies: 1983, Chapter 3, pp 100.
3. Wilson, J. L. *Carbonate Facies in Geologic History*; Springer-Verlag: 1975, pp 471.
4. Oomori, T.; Kaneshima, K.; Nakamura, Y. *Galaxea* **1982**, *1*, 77-86.
5. Volkov, I. I.; Rozanov, A. G. *The Global Biochemical Sulfur Cycle*; Ivanov, M.V.; Freney, J. R., Ed.; John Wiley & Sons: 1983, pp 357-439.
6. Busenberg, E.; Plummer, L. N. *Geochim. Cosmochim. Acta* **1985**, *49*, 713-725.
7. Sass, E.; Bein, A. *Sedimentology and Geochemistry of Dolostones*; Shukla, V.; Baker, P. A., Eds.; Soc. Econ. Paleont. Mineral. Spec. Publ. 43: 1988, 223-233
8. Staudt, W. J.; Oswald, E. J.; Schoonen, M. A. A. *Chemical Geology* **1993**, *107*, 97-109.
9. Crenshaw, M. A. *Biomineralisation* **1972**, *6*, 6-11.
10. Burdett, J. W.; Arthur, M. A.; Richardson, M. *Earth and Planet. Sci. Let.* **1989**, *94*, 189-198.
11. Takano, B. *Chemical Geology* **1985**, *49*, 393-403.
12. Takano, B.; Asano, Y.; Watanuki, K. *Contrib. Mineral. Petrol.* **1980**, *72*, 197-203.
13. Staudt, W. J.; Reeder, R. J.; Schoonen, M. A. A. *Geochim. Cosmochim. Acta* **1994**, 2087-2098.
14. Reeder, R. J.; Lamble, G. M.; Lee, J.-F.; Staudt, W. J. *Geochim. Cosmochim. Acta* **1994**, 58, 5639-5646.
15. Pingitore, N.E., Meitzner, G., Love, K.M. *Geochim. Cosmochim. Acta* **in press.**
16. Morse, J. W.; Mackenzie, F. T. *Geochemistry of Sedimentary Carbonates*; Developments in Sedimentology 48; Elsevier: 1990, pp 707.
17. Bathurst, R. G. C. *Carbonate Sediments and Their Diagenesis;* Elsevier: 1975, pp 658.
18. Milliman, J. D. *Marine Carbonates*; Springer-Verlag: 1974, pp 375.
19. Milliman, J. D. *Global Geochemical Cycles* **1993**, 7, 927-957.
20. Drever, J. I. *The Geochemistry of Natural Waters*; Prentice Hall: 1988, pp 437.

21.   Charlson, R. J.; Anderson, T. L.; McDuff, R.E. *Global Geochemical Cycles,* Butcher, S. S.; Charlson, R. J.; Orians, G. H.; Wolfe, G.V., Eds.; Academic Press: 1992, pp 285-299.

22.   Ivanov, M.V. *The Global Biochemical Sulfur Cycle,* Ivanov, M.V., Freney, J.R., Eds; John Wiley & Sons: 1983, pp 297-356.

23.   Okumura, M.; Kitano, Y.; Idogaki, M. *Geochem. Journal.* **1983,** *17,* 105-110.

24.   Holser, W.T., Schidlowski, M., Mackenzie, F.T., Maynard, J.B. *Chemical Cycles in the Evolution of the Earth,* Gregor, C.B.,Garrels, R.M., Mackenzie, F.T., Maynard, J.B., Eds., Wiley-Interscience: 1988, pp 105-174.

25.   Berner, R.A. *American Journal Sci.* **1982,** *282,* 451-473.

26.   Berner, R. A.; Raiswell, R. *Geochim. Cosmochim. Acta* **1983,** *47,* 855-862.

27.   Zharkov, M. A. *History of Paleozoic Salt Accumulation,* Springer-Verlag: 1981, pp. 308.

28.   Berner, E. K.; Berner, R. A. *The Global Water Cycle,* Prentice-Hall: 1987, pp. 397.

29.   Harris, D. C.; Meyers, W. J. *Diagenesis of Sedimentary Sequences,* Marshall, J. D., Ed., Geol. Society of London, Special Publ. 36: 1987, 237-238.

30.   Banner, J. L.; Hanson, G. N.; Meyers, W. J. *Journal of Sedimentary Petrology* **1988,** *58,* 415-432.

31.   Banner, J. L.; Hanson, G. N.; Meyers, W. J. *Journal of Sedimentary Petrology* **1988,** *58,* 673-688.

32.   Cander, H. S.; Kaufman, J.; Daniels, L. D.; Meyers, W. J. *Sedimentology and Geochemistry of Dolostones,* Shukla, V.; Baker, P. A., Eds., SEPM Special Publication 43, 1988, 129-144.

33.   Staudt, W.J., Meyers, W.J., Schoonen, M.A.A. Submitted to Journal of Sedimentary Research.

34.   Skinner, B.J. *Earth Resources,* Prentice-Hall: 1986, pp. 184.

35.   Oswald, E. J. Ph.D. Dissertation, SUNY Stony Brook, N.Y.: 1992, pp. 424.

36.   Leaver, J. Master Thesis, Duke University, Durham, N.C.: 1985.

37.   Ward, W. C.; Halley, R. B. *J. Sediment. Petrol.* **1985,** *55,* 407-420.

38.   *Fouke, B. W.* Ph.D. Dissertation, SUNY Stony Brook, N.Y.: 1992.

39.   Sibley, D. F. *Concepts and Models of Dolomitization,* Zenger, D. H.; Dunham, J. B.; Ethington, R.L., Eds., Soc. Econ. Paleont. Mineral., Spec. Publ. 28: 1980, 247-258.

40.   Kaufman, J.; Hanson, G. N.; Meyers, W. J. *Sedimentology* **1991,** *38,* 41-66.

41.   Ward, W. B. Ph.D. Dissertation, SUNY Stony Brook, N.Y.: 1993, pp.

42.   Hoff, J. A. Ph.D. Dissertation, SUNY Stony Brook, N.Y.: 1992, pp. 156.

43.   Douthit, T. L.; Meyers, W. J.; Hanson, G. N. *Jour. of Sediment. Petrol.* **1993,** *63,* 539-549.

RECEIVED June 28, 1995

BIOGEOCHEMICAL TRANSFORMATIONS

Chapter 19

^{35}S-Radiolabeling To Probe Biogeochemical Cycling of Sulfur

Henrik Fossing

Department of Biogeochemistry, Max Planck Institute for Marine Microbiology, Fahrenheitstrasse 1, D–28359 Bremen, Germany

This paper focuses on the precision of rate measurements with radiolabeled sulfur, for tracking intermediate sulfur compounds and their consumption pathways in biogeochemical systems involving sulfur turnover. The discussion is based on models and examples from sulfate reduction and thiosulfate turnover. Complications arising from the oxidation of formed reduced sulfur compounds and isotopic exchange are discussed. Finally, theoretical and practical aspects of amendment studies are presented using examples from thiosulfate turnover.

The use of radioactive ^{35}S-compounds is a valuable tool in determining rates, tracing pathways, and documenting intermediate S-compounds in the sulfur cycle. The use of $^{35}SO_4^{2-}$ to measure rates of sulfate reduction is particularly well documented (see for example *1-7*) but the radiolabeled compounds $H_2^{35}S$, $^{35}S^o$, $Fe^{35}S$, and $^{35}S_2O_3^{2-}$ have also been used in biogeochemical studies (*8-15*).

The use of ^{35}S has the advantage that chemical processes can be studied without disturbing chemical and physical equilibria and even minute transfers of radioactivity between various sulfur compounds can be detected. However, planning a radiotracer experiment can be difficult due to, for example, isotope exchange, undetectable intermediates, or unknown pathways which may all complicate the interpretation of the observations. In this paper I will discuss some of the problems that are encountered when ^{35}S is used in studies of inorganic sulfur cycling. I will focus on rate measurements, isotope exchange, and amendment experiments by which intermediate sulfur compounds can be documented and pathways elucidated. The mathematical models that are presented have all been generated using "Stella II. 2.2.2." (High Performance Systems) for Macintosh.

Rate Measurements at Dynamic Equilibrium Conditions: Theory

At dynamic equilibrium, the turnover of two compounds, A and B, may be expressed by the reaction

$$------> X \xrightarrow{x_t} A \xrightarrow{a_t} B \xrightarrow{b_t} ------> \tag{1}$$

where x_t, a_t, and b_t are the reaction rates, at time t, by which the compounds X, A, and B are continuously produced and consumed. If chemical reactions are of first order, turnover rates are proportional to concentrations expressed by

$$a_t = k_a \cdot [A]_t \tag{2}$$

where k_a is the rate constant and $[A]_t$ is the concentration of A at time t. However, when a reaction is enzyme dependent, as for example a biological reaction, it is expected to follow Michaelis-Menten kinetics and the rate is expressed by

$$a_t = a_{max} \cdot \{[A]_t / ([A]_t + K_a)\} \tag{3}$$

with $[A]_t$ as above and additionally K_a is the Michaelis constant (i.e., the concentration of A at which $a_t = a_{max}/2$) and a_{max} is the maximum reaction rate (achieved when $[A]_t \gg K_a$). The reaction rate b_t is obtained by substituting a with b and A with B in equation 2 and 3.

Figure 1 shows the theoretical observations after an addition of a radiolabeled compound A* to a system with two compounds (A and B) in dynamic equilibrium, i.e., constant concentrations and thus $x_t = a_t = b_t$, see equation 1. An insignificant amount of A* is added at time t = 0 which does not change the concentration of A ([A] $\approx$ [A] + [A*]) and all rates therefore remain unchanged. This kind of radiolabeled experiment is named a *tracer experiment*. The turnover of the added *tracer*, A*, with a radioactivity of a*, follows the same kinetics as does the non-radioactive compound, A. Thus, at constant concentrations of compound A and B the change in radioactivity of these compounds are expressed as

$$\partial a^*/\partial t = a_t \cdot a^*_t/[A] \text{ and } \partial b^*/\partial t = b_t \cdot b^*_t/[B] \tag{4}$$

where $a^*_t / [A]$ and $b^*_t / [B]$ are the specific radioactivities of compound A and B, respectively, at time t. The change in radioactivity of A and B, defined by equation 4, is shown on Figure 1a from the initial labeling until the radioactivity is no longer detectable in A or B. At this time compound A has been turned over more than five times, i.e., the amount of A consumed equals five fold the concentration.

Initially when < 1 % of A has been turned over, the specific radioactivity of A can be considered constant and equal to a*$_0$/[A], where a^*_0 is the initial radioactivity of A (Figure 1b). Thus, the change in radioactivity of A (= Δa^*) during Δt (= t - t$_0$) is

$$\Delta a^* = a_t \cdot a^*_0/[A] \cdot \Delta t \tag{5a}$$

As long as the specific radioactivity of A is unchanged the rate is a linear function of time and calculated as

$$a_t = \Delta a^* / a^*_0 \cdot [A] \cdot 1/\Delta t = a_{calc} \tag{5b}$$

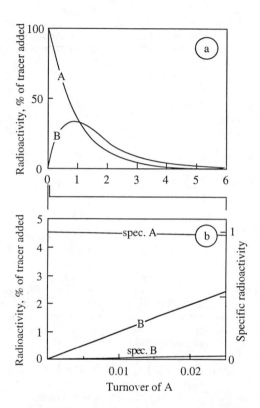

Figure 1. a) Distribution of radioactivity between two compounds A and B in dynamic equilibrium, i.e. no change in concentrations, expressed relative to the turnover of the A pool. b) The initial period during which up to 2.5 % of A is turned over seen as an equivalent increase of radioactivity of B (Note the expanded scales). The specific radioactivities of A and B (i.e., radioactivity/ concentration) are normalized and [A] : [B] = 4 : 3.

It should be noted that isotopic discrimination may often occur between radioactive and non-radioactive compounds due to their mass difference, and that the rate, to be correct, should be multiplied by this fractionation factor; for sulfate reduction, for example, a value of 1.06 is often used (*16*). In most of the examples presented here, I have neglected this factor to keep the equations as simple as possible.

At dynamic equilibrium, however, the consumption of compound A is balanced by production and the product A will be unlabeled. The radioactivity of A, therefore, decreases with time and so does the specific radioactivity. As continuously more of A is turned over, the decline in radioactivity of A deviates from the linear decrease prescribed by equation 5a (see Figure 1a) and the calculated reaction rate (a_{calc}) will accordingly differ from the true rate, a_t, as shown in Table I. Hence, to be able to calculate reaction rates by equation 5b, the incubation length of any tracer experiment should be kept short enough to avoid any significant change in specific radioactivity of the initially labeled compound (A), and thus ensure a linear relationship between the length of incubation and decrease in radioactivity. I, therefore, define *a short time incubation experiment* as an experiment in which less than 1 % of the initial radioactive pool is turned over during the incubation. In such an experiment, however, it is not possible to detect a radioactive decrease of < 1 % and, therefore, the radioactive *increase* of the reaction product (B) is measured instead (Figure 1b). Prior to ^{35}S labeling the product B is not radioactive, thus after ^{35}S-addition an increase of even less than 1 % of B is detected without problems. This concept is used, for example, when sulfate reduction rates are measured. Thus, the reaction rate for compound A, a_{calc}, is expressed as

$$a_{calc} = \Delta b^* / a^*_0 \cdot [A] \cdot 1/ \Delta t \tag{5c}$$

where Δb^* is the radioactive content of product B after an incubation length of Δt, a^*_0 is the initial radioactivity of compound A, and *[A]* the concentration of A.

Application to Measurements of Sulfate Reduction Rates

Sulfate reduction rates are calculated based on the reduction of $^{35}SO_4^{2-}$ to sulfide, some of which may have partly oxidized. Overall, reduced label ($^{35}S_{red}$) may be found as $H_2^{35}S$, $^{35}S_0$, $^{35}S_n^{2-}$, $Fe^{35}S$ and $Fe^{35}S_2$ (*2, 3, 5-7, 17-19*). After sediments have been incubated for a short time (4 - 8 h) with trace amounts of $^{35}SO_4^{2-}$, sulfate reduction may be terminated when the sediment is preserved, for example, in 20 % zinc acetate (ZnAc). The zinc precipitates with the dissolved compounds of H_2S and S_n^{2-} and exchanges with FeS to form ZnS. In this way all $^{35}S_{red}$ is preserved as solid compounds (= ZnS, S^0, and FeS_2). The $^{35}SO_4^{2-}$ is separated from its radioactive products by centrifugation, and a pure sample of $^{35}SO_4^{2-}$ can then be taken from the supernatant. The solid phase of $^{35}S_{red}$ is dissolved or reduced to gaseous $H_2^{35}S$ by an acid solution of reduced chromium (1 N Cr^{2+} in 5 HCl) and stripped from the sediment by a flow of N_2 into a 5 % ZnAc-trap (*20-22*). The Cr^{2+} does not reduce $^{35}SO_4^{2-}$ to $H_2^{35}S$ (*17*). Based on the radioactive content of $^{35}SO_4^{2-}$ and $^{35}S_{red}$, the sulfate reduction rate (SRR) is calculated using equation 5d,

$$SRR = {}^{35}S_{red}/ ({}^{35}SO_4^{2-} + {}^{35}S_{red}) \cdot [SO_4^{2-}] \cdot 24/ t \cdot 1.06 \quad nmol \ SO_4^{2-} \ cm^{-3} \ day^{-1} \tag{5d}$$

where $^{35}S_{red}$ is the radioactivity of the reduced sulfur (per unit volume), $^{35}SO_4^{2-}$ is the radioactivity of the sulfate *after* incubation (per unit volume), $[SO_4^{2-}]$ is the sulfate concentration (nmol cm^{-3}), t is the incubation time (hours), and *1.06* is the isotopic fractionation factor (*16*). Note that generally $^{35}SO_4^{2-} + {}^{35}S_{red} \approx {}^{35}SO_4^{2-}$.

Concomitant $^{35}SO_4^{2-}$ Reduction and Oxidation of Reduced ^{35}S Compounds. A complicating factor in sulfate reduction rate measurements is reoxidation of the formed $H_2^{35}S$. The importance of reoxidation can be visualized by a model based on the dynamic equilibrium

$$^{35}SO_4^{2-} \overset{r}{\underset{o}{\rlap{\text{<}}===}} \text{>} (H_2^{35}S, {}^{35}S^o) \overset{p}{------} \text{>} (Fe^{35}S, Fe^{35}S_2) \qquad (6)$$

where r is the sulfate reduction rate, o is the combined reoxidation rate of $H_2S + S^o$, and p is the combined precipitation rate of $H_2S + S^o$ to $FeS + FeS_2$. Thus, at constant concentrations of H_2S and S^o, $r = p + o$ (see also *23*).

In a ^{35}S labeled sulfate reduction rate experiment $H_2^{35}S$ is the first product formed, whereas secondary radiolabeled products, like $^{35}S^o$, $Fe^{35}S$, and $Fe^{35}S_2$ must be produced from $H_2^{35}S$ either by chemical reactions or by isotope exchange. The isotope exchange between H_2S and S^o is complete within minutes which leads to equal specific radioactivities of these two pools, whereas $Fe^{35}S$ and $Fe^{35}S_2$ are not significantly involved in isotope exchange reactions (*15, 18*; see below). When the concentrations of H_2S and S^o are small compared to FeS and FeS_2 the specific radioactivity of the H_2S and S^o pools will far exceed the specific radioactivity of the secondary products, $Fe^{35}S$ and $Fe^{35}S_2$. Thus, with a high specific radioactivity of $H_2^{35}S$ and $^{35}S^o$ a significant amount of ^{35}S will be recycled to the sulfate pool, as $^{35}SO_4^{2-}$, if reoxidation occurs. In this case no significant loss of $Fe^{35}S$ and $Fe^{35}S_2$ will be observed due to an eventually oxidation of FeS and FeS_2 because of the low specific radioactivity of these pools.

In theory it is possible to separate the $^{35}SO_4^{2-}$ reduction products in the two pools comprising $H_2^{35}S + {}^{35}S^o$ and $Fe^{35}S + Fe^{35}S_2$. However, in practice the ^{35}S-content of these compounds cannot be separated as H_2S and FeS, preserved as ZnS, are dissolved simultaneously when acidified and degassed as H_2S (*22*). Thus, total $^{35}S_{red}$ is expressed as $H_2^{35}S + {}^{35}S^o + Fe^{35}S + Fe^{35}S_2$. Figure 2 shows the amount of $^{35}S_{red}$ produced when $[H_2S] + [S^o] << [FeS] + [FeS_2]$ and reoxidation of the $H_2S + S^o$ pool is anticipated. A curvature of the graph is observed for $o > 0$ due to the gradual ^{35}S saturation of the H_2S and S^o pool and is most pronounced when all of the produced $H_2^{35}S$ and $^{35}S^o$ are rapidly reoxidized and thus no precipitation of $Fe^{35}S$ and $Fe^{35}S_2$ occur (i.e. 100 % reoxidation, Figure 2). Initially the slope, and hence rate, is identical for all curves and equal to r, the real (gross) sulfate reduction rate, also observed when $o = 0$, i.e. when no $^{35}S_{red}$ is reoxidized. Within time, however, the curves deviate depending on reoxidation and when the combined pool of H_2S and S^o becomes saturated with ^{35}S, a linear relation between *time* and $^{35}S_{red}$ is observed, equal to the combined precipitation rate of FeS and FeS_2 (i.e. the net sulfate reduction rate, $p = r - o$). In sediments where reoxidation is of no importance ($o = 0$) gross and net sulfate reduction rates equal each other.

Table I. Consumption rates calculated (a_{calc})
in % of true rate (a_t) relative to turnover of
labeled pool

% of A pool turned over	a_{calc} in % of a_t
0.1 %	100.0 %
1.0 %	99.5 %
2.0 %	99.0 %
5.0%	97.6 %
10.0 %	95.3 %
50.0 %	79.3 %
100.0 %	63.6 %
200.0 % [1]	43.4 %

[1] i.e. turned over two times.

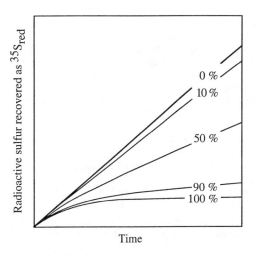

Figure 2. Radioactivity of the reduced sulfur pool ($^{35}S_{red}$ = $H_2{}^{35}S$ + $^{35}S^o$ + $Fe^{35}S$ + $Fe^{35}S_2$) produced by $^{35}SO_4{}^{2-}$ reduction at various reoxidation intensities (%) of the formed $H_2{}^{35}S$ + $^{35}S^o$. No reoxidation (0 %) equals $o = 0$, whereas $o = 1$ when all formed $H_2{}^{35}S$ + $^{35}S^o$ is reoxidized (100 %), see text. Radioactivities are normalized.

An Example. The application of equation 5d to sulfate reduction rate determination normally assumes no reoxidation of the reduced products to SO_4^{2-}, although the production of radiolabeled S^o, $S_2O_3^{2-}$, and FeS_2 implies some reoxidation may occur. We (in cooperation with Ferdelman, T., Max Planck Institute for Marine Microbiology) performed high time resolution sulfate reduction experiments to explore the possible reoxidation of reduced products in more detail. Figure 3a shows the time course of a $^{35}SO_4^{2-}$ labeled sulfate reduction experiment performed over 4 hours in an anoxic marine sediment slurry. Surface sediment (0 - 2 cm) from Skagerrak (Denmark) at 400 m water depth was used. Sulfate concentration was 28 mM with no detectable H_2S. There was an apparent linear correlation between incubation time and sulfate reduced ($r^2 = 0.998$) and a SRR of 2.70 μM h^{-1} was calculated. Hence, in 4 hours < 0.04 % of the sulfate pool was turned over. However, when the first 30 minutes of the experiment were studied in higher resolution, a curvature was observed which implied a higher initial sulfate reduction rate that decreased with time (Figure 3a). Similar observations, albeit with fewer data points, have been reported by Moeslund and co-workers (*23*), and were explained by reoxidation of reduced ^{35}S-compounds, particularly $H_2^{35}S$, back to $^{35}SO_4^{2-}$.

Based on equation 5d, the sulfate reduction rates were calculated from each data point on Figure 3a. Due to reoxidation, as argued above, the measured sulfate reduction rates deviated with time from a gross sulfate reduction rate of approximately 7.81 μM h^{-1} and approached a net reduction rate of 2.56 ± 0.22 μM h^{-1} averaged over 16 - 240 min (Figure 3b). Thus, because of reoxidation the measured SRR decreased by ((7.8-2.6)/ 7.8 =) 67% of the gross sulfate reduction rate after 15 min.

Routinely, sulfate reduction rates are calculated following one defined incubation length, i.e., based on only one data point. Therefore, if reoxidation is of importance, the sulfate reduction rate will be underestimated to a lesser or greater degree, dependent on the length of incubation due to the non linearity between sulfate reduced and length of incubation. Such an observation has been made in undisturbed marine surface sediment where the aerial sulfate reduction rate (0 to 2 cm) decreased by up to 80 % when incubation time was increased from 1.5 h to 6 h (unpublished data) .

Isotope Exchange as a complicating factor in rate measurements

Isotope exchange is defined as a *chemical reaction in which the atoms of a given element interchange between two or more chemical forms of this element* (*24*). Such isotope exchange reactions have been studied by Fossing and Jørgensen (*18*) in solutions of H_2S and S^o in equilibrium with S_n^{2-}, which is a complex mixture of polysulfides with S-chain lengths from n = 2 to 5 sulfur atoms but with S_4^{2-} and S_5^{2-} as the dominating species at the experimental pH (*25, 26*)

$$8\ HS^- + 3S_8 \iff 8\ S_4^{2-} + 8H^+ \tag{7a}$$
$$2\ HS^- + \ S_8 \iff 2\ S_5^{2-} + 2H^+ \tag{7b}$$

The isotope exchange experiment of Fossing and Jørgensen (*18*) was performed at room temperature in a solution of anoxic, sterile sea water at pH 7.6. The three sulfur compounds H_2S (0.99 mM), S_n^{2-} (0.53 mM), and S^o (0.43 mM) were all in equilibrium when trace amounts of $^{35}S^o$ was added. Initially, the added $^{35}S^o$ tracer exchanged with S_n^{2-} at a rate of approximately 8300 μM h^{-1} (Figure 4). Thus, more than half of the tracer from $^{35}S^o$ had exchanged with S_n^{2-} in < 2 min. The ^{35}S exchange between S^o and S_n^{2-} ceased when the specific activities of these

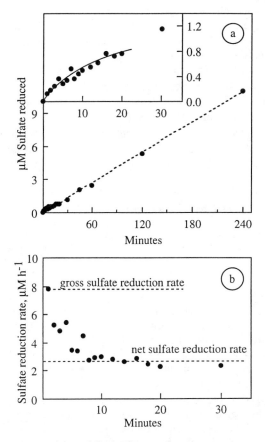

Figure 3. Time course of a $^{35}SO_4{}^{2-}$ labeled sulfate reduction experiment. a) The observed amount of $SO_4{}^{2-}$ reduced over 240 min (calculated as $^{35}S_{red}/ ^{35}SO_4{}^{2-} \cdot [SO_4{}^{2-}]$; μM) with the first 30 min period inserted. The regression line over 240 min is also shown (2.70 μM $SO_4{}^{2-}$ h^{-1}, $r^2 = 0.998$). b) The sulfate reduction rate (μM h^{-1}) calculated after each observation based on equation 5d, see text.

compounds approached each other. The increase in $H_2^{35}S$ radioactivity was delayed because of a slower exchange rate between S_n^{2-} and H_2S. Accordingly, a $^{35}S_n^{2-}$ peak emerged before the ^{35}S-atoms were exchanged from $^{35}S^o$ to H_2S. After approximately 30 min the specific radioactivity of all compounds were equal, meaning that the total radioactivity was distributed between the three compounds in proportion to their concentrations, which remained unchanged during the experiment. Other exchange experiments with FeS, S^o, and H_2S demonstrated an incomplete isotope exchange of 7 - 83 % between FeS and the compounds of S^o and H_2S, mediated through polysulfide (*15, 18*).

In sediments with low H_2S concentration, added $H_2^{35}S$ will react immediately with iron (or manganese) compounds and precipitate as $Fe^{35}S$, or be oxidized to $Fe^{35}S_2$, $^{35}S^o$ or even $^{35}SO_4^{2-}$ (*27-31*) and exchange reactions will further change the specific radioactivities of H_2S, S^o and FeS within minutes after labeling with $H_2^{35}S$ (or $^{35}S^o$). Thus, isotope exchange between the reduced sulfur compounds of H_2S, S_n^{2-}, S^o, and FeS adds to the problem of controlling $H_2^{35}S$ specific radioactivity and measure H_2S oxidation rates successfully (*15, 18*).

It should be mentioned that there presently is no evidence for isotope exchange between the reduced sulfur compounds (H_2S, S^o, and FeS) and their more oxidized forms ($S_2O_3^{2-}$, SO_3^{2-} and SO_4^{2-}) nor is isotope exchange observed between the latter forms at or below room temperature (*18, 32-34*). Thus, it appears that radiolabeled compounds of thiosulfate, sulfite, and sulfate can be used for rate measurements in aquatic systems without problems related to isotope exchange.

Amendment Studies

An amendment study is a useful approach to show the importance of an intermediate compound that otherwise would not have been detected in a radiolabeled experiment due its low concentration. For example $S_2O_3^{2-}$ and SO_3^{2-} have been reported in marine sediments from nano molar to sub-milli molar concentrations (*30, 35*) or have sometimes not even been detected (< 50 nM, Ferdelman, T., Max Planck Institute for Marine Microbiology, personal communication, 1994). The importance of $S_2O_3^{2-}$ and SO_3^{2-} as intermediates in H_2S oxidation has been documented from sediments amended with $S_2O_3^{2-}$ (*9-12, 14*) or SO_3^{2-} (Habicht, K., Max Planck Institute for marine Microbiology, personal communication, 1994) and consecutively labeled with $H_2^{35}S$ or $^{35}S^o$. In these amendment experiments, ^{35}S during $H_2^{35}S$ oxidation was transferred into and retarded in the thiosulfate or sulfite pools and hence documented that these compounds are potential intermediates. The concept of increasing the concentration of a compound significantly above the detection limit in order to document its potential importance as an intermediate in a reaction pathway is discussed below.

Theoretical Considerations. Consider three compounds, A, B, and C that are consumed and produced, in a closed system, according to equation 8 with rates $a_t = b_t = c_t$

$$A \xrightarrow{a_t} B \xrightarrow{b_t} C \xrightarrow{c_t} \qquad (8)$$

Compound A and C are present in detectable amounts whereas the concentration of B is below or near its detection limit. At time 0, A is labeled with trace amounts of its corresponding radioactive compound (A*).

In *theory*, due to the small pool size of B, the specific radioactivity of this pool quickly approaches the specific radioactivity of A. Concurrently, the B pool becomes saturated with radioactivity and no further radioactive increase in this pool is observed (Figure 5). The radioactivity that flows to C passes through B and therefore radioactivity first appears in C after a lag period, seen as an exponential increase, before a linear increase is observed at the time when the B pool is saturated with radioactivity (Figure 5).

Practically, in this example, the radioactivity of B is too close to the back ground level to be detected. Thus, radioactivity is first and only observed in the C pool and therefore it is most likely that compound C will be taken as the first intermediate, unless the lag period in radioactive increase of C is detected. A recognition of such a lag implies that there must be an intermediate prior to C. The intermediate B, until now undetectable, can only be identified by a radioactive labeling experiment if its concentration is increased, by addition, significantly above its detection limit. Assuming that this amendment does not change the turnover rates of the enriched pool, B, significantly, the radioactivity of B is expected to increase until B becomes saturated with radioactivity (Figure 5). Thus, not only does a linear increase in radioactivity of B persist much longer in the amended experiment relative to the unamended experiment but also the radioactivity of this pool exceeds the back ground level and is thus detected. In these amendment experiments, however, radioactivity will not be detected in compound C before a much longer lag period has passed, relative to the unamended experiment.

An Example. Elsgaard and Jørgensen (*14*) used amendment with $S_2O_3^{2-}$ to show its importance as an intermediate in H_2S oxidation. The H_2S oxidation was studied in a 1 mM H_2S anoxic, reduced slurry of fresh water sediment. Prior to $H_2^{35}S$ labeling the sediment slurry was amended with $S_2O_3^{2-}$ and SO_4^{2-} to final concentrations of 500 μM and 750 μM, respectively. The experiment was performed over 5 days and showed an immediate increase of ^{35}S in the $S_2O_3^{2-}$ followed by a decrease concomitant with the gradual depletion of thiosulfate (Figure 6). Thus, despite net thiosulfate consumption, a production of thiosulfate occurred simultaneously. Also, $^{35}SO_4^{2-}$ increased linearly immediately after $H_2^{35}S$ addition and Elsgaard and Jørgensen therefore concluded that a direct pathway of H_2S oxidation to SO_4^{2-} also existed in addition to the $^{35}SO_4^{2-}$ formation through $^{35}S_2O_3^{2-}$.

Amendment Disturbs Equilibrium. In Elsgaard and Jørgensen's $S_2O_3^{2-}$ amendment experiment the system was obviously brought out of equilibrium when $S_2O_3^{2-}$ was added to the sediment and the consumption rate of $S_2O_3^{2-}$ thus, exceeded the production rate. This non-equilibrium state resulted in a linear net decrease of thiosulfate concentration and a peak in radioactive $S_2O_3^{2-}$ from the oxidation of $H_2^{35}S$ (or $^{35}S^o$ or $Fe^{35}S$ formed by isotope exchange immediately after $H_2^{35}S$ labeling; Figure 6). As discussed above, the consumption of $S_2O_3^{2-}$ may either be of first (or higher) order, or follow Michaelis-Menten kinetics (see equation 2 and 3, respectively). Figure 7 shows the modeled changes in $S_2O_3^{2-}$ radioactivity after thiosulfate amendment provided this addition does not change the *production rate* of $S_2O_3^{2-}$. The increase in $S_2O_3^{2-}$ radioactivity is most pronounced when $S_2O_3^{2-}$ consumption rate is unaffected by the thiosulfate amendment and hence no

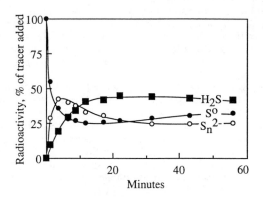

Figure 4. Isotope exchange experiment with labeled $^{35}S^o$. Distribution of ^{35}S between S^o, S_n^{2-}, and H_2S at constant concentrations of 0.43 mM, 0.53 mM, and 0.99 mM, respectively. (Adapted from ref. 18).

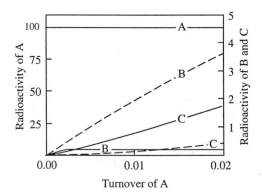

Figure 5. Distribution of radioactivity between three compounds A, B, and C, in which the radioactivity flows from the labeled pool A, via the B pool, to C, expressed relative to the turnover of the A pool. The solid lines show the radioactive increase in B and C when [B] << [A] and [C] and the B pool is not detectable. When the concentration of B is increased by amendment the broken line of B and C is observed. The radioactive content of A is unaffected ± amendment. Note the different radioactivity scale for A, and B and C, respectively.

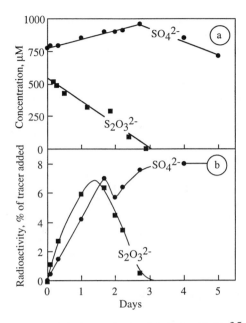

Figure 6. Thiosulfate amended experiment labeled with $H_2{}^{35}S$. a) $S_2O_3^{2-}$ and SO_4^{2-} concentrations. b) Radioactivity of $S_2O_3^{2-}$ and SO_4^{2-} in % of ^{35}S added. (Adapted from ref. 14).

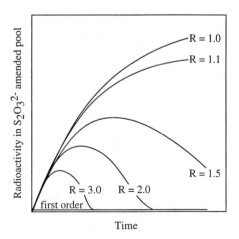

Figure 7. Expected radioactivity in a $S_2O_3^{2-}$ pool after $H_2{}^{35}S$ labeling and thiosulfate amendment, resulting in non-equilibrium (i.e. $S_2O_3^{2-}$ consumption exceeds production). The evolution of $^{35}S_2O_3^{2-}$ is shown when consumption follows first order or Michaelis-Menten kinetics, the latter depending on the increase in thiosulfate consumption rate expressed as $R = S_2O_3^{2-}{}_{max}/ S_2O_3^{2-}{}_{deq}$, see text.

net consumption is observed. This situation is obtained for zero order kinetics. The same observation is also true for reactions driven by Michaelis-Menten kinetics when the maximum reaction rate (after amendment) equals the reaction rate at dynamic equilibrium, i.e. $S_2O_3^{2-}{}_{max}/\ S_2O_3^{2-}{}_{deq} = R = 1$. The situation is quite different when thiosulfate consumption follows first order kinetics. In this case no $^{35}S_2O_3^{2-}$ peak develops and the increase in $^{35}S_2O_3^{2-}$ radioactivity is thus the same whether or not the system has been amended with $S_2O_3^{2-}$ prior to radiolabeling. Between these two extremes more or less pronounced $^{35}S_2O_3^{2-}$ peaks can be observed when $S_2O_3^{2-}$ depletion follows Michaelis-Menten kinetics provided that $S_2O_3^{2-}{}_{max} >$ $S_2O_3^{2-}{}_{deq}$. The peak intensity depends on how much the consumption rate is increased by thiosulfate addition, i.e. the ratio $R = S_2O_3^{2-}{}_{max}/\ S_2O_3^{2-}{}_{deq}$, as peaks become less pronounced with increasing R (Figure 7).

Consequently, in the above experiment (Figure 6) the observed peak in thiosulfate radioactivity and the concurrent decrease in concentration revealed that thiosulfate was turned over by an enzyme dependent reaction and therefore was of biological (bacterial) origin (i.e., followed Michaelis-Menten kinetics). Opposite, a non-biological reaction would, due to first order kinetics, appear with no radioactive $^{35}S_2O_3^{2-}$ peak, but with a decline in concentration.

Amendment Experiments and Rate Measurements

Turnover of Thiosulfate. Amendment also provides a useful approach for studying the continuous turnover of an added compound. Thiosulfate in aquatic sediments has been shown to follow concurrently three various pathways including oxidation (production of SO_4^{2-}), reduction (production of H_2S) and disproportionation (concurrent production of SO_4^{2-} and H_2S, 1:1; 36-37). Thiosulfate sulfur exists in two different oxidation states, -1 for the "outer" (sulfonate) sulfur and +5 for the "inner" (sulfane) sulfur (38) and Jørgensen and co-workers (9-12) have shown that the *radioactive* outcome of a thiosulfate consumption experiment varies dependent on which of the sulfur atoms are radiolabeled (Figure 8).

In two analogous thiosulfate amended experiments (10), $S_2O_3^{2-}$ was turned over with identical rates. However, the two experiment were labeled with, respectively, "outer" and "inner" labeled thiosulfate. A production of $H_2^{35}S$ was observed only when "inner" labeled thiosulfate was reduced or "outer" labeled thiosulfate was disproportionated or reduced

$$^-S\text{-}^{35}SO_3^- \ \text{------>}\ HS^- + H^{35}S^- \qquad \text{(reduction)} \qquad (9a)$$

$$^{-35}S\text{-}SO_3^- \ \text{------>}\ H^{35}S^- + SO_4^{2-} \qquad \text{(disproportionation)} \qquad (9b)$$

$$^{-35}S\text{-}SO_3^- \ \text{------>}\ H^{35}S^- + HS^- \qquad \text{(reduction)} \qquad (9c)$$

By contrast, the oxidation of "outer" labeled thiosulfate or the disproportionation or oxidation of "inner" labeled thiosulfate produced $^{35}SO_4^{2-}$

$$^{-35}S\text{-}SO_3^- \ \text{------>}\ ^{35}SO_4^{2-} + SO_4^{2-} \qquad \text{(oxidation)} \qquad (9d)$$

$$^-S\text{-}^{35}SO_3^- \ \text{------>}\ HS^- + ^{35}SO_4^{2-} \qquad \text{(disproportionation)} \qquad (9e)$$

$$^-S\text{-}^{35}SO_3^- \ \text{------>}\ SO_4^{2-} + ^{35}SO_4^{2-} \qquad \text{(oxidation)} \qquad (9f)$$

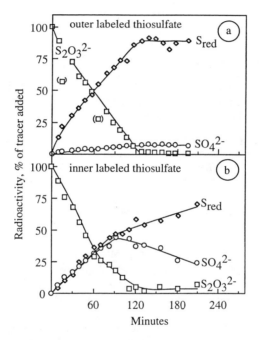

Figure 8. Thiosulfate oxidation, reduction and disproportionation in a $S_2O_3^{2-}$ amended sediment slurry, labeled with respectively "inner" and "outer" labeled $S_2O_3^{2-}$. a) Radioactivity of "outer" labeled $S_2O_3^{2-}$, SO_4^{2-}, and S_{red} (= H_2S + S^0 + FeS + FeS2). b) Radioactivity of "inner" labeled $S_2O_3^{2-}$, SO_4^{2-}, and S_{red}. (Adapted from ref. 10).

Thus, from the "outer" labeled thiosulfate experiment (Figure 8a) it was calculated that 6% of the thiosulfate was oxidized, with the rest (i.e. 94%) either reduced or disproportionated. From the "inner" labeled thiosulfate experiment (Figure 8b) 50% of the thiosulfate was reduced, whereas the other half was oxidized or disproportionated. Consequently, 44% (= 100% - 6% - 50%) of the thiosulfate was turned over by disproportionation. The radiolabeled $^{35}SO_4^{2-}$ that was formed by the turnover of "inner" labeled thiosulfate was continuously reduced to $H_2^{35}S$ by sulfate reduction. Thus $^{35}S_{red}$, comprising $H_2^{35}S$ and its precipitation products, continued to increase even after thiosulfate consumption had ceased.

Amendment Increases Turnover Rates. It is important to note that an increase of thiosulfate concentration will, when $S_2O_3^{2-}_{deq} < S_2O_3^{2-}_{max}$, increase the turnover rate of thiosulfate in accordance to Michaelis-Menten kinetics as discussed above. Therefore, observations done in experiments with $S_2O_3^{2-}$ addition may not always reveal the "true" turnover rates at dynamic equilibrium and even the importance of various pathways may well be affected. Experiments should therefore be performed under various concentration regimes of $S_2O_3^{2-}$ in order to estimate the impact that thiosulfate addition has on reaction rates and pathways.

Elemental Sulfur and Sulfite Turnover. Elemental sulfur is known to be an important intermediate in H_2S oxidation in aquatic sediments and is currently reduced or oxidized. Recently, it has been shown that also disproportionation is of major importance for elemental sulfur turnover (*39, 40*), producing SO_4^{2-} and H_2S, 1 : 3, as

$$4\ S^o + 4\ H_2O \longrightarrow SO_4^{2-} + 3\ H_2S + 2H^+ \qquad (10)$$

The turnover of SO_3^{2-} in aquatic sediments has to my knowledge not been published. However, experiments with pure cultures have showed that sulfite is disproportionated (*36, 41, 42*) and yields SO_4^{2-} and H_2S, 3 : 1, according to

$$4\ SO_3^{2-} + 2\ H^+ \longrightarrow 3\ SO_4^{2-} + H_2S \qquad (11)$$

However, it is not possible to perform a sediment experiment with $^{35}SO_3^{2-}$ or $^{35}S^o$ in order to discriminate disproportionation from oxidation or reduction like it was done with "inner" and "outer" labeled thiosulfate. Disproportionation of $^{35}SO_3^{2-}$ or $^{35}S^o$ produces $H_2^{35}S$ and $^{35}SO_4^{2-}$ which experimentally cannot be distinguished from the $H_2^{35}S$ and $^{35}SO_4^{2-}$ produced by reduction and oxidation, respectively.

Summary

Rate measurements with radiolabeled compounds must be done with addition of the radioactive compound in trace amounts in order not to change the chemical and physical equilibrium of the system. Further, the specific radioactivity of the added tracer must not change during incubation and thus, the rate measurement should be done with short time incubations, defined as the length of time in which <1% of the added tracer is turned over. However, even with these precautions, correct rates may be difficult to obtain as discussed above. The S-tracer may proceed through unknown pathways to compounds that are or cannot be analyzed, reverse reaction may return the radiolabeled sulfur atom to the tracer, or the radioactive product may the saturated with the tracer during incubation which all tends to underestimate the turnover rate.

Isotope exchange reactions between H_2S, S_n^{2-}, S^o, and FeS confound any attempt to use these radiolabeled compounds to measure turnover rates. Within minutes the ^{35}S-atoms will be transferred to $H_2^{35}S$, $^{35}S_n^{2-}$, $^{35}S^o$, and ^{35}FeS and the specific radioactivity will therefore change during incubation. However, these reduced compounds do not exchange with the more oxidized sulfur compounds like $S_2O_3^{2-}$, SO_3^{2-}, and SO_4^{2-}. Thus, they can be used qualitatively in amendment studies to show pathways and intermediate compounds in the sulfur cycle.

In amendment studies the concentration of a compound, initially low, is artificially increased by addition. Thus, amendment delays the depletion of the compound and hence retards any loss of radioactivity from this pool, compared to non-amended experiments. Therefore, by amendment studies it is possible to show the potential importance of a particular compound, to document transformation pathways, and measure reaction rates. However, the reaction rate may change by amendment, and even the importance of pathways may be incorrectly interpreted.

Acknowledgment

Thanks are due to Lars Elsgaard and Bo Barker Jørgensen for sharing their data from studies on thiosulfate turnover. Donald E. Canfield and Timothy G. Ferdelman are acknowledged for valuable discussions and suggestions which together with careful reviews and critical comments by Anne Giblin, John Parkes, and Murthy A. Vairavamurthy improved an earlier version of this paper.

Literature Cited

1. Jørgensen, B. B. *Limnol. Oceanogr.* **1977**, *22*: 814.
2. Howes, B. L.; Dacey, J. W.; King, G. M. *Limnol. Oceanogr.* **1984**, *29*: 1037.
3. King, G. M.; Howes, B. L.; Dacey, J. W. H. *Geochim. Cosmochim. Acta* **1985**, *49*, 1561.
4. Skyring, G. W. *Geomicrobiol. J.* **1987**, *45*, 1956.
5. Thode-Andersen, S.; Jørgensen, B. B. *Limnol. Oceanogr.* **1989**, *34*, 793.
6. Canfield, D. E.; Des Marais, D. J. *Science* **1991**, *251*, 1471.
7. Canfield, D. E.; Thamdrup, B.; Hansen, J. W. *Geochim. Cosmochim. Acta* **1993**, *57*, 3867.
8. Jørgensen, B. B.; Kuenen, J. G.; Cohen, Y. *Limnol. Oceanogr.* **1979**, *24*, 799.
9. Jørgensen, B. B. A *Science* **1990**, *249*, 152.
10. Jørgensen, B. B. *Limnol. Oceanogr.* **1990**, *35*: 1329.
11. Fossing, H.; Jørgensen, B. B. *Geochim. Cosmochim. Acta* **1990**, *54*, 2731.
12. Jørgensen, B. B.; Bak, F. *Appl. Environ. Microbiol.* **1991**, *57*, 847.
13. Jørgensen, B. B.; Fossing, H.; Wirsen, C. O.; Jannasch, H. W. *Deep-Sea Res.* **1991**, *38*, 1083.
14 Elsgaard, L.; Jørgensen, B. B. *Geochim. Cosmochim. Acta* **1992**, *56*, 2435.
15. Fossing, H.; Thode-Andersen, S.; Jørgensen, B. B. *Mar. Chem.* **1992**,*38*, 117.
16. Jørgensen, B. B.; Fenchel, T. *Mar. Biol.* **1974**, *24*, 189.
17. Howarth, R. W.; Jørgensen, B. B. *Geochim. Cosmochim. Acta* **1984**, *48*, 1807.
18. Fossing, H.; Jørgensen, B. B. *Biogeochemistry* **1990b**, *9*, 223.
19. Fossing, H. *Continental Shelf Res.* **1990**, *10*, 355.
20. Zhabina, N. N.; Volkov, I. I. In *Environmental Biogeochemistry and Geomicrobiology*; Editor, W. E. Krumbein; Ann Arbor Science Publishers: Michigan, **1978**, Vol. 3; pp 735-746.
21. Canfield, D. E.; Raisewll, R.; Westrich, J. T.; Reaves, C. M.; Berner, R. A. *Chem. Geol.* **1986**, *54*, 149.
22. Fossing, H.; Jørgensen, B. B. *Biogeochemistry* **1989**, *8*, 205.
23. Moeslund, L; Thamdrup, B.; Jørgensen, B. B. *Biogeochemistry* **1994**, *27*, 129.

364   GEOCHEMICAL TRANSFORMATIONS OF SEDIMENTARY SULFUR

24. Stranks, D. R.; Wilkins, R. G. *Chem. Rev.* **1957**, *57*, 743.
25. Schwarzenbach, G.; Fischer, A. *Helv. Chem. Acta* **1960**, *169*, 1365.
26. Giggenbach, W. *Inorg. Chem.* **1974**, *6*, 1201.
27. Aller, R. C.; Rude, P. D. *Geochim. Cosmochim. Acta* **1988**, *52*, 751.
28. Pyzik, A. J.; Sommer, S. E. *Geochim. Cosmochim. Acta* **1981**, *45*, 687.
29. Burdige, D. J.; Nealson, K. H. *Geomicrobiol. J.* **1986**, *4*, 361.
30. Thamdrup, B.; Finster, K.; Fossing, H.; Hansen, J. W.; Jørgensen, B. B. *Geochim. Cosmochim. Acta* **1994**, *58*, 67.
31. Thamdrup, B.; Fossing, H.; Jørgensen, B. B. *Geochim. Cosmochim. Acta* **1994**, *58*, 5115.
32. Voge, H. H. *J. Amer. Chem. Soc* **1939**, *61*, 1031.
33. Uyama, F.; Chiba, H.; Kuusakabe, M.; Sakai, H. *Geochem. J.* **1985**, *19*, 301.
34. Chu, X. L.; Ohmoto, H. *Geochim. Cosmochim. Acta* **1991**, *55*, 1953.
35. Luther III, G. W.; Church, T. M. *Mar. Chem.* **1988**, *23*, 295.
36. Bak, F.; Cypionka, H. *Nature* **1987**, *326*, 891.
37. Krämer, M.; Cypionka, H. *Arch. Microbiol.* **1989**, *151*, 232.
38. Vairavamurhty, A.; Manowitz, B.; Luther III, G.W.; Jeon, Y. *Geochim. Cosmochim. Acta* **1993**, *57*, 1619.
39. Thamdrup, B.; Finster K.; Hansen, F.; Bak, F. *Appl. Environ. Microbiol.* **1993**, *59*, 101.
40. Lovley, D. R.; Phillips, E. J. P. *Appl. Environ. Microbiol.* **1994**, *60*, 2394.
41. Bak, F.; Pfennig, N. *Arch. Microbiol.* **1987**, *147*, 184.
42. Widdel, F. In *Biology of anaerobic microorganisms*; Zehnder, J. B., Ed.; John Wiley & Sons, Inc., New York, 1988; pp 469-585.

RECEIVED April 17, 1995

Chapter 20

Microbial Assimilation and Dissimilation of Sulfonate Sulfur

A. P. Seitz and E. R. Leadbetter

Department of Molecular and Cell Biology, University of Connecticut, Storrs, CT 06269–2131

A wide variety of microbes use sulfonate-sulfur as the sole sulfur source for biosynthesis even when the carbon of that sulfonate cannot be used as an energy source for growth. Our studies of bacteria, including members of the genera *Comamonas* and *Escherichia*, as well as ascomycetous and basidiomycetous yeasts indicate that the sulfur of many naturally-occurring sulfonates can be reduced and assimilated into cellular sulfur compounds during aerobic, respiratory growth. Other unrelated bacteria (e.g. members of the genera *Clostridium, Klebsiella*) are able to use sulfonate-sulfur for biosynthesis under anaerobic conditions. Sulfonate can also serve as the terminal electron acceptor for *Desulfovibrio*'s anaerobic respiration. The breadth of microbial participation in sulfonate-sulfur transformations in the natural sulfur cycle is thus established.

Sulfonates, organosulfur compounds containing the $R-CH_n-SO_3H$ moiety, occur in the biosphere as a result of synthesis by diverse organisms and introduction by human activity. Naturally-produced sulfonates (see Table I for some examples) include taurine in the hearts and/or eyes of vertebrates (*1, 2*), coenzyme M in the methanogenic *Archaea* (*3*), aeruginosin, a pigment in some pseudomonads (*4*), the sulfonolipids of gliding bacteria (*5*) and diatoms (*6*), taurocholic acid in the digestive system of many mammals (*7*), sulfolactate in *Bacillus subtilis* spores (*8*), and isethionate in the axoplasm of squid (*9*). Some natural sulfonates are secondary products of the breakdown of other sulfur compounds. Methanesulfonate, for instance, is a product of the chemical oxidation in the atmosphere of dimethylsulfide (*10, 11*) which is produced by phytoplankton (*12-14*) and marsh grass (*15*). Thus, these biosynthesized sulfonates range from very simple short chains to aromatic structures. Examples of commercially-produced sulfonates include laboratory buffers such as HEPES, MOPS, and MES, the aminobenzenesulfonates which are used in the manufacture of dyes (*16*) and optical brighteners, the detergent-additive toluenesulfonate (*17*), and linear alkylsulfonate surfactants (*18*).

It is to be expected that the naturally-occurring sulfonates, at least, would be

0097–6156/95/0612–0365$12.00/0

Table I. Structure, origin and function or use of sulfonates

Some Naturally-Occuring Sulfonates

Sulfonate	Source	Function
Taurine (2-aminoethanesulfonate)	Vertebrate eyes, heart	Uncertain
Isethionate (2-hydroxyethanesulfonate)	Squid giant axon	Uncertain
2-Sulfolactate	*Bacillus* spores	Osmolyte?
Coenzyme M (2-mercaptoethanesulfonate)	Methanogenic archaea	Methyl group carrier
Sulfoquinovosyl diglyceride (6-deoxy-6-sulfoglucosyl diglyceride)	All phototrophs examined, pro- and eukaryotic	Implicated in photosynthesis
Methanesulfonate	Atmosphere, rain	Not applicable
Sulfonolipids (2-fattyacylamino-3-hydroxy-15-methylhexadecane-1-sulfonic acids)	Gliding bacteria (cytophagas, flexibacters)	Necessary for motility
Aeruginosin (2-amino-6-carboxy-10-methyl-8-sulfophenazium betaine)	*Pseudomonas aeruginosa*	Unknown

Some Synthetic Sulfonates

Sulfonate	Source/Use
Sodium O-cocoyl isethionate	Household soaps
Ammonium xylenesulfonate	Shampoos
Sodium dodecylsulfonate	Detergents
HEPES (N-[2-hydroxyethyl]piperazine-N'-[2-ethanesulfonate]), MOPS (3-[N-morpholino]propanesulfonate)	Biological buffers

attacked and degraded by microbes since, as recognized since the 1920's (*19*), for all naturally-synthesized organic compounds occuring in nature there are organisms which have evolved methods of degrading and mineralizing them.   Not all products of biosynthesis, however, are degraded at identical rates, as is well demonstrated by the slow rate of lignin disappearance, and thus its continued presence in significant quantities in forests, for example.   Some sulfonates may be in such a category as well, for as two recent reports indicate, sulfonates may accumulate in some habitats.   Autry and Fitzgerald (*20*) evaluated forest soils for the relative contributions of various organic sulfur compounds and found that sulfonate-sulfur comprised the major fraction of organic sulfur at all depths sampled; in forest litter greater than 40% of the total sulfur was sulfonate-sulfur in the majority of sites examined.   Vairavamurthy et al. (*21*) reported, in a study of three marine sediments, that sulfonates comprised 20-40% of the total organic sulfur in near-surface layers.

Several short chain, simple sulfonates were shown to be degraded by microorganisms when the compounds served as the sole source of carbon, nitrogen and, in some cases also sulfur, for growth.   Reports of taurine utilization suggested that the first step in the biodegradation of this compound can be either deamination (*22-24*) or transamination (*25-27*) to sulfoacetaldehyde, followed by release either of sulfite or sulfate (*22*), as measured by the appearance first of sulfate, then later, ammonia, in the culture medium.   The sulfoacetaldehyde was cleaved by a sulfolyase to form acetate and sulfite (*28, 29*).   Deamination and subsequent desulfuration was also suggested for cysteate (*23*).   Methanesulfonate degradation occurred via action of a monooxygenase to produce formaldehyde and sulfite (*30*) .

Longer chain sulfonates can also serve as a source of carbon for growth.   Thysse and Wanders (*31*) isolated pseudomonads which hydroxylated *n*-alkane-1-sulfonates, forming 1-hydroxy-*n*-alkanesulfonates, from which the corresponding aldehyde and bisulfite were formed.   As in the case of methanesulfonate, intital attack on the sulfonates was presumed to involve participation of molecular oxygen.

Alkylbenzenesulfonate-containing detergents caused serious foaming problems in treatment plants, on waterways and in tap water beginning in the late 1940's (*32*).   Because of this pollution problem much emphasis has been placed on the degradation of toluene and benzene sulfonic acids as models of the breakdown of these detergent components. Aromatic sulfonates themselves are major pollutants because they are by-products of the dyestuff industry and used as optical brighteners in detergent formulations (*17*).   At one point in the 1970's about 10% of the organic matter in the Rhine River was reported to be alkyl sulfonates (*33*).   Studies on the biodegradability of these compounds indicated that they could be utilized as the sole source of carbon by bacteria; two possible modes of attack were suggested: 1) initial dioxygenation of the carbon-sulfur bond with release of sulfite (*34-36*) or 2) desulfonation after hydroxylation (*37*) or side chain oxidation (*38*).

In spite of the number of reports of organisms, mostly bacteria, able to use sulfonates as the sole source of carbon for growth, the variety of bacteria known to be capable of mediating this breakdown is surprisingly narrow.   The vast majority were described as pseudomonads although a few others (species of *Bacillus*, *Agrobacterium*, *Alcaligenes* and *Achromobacter*) have been reported (*22, 25, 36, 39, 40*).

Transformation of sulfonate-sulfur is not restricted to instances in which a sulfonate is consumed as a carbon and energy source.   Several reports, from our laboratory and others, have shown that a wide variety of bacteria was able to use the sulfur of several chemically diverse sulfonates for assimilation into the sulfur-containing molecules of the cell even though they were unable to use the sulfonates as a carbon and energy source (41-50). Among the organisms with this trait are the fungus *Aspergillus niger* (43), the alga *Chlorella fusca* (45), several genera of yeasts (42, 48) as well as numerous bacterial genera (41, 46, 47, 49, 50).

Assimilation of sulfonate-sulfur under aerobic conditions

Probably the first extensive description of bacterial utilization of the sulfur of a wide variety of sulfonates was a 1987 report by Zurrer et al. (46) who demonstrated the desulfonation of several naphthalene- and benzenesulfonic acids by three types of bacteria. A pseudomonad, an *Arthrobacter*, and an unidentified bacterium all could use at least 16 aromatic sulfonates as a source of sulfur, even though none of these could serve as a carbon and energy source. Desulfonation occurred via a monooxygenolytic cleavage of the carbon-sulfur bond.

At approximately the same time, our laboratory began a search for sulfonate-utilizing bacteria, employing the enrichment culture approach (51). We isolated soil bacteria able to assimilate short-chain sulfonates such as cysteate, taurine, or isethionate as the sole carbon and energy source (50). These strains subsequently lost the ability to use these sulfonates as a carbon source and energy source, but retained the ability to use them as a source of sulfur. The organism showing the most vigorous growth on sulfonates was identified as the strictly aerobic soil bacterium *Comamonas acidovorans;* it was tested for its ability to use a range of sulfonates. In addition to the three sulfonates noted above, it was able to utilize methanesulfonate, *p*-toluenesulfonate, or 1-dodecanesulfonate as a sole source of sulfur but not of carbon and energy; metanilate, *m*-nitrobenzenesulfonate, sulfanilamide, or sulfamate did not serve either as sole sulfur or sole carbon and energy source (see Table II).

To determine whether the ability to use sulfonate-sulfur extended to additional, unrelated bacteria that had not been specifically isolated on the basis of the ability to utilize sulfonate-sulfur, laboratory strains of the facultatively anaerobic enterobacteria, which included *Enterobacter aerogenes*, *Serratia marcescens*, *Escherichia coli* K12 and others, were tested (47). These organisms also used sulfonate-sulfur but could not utilize sulfonates as a carbon and energy source. This assimilation of sulfonate-sulfur was restricted to growth under aerobic, respiratory conditions, and did not occur during anaerobic (fermentative) growth, although growth under the latter conditions did occur when sulfate-sulfur was supplied. As indicated in Table II, several of the other sulfonates tested could serve as a source of sulfur for aerobic growth.

The many reports of sulfonate-sulfur release as sulfite or sulfate by organisms using sulfonates as a source of carbon and energy (22, 23, 25, 28, 29) led to the assumption that sulfur in one of these inorganic forms is utilized for biosynthesis by the well-established assimilatory sulfate reduction pathway (Figure 1). In order to determine whether oxidative release of sulfate was a mandatory aspect of sulfonate-sulfur assimilation, a series of *E. coli* K12 strains, each mutated in one of the enzymes of the assimilatory sulfate reduction (ASR) pathway, were studied (47). Each mutant was tested for the ability to grow using a sulfonate as the sulfur source. Sulfonate-sulfur could be assimilated for growth by

Table II. Growth of some bacterial strains with selected sulfonates as sole sulfur source[a].

	Strains									
	Comamonas acidovorans	*Enterobacter aerogenes*	*Serratia marcescens*	*Escherichia coli K12*	*Escherichia coli*[c]	*Shigella flexneri*	*Proteus vulgaris*	*Salmonella typhimurium*	*Klebsiella* species	*Clostridium pasteurianum*
Taurine	+[b]	+	+	+	-	-	-	-	+/-[d]	+
Isethionate	+	+	+	+	-	-	-	-	+/-	+
Cysteate	+	+	+	+	-	-	-	-	+/+	-
Methanesulfonate	+	+	+	+	ND[e]	ND	ND	ND	+/-	-
HEPES	ND	+	+	+	ND	-	-	ND	+/-	-
Dodecanesulfonate	+	-	ND	+	ND	-	-	ND	ND	ND

[a] Zurrer et al. provide an extensive list of aromatic sulfonates utilized as sole sulfur source.
[b] + means the organism grew using the compound, - means the organism did not grow
[c] Fecal isolate
[d] Results are reported for aerobic/anaerobic conditions
[e] Not determined

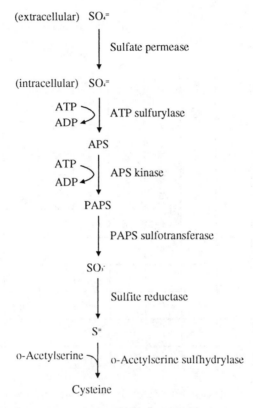

Figure 1. The assimilatory sulfate reduction pathway. The name of the enzyme catalyzing each step is shown to the right of the arrow. ATP: adenosine triphosphate, ADP: adenosine diphosphate, APS: adenylyl sulfate (adenosine 5'-phosphosulfate), PAPS: adenosine 3'-phosphate 5'-phosphosulfate.

mutant strains deficient in any portion of the ASR pathway prior to formation of sulfite, but sulfite-reductase deficient mutants were unable to utilize sulfonate-sulfur. Sulfate formation, thus, is not a necessary step in sulfonate-sulfur assimilation. However, these results suggested that sulfite reductase was a part of the pathway for sulfonate-sulfur assimilation and that sulfite was an intermediate of the pathway. This latter notion was supported by the demonstration that when supplied with sulfonate-sulfur, sulfite-reductase-less mutants accumulated sulfite.

Additional experiments with these mutant strains demonstrated sulfide formation from sulfonate-sulfur, except when the enzymatic deficiency was the lack of sulfite reductase, again lending credence to the idea that sulfonate-sulfur assimilation involves release of sulfite and its subsequent reduction to sulfide which is then used for biosynthesis of the amino acid cysteine.

An early study by Margolis and Block (*42*) examining cysteate and taurine utilization with five strains of yeast representing three different species was one of the few, and probably the first, studies of sulfonate-sulfur utilization in eukaryotes. All species tested were able to use taurine-sulfur but only three of the five strains showed moderate to good growth with cysteate. We have examined four different yeasts, both ascomycetes and basidiomycetes, for the ability to use cysteate-, taurine-, or isethionate-sulfur (*48*). These compounds served as sole sulfur source for all strains examined although the final cell yield in some cases was less than that obtained with an equimolar amount of sulfate. Krauss and Schmidt (*45*) showed that the alga *Chlorella fusca* was capable of using taurine, isethionate, cysteate or other short chain sulfonates as a sulfur source, but did not use several aromatic sulfonates or longer chain buffers such as HEPES.

Assimilation of sulfonate-sulfur under anaerobic conditions

Because of the inability, noted above, of enterobacteria to use sulfonates for anaerobic growth, we employed enrichment cultures to determine if anaerobic sulfonate utilization might occur. Two organisms, the strictly fermentative (anaerobic) *Clostridium pasteurianum* and a strain of the facultatively anaerobic *Klebsiella*, were isolated and utilization of sulfonate-sulfur studied (see Table II) (Chien, C. C., Leadbetter, E. R., Godchaux, W., III, in press). *C. pasteurianum* could assimilate taurine, isethionate, and *p*-toluenesulfonate sulfur, but not that of cysteate, for its obligatorily anaerobic growth. In contrast, anaerobically grown *Klebsiella* could use cysteate, but not taurine and isethionate, as a sulfur source. Under aerobic conditions *Klebsiella* used the sulfur of cysteate, isethionate, taurine, HEPES, or methanesulfonate (see Table II).

Selectivity in assimilation of sulfur sources

Because in most natural habitats a variety of sulfur sources is available to organisms, we examined the ability of sulfonate-sulfur to be assimilated under conditions when more than a single sulfur-containing compound was present for growth of *C. acidovorans*, *E. coli* K12, *C. pasteurianum*, *Klebsiella*, or the yeasts *Rhodotorula*, *Hansenula*, *Trigonopsis*, and *Saccharomyces*. The dogma on this subject implies (*52-56*) that the more reduced sulfur sources are assimilated preferentially over those more oxidized; thus, in experiments in which equimolar concentrations of both a sulfonate (oxidation state +5) and sulfate (oxidation state +6) were present, we expected sulfonate to be used either preferentially or exclusively. In each instance, however, the result was the opposite: sulfate-sulfur competed with, and, in fact, excluded the use of cysteate-, isethionate-, or taurine-sulfur (*47, 48, 50*), (Chien, C. C., Leadbetter, E. R., Godchaux, W., III, in press). Zurrer et al. had also noted similar competition during study of a pseudomonad able to use sulfur of different naphthalenesulfonates (*46*) as has, more recently, Kertesz et al. with a culture of *Pseudomonas putida* able to use alkylbenzenesulfonates as sole sulfur source (*49*). Additional tests in our laboratory in which the sulfur sources available to the cells were both a sulfonate and the more reduced sulfur of cysteine fit the expected pattern: cysteine-sulfur, not that of a sulfonate, was used (*50, 57*). In sharp contrast are the results of Biedlingmaier and Schmidt (*58*) who showed that uptake of ethanesulfonate by *Chlorella fusca* was inhibited neither by the oxidized sulfur of sulfate nor by the more reduced compounds such as cysteine.

Dissimilation of sulfonate sulfur

Our discussion up to this point has been concerned solely with sulfur assimilation (the incorporation of sulfur into the cell's organic molecules). Dissimilation of sulfur refers to its reduction, usually to sulfide, which is mainly exported from the cells, rather than being incorporated into cellular components. This process represents a class, or type, of anaerobic respiration in which oxidized compounds serve as an electron acceptor for anaerobic, non-fermentative growth. There are several types of anaerobic respiration known in prokaryotes, an example being sulfate respiration which is common in marine sediments; carbonate and nitrate reduction are other types of anaerobic respiration, known also by the terms methanogenesis and denitrification, respectively.

To determine if anaerobic respiration of sulfonate-sulfur--analogous perhaps to sulfate-sulfur reduction--existed, we set up enrichment cultures for bacteria able to reduce (respire) sulfonates to support anaerobic growth. A bacterium identified as *Desulfovibrio desulfuricans*, and given strain designation IC1, was isolated from marine sediment. The organism was able to reduce the sulfur of either isethionate or cysteate (serving as electron acceptors) with lactate as carbon and energy source (electron donor); sulfide and acetate were detected as metabolic end products (Lie, T. J., Pitta, T. P., Leadbetter, E. R., Godchaux, W., III, Leadbetter, J. R., in press). A variety of other sulfonates were not used as electron acceptors to support growth; this latter category included methane- and ethanesulfonate, taurine, 3-hydroxypropanesulfonate, the buffers HEPES and MOPSO, sulfanilate, metanilate, *p*-toluenesulfonate and *m*-nitrobenzenesulfonate.

Well studied laboratory cultures of other known sulfate-reducing bacteria that had not been isolated specifically by virtue of the ability to reduce sulfonate-sulfur were assessed for this trait. A strain each of *D. desulfuricans*, *D. vulgaris*, *D. sulfodismutans*, *Desulfomicrobium baculatum*, *Desulfobulbus propionicus*, *Desulfobacterium autotrophicum*, and *Desulfobacter postgatei*, were tested for their ability to reduce isethionate-sulfur; only *D. desulfuricans* and *D. baculatum* were able to reduce this organosulfur compound in anaerobic respiration.

Discussion

While the reported ability to use sulfonates as a carbon and energy source for growth seems to be confined mainly to relatively few bacterial species, the utilization of sulfonate-sulfur appears to be a fairly widespread trait as demonstrated in aerobic, facultative, and anaerobic organisms, prokaryotic as well as some eukaryotic. This finding may indicate a greater potential for the biodegradability of these compounds since, after removal of the sulfonate-sulfur, the resulting compounds may be of a type more readily subject to attack by a greater variety of microorganisms than those able to attack and cleave the rather stable carbon-sulfur bond.

Past studies of sulfonate utilization as a source of carbon and energy either suggested or established that sulfite or sulfate was released during sulfonate oxidation and degradation. However, this does not necessarily indicate that sulfate or sulfite was the form in which sulfonate-sulfur was assimilated. Clearly, an organism gaining its energy from oxidation of sulfonate-carbon needs to utilize a considerably greater number of sulfonate molecules for this purpose than it needs for synthesis of sulfur-containing cell components. The sulfite or sulfate released by virtue of sulfonate oxidation would indeed be expected to be in excess of biosynthetic needs and to accumulate in the growth medium; this has been regularly observed. Studies of the form in which sulfonate-sulfur is assimilated when sulfonates served as sole source of carbon, energy and sulfur for growth have not been reported.

The studies with mutants of the bacterium *E. coli* and the yeast *S. cerevisiae* (*48, 57*) provide a clear indication that sulfonate-sulfur assimilation can occur without prior oxidation of this sulfur to sulfate, and indicate that sulfite is a likely intermediate in the pathway of sulfonate-sulfur assimilation. Also consistent with this may be the report of sulfite production from toluenesulfonate by cell extracts of *Pseudomonas testosteroni* T-2 (*38*).

The contrasting results of Biedlingmaier and Schmidt (*58*), reporting sulfate in extracts of *Chlorella fusca* grown with ethanesulfonate as the sulfur source, suggest that an oxidation of sulfonate-sulfur to sulfate may be obligatory in this alga; such an event clearly necessitates the expenditure of an additional ATP for each sulfate-sulfur atom assimilated over that required for assimilation of sulfonate-sulfur by a more direct (release as sulfite) pathway.

The curious results indicating that incorporation of sulfonate-sulfur into cell components was inhibited when sulfate was also present is apparently a trait common both to pro- and eukaryotes, and for which the responsible mechanisms remain unclear. On the other hand, the apparently consititutive nature of the assimilatory processes in the Bacteria seems not to be characteristic of the eukaryote *C. fusca*. In this regard, it may be significant that in *C. fusca* sulfate appeared to be the first demonstrable sulfur product of carbon-sulfur bond cleavage; neither sulfide nor sulfite was detected.

The great differences in the metabolism of sulfonate-sulfur for assimilation in different biota is reflected in the constitutive vs. non-constitutive feature just noted, as well as by the observations that the ability of different organisms to metabolize different sulfonates is not identical. What is, not surprisingly, apparent is that possession of the sulfonate moiety *per se* does not insure identical, or indeed any, metabolism of a given sulfonate; thus, as for so many other classes of molecules, biological specificity, undoubtedly at the enzyme level, plays a significant discriminatory, determinative role.

Among the many aspects of sulfonate metabolism that remain to be clarified is why the sulfur of isethionate or taurine, for example, can be assimilated under respiratory conditions but that the sulfonate is unable to serve as a source of carbon and energy for growth; the carbon compounds likely to result from desulfonation normally are utilized as carbon and energy sources by these bacteria, and yet no evidence has accrued to indicate that these sulfonates are themselves toxic. Equally perplexing is the inability of sulfonate-sulfur to be used under fermentative growth conditions by several enteric bacteria able to ultilize these sulfonates under respiratory conditions; the involvement of an oxygenolytic attack on the C-S bond is possible, but not yet demonstrated, in this instance. This explanation seems less likely given the ability of other related enteric bacteria to use a sulfonate under both the anaerobic and aerobic conditions.

In addition to the attack on sulfonate-sulfur by bacteria, yeasts, other fungi, and algae for assimilatory purposes, the demonstration of dissimilatory reduction to sulfide by anaerobic respiratory processes further substantiates consideration of sulfonates as part of the natural sulfur cycle. Whether a greater variety of bacteria will be implicated in this anaerobic respiration, and whether more-reduced organosulfur compounds may be the, or among the, products of this process in other of these bacteria remains for study.

There are many reports of natural and synthetic sulfonates in water, soil and sediments (20, 21, 59-63) but the degree to which sulfonates are actually used as a sulfur source in natural habitats has yet to be determined. The demonstration that in some organisms sulfate-sulfur can be assimilated to the exclusion of sulfonates (46-50) may well have important ecological implications. In environments where sulfate rarely accumulates, such as soils of humid and semi-humid regions (64), sulfonates may be incorporated into cellular sulfur compounds while in environments with available sulfate, sulfonates may not be used as a sulfur source, resulting in the sulfonate accumulation reported in recent years (20, 21). It is also possible that sulfonate accumulation simply reflects the fact that these compounds are biodegraded much more slowly than they are biosynthesized or introduced by societal activities.

Acknowledgments

This research was funded in part by the the the Institute of Water Resources (U.S. Geological Survey - Department of Interior, 14-08-0001-G2009), and the University of Connecticut Research Foundation. A. P. Seitz was also supported by a Connecticut High Technology Scholarship.

Literature cited

1.   Jacobsen, J. G.; Smith, L. H. *Physiol. Rev.* **1968,** *48,* 424-511.
2.   Huxtable, R. J. *Physiol. Rev.* **1992,** *72,* 101-163.
3.   Taylor, C. D.; Wolfe, R. S. *J. Biol. Chem.* **1974,** *249,* 4879-4885.
4.   Herbert, R. B.; Holliman, F. G. *Proc. Chem. Soc.* **1964,** 19.
5.   Godchaux, W., III; Leadbetter, E. R. *J. Bacteriol.* **1980,** *144,* 592-602.
6.   Anderson, R.; Kates, M.; Volcani, B. E. *Biochim. Biophys. Acta.* **1978,** *528,* 89-106.

7.   Haslewood, G. A. D.; Wootton, V. *Biochem. J. 1950*, *47*, 584-597.
8.   Bonsen, P. P. M.; Spudich, J. A.; Nelson, D. L.; Kornberg, A. *J. Bacteriol. 1969*, *98*, 62-68.
9.   Koechlin, B. A. *Proc. Natl. Acad. Sci. 1954*, *40*, 60-62.
10. Hatakeyama, S.; Okuda, M.; Akimoto, H. *1982*, 583-586.
11. Grosjean, D.; Lewis, R. *Geophys. Res. Lett. 1982*, *9*, 1203-1206.
12. Turner, S. M.; Malin, G.; Liss, P. S.; Harbour, D. S.; Holligan, P. M. *Limnol. Oceanogr. 1988*, *33*, 364-375.
13. Iverson, R. L.; Nearhoof, F. L.; Andreae, M. O. *Limnol. Oceanogr. 1989*, *34*, 53-67.
14. Kelly, D. P.; Smith, N. A., In *Advances in Microbial Ecology*; K. C. Marshall, Ed.; Plenum Press: New York, 1990, Vol. 11; pp 345-385.
15. Dacey, J. W. H.; King, G. M.; Wakeham, S. G. *Nature 1987*, *330*, 643-645.
16. *The Merck Index*; Windholz, M.; Budavari, S., Eds.; Merck and Co., Inc.: Rahway, New Jersey, 1983.
17. Woollatt, E., *The Manufacture of Soaps, Other Detergents, and Glycerine*; Halsted Press: New York, 1985.
18. Porter, M. R., *Handbook of Surfactants*; Chapman and Hall: New York, 1991.
19. den Dooren de Jong, L., *Bijdrage tot de kennis van het mineralisatie-proces*; Nijgh and Van Ditmar: Rotterdam, 1926.
20. Autry, A. R. *Biol. Fertil. Soils 1990*, *10*, 50-56.
21. Vairavamurthy, A.; Zhou, W.; Eglinton, T.; Manowitz, B. *Geochim. Cosmochim. Acta 1994*, *58*, 4681-4687.
22. Ikeda, K.; Yamada, H.; Tanaka, S. *J. Biochem. 1963*, *54*, 312-316.
23. Stapley, E.; Starkey, R. *J. Gen. Microbiol. 1970*, *64*, 77-84.
24. Kondo, H.; Anada, H.; Ohsawa, K.; Ishimoto, M. *J. Biochem. 1971*, *69*, 621-623.
25. Toyama, S.; Soda, K. *J. Bacteriol. 1972*, *109*, 533-538.
26. Toyama, S.; Miyasato, K.; Yasuda, M.; Soda, K. *Agr. Biol. Chem. 1973*, *37*, 2939-2941.
27. Shimamoto, G.; Berk, R. *Biochim. Biophys. Acta. 1979*, *569*, 287-292.
28. Kondo, H.; Ishimoto, M. *J. Biochem. 1975*, *78*, 317-325.
29. Shimamoto, G.; Berk, R. S. *Biochim. Biophys. Acta 1980a*, *632*, 121-130.
30. Kelly, D. P.; Baker, S. C.; Trickett, J.; Davey, M.; Murrell, J. C. *Microbiol. 1994*, *140*, 1419-1426.
31. Thysse, G. J. E.; Wanders, T. H. *Antonie van Leeuwenhoek 1974*, *40*, 25-37.
32. McGucken, W., *Biodegradable Detergents and the Environment*; Texas A&M University Press: College Station, Texas, 1991.
33. Malle, K.-G. *Chemie Unserer Zeit 1978*, *12*, 111-122.
34. Cain, R. B.; Farr, D. R. *Biochem. J. 1968*, *106*, 859-877.
35. Brilon, C.; Beckman, W.; Knackmuss, H.-J. *Appl. Environ. Microbiol. 1981*, *42*, 44-55.
36. Thurnheer, T.; Zurrer, D.; Hoglinger, O.; Leisinger, T.; Cook, A. M. *Biodeg. 1990*, *1*, 55-64.
37. Focht, D. D.; Williams, F. D. *Can. J. Microbiol. 1970*, *16*, 309-316.
38. Locher, H. H.; Leisinger, T.; Cook, A. M. *J. Gen. Microbiol. 1989*, *135*, 1969-1978.

39.   Martelli, H. L.; Souza, S. M. *Biochim. Biophys. Acta* **1970**, *208*, 110-115.
40.   Willetts, A. J.; Cain, R. B. *Biochem. J.* **1972**, *129*, 389-402.
41.   Roberts, R. B.; Abelson, P. H.; Cowie, D. B.; Bolton, E. T.; Britten, R. J., *Studies of Biosynthesis in Escherichia coli*; Carnegie Institution of Washington: Washington, D.C., 1957.
42.   Margolis, D.; Block, R. J. *Contrib. Boyce Thompson Inst. Plant Res.* **1958**, *19*, 437-443.
43.   Braun, R.; Fromageot, P. *Biochim. Biophys. Acta* **1962**, *62*, 548-555.
44.   Benarde, M. A.; Koft, B. W.; Horvath, R.; Shaulis, L. *Appl. Microbiol.* **1965**, *60*, 2296-2303.
45.   Krauss, F.; Schmidt, A. *J. Gen. Microbiol.* **1987**, *133*, 1209-1219.
46.   Zurrer, D.; Cook, A. M.; Leisinger, T. *Appl. Environ. Microbiol.* **1987**, *53*, 1459-1463.
47.   Uria-Nickelsen, M. R.; Leadbetter, E. R.; Godchaux, W., III *J. Gen. Microbiol.* **1993**, *139*, 203-208.
48.   Uria-Nickelsen, M. R.; Leadbetter, E. R.; Godchaux, W., III *FEMS Microbiol. Lett.* **1993**, *114*, 73-78.
49.   Kertesz, M. A.; Kolbener, P.; Stockinger, H.; Beil, S.; Cook, A. M. *Appl. Environ. Microbiol.* **1994**, *60*, 2296-2303.
50.   Seitz, A. P.; Leadbetter, E. R.; Godchaux, W., III *Arch. Microbiol.* **1993**, *159*, 440-444.
51.   Poindexter, J. S.; Leadbetter, E. R., In *Bacteria in Nature;* J. S. Poindexter, E. R. Leadbetter, Eds.; Plenum Press: New York, 1985, Vol. 2; pp 229-260.
52.   Dreyfuss, J.; Monty, K. J. *J. Biol. Chem.* **1963**, *238*, 3781-3783.
53.   Pasternak, C. A.; Ellis, R. J.; Jones-Mortimer, M. C.; Crichton, C. E. *Biochem. J.* **1965**, *96*, 270-275.
54.   Burnell, J. N.; Whatley, F. R. *J. Gen. Microbiol.* **1980**, *118*, 73-78.
55.   Imhoff, J. F.; Then, J.; Hashwa, F.; Truper, H. G. *Arch. Microbiol.* **1981**, *130*, 234-237.
56.   Imhoff, J. F.; Kramer, M.; Truper, H. G. *Arch. Microbiol.* **1983**, *136*, 96-101.
57.   Uria-Nickelsen, M. R.; Leadbetter, E. R.; Godchaux, W., III *Arch. Microbiol.* **1994**, *161*, 434-438.
58.   Biedlingmaier, S.; Schmidt, A. *Biochim. Biophys. Acta* **1986**, *861*, 95-104.
59.   Stevenson, F. J. *Proc. Soil Sci. Soc. Amer.* **1956**, *20*, 201-.
60.   McKinney, R. E.; Symons, J. M. *Sewage Indus. Wastes* **1959**, 549-556.
61.   Sullivan, W. T.; Swisher, R. D. *Environ. Sci. Technol.* **1969**, *3*, 481-483.
62.   Freney, J. R.; Stevenson, F. J.; Beavers, A. H. *Soil Sci.* **1972**, *114*, 468-476.
63.   Chae, Y. M.; Tabatabai, M. A. *Soil Sci. Soc. Amer. J.* **1981**, *45*, 20-25.
64.   Stevenson, F. J., *Cycles of soil: Carbon, nitrogen, phosphorus, sulfur, micronutrients*; John Wiley & Sons: New York, 1986.

RECEIVED April 17, 1995

ISOTOPIC EFFECTS
DURING SULFUR TRANSFORMATIONS

Chapter 21

Isotopic Evidence for the Origin of Organic Sulfur and Elemental Sulfur in Marine Sediments

Thomas F. Anderson[1] and Lisa M. Pratt[2]

[1]Department of Geology, University of Illinois, Urbana, IL 61801
[2]Biogeochemical Laboratories and Department of Geological Sciences,
Indiana University, Bloomington, IN 47405

Organic sulfur and elemental sulfur have nearly the same isotopic composition but are enriched substantially in ^{34}S relative to co-existing pyrite in marine sediments. Isotope mass-balance budgets together with evidence on the timing of sulfur transformations suggest that ^{34}S-enriched species produced by sulfide oxidation, such as polysulfides or sulfoxyanions, during the earliest stages of diagenesis are the predominant precursors of organic sulfur and elemental sulfur. Early-diagenetic H_2S contributes to the formation of both organic and elemental sulfur but is not the principal source. Model results also suggest that sulfur species derived from primary organic matter make up 20 % of neoformed organic-sulfur compounds.

Organically-bound sulfur and elemental sulfur are enriched in ^{34}S by up to 30 permil relative to co-existing pyrite in virtually all modern marine sediments and sedimentary rocks for which sulfur isotopic data on those species have been reported *(e.g., 1-7)*. This pattern of isotopic fractionation thus appears to be a characteristic feature of the marine biogeochemical cycle of sulfur, and one that has been neither widely recognized nor satisfactorily addressed. The objective of this paper is to provide new insights on the origin of organic sulfur and elemental sulfur in marine sediments based on the isotopic evidence. First, we derive mass-balance budgets based on isotopic correlation trends for organic sulfur and elemental sulfur relative to pyrite. Those calculations place limits on the proportions and isotopic compositions of likely sources of organic sulfur and elemental sulfur. We then consider the implications of the isotope budgets for possible pathways of organic-sulfur and elemental-sulfur formation in marine sediments.

Marine Biogeochemistry of Sulfur and Sulfur Isotopes

The biogeochemistry of sulfur in marine sediments is dominated by dissimilatory bacterial sulfate reduction to H_2S *(8)*. Because sulfate-reducing bacteria are obligate anaerobes *(9-11)*, this process can only occur in anoxic macro-environments or in anoxic micro-environments within suboxic macro-environments. The highest rates of sulfate reduction commonly occur in the surficial zone of bioturbation, where biological advection provides a continuous supply of reactive organic matter *(12-15)*. Although the complex series of reactions that control the fate of H_2S in marine sediments is not well understood in detail, the following general pathways are involved: (i) Chemical and biological oxidation to sulfate and various partial-oxidation products, e.g. elemental sulfur, polysulfides, and sulfoxyanions such as sulfite, thiosulfate, and polythionates *(13, 16, 17)*. Oxidation should be especially vigorous at oxic-anoxic interfaces in the zone of bioturbation, where sulfide encounters oxidants such as O_2 and oxidized metal-bearing sediments. Anaerobic microbial oxidation of sulfide can also occur *(17)*. (ii) Reaction with iron-bearing minerals to precipitate metastable iron sulfide phases and eventually pyrite *(8)*. The transformation of iron sulfides to pyrite involves reaction with partially reduced sulfur species, such as elemental sulfur or polysulfides. Most of the pyrite in modern marine sediments is precipitated within or slightly below the zone of maximum sulfate reduction during the earliest stages of diagenesis *(4, 8, 18, 19)*. (iii) Reaction with sedimentary organic matter to yield organic sulfur compounds, either directly or through intermediate oxidation products *(1-7, 20-26)*. Organic-sulfur formation also appears to be a feature of early diagenesis *(4, 20-28)*. The amounts of sulfur retained in diagenetic phases invariably exceed the amounts present initially as sulfate in pore waters *(e.g., 1, 18)*. This situation requires the transport of sulfate into sediments via diffusion and/or advection during early diagenesis and implies that the rate of sulfate reduction equaled or exceeded rates of sulfate transport *(1, 29-31)*.

Dissimilatory bacterial sulfate reduction is also the dominant process controlling the isotopic composition of sulfur species in marine sediments. The isotopic composition of a sulfur sample is expressed as $\delta^{34}S$, the deviation (in permil, or parts per thousand) of the $^{34}S/^{32}S$ ratio of the sample relative to that of Canyon Diablo troilite (CDT):

$$\delta^{34}S = \{[(^{34}S/^{32}S)_{sample} - (^{34}S/^{32}S)_{CDT}] / (^{34}S/^{32}S)_{CDT}\} \times 10^3 \text{ (permil)} \quad (1)$$

Kinetic isotope effects in bacterial sulfate reduction favor the reduction of the $^{32}SO_4$ species over that of the $^{34}SO_4$ species *(32-35)*. In experimental culture studies using organic electron donors, differences in $\delta^{34}S$ between product H_2S and reactant SO_4, i.e., isotopic fractionation between H_2S and SO_4, of from -9 to -46 permil have been obtained; the magnitude of fractionation was inversely related to the rate of sulfate reduction per cell *(33-37)*. However, maximum ^{34}S depletions in sediment or water-column sulfides in modern natural systems are larger, typically -40 to -60 permil *(33)*. Large apparent fractionations in natural environments may be due to much slower specific rates of reduction compared to experimental systems. In addition, recent experimental evidence has suggested that fractionation accompanying cycles of sulfide oxidation to elemental sulfur and subsequent bacterial disproportionation can generate the large ^{34}S depletions often observed in natural systems *(38)*. The cyclic formation and disproportionation of thiosulfate may also contribute to ^{34}S depletions in sedimentary sulfides *(39)*.

Isotopic differences between sedimentary pyrite and contemporaneous seawater sulfate are often less than maximum fractionations observed in modern systems. This attenuation is most likely a consequence of rates of sulfate reduction (and sulfide reoxidation) exceeding rates of sulfate transport, even in near-surface sediments.

Because of the kinetic isotope effect in bacterial sulfate reduction, the difference in rates results in enrichment of ^{34}S in pore-waters sulfate and its reduction products *(30, 31, 40, 41)*. This transport-related phenomenon is enhanced below surficial sediments in the so-called "zone of diffusion," where $\delta^{34}S$ values of sulfate and sulfide increase systematically with depth *(4, 18, 20, 30,40-42)*.

The isotopic composition of sedimentary sulfur species reflects the composition of the pools from which they form and isotopic fractionation in those reactions Experimental studies *(43)* have shown that pyrite has the same isotopic composition as the H_2S pool from which it forms, implying that the partially reduced species involved in the pyritization of iron sulfides must also have the same isotopic composition as H_2S. The isotopic composition of pyrite in any given sedimentary horizon is not homogeneous *(e.g., 42, 43)* and thus reflects isotopic variations of its precursor H_2S pool in both time and space (depth) during pyrite formation.

Organic sulfur and elemental sulfur that form directly from H_2S should also have the same isotopic composition as this precursor because isotope effects in transformations between fully and partially reduced sulfur species are small *(43, 45-47)*. On the other hand, there is some experimental evidence to suggest that microbial oxidation of H_2S produces intermediate products that are significantly enriched in ^{34}S relative to the H_2S substrate *(34, 36)*. If isotopically enriched intermediates of sulfide oxidation are produced in natural settings, then organic sulfur and elemental sulfur that form via more complex oxidative pathways may have higher $^{34}S/^{32}S$ ratios than their ultimate H_2S source.

The enrichment of ^{34}S in organic sulfur and elemental sulfur relative to co-existing pyrite suggests that reservoirs in addition to (or other than) the source of pyritic sulfur are involved in the formation of the former species. As noted previously, the dominant source of pyrite is the dissolved sulfide pool in near-surface sediments, referred to hereafter as "early-diagenetic H_2S." Potential reservoirs with higher $^{34}S/^{32}S$ ratios include (but are not necessarily limited to) the following: (i) Dissolved sulfide species present deeper in the sediment-pore water column ("later-diagenetic H_2S "). Several investigators have emphasized this source for the origin of organic sulfur *(5, 6)*. (ii) Reactive intermediate products of H_2S oxidation (see above). This source must be distinct from the partially reduced sulfur pool that participates in pyrite formation. (iii) Sulfur-containing compounds in primary organic matter. Almost all microorganisms and plants incorporate cellular sulfur by the process of assimilatory sulfate reduction *(9)*. Because there are no significant isotope effects involved in that process *(1, 36)*, sulfur in primary marine organic matter has about the same isotopic composition as seawater sulfate. The contribution of assimilatory sulfur to sedimentary organic matter is not known, although it is generally considered not to be significant. The molecular structures of biosynthetic sulfur compounds are quite different from sulfur-containing compounds in sedimentary organic matter *(22, 23, 48, 49)*. Also, sulfur-containing molecules in organisms (proteins, coenzymes, vitamins, etc.) are readily metabolized by microorganisms *(50)*. If assimilatory sulfur is incorporated into sedimentary organic matter to any significant extent, then reactions involving organic or inorganic degradation products of the former may be involved.

Isotopic Compositions

Table I describes the age, lithology, and sulfur-carbon geochemistry of the samples utilized in this study *(1-7)*. The samples range in age from Jurassic to Recent, and represent a variety of lithologies and depositional environments, as well as organic carbon and reduced sulfur contents. The ranges in isotopic compositions of co-existing pyrite, organic sulfur, and elemental sulfur as well as fractionations (Δ) between species are shown in Table II. Note that the isotopic composition of organic sulfur is reported as solvent-extractable bitumens, residual kerogen, or total organic sulfur. The use of sulfur isotopic data from different organic fractions is unlikely to bias our results significantly because sulfur isotopic fractionation between bitumens and kerogen extracted from the same rock are minimal *(3)*. The overall ranges in isotopic compositions are large: 60 permil for pyrite and 45 permil for organic sulfur and elemental sulfur. We found no systematic correlation between $\delta^{34}S$ values of species (or fractionation between species) and the sulfur-carbon geochemical parameters listed in Table I. In all but 2 of the 110 samples in Table II, the $\delta^{34}S$ value of organic sulfur is higher than that of pyrite, i.e. Δ(org-pyr) > 0. In the 20 samples analyzed for elemental sulfur, the $\delta^{34}S$ values of elemental sulfur and organic sulfur are generally within ±5 permil (mean Δ(elem-org) = -1.0 permil). This isotopic similarity suggests strongly that elemental sulfur and organic sulfur were derived from the same source(s).

Figure 1 shows the covariance between δ(org) and δ(pyr) for all samples considered in this study. The correlation is linear over the entire isotopic range with a slope of about 0.8 and an intercept of about +10 permil. (Correlation within many of the sample suites in Tables I and II is less significant because of the relatively small number of data points and restricted isotopic ranges.) Figure 2 shows the covariance between δ(elem) and δ(pyr). This correlation is also linear, corresponding to a slope of about 1 and an intercept of about +12 permil. The slopes and intercepts of the two correlation lines are virtually identical at the 95 % confidence level. The excellent isotopic correlations in Figs. 1 and 2 suggests that the relative proportions and isotopic properties of the sulfur pools involved in organic-sulfur and elemental-sulfur formation were essentially constant over the range of depositional and diagenetic environments represented by the samples.

Isotope Mass-Balance Budgets

In this section, we use isotope mass-balance calculations to estimate theoretical limits on the proportions of likely precursors/sources of organic sulfur and elemental sulfur. The slope and intercept of the isotopic correlation lines serve as constraints for the mass-balance calculations. For example, the fact that the δ(org) - δ(pyr) correlation lies above the δ(org) = δ(pyr) line and has a slope < 1 (Fig. 1) suggests that at least one source of organic sulfur has a constant $\delta^{34}S$-value that is higher than that of pyritic sulfur. The displacement of the δ(elem) - δ(pyr) correlation above the δ(elem) = δ(pyr) reference line by a constant amount (i.e., slope $\approx$ 1; Fig. 2) requires that the source of elemental sulfur is pyritic sulfur and/or a pool that is isotopically enriched relative to pyritic sulfur by a constant amount.

From these qualitative considerations and the previous discussion, we assume that one or a combination of three components with the following isotopic properties are sources for organic sulfur and elemental sulfur: (i) Component p, with the same isotopic composition as pyritic sulfur, i.e. early-diagenetic H_2S. (ii) Component b, whose $\delta^{34}S$-value is constant and higher than the $\delta^{34}S$ of pyritic sulfur. This source is a proxy for species derived from biosynthetic sulfur compounds. We assume a $\delta^{34}S$ value for component b of +20 permil, the value for modern seawater sulfate. Although the isotopic composition of seawater sulfate has varied over the geologic

Table I. Description and Ranges in Chemical Composition of Samples used in this Study

Age, Fm.[a]	Location	Lithology	TOC (wt. %)[b]	TRS (wt. %)[c]	f (org. S)[d]	f (elem. S)[e]	S(org)/TOC[f]	Ref
Recent [3]	Newport Marsh, CA	Clastic sands, clays, and silts	0.1 to 1.9	0.2 to 1.0	0.01 to 0.03	0.01 to 0.02 [3]	0.005 to 0.02	1
Recent and Quaternary [4]	Santa Barbara Basin, offshore CA	Calcareous muds	2.4 to 3.0	0.6 to 0.9	0.01 to 0.03	0.008 to 0.04 [4]	0.002 to 0.01	1
Quaternary to U. Pliocene [16]	E. Pacific, Peru margin (ODP Sites 680 and 686)	Diatomaceous, phosphatic, and calcareous muds	1.1 to 9.0	1.0 to 2.6	0.1 to 0.5	0 to 0.08 [9]	0.03 to 0.15	4
Miocene Monterey Fm. [28]	Santa Maria Basin, onshore CA (2 cores)	Siliceous, phosphatic, and calcareous mudstones	0.7 to 6.3	0.3 to 2.6	0.1 to 0.8	n.d.	0.05 to 0.18	3
Miocene Monterey Fm. [12]	Santa Maria Basin, onshore CA (outcrops)	Siliceous, phosphatic, and calcareous mudstones	1.6 to 10.3	0.2 to 2.6	0.1 to 0.7	0.001 to .007 [4]	0.03 to 0.17	5
U. Cret. Gareb Fm. [4]	Israel	Chalks	6.8 to 12.5	1.1 to 2.6	0.5 to 0.8	n.d.	0.09 to 0.17	2
M. Jurassic Oxford Clay Fm. [28]	England	Mudstones	0.9 to 16.1	0.4 to 4.7	0.01 to 0.5	n.d.	0.02 to 0.11	7

L. Jurassic Jet Rock [11]	England	Mudstones	4.6 to 17.0	5.0 to 6.4	0.06 to 0.22	n.d.	0.02 to 0.09	6
L. Jurassic Posidonien-scheifer [6]	Germany	Mudstones	6.6 to 12.4	2.7 to 8.0	0.02 to 0.17	n.d.	0.05 to 0.06	6
OVERALL [110]			0.1 to 17.0	0.2 to 8.0	0.01 to 0.8	0.001 to 0.08 [20]	0.002 to 0.17	

aBracketed value is total number of samples in data set; all samples were analyzed for TOC, TRS, and organic sulfur content [S(org)].
bTotal organic carbon.
cTotal reduced sulfur.
dFraction of total reduced sulfur as organic sulfur.
eFraction of total reduced sulfur as elemental sulfur; bracketed value is the number of samples analyzed for elemental sulfur.
fOrganic sulfur content/TOC, weight ratio.
n.d. = no data reported.

Table II. Ranges in Isotopic Compositions of Co-Existing Pyrite, Organic Sulfur, and Elemental Sulfur for Samples Described in Table I. δ34S Values are in Permil Relative to the CDT standard. Also Shown are Ranges in Fractionations (Δ) Between Co-Existing Sulfur Species.

Samples[a]	δ34S(pyr)	δ34S(org)	δ34S(elem)[e]	Δ(org-pyr)	Δ(elem-pyr)	Δ(elem-org)	Ref
Newport Marsh [3]	-27.7 to -20.0	-19.8 to -9.8[b]	-21.1 to -11.7 [3]	+7.9 to +10.2	+6.5 to +8.3	-2.1 to -1.4	1
Santa Barbara Basin [4]	-26.6 to -17.3	-23.1 to +6.5[b]	-25.2 to +8.0 [4]	+3.5 to +26.4	+1.4 to +27.9	-2.1 to +1.5	1
Peru Margin [16]	-35.1 to -5.7	-17.6 to +5.7[b]	-22.3 to -0.3 [9]	+5.9 to +23.7	+5.3 to +18.8	-5.7 to +3.2	4
Monterey Fm. (cores) [28]	-32.1 to +14.4	-16.1 to +22.8[c]	n.d.	+2.9 to +21.2	n.d.	n.d.	3
Monterey Fm. (outcrops) [12]	-18.5 to +15.1	+1.8 to +21.7[c]	+9.5 to +20.3[f] [4]	+3.4 to +20.3	+8.7 to +18.7	-5.3 to +7.9	5
Gareb Fm. [4]	-23.8 to +8.3	-10.7 to +8.3[c]	n.d.	+2.0 to +14.5	n.d.	n.d.	2
Oxford Clay Fm. [30]	-44.8 to -11.7	-21.4 to -4.6[b]	n.d.	-4.3 to +32.7	n.d.	n.d.	7
Jet Rock [11]	-28.4 to -19.8	-13.6 to +0.8[d]	n.d.	+14.9 to +24.5	n.d.	n.d.	6
Posidonien-scheifer [6]	-28.7 to -15.1	-10.8 to -4.1[d]	n.d.	+11.0 to +20.7	n.d.	n.d.	6

OVERALL -44.8 to +15.1 -23.1 to +21.7 -25.2 to +20.3 -4.3 to +32.7 +1.4 to +27.9 -5.7 to +7.9
[110]   [20]

[a]Bracketed value is the number of samples in the data set; all samples were analyzed isotopically for pyritic sulfur and organic sulfur.

[b]Analyses on total organic sulfur.

[c]Analyses on kerogen sulfur, i.e. sulfur remaining in organic matter after removal of solvent-extractable bitumens.

[d]Analyses on bitumen sulfur.

[e]Bracketed value is the number of samples in the data set that were analyzed isotopically for elemental sulfur.

[f]Excludes one sulfate-rich sample with an anomalously low $\delta^{34}S$(elem) of -20.9 permil.

n.d. = no data reported.

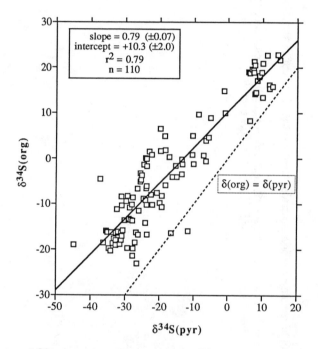

Figure 1. δ^{34}S(organic sulfur) versus δ^{34}S(pyrite) for the samples summarized in Tables I and II. δ^{34}S-values are in permil relative to the CDT standard. The best-fit linear regression line is drawn through the data. Uncertainties in the slope and intercept are at the 95% confidence limit. Also shown is the line for δ^{34}S(organic sulfur) = δ^{34}S(pyrite).

time interval represented by the samples *(51)*, this variation is relatively small ($< \pm 5$ permil) and +20 permil is an acceptable mean value. (iii) Component h, having a $\delta^{34}S$ value that is also higher than that of pyritic sulfur, but that may or may not be dependent on the $\delta^{34}S$ value of pyritic sulfur. (No relationship between the isotopic compositions of components h and b is specified, i.e., the $\delta^{34}S$ value of component h may higher or lower than that of component b.) Component h would be a proxy for later-diagenetic H_2S or reactive intermediate products of H_2S oxidation.

The proportion of source components in organic sulfur or elemental sulfur is expressed as the sum of their respective mole fractions (X) as:

$$X_p + X_b + X_h = 1 \tag{2}$$

The isotopic composition of organic sulfur or elemental sulfur, δ^*, is given as the sum of the products of mole fraction and δ value of source components:

$$\delta^* = X_p\delta_p + X_b\delta_b + X_h\delta_h \tag{3a}$$

Note from equation (2) that specifying the mole fraction of two components fixes the value of the third. Thus, the general isotope mass-balance statement in equation (3a) can be expressed in terms of the mole fractions of only two components. For example, by substituting for X_p ($= 1 - X_b - X_h$) in equation (3a), we can derive that:

$$\delta^* = (1 - X_b)\delta_p + X_b\delta_b + X_h\Delta_{h\text{-}p} \tag{3b}$$

where $\Delta_{h\text{-}p} = \delta_h - \delta_p$, fractionation between component h and component p.

The parameters of the isotope correlation lines in Figs. 1 and 2 constrain the value of variables in the isotope mass-balance expressions. The slope (m) determines the coefficient of δ_p, and the intercept (I) fixes the sum of the other terms on the right-hand side of the expressions. The slope and intercept yield only "mean" values; the scatter of data points in Figs. 1 and 2 are considered to reflect deviation of the variables from their budget-derived mean values. Nonetheless, because the slope and intercept are constant, the mean values of the specific variables and terms that are determined by those parameters must also be constant.

We have evaluated all possible isotope mass-balance expressions of one-, two-, and three-component systems for organic sulfur and elemental sulfur. The isotope trends for organic sulfur and elemental sulfur are considered separately in order to emphasize the differences in source components for those two phases. In reality, the fact that the parameters of their respective correlation lines are identical within statistical uncertainty suggests strongly that they share the same components. Most expressions require either unreasonably high δ values for component h ($\geq +50$ permil) or a complex relationship between δ_h and δ_p. The only mass-balance expression that yields acceptable results based on isotopic criteria and model assumptions is equation (3b). The results are described below:

(A) Organic sulfur: The mean value of X_b is determined by the slope (m = 0.8) of the $\delta(\text{org}) - \delta(\text{pyr})$ correlation line in Fig 1:

$$X_b = 1 - m = 0.2 \ (\pm 0.07) \tag{4}$$

Since X_b is fixed by the slope and δ_b is assumed constant (= +20 permil), then the intercept (I = +10 permil) determines the mean value of the product $X_h\Delta_{h-p}$:

$$X_h\Delta_{h-p} = I - X_b\delta_b = +10 - (0.2)(+20) = +6 \ (\pm 3) \text{ permil} \qquad (5)$$

(The uncertainty is estimated from the uncertainties in the slope and intercept at the 95 % confidence limit; see Fig. 1). Although the values X_h and Δ_{h-p} are not constrained independently by isotope mass-balance (other than $0 < X_h < m$), the product of those two variables is fixed at a constant mean value.

(B) Elemental sulfur: Since the slope of the $\delta(\text{elem})$ - $\delta(\text{pyr})$ correlation line is ≈ 1 (Fig. 2), the contribution of a component with a constant δ value (i.e., component b) to elemental sulfur is insignificant. Thus, $X_b = 0$ and $X_p + X_h = 1$. The intercept (I = +12 permil, Fig. 2) also fixes the mean value of the product $X_h\Delta_{h-p}$ for elemental sulfur:

$$X_h\Delta_{h-p} = I = +12 \ (\pm 5) \text{ permil} \qquad (6)$$

Table III shows the isotope budget results for organic sulfur and elemental sulfur for the possible range of component mole fractions. The results were obtained by calculating X_p and X_h for fixed values of Δ_{h-p} (X_b is fixed by the slope of the correlation lines, and $\delta_b = +20$ permil is assumed). Those results show the proportions of pyritic sulfur and component h if a common pool of component h with a fixed Δ_{h-p} value contributes to both organic sulfur and elemental sulfur.

Discussion

Mass-balance calculations based on the linear isotopic correlations of Figures 1 and 2 define some relationships between diagenetic and primary species that contribute to organic sulfur and elemental sulfur. For example, the proportion of a component with a constant δ value (component b) is determined by the slope of the lines. However, isotope mass-balance budgets do not specify values of the variables X_p, X_h, and Δ_{h-p}. On the other hand, if we fix one of those variables, we specify the values of the other two (e.g., Table III). We must use evidence from sources other than the isotopic correlations to constrain one of the three variables and, thereby, to provide a more quantitative assessment of the origin of organic sulfur and elemental sulfur. For example, we might assume that early-diagenetic H_2S was the dominant source for both phases because it was probably the largest pool of reactive sulfur available during diagenesis. However, results in Table III show that Δ_{h-p} values are unrealistically high when $X_p \gg X_h$. Thus, although early-diagenetic H_2S may have been the initial diagenetic component, secondary diagenetic components such as ^{34}S-enriched later-diagenetic H_2S or oxidation intermediates were the principal immediate precursors for organic-sulfur and elemental-sulfur formation. Rather than setting arbitrary limits on X_p or X_h, a better approach is to constrain Δ_{h-p} values from the isotopic properties of likely sources for component h. An important consideration in this approach is the timing of organic-sulfur and elemental-sulfur formation.

Later-Diagenetic H_2S as the Source of Component h. In this case, organic sulfur and elemental sulfur would continue to form after pyrite formation ceased. This would be consistent with the consensus view that sulfidization of fine-grained iron

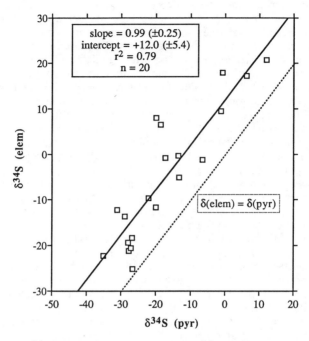

Figure 2. δ^{34}S(elemental sulfur) versus δ^{34}S(pyrite) for the samples summarized in Tables I and II. See Figure 1 for explanation.

Table III.   Isotope mass-balance results calculated for fixed $\Delta_{h\text{-}p}$ values

	Organic sulfur $[\delta_b = +20]$			Elemental sulfur		
$\Delta_{h\text{-}p}$	X_b	X_p	X_h	X_b	X_p	X_h
+60	0.2	0.70	0.10	0.0	0.80	0.20
+50	0.2	0.68	0.12	0.0	0.76	0.24
+40	0.2	0.65	0.15	0.0	0.70	0.30
+30	0.2	0.60	0.20	0.0	0.60	0.40
+20	0.2	0.50	0.30	0.0	0.40	0.60
+15	0.2	0.40	0.40	0.0	0.20	0.80
+12	0.2	0.30	0.50	0.0	0.00	1.00
+10	0.2	0.20	0.60	0.0	-	-
+7.5	0.2	0.00	0.80	0.0	-	-

oxide and oxyhydroxide minerals is the dominant sedimentary sink for early-diagenetic sulfide species *(8, 13-15, 20)*. Iron sulfidization reactions buffer dissolved sulfide concentrations in pore-waters to very low levels, even in zones of rapid sulfate reduction *(14)*. Only when reactive iron phases are depleted does H_2S accumulate in pore waters and become available for transformation to organic sulfur and elemental sulfur. As described previously, $\delta^{34}S$ of dissolved sulfide increases progressively downward in modern marine sediments, with gradients ranging from 1 to 100 permil/meter and maximum ^{34}S enrichments relative to early-diagenetic H_2S of +20 to +60 permil *(4, 18, 30, 41, 42)*. For Δ_{h-p} values corresponding to those maximum observed enrichments, isotope budget calculations predict minimum X_h values of 0.1 to 0.6 (Table III). Thus, a significant fraction of organic sulfur and elemental sulfur would have to form from isotopically enriched, late-diagenetic H_2S.

However, there are several troubling implications in this scenario. First, organic sulfur and elemental sulfur should be more isotopically heterogeneous than pyrite because they formed in part from a sulfide pool of increasing $\delta^{34}S$. Sulfur isotopic variations between different fractions of bitumens and crude oils provide a partial test of this implication for organic sulfur (although different fractions may have formed at different times). In most cases, those variations are less than a few permil *(e.g. 52, 53)*. Larger variations between fractions *(6, 53)* may be the consequence of isotope effects related to differences in sulfur bonding strengths *(53)* rather than to differences in time of formation. Second, the continuous formation of organic sulfur and elemental sulfur should have produced continuous $\delta^{34}S$ ranges that were higher than but overlapped with the $\delta^{34}S$ values of co-existing pyrite. However, organic sulfur and elemental sulfur are always isotopically enriched relative to pyrite, i.e., there is virtually no overlap (Table II; Figs. 1 and 2). Those systematic fractionations are also illustrated by the fact that minimum Δ_{h-p} values from isotope mass-balance models are significantly greater than zero (Table III). Such strong partitioning in the isotopic record would seem to imply that organic-sulfur and elemental-sulfur formation was not continuous with depth, but "turned-off" after the initial stage of formation in surficial sediments, then "turned-on" again at a depth range where the $\delta^{34}S(H_2S)$ gradient produced a Δ_{h-p} value consistent with isotope mass-balance. This is clearly implausible. Other explanations involving a specific dependence of the rates of organic-sulfur and elemental-sulfur formation on the $\delta^{34}S(H_2S)$ gradient and sediment accumulation rate would require similarly fortuitous and unsupported assumptions.

Direct evidence on the timing of organic-sulfur and elemental-sulfur formation provide the most convincing argument against an exclusively late-diagenetic origin for those phases. Studies on the abundance of sulfur phases in modern marine sediments from a variety of depositional environments suggest that organic sulfur and elemental sulfur form in surficial sediments during the earliest stages of diagenesis *(1, 4, 20, 27, 28)*, hence in diagenetic settings where H_2S is recycled and intermediate oxidation products, such as elemental sulfur, polysulfides, sulfite, thiosulfate, and polythionates are present. The formation of organic sulfur may be less significant in sediments deposited under euxinic conditions or in contact with highly reducing pore waters where the concentration of partially oxidized intermediate species may be suppressed *(6, 54)*. In addition, detailed structural characterization of organic-sulfur compounds in sediments and crude oil provide important information on the timing and mechanisms of sulfur incorporation in biogenic materials *(e.g., 22-24, 55, 56)*. The general model for sulfur incorporation based on those studies proposes that H_2S and/or partially oxidized species are added to double bonds or other reactive functionalities in precursor molecules *(22-24, 55)*. Intramolecular sulfur incorporation and intermolecular sulfur bridges preserve the original carbon skeleton of labile functionalized lipids from microbial alteration. The preservation of otherwise reactive molecules implies that the formation of organic-sulfur compounds must occur very early in the diagenetic sequence. Experimental studies have confirmed that H_2S and polysulfides are important agents in the sulfidization of organic matter *(21, 24, 25, 57-59)*. The presence in marine sediments of organic-sulfur compounds with a broad range of oxidation states *(27, 60)* suggests that reactions involving sulfoxyanions may also be involved.

The various lines of evidence described above argue strongly that organic sulfur and elemental sulfur are primarily the products of early diagenesis and thus form contemporaneously with pyrite. Reactions that consume and produce sulfur phases may continue at depth *(e.g., 4, 20, 27, 28)*. Those processes may contribute to the scatter in the isotopic data in Figures 1 and 2. Nonetheless, we conclude that the isotopic signatures of organic sulfur and elemental sulfur (as well as pyrite) are generally established during the earliest stages of diagenesis.

Early-Diagenetic Polysulfides as the Source of Component h. Considerable attention has been focused on the role of polysulfides in the formation of organic-sulfur compounds in sediments. The occurrence of polysulfide linkages in sedimentary macromolecules and of both cyclic and acyclic organopolysulfides is evidence for reactions between inorganic polysulfide species and organic matter *(23, 55, 56)*. Theoretical calculations suggest that polysulfides are more reactive than (mono)sulfides to appropriate functionalized lipids *(21)*. Inorganic polysulfides in natural environments are formed by either reaction between H_2S and elemental sulfur or by both biotic and abiotic oxidation of H_2S *(61, 62)*. Polysulfides are the first products in the oxidation of dissolved sulfide species to elemental sulfur, intermediate sulfoxyanions, and sulfate in oxic seawater *(63-66)*, thus establishing a link between polysulfides and elemental sulfur. However, the isotopic composition of polysulfides generated during early diagenesis is largely unknown. Mossmann *et al.* *(4)* have proposed that oxidation of upward-diffusing, late-diagenetic [34]S-enriched H_2S in the topmost layer of sediments may yield [34]S-enriched elemental sulfur and polysulfides. Also, as noted below, isotopically enriched polysulfides may be intermediate products of bacterial sulfide oxidation *(34)*. On the other hand, the results of isotope exchange experiments with ^{35}S-radiolabeled species in anoxic seawater indicate that isotopic equilibration in the system H_2S-H_2S_n-S^0-FeS is rapid on laboratory time-scales *(66)* . Polysulfide was an essential mediator for isotopic exchange between dissolved sulfide and elemental sulfur in those experiments. Stable isotopic equilibrium fractionations between H_2S, S^0, and FeS are <2 permil *(67)*; polysulfide species probably have similar isotopic properties. Thus, H_2S, H_2S_n, and S^0 may be considered an isotopically homogeneous pool whose composition is controlled by and essentially identical to H_2S, the ultimate source of the other species. If isotopically enriched polysulfides are a precursor of organic sulfur and elemental sulfur (i.e., component h), then the rates of transformation to those phases must exceed the rate of isotopic exchange with H_2S. Alternatively, there may be kinetic inhibitors to equilibration in natural sediment-pore water systems that are not present in experimental systems *(see 66)*. Otherwise, organic sulfur, elemental sulfur, and pyrite that form contemporaneously by reactions involving dissolved sulfide and polysulfide ions should have the same isotopic composition.

Early-Diagenetic Sulfoxyanions as the Source of Component h. Sulfites thiosulfate, and polythionates are also intermediates in sulfide oxidation. Although reported concentrations in marine pore-waters are typically low, those sulfoxyanions appear to be involved to varying extents in biogeochemical transformations of sulfur. For example, thiosulfate is a major product of anoxic sulfide oxidation and is consumed rapidly by reduction, oxidation, and disproportionation from oxic surface sediments to the zone of sulfide accumulation *(39, 62)*. Thiosulfate reduction yields elemental sulfur as one of the products *(62)*. In addition, experimental studies have shown that thiosulfate can react with simple unsaturated organic compounds to yield organic sulfonates, although the rate of reaction is much slower than that for sulfite *(60)*. Polythionates are closely linked to thiosulfate in redox reactions catalyzed by microbial processes. Polythionates are apparently produced by bacterial oxidation of thiosulfate under anaerobic conditions *(68)* but are also rapidly reduced to thiosulfate in anoxic but non-sulfidic sediments *(69)*. In addition, polythionates are able to react with a thiol group in organic compounds, leading to the formation of a sulfane group *(70)*. Polythionates have been detected in many sulfur bacteria *(71)*; their release from bacterial cells is the probable source of elemental sulfur in some marine sediments *(28)*. Polythionates can also produce elemental sulfur (plus thiosulfate) by reaction with sulfide.

Polythionates and thiosulfate merit consideration as a source of component h primarily on isotopic grounds. H_2S oxidation by the acidophilic chemoautotroph *Thiobacillus concretivorus* under aerobic conditions and by the photoautroph *Chromatium sp.* under anaerobic conditions produced (in addition to sulfate and elemental sulfur) dissolved species that were enriched in ^{34}S with respect to the initial sulfide by up to +19 permil (ave. of 9 experiments: +8 permil) *(34)*. Those species were identified as polythionates, although from the analytical procedure used other species, including polysulfides, may have been present as well. Isotopically enriched polythionate species (trithionate and tetrathionate) were also intermediates in the oxidation of thiosulfate to sulfate by *Thiobacillus neapolitanus (see 36)*. Isotopic enrichment in polythionates and residual thiosulfate relative to initial thiosulfate increased with the extent of oxidation to maximum values of +6 to +11 permil as calculated by mass-balance. In the late stages of oxidation, when sulfate and polythionates were the main constituents (>98%) of the medium, sulfonate-sulfur in polythionates was calculated to be +18 permil relative to initial thiosulfate. Assuming no isotopic fractionation in the oxidation of H_2S to thiosulfate *(45-47)*, Δ_{h-p} values corresponding to the enrichments described above predict minimum X_h values of 0.3 to 0.6 (Table III). Thus, polythionates (and/or other intermediate species) with isotopic properties similar to those reported in the above experiments could be the predominant ^{34}S-enriched precursors of organic sulfur and elemental sulfur.

There are serious problems and uncertainties with this interpretation. Foremost is the fact that the production of isotopically enriched polythionates or other intermediate species by bacterial and chemical oxidation of H_2S has not been duplicated *(45-47)*. Also, polythionates are relatively labile compounds that participate in a variety of biological and chemical reactions. In order for polythionates (or similar intermediate species) to be an important precursor for organic sulfur and elemental sulfur, the rates of transformations to those phases must be comparable to the rates at which polythionates are consumed by other processes. In addition, ^{34}S is preferentially sited in oxidized sulfonate-sulfur relative to reduced sulfane-sulfur in both thiosulfate and polythionates *(36, 46, 72)*. Consequently, sulfonate-sulfur must be preferentially converted to ^{34}S-enriched organic sulfur and elemental sulfur. But in order to produce those relatively reduced phases from sulfonate-sulfur, the overall transformations must involve a series of electron-transfer reactions. Thus, even though there is some isotopic evidence to suggest that polythionates (or other intermediate products of sulfide oxidation) are precursors of organic sulfur and elemental sulfur, the problems associated with those proposed transformations are formidable.

Preservation of Biosynthetic Sulfur in Sedimentary Organic Matter. According to our isotope mass-balance calculations, assimilatory sulfur from primary organic matter is a significant source of sedimentary organic sulfur. Sulfur-bearing biomolecules are among the most readily degradable compounds in sedimentary environments *(50)*. Thus, the prediction that organic sulfur in the sediments of our data set is made up of a constant and rather high proportion of biosynthetic sulfur is surprising. Moreover, S(org)/TOC ratios in those sediments (0.002-0.17, Table I) have a much wider range that the S/C wt. ratios in primary marine organic matter (0.01-0.03; *1, 20*). This contrast may be due in part to large differences in the proportions of terrestrial organic matter in the sediments. However, the contrast is more likely related to differences in the preservational (burial) efficiencies of biosynthetic sulfur and primary organic carbon, i.e., the fraction of sulfur or carbon in primary organic matter that is preserved in sedimentary organic matter. A detailed discussion of the relationships between preservational efficiencies for sulfur and carbon, the fraction of biosynthetic sulfur in sedimentary organic sulfur (X_b), and S/C ratios in primary and sedimentary organic matter is beyond the scope of this paper. A preliminary assessment suggests that S(org)/TOC ratios for sedimentary organic matter of constant X_b reflect the relative preservational efficiencies of sulfur and carbon. For sediments in our data set having S(org)/TOC $\geq$ 0.1 (35 of the 110 samples in Tables I and II), the preservational efficiency of biosynthetic sulfur must have exceeded that of carbon, with values approaching 100 %.

Isotope mass-balance budgets suggest that the retention of biosynthetic sulfur in sedimentary organic matter can be significant. Accordingly, the break-down of sulfur biomolecules (a process that must occur during the earliest stages of diagenesis) releases species that may be rapidly and efficiently incorporated into residual organic constituents. The identity of those reactive species is unknown. They are probably not inorganic species, such as sulfides, polysulfides, or sulfoxyanions, because the production of such species by other processes should exceed production from biomolecular decomposition, thereby diluting the isotopic signature of biosynthetic sulfur. The reactive species are more likely to be low molecular-weight organic compounds that are reincorporated into organic matter close to their sites of production. Thiols formed from the breakdown of biosynthetic sulfur compounds are probably involved *(73)*. The pathways of biosynthetic-sulfur degradation and recycling requires further study.

Conclusions

Organic sulfur and elemental sulfur are systematically enriched in [34]S relative to co-existing pyrite in marine sediments and sedimentary rocks representing a wide variety of depositional and diagenetic environments. A mass-balance budget based on isotopic correlations between sulfur phases suggests a significant contribution of biosynthetic sulfur to neoformed organic-sulfur compounds. This result is consistent with the emerging view of rapid and irreversible reactions at reactive sites in organic matter. Mass-balance calculations also suggest that early-diagenetic H_2S, the source of pyrite, was not the dominant immediate precursor of organic sulfur and elemental sulfur. Another component that was enriched isotopically by a constant amount relative to early diagenetic H_2S was probably the predominant source of those phases. The contemporaneous formation of pyrite, organic sulfur, and elemental sulfur in modern marine sediments plus additional isotopic constraints argue that the ^{34}S-enriched component was not dissolved sulfide species present deeper in the sediment-pore water system. The isotopically enriched component may be intermediate products of H_2S oxidation, such as polysulfides or sulfoxyanions. However, based on the available evidence, it is unclear that those species have the necessary isotopic and chemical properties. Additional field and laboratory investigations on the reactivity and isotopic composition of intermediate species in H_2S oxidation are needed. Also, comprehensive isotopic data sets from additional modern marine sediments are needed to test the general applicability of the isotopic correlations and budget calculations reported here.

Our work demonstrates that the isotopic record of sedimentary sulfur phases contains important and unique information on biogeochemical transformations of sulfur. Diagenetic partitioning of sulfur isotopes provides compelling evidence for unrecognized (or unappreciated) isotope fractionation during sulfide oxidation. Sulfidization of sedimentary organic matter sequesters labile organic compounds and enhances the overall preservation of organic matter. Sulfidization is most effective in suboxic depositional settings with persistent recycling of sulfide and sulfate across an oxic-anoxic boundary located near the top of the sediment column. Anoxic micro-niches in suboxic settings will be sites of pyrite formation, while the formation of diagenetic organosulfur compounds and elemental sulfur (as well as oxidized metal species such as Fe^{3+} and Mn^{4+}) will take place in the surrounding weakly oxidized environments.

Acknowledgments

We thank ACS reviewers M. B. Goldhaber and A. Vairavamurthy for thoughtful critiques of this paper. Financial support was provided by NSF EAR 90-17462 and NSF EAR 93-04401 grants to TFA and by NSF EAR 88-16371 AND NSF EAR 93-16279 to LMP.

References

1. Kaplan, I.R.; Emery, K.O.; Rittenberg, S.C. *Geochim. Cosmochim. Acta* **1963**, *27*, pp 297-331.
2. Dinur, D.; Spiro, B.; Aizenshtat, Z. *Chem. Geol.* **1980**, *31*, pp 37-51.
3. Orr, W.L. *Org. Geochem.* **1986**, *10*, pp 499-516.
4. Mossmann, J.; Aplin, A.C.; Curtis, C.D.; Coleman, M.L. *Geochim. Cosmochim. Acta* **1991**, *55*, pp 3581-3595.
5. Zaback, D.A.; Pratt, L.M. *Geochim. Cosmochim. Acta* **1992**, *56*, pp 763-774.
6. Raiswell, R.; Bottrell, S.H.; Al-Biatty, H.J.; Tan, M. Md. *Amer. J. Sci.* **1993**, *293*, pp 569-596.
7. Chu, T.H.; Bonnell, L.M.; Anderson, T. F. *Chem. Geol.* **1993**, *107*, pp 443-445.
8. Berner, R.A. *Amer. J. Sci.* **1970**, *268*, pp 1-23.
9. Postgate, J. R. *Ann. Rev. Microbiol.* **1959**, *13*, pp 505-520.
10. Trudinger, P.A. *Adv. Microbial Physiol.* **1969**, *3*, pp 111-158.
11. Le Gall, J.; Postgate, J.R. *Adv. Microbial Physiol.*, **1973**, *10*, pp 82-133.

12. Jorgensen, B.B. *Nature* **1990**, *296*, pp 643-645.
13, Berner, R.A.; Westrich, J T. *Amer. J. Sci.* **1985**, *285*, pp 193-206.
14. Canfield, D. E. *Deep-Sea Res.* **1989**, *36*, pp 121-138.
15. Canfield, D. E. *Amer. J. Sci.* **1991**, *291*, pp 177-188.
16. Jorgensen, B. B. *Limnol. Oceanogr.* **1977**, *22*, pp 814-832.
17. Thamdrup, B.; Finster, K.; Fossing, H.; Hansen, J.W.; Jorgensen, B.B. *Geochim. Cosmochim. Acta* **1994**, *58*, pp 67-73.
18. Hartmann, U. M.; Nielsen, H. *Geol. Runsch.* **1969**, *58*, pp 621-655.
19. Canfield, D.E.; Raiswell, R. In *Taphonomy: Releasing the Data Locked in the Fossil Record;* Allison, P.A.; Briggs, D.E.G., Eds., *Topics in Geomicrobiology;* Plenum Press: New York, NY, 1992, Vol. 9; pp 337-387.
20. Francois, R. *Geochim. Cosmochim. Acta* **1987**, *51*, pp 17-27.
21. Lalonde, R. T.; Ferrara, L. M.; Hayes, M. P. *Org. Geochem.* **1987**, *11*, pp 563-571.
22. Sinninghe Damsté, J. S.; Rupstra, W. I. C.; Kock-Van Dalen, A. C.; de Leeuw, J. W.; Schenck, P. A. *Geochim. Cosmochim. Acta* **1989**, *53*, pp 1343-1355.
23. Kohnen, M. E. L.; Sinninghe Damsté, J. S.; ten Haven, H.L.; Kock-Van Dalen, A. C.; Schouten, S.; de Leeuw, J. W. *Geochim. Cosmochim. Acta* **1991**, *55*, pp 3685-3695.
24 Vairavamurthy, A.; Mopper, K. *Nature* **1987**, *329*, pp 623-627.
25 Vairavamurthy, A.; Mopper, K. In *Biogenic Sulfur in the Environment;* Saltzman, E.S.; Cooper, W. J., Eds.; ACS Symp. Ser. No. 393; American Chemical Society: Washington, D.C., 1989, pp 231-242.
26 Vairavamurthy, A.; Mopper, K.; Taylor, B.F. *Geophys. Res. Lett.* **1992**, *19*, pp 2043-2046.
27. Ferdelman, T. G.; Church, T. M.; Luther, G. W., III. *Geochim. Cosmochim. Acta* **1991**, *55*, pp 979-988.
28. Schimmelmann, A.; Kastner, M. *Geochim. Cosmochim. Acta* **1993**, *57*, pp 67-79.
29. Goldhaber, M. B.; Kaplan, I. R. In *The Sea, Vol. 5;* Goldberg, E.D., Ed.; John Wiley & Sons: New York, 1974; pp 569-655.
30. Goldhaber, M.B.; Kaplan, I R. *Mar. Chem.* **1980**, *9*, pp 95-143.
31. Zaback, D.A.; Pratt, L.M.; Hayes, J.M. *Geology* **1993**, *21*, pp 141-144.
32. Thode, H.G.; Kleerekoper, H.; McElcheran, D.E. *Research* **1951**, *4*, pp 581-582.
33. Harrison, A.G.; Thode, H. G. *Trans. Faraday Soc.* **1958**, *54*, pp 84-92.
34. Kaplan, I.R.; Rittenberg, S. C. *J. Gen. Microbiol.* **1964**, *34*, pp 195-212.
35. Kemp, A.L.W.; Thode, H.G. *Geochim. Cosmochim. Acta* **1968**, *32*, pp 71-91.
36. Chambers, L.A.; Trudinger, P.A. *Geomicrobiol J.* **1979**, *1*, pp 249-293.
37. Chambers, L.A.; Trudinger, P.A.; Smith, J.W.; Burns, M. S. *Can. J. Microbiol.* **1975**, *21*, pp 1602-1606.
38. Canfield, D.E.; Thamdrup, B. *Science* **1994**, *266*, pp 1973-1975.
39. Jorgensen, B.B. *Science* **1990**, *249*, pp 152-154.
40. Jorgensen, B.B. *Geochim. Cosmochim. Acta* **1979**, *43*, pp 363-374.
41. Chanton, J.P.; Martens, C.S.; Goldhaber, M.B. *Geochim. Cosmochim. Acta* **1987**, *51*, pp 1201-1208.
42. Canfield, D.E.; Raiswell, R.; Bottrell, S. *Amer. J. Sci.* **1993**, *292*, pp 649-683.
43. Price, F.T.; Shieh, Y.N. *Chem. Geol.* **1979**, *27*, pp 245-253.
44 Sweeney, R.E.; Kaplan, I.R. *Mar. Chem.* **1973**, *9*, pp 165-174.
45. Fry, B.; Gest, H.; Hayes, J.M. *FEMS Microbiol. Lett.* **1984**, *22*, pp 283-287.
46. Fry, B.; Cox, J.; Gest, H.; Hayes, J.M. *J. Bacteriol.* **1986**, *165*, pp 328-330.
47. Fry, B.; Ruff, W.; Gest, H.; Hayes, J.M. *Chem. Geol.* **1988**, *73*, pp 205-210.
48. Sinninghe Damsté, J.S. ; de Leeuw, J.W. *Org. Geochem.* **1990**, *16*, pp 1077-1101.

49. Orr, W.L.;Sinninghe Damsté, J.S. In *Geochemistry of Sulfur in Fossil Fuels;* Orr, W.L.; White, C.M., Eds; ACS Symp. Ser. No. 429; American Chemical Society: Washington, D.C., 1990, pp 2-29.
50. Orr, W.L. In *Handbook of Geochemistry;* Wedepohl, K.H., Ed.; Springer-Verlag: Berlin, 1974, Vol. II-I, Sect. 16-L.
51. Claypool, G.E.; Holser, W.T.; Kaplan, I.R.; Sakai, M. *Chem. Geol.* **1980,** *28,* pp 199-260.
52. Krouse, H.R. *J. Geochem. Explor.* **1977,** *7,* pp 189-211.
53. Hirner, A.V.; Graf, W.; Treibs, R.; Melzer, A.N.; Hahn-Weinheimer, P. *Geochim. Cosmochim. Acta* **1984,** *48,* pp 2179-2186.
54. Aplin, A.C.; Macquaker, J.H.S. *Phil. Trans. Royal Soc.* **1993,** *344,* pp 89-100.
55. Kohnen, M.E.L.; Sinninghe Damsté, J.S.; Kock-van Dalen, A.C.; ten Haven, H.L.; Rullkotter, J.; de Leeuw, J.W. *Geochim. Cosmochim. Acta* **1990,** *54,* pp 3053-3063.
56. Kohnen, M.E.T.; Sinninghe Damsté, J.S.; Bass, M.; Kock-van Dalen, A.C.; de Leeuw, J. W. *Geochim. Cosmochim. Acta* **1993,** *57,* pp 2515-2528.
57. Fukishima, K.; Yasukawa, M.; Muto, N.; Uemura, H.; Ishiwatari, R. *Org. Geochem.* **1992,** *18,* pp 93-91.
58. de Graaf, W.; Sinninghe Damsté, J.S.; de Leeuw, J.W. *Geochim. Cosmochim. Acta* **1992,** *56,* pp 4321-4328.
59. Rowland, S; Rockey, C.; Al-Lihaibi, S.S.; Wolff, G.A. *Org. Geochem.* **1993,** *20,* pp 1-5.
60. Vairavamurthy, A.; Zhou, W.; Eglinton, T.; Manowitz, B. *Geochim. Cosmochim. Acta* **1994,** *58,* pp 4681-4687.
61. Boulegue, J.; Lord, C.L.; Church, T.M. *Geochim. Cosmochim. Acta* **1982,** *46,* pp 453-464.
62. Fossing, H.; Jorgensen, B.B. *Geochim. Cosmochim. Acta* **1990,** *54,* pp 2731-2742.
63. Chen, K.Y.; Gupta, K. *Environ. Lett.* **1973,** *4,* pp 187-200.
64. Chen, K.Y.; Morris, J.C. *Environ. Sci. Tech.* **1972,** *6,* pp 529-537.
65. Milero, F.J. *Mar. Chem.* **1986,** *18,* pp 121-147.
66. Fossing, H.; Jorgensen, B.B. *Biogeochem.* **1990,** *9,* pp 223-245.
67. Ohmoto, H.; Rye, R. O. In *Geochemistry of Hydrothermal Ore Deposits*; *2nd Edit.;* Barnes, H. L., Ed.; Holt Rinehart and Winston: New York, NY, 1979, pp 509-567.
68. van Gemerden, H.; Tughan, C. S.; de Wit, R.; Herbert, R.A. *FEMS Microbiol. Ecol.* **1989,** *62,* pp 87-102.
69. Ferdelman, T.G.; Fossing, H. *208th ACS Natl. Mtg.,* **1994.**
70. Roy, A.B.; Trudinger, P.A. *The Biochemistry of Inorganic Compounds of Sulfur;* Pergamon Press: Oxford, 1970, 400 p.
71. Steudel, R.; Holdt, G.; Gobel, T.; Hazeu, W. *Angew. Chemie, Intl. Ed., Engl.* **1987,** *26,* pp. 151-153.
72. Uyama, F.; Chiba, H., Kusakabe, M.; Sakai, H. *Geochem. J.* **1985,** *19,* pp 301-315.
73. Vairavamurthy, A.; Zhou, W.; Manowitz, B. *208th ACS Natl. Mtg.,* **1994.**

RECEIVED June 7, 1995

Chapter 22

Stable Sulfur Isotopic Compositions of Chromium-Reducible Sulfur in Lake Sediments

Brian Fry[1], Anne Giblin, Mark Dornblaser, and Bruce Peterson

Ecosystems Center, Marine Biological Laboratory,
Woods Hole, MA 02543

We measured $\delta^{34}S$ values of water column sulfates and reduced sedimentary sulfides (chromium reducible sulfur) in twelve low sulfate lakes from the northern United States and Nova Scotia. The results show that the net isotopic fractionation occurring during sulfate reduction is smaller than that observed in marine environments, with maximum differences ($\Delta\delta^{34}S$) between sulfate and CRS ranging from 0.4 to 18.8 ‰ vs. the 40-70 ‰ values typical of marine sediments. In lakes with 12-83 μM sulfate, the observed fractionation was not well correlated with sulfate level, but was inversely correlated with two measures of carbon supply available to sulfate reducing bacteria. The data are in agreement with theoretical models that predict smaller isotopic fractionation when increasing carbon availability leads to increased demand for very limited sulfate supplies. We suggest that the observed $\Delta\delta^{34}S$ estimate of isotopic fractionation in low sulfate lakes reflects the ratio of sulfate to carbon supply available to sulfate reducing bacteria in sediments.

Sulfate reducing bacteria produce sulfides that are stored in sediments as iron sulfides or organic sulfur. These sulfides are typically depleted in ^{34}S vs. the initial sulfate because of isotopic fractionation during the process of dissimilatory sulfate reduction. This fractionation is measured most simply as the isotopic difference between sulfate and sulfides, i.e., $\Delta\delta^{34}S$, and the magnitude of this isotopic fractionation ranges from about 2-70 ‰ in marine and freshwater systems. Laboratory and field studies of sulfate-rich systems have repeatedly shown that this large range in fractionation values is controlled by the rate of sulfate reduction (*1, 2*), with fractionation decreasing as carbon supply or temperature increases the overall reduction rate. This inverse relationship applies when the reduction rate is

[1]Current address: Department of Biology, Florida International University, University Park, FL 33199

expressed on a per cell basis (*3*) or on an areal basis that sums over the community of sulfate-reducing bacteria (*4*). Particularly in marine systems where sulfate levels are high (28mM in full strength seawater) and carbon supply usually limits sulfate reduction rates, maximal fractionations of 40-70 ‰ are observed, e.g., Chanton (*4*), Hartman and Neilsen (*5*), Goldhaber and Kaplan(*6*). In sharp contrast to marine systems, lakes have generally much lower sulfate levels (usually less than 100µM), and smaller fractionations in the 2-20 ‰ range are observed (*7, 8, 9, 10*). We began the present study to determine whether the difference in sulfate levels between freshwater and marine systems was sufficient to explain the large variation in observed isotopic fractionations, or whether other factors such as carbon supply were involved.

We have surveyed sediments of 12 North American lakes for amounts and isotopic compositions of sulfidic, chromium reducible sulfur (CRS). Previous studies have shown that CRS is primarily composed of sulfide-derived sulfur in most lake sediments, with little detectable contribution from organic sulfur compounds (*11, 12, 13*). Using the CRS assay that is specific for the sulfide products of sulfate reduction, we investigated sulfate concentrations, temperature, bottom water anoxia, and carbon availability as possible controls of isotopic fractionation. We advance the hypothesis that the magnitude of sulfur isotopic fractionation observed in lake sediments is not solely governed by sulfate levels, but rather records the balance of sulfate and carbon supplies available to benthic sulfate reducing bacteria.

Methods

The twelve lakes studied were diverse in terms of their geography, trophic status, sulfate concentrations, and hypolimnetic oxygen deficits (Table I). The lakes were sampled for water and sediment chemistry at various times during the period between 1987-1991. Oxygen concentrations and temperature were measured with an Orbisphere 2714 meter and stirring probe. Dissolved SO_4^{2-} analysis was analyzed by ion chromatography (Dionex 2010): precision was 0.3 %. Sediment cores were taken by SCUBA divers with small (6.5 cm) diameter piston corers (*14*). The cores were sectioned under N_2 and frozen until analysis (*15*). Lake sediments did not contain appreciable amounts of carbonates , and organic carbon content was analyzed on a Perkin-Elmer 240C CHN elemental analyzer. CRS (chromium-reducible sulfur), which includes all monosulfides, pyrite (FeS_2), and elemental S, was determined with freeze-dried mud by heating in an acid $Cr(II)Cl_2$ solution (*16*).

Sulfate in water samples was precipitated with barium chloride, filtered, ashed, and decomposed to SO_2 with V_2O_5 for mass spectrometric analysis (*17*). Zinc sulfide precipitates captured from CRS analyses were similarly decomposed to SO_2 (for a complete description of isotopic analyses, see (*14*)). All S isotopic determinations were measured on a Finnigan MAT 251 isotope ratio mass spectrometer. Results are reported as $\delta^{34}S$ values versus the Canyon Diablo Troilite standard using conventional notation, e.g. Fry (*8*). Duplicate samples varied by < 0.2 ‰ $\delta^{34}S$.

Table I. Characteristics of Study Lakes

Lakes	Latitude/Longitude	Anoxia[1]	SO$_4^{2-}$ (µM) Epilimnetic	Hypolimnetic	Sediments % C[2]	µmol CRS/g[3] dry wt	Epilimnetic SO$_4^{2-}$	δ^{34}S CDT CRS[4]	Δδ^{34}S$_{SULFATE-CRS}$
ELA									
Lake 240	N49.65W93.73	+/-	48	40	14.6	86.5	4.2	0.7	3.5
WISCONSIN									
Bird	N45.48W89.40	+	20	17	25.2	167.9	6.3	1.4	4.9
Trout Basin 2	N46.02W89.40	-	32	25	14.2	78.8	6.1	-4.7	10.8
Trout Basin 3	N46.02W89.40	+	32	14	17.2	126.0	7.2	2.3	4.8
Trout North		+/-	33	28	18.3	208.6	7.0	0.7	6.3
Crystal	N46.00W89.37	-	35	32	20.2	107.7	6.7	-2.4	9.1
FINGER LAKES									
Canadice	N42.44W77.34	-	160	146	3.1	112.9	3.1	-15.7	18.8
Conesus	N42.47W77.43	+	230	144	5.4	297.6	3.6	-14.1	17.7
ADIRONDACKS									
Heart	N44.20W74.08	+	12	1	27.9	269.0	8.3	7.9	0.4
Dart	N43.47W74.52	+/-	62	52	20.3	90.7	5.4	2.4	3.0
CAPE COD									
Cliff	N41.45W70.02	-	67	51	8.8	48.8	7.6	-4.0	11.6
Mares	N41.35W70.36	+/-	73	45	12.5	15.7	8.9	-4.1	13.0
Gull	N41.58W69.71	+	83	49	13.9	226.3	13.7	8.6	5.1
NOVA SCOTIA									
Mountain	N44.22W65.12	-	22	19	12.9	8.5	9.4	-2.5	11.9

[1] + = Recurrent anoxia, +/- = anoxic bottom water in some years, - = anoxic bottom waters never observed

[2] Total Carbon, 0-1 cm

[3] CRS concentration at depth of δ^{34}S minimum

[4] Lowest δ^{34}S value of CRS in upper 10 cm of core

Results

We generally found different isotopic patterns for CRS-δ^{34}S in lakes with anoxic vs. well-oxygenated bottom waters. Sediments from lakes with anoxic bottom waters usually accumulated more CRS with less overall isotopic fractionation (smaller difference between epilimnetic sulfate δ^{34}S and CRS-δ^{34}S) than did sediments from lakes with well-oxygenated bottom waters. This contrast was evident even in a single lake where we collected cores from oxic and anoxic subbasins (Figure 1). In Trout Lake, Wisconsin, CRS concentrations were higher in sediments underlying anoxic bottom waters, while CRS accumulation in sediments of the oxic basin was lower, even though organic C contents were similar in both cores analyzed. The isotopic values of CRS in the anoxic basin were uniform with depth and close to that of starting sulfate (Figure 1). The CRS isotopic profile in sediments of the oxic basin was more complex, and showed lowest values at 3 cm, a point just above the CRS concentration maximum, followed by increasing values with depth (Figure 1). The contrasts in the isotopic profiles might indicate reduction of most available sulfate in the anoxic basin, with little regard to isotopic composition, but less rapid sulfate consumption that allows larger fractionation in near-surface sediments of the oxic basin.

   To investigate these possibilities, we measured sulfate concentrations in hypolimnetic waters of Trout Lake. The results show steep concentration decreases in the hypolimnion of the regularly anoxic basin, with a further decline in porewater sulfate to near-zero levels in the top 5 cm of sediments (Figure 2). This very limited sulfate availability for benthic sulfate reducing bacteria led to a low fractionation, calculated either from the increase of δ^{34}S values in hypolimnetic sulfate (3.6 ‰ fractionation factor calculated using closed system assumptions, following Mariotti et al. (*18*)) or as the 4.8 ‰ $\Delta\delta^{34}$S difference between epilimnetic sulfate and minimum δ^{34}S-CRS value in the upper 10 cm of sediments. We observed a two-fold larger $\Delta\delta^{34}$S value (10.8 vs. 4.8 ‰) in the oxic basin of Trout Lake, where sulfate reduction in bottom sediments was not sufficiently strong to deplete sulfate from the hypolimnion (Basin 2, Figure 2).

   We also examined sediments from several other lakes with varying degrees of bottom water hypoxia. In lakes with recurrent or occasional anoxia, the CRS-δ^{34}S values were usually close to those of epilimnetic sulfate, and CRS isotopic values were fairly constant with depth or increased (Figure 3, top 3 left panels). In contrast, sediments from lakes with well-oxygenated bottom waters usually showed an isotopic minimum in the upper five cm and a larger overall isotopic difference between epilimnetic sulfate and CRS values (Figure 3, top 3 right panels). There were exceptions to these generalized patterns of CRS isotopic composition. For example, two lakes that had occasionally anoxic bottom waters (Dart Lake and Lake 240, Figure 3) showed contrasting CRS patterns with Dart Lake lacking a sub-surface isotopic minimum as do other lakes with strongly anoxic bottom waters, while Lake 240 showed a sub-surface isotopic minimum that is usually seen in lakes with well-oxygenated bottom waters (Figure 3). In other lakes such as Mountain Lake, we lacked detailed CRS profiles, but the trend of the existing data are consistent with a subsurface isotopic minimum (Figure 3).

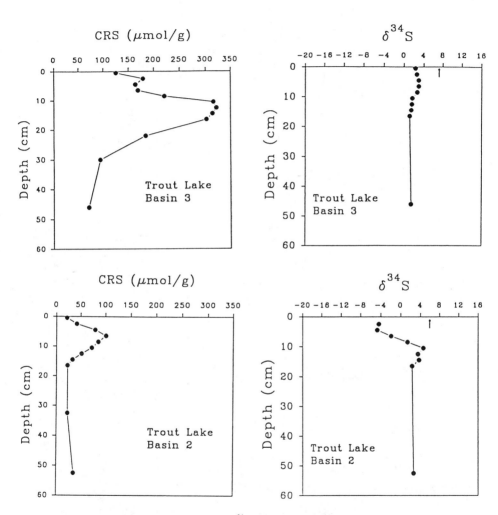

Figure 1. CRS concentrations and $\delta^{34}S$ of epilimnetic sulfate (arrows) and sedimentary CRS (dots) in two basins of Trout Lake, Wisconsin. Top panels: basin with recurrently anoxic bottom waters. Bottom panels: basin with well-oxygenated bottom waters.

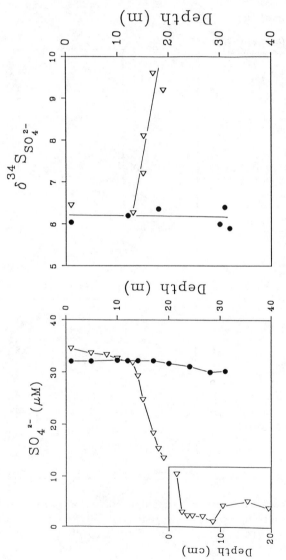

Figure 2. Concentrations and isotopic compositions of sulfate in Trout Lake in the main oxic basin (•) and in a smaller anoxic basin (Δ) from the summer of 1987. Inset shows porewater sulfate concentrations in a core collected in August 1988 from the anoxic basin.

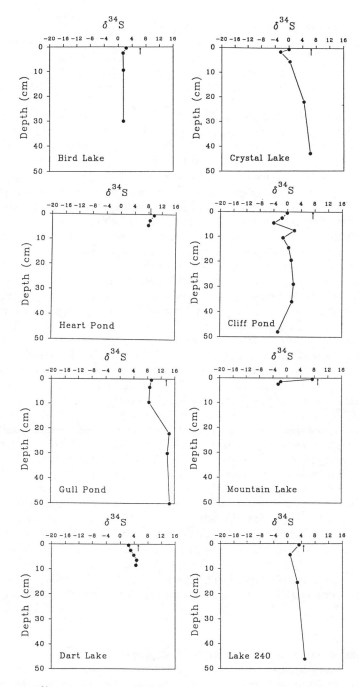

Figure 3. δ^{34}S of epilimnetic sulfate (arrows) and sedimentary CRS (dots) in lakes with recurrently anoxic bottom waters (top 3 panels, left) well-oxygenated bottom waters (top 3 panels, right) and occasionally anoxic bottom waters (left and right panels at bottom).

We explored several possible reasons for the observed range in $\Delta\delta^{34}S$. In general, we expected small fractionation when sulfate is relatively unavailable and sulfate reducing bacteria use most available sulfate. Thus, we expected small fractionation where sulfate concentrations were low, or where high inputs of labile organic carbon resulted in a strong demand for limited sulfate by sulfate reducing bacteria.

Although the smallest fractionation was observed at the lowest sulfate levels, neither epilimnetic nor hypolimnetic sulfate concentrations appeared generally predictive of $\Delta\delta^{34}S$ (Table I, Figure 4a). Isotopic fractionation can also be controlled by the rate of sulfate reduction, which is in turn influenced by temperature and carbon supply, e.g., Harrison and Thode (1), Kaplan and Rittenberg (2). Correlations between $\Delta\delta^{34}S$ and temperature were not strong (Figure 4b). We also found that several overall indicators of hypolimnetic carbon metabolism (including total hypolimnetic CO_2 and alkalinity generation, hypolimnetic oxygen and sulfate consumption), did not show strong correlations with the magnitude of isotopic fractionation (data not shown).

However, two core-specific proxies for labile C supply did show an inverse correlation with $\Delta\delta^{34}S$. The amount of CRS measured at the isotope minimum (typically occurring between 3-5cm) partly reflects integrated sulfate reduction at that depth, and thus also carbon supply, although bottom water oxygenation exerts a strong control on CRS accumulation (19). Fractionation was inversely correlated with CRS storage (Figure 4c). The second proxy for labile carbon, % organic C at 0-1cm, also showed an inverse correlation with fractionation (Figure 4d).

In evaluating this field data, we performed simple and stepwise linear regressions (20) to test the overall strength of correlations between $\Delta\delta^{34}S$ and the parameters shown in Figure 4. The r^2 values for simple linear regressions were 0.06, 0.03, 0.55 and 0.52 for sulfate, temperature, CRS concentration and %C, respectively, and significant at the $p < 0.05$ level for CRS concentration and %C. Stepwise regression yielded significant ($p < 0.05$) r^2 values of 0.56 for sulfate, temperature and CRS concentration and 0.61 for sulfate, temperature and %C. CRS concentration and %C were significantly correlated ($r^2 = 0.47$; $p < 0.05$) so that we did not include both variables in the stepwise regressions and temperature and sulfate contributed only in a minor way to these stepwise correlations (0.09 r^2 for the combined effects of temperature and sulfate). Including data for two Finger Lakes district of New York state (Table I, Figure 5) which had much higher sulfate concentrations ($> 140\mu M$), did not significantly improve r^2 values vs. those obtained using only the ten lakes with low ($< 83\mu M$) sulfate (i.e., sulfate and temperature still showed poor correlations with fractionation, while the correlation with CRS concentration was weaker, and the correlation with %C stronger).

Discussion

The detailed vertical structure of the $\delta^{34}S$ isotopic profiles in lake sediments has been previously discussed in terms of recent anthropogenic-influenced sulfur deposition in the upper 5-15 cm, and pre-industrial sulfur deposited in deeper sediments (8, 9, 21, 22). The sulfur profiles are not at steady-state in these systems because of increased sulfate loading over the last century that has changed both the

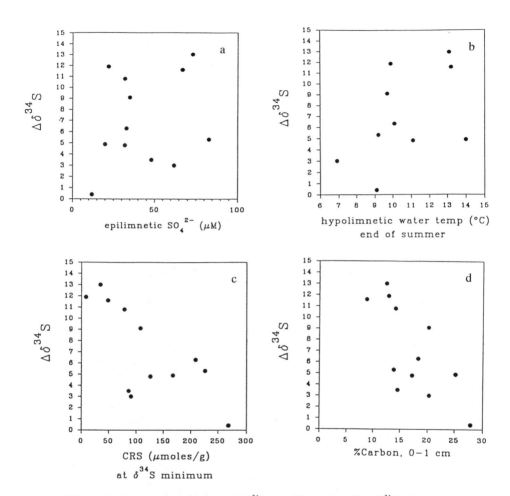

Figure 4. Isotopic fractionation ($\Delta\delta^{34}S$ = epilimnetic sulfate $\delta^{34}S$ - lowest CRS- $\delta^{34}S$ occurring in the upper 5 cm of sediments) as a function of a) epilimnetic sulfate levels, b) end of summer hypolimnetic temperature, c) concentration of CRS at the $\delta^{34}S$ minimum in the upper 5cm, and d) % organic C in surface sediments (0-1cm) in the eleven study lakes.

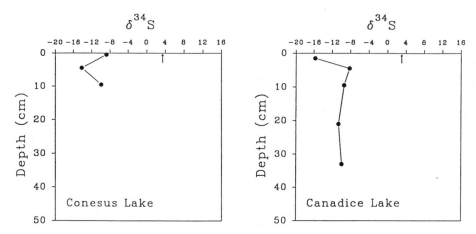

Figure 5. δ^{34}S of epilimnetic sulfate (arrows) and sedimentary CRS (dots) in two high sulfate (160-230μM) Finger Lakes of New York State. Conesus has strongly anoxic bottom waters, while bottom waters in Canadice are well-oxygenated.

availability of sulfate to benthic bacteria and the average isotopic composition of sulfate (*10*). Here we focus on results for the upper 5 cm that lie within the zone of current, active sulfate reduction, and especially deal with the isotopic fractionation occurring in the lake-sediment system.

There are three principal controls on the magnitude of sulfur isotope fractionation in lake sediments: sulfate supply, carbon supply and oxygen supply. Sulfate and carbon supplies together control the rate and amount of sulfate reduction, while oxygenation of sediments limits the net storage of produced sulfides (*15*, 23, 24). The net fractionation expressed in sedimentary CRS is the balance between isotope effects occurring during sulfide formation and oxidative loss. We briefly discuss how each of these factors - sulfate, carbon and oxygen - may have affected the observed CRS isotopic distributions.

Sulfate concentrations were low, < 83μM in surface and bottom waters of most lakes studied, and sulfate levels further declined in porewaters. Porewater sulfate measurements reported here (Figure 2) and by others , e.g. Kelly and Rudd (*25*), Lovely and Klug (*26*), Rudd et al. (*27*) for low-sulfate lakes typically show a strong decrease in sulfate concentration of 50-95% in the top 5 cm of sediments, e.g. from 70 μM to 20μM in sediments of Big Moose Lake (*28*). Resulting low sulfate concentrations at depth could lead to a diffusional limitation on sulfate reduction, with small isotopic fractionations expected when most or all sulfate is consumed. However, low sulfate levels did not correlate well with observed fractionations (Figure 4a), with some of the lowest sulfate lakes (e.g. Mountain Lake, 20μM sulfate) having the largest fractionations. This lack of close correspondence between sulfate levels and fractionation was surprising, and suggested other factors may be involved.

Oxygenation of sediments is a second possible control of isotopic fractionation. The top few cm of sediments are known sites of intense sulfide oxidation (*14, 27, 29*), and oxidation of sulfides can increase $\delta^{34}S$ values (*30*). Some of the field evidence was consistent with an important role of oxygenation, especially the observation that CRS isotopic values in surface sediments of lakes with oxygenated bottom waters were relatively high compared to CRS-$\delta^{34}S$ values a few cm deeper at the isotopic minimum (Figure 3, right panels), even though sulfate levels are actually higher at the sediment surface, and more fractionation would be expected. Extensive oxidation of H_2S prior to CRS formation could cause the higher surface CRS-$\delta^{34}S$ values.

We note that another oxidation-related hypothesis, i.e., oxidation of solid-phase CRS, does not seem sufficient to explain our results. If oxidation after rather than before CRS formation resulted in strong fractionation, we might expect ^{34}S-enriched CRS in surface sediments of all lakes, since sediments from anoxic basins also experience oxygenated bottom waters in the winter. The lack of strong ^{34}S enrichment in CRS from these anoxic basin cores (Figure 1, top and Figure 3, left panels) and the fairly small fractionations associated with oxidation of solid sulfides (*31*) probably indicates that sulfide oxidation <u>after</u> CRS formation exerts a relatively minor effect on CRS isotopic compositions.

Carbon supply is the third major control of sulfate reduction. There is limited correlational evidence that carbon supplies in sediments might influence CRS isotopic values (Figure 4c, d). The correlations of $\Delta\delta^{34}S$ with %C and CRS content are more suggestive than conclusive, however, because we lacked a good, specific measure of labile carbon available to sulfate reducing bacteria. For example, some of the bulk carbon in surface sediments (%C, Figure 4d) can be fairly inert to bacterial metabolism; we do not know what proportion of the total carbon is readily labile, and available to sulfate reducing bacteria. Similarly, the concentration of accumulated CRS (Figure 4c), is affected by bottom water anoxia and oxidation losses (*32*), and thus not truly specific for labile carbon. Overall, we admit that our current proxies for estimating carbon availabilities are crude and sometimes conflicting, so that arguing from field evidence about a carbon control of isotopic fractionation cannot be totally convincing at present. However, modeling efforts suggest that the balance between sulfate uptake and sulfate reduction steps is likely to control the observed isotopic fractionation (*33*), so that carbon supply which especially influences reduction steps is likely to be a key factor in governing observed isotopic fractionations.

We summarize our current qualitative understanding of isotopic results from lake sediments as follows. In low-sulfate lakes with bottom waters that are regularly anoxic, isotopic results are consistent with the idea that most sulfate reduction is focused into a narrow near-surface zone where most sulfate is consumed. If this interpretation is correct, this focusing should make sediments from anoxic basins attractive for studying the onset of sulfate loading from industrial sources (*9*), since the sulfate reduction record should not be smeared broadly downward over the pre-industrial sulfur record.

More complex isotopic profiles are found in lakes with well-oxygenated bottom waters. In near-surface sediments, isotopic compositions decline towards a minimum value in the 3-5 cm depth range, which is probably the transition point between more oxidized surface conditions above and a more closed anaerobic system below. White et al. (28) showed that in one lake with well-oxygenated bottom waters, the isotopic minima in several cores typically occurred just above the point of maximum CRS accumulation, which makes sense if storage is maximized as oxidation becomes unimportant below the isotopic minimum. Below the isotopic minimum, a few detailed studies of CRS and total S isotopic compositions show that there is often a deep $\delta^{34}S$ maximum in the 8-20 cm depth range (21, 28, Figure 1, Basin 2 profile of this study), after which isotopic values approach background, pre-industrial levels. Deep isotope maxima probably arise because sulfate reduction has cumulatively removed ^{34}S-depleted sulfur from downward diffusing sulfate, so that sulfides forming CRS towards the end of the active reduction zone become quite enriched in ^{34}S. This cumulative, depth-related ^{34}S enrichment in sulfides is well-known from studies of marine sediments (e.g., Goldhaber and Kaplan 6), and is likely to occur in lake sediments, even though low porewater concentrations have thus far prevented direct measurement of porewater sulfate and sulfide isotopic compositions from low sulfate lakes. Thus, our overall view of the isotopic profiles in lakes with well oxygenated bottom waters is that fractionation during sulfate reduction leads to first low then high values measured downcore, a pattern that is consistent with sufficient sulfate surviving downcore to fuel sulfate reduction occurring over a rather broad interval.

In a more quantitative sense, we have cited the theoretical models of Rees (33) that indicate the likely importance of the ratio of sulfate to labile carbon supplies in determining isotopic profiles. We recommend that future laboratory and field studies directly examine the effects of this ratio on isotopic fractionation. Laboratory studies should test how sulfate/carbon supply ratios affect fractionation under low-sulfate conditions (20-200 micromolar sulfate), as well as the effects of oxidation on CRS isotopic composition. Field studies might most profit from examining the relationship between the isotopic profiles and porewater sulfate. The steepness of the porewater sulfate gradient in sediments and the overall length of the sulfate reduction zone should be good indicators of the sulfate/carbon supply ratios, with steep gradients and compressed sulfate reduction zones indicating low sulfate/carbon supply ratios. It is likely that the solid-phase isotopic profiles reflect the time-integrated dynamics of sulfate reduction, so that the steepness and compression of the isotopic profiles could be similarly examined as indicators of the sulfate/carbon supply ratios.

Finally, our isotopic investigations in lakes show that smallest fractionations are observed in anoxic, low sulfate conditions that are in principle similar to conditions prevailing in the Archean Ocean where small (< 5 ‰) sulfur isotopic differences between sulfate and sulfides typically existed (34). The ancient sulfur isotope record has been scrutinized for evidence of larger fractionations that would unambiguously indicate bacterial sulfate reduction, establishing the overall antiquity of this process (35). We suggest that lake sediments may provide modern laboratories to investigate the controls of sulfur isotopic fractionation under these low sulfate conditions. For example, examining lakes with bottom water anoxia and sulfate levels in the 100-1000μM range might help establish minimum sulfate levels expected for significant fractionations to occur. The current study is a start in this direction, and showed small fractionations, < 5 ‰, for all lakes with recurrently anoxic bottom waters and sulfate levels of 83μM or less. We have some evidence that larger fractionations occur at only moderately higher sulfate levels. Conesus Lake, one of the Finger Lakes of New York state, has anoxic bottom waters containing 144μM sulfate and a $\Delta\delta^{34}S$ value of 17.7 ‰ in sediments of relatively low carbon content (Table I). This preliminary datum indicates that the transition between 5 and ≥15 ‰ fractionations could be related to low carbon supply rates and could occur rather abruptly at sulfate concentrations considerably less than the approximately 1mM threshold sulfate concentration currently under discussion in the geological literature (e.g., Ohmoto et al. *36*).

Acknowledgments

We thank Kris Tholke and Robert Garritt for assistance with the isotopic determinations. This work was supported by NSF grants BSR 86-15191 and 89-18273. Tom Anderson provided helpful comments during the review process.

Literature Cited

1. Harrison, A.G.; Thode, H.G. *Faraday Soc. Trans.* **1958**, *58*, 84-92.
2. Kaplan, I.R.; Rittenberg, S.C. *Microbiology,* **1964**, *34*,195-212.
3. Goldhaber, M.B.; Kaplan, I.R. *Soil Sci.* **1975**, *119*, 42-55.
4. Chanton, J.P. Ph.D. Dissertation. **1985**, The University of North Carolina. 406p.
5. Hartmann, V.M.; Neilsen, H. *Geol. Rundschau.* **1969**, *58*, 621-655.
6. Goldhaber, M.B; Kaplan, I.R. *Mar. Chem.* **1980**, *9*, 95-143.
7. Matrosov, A.G.; Cehbotarev, G.N.; Kudryavtseva, A.J.; Zyakun, J.M.; Ivanov, M.V. *Geochem.* **1975**, *12*, 217-221.
8. Fry, B. *Biogeochem.* **1986,** *2*, 329-343.
9. Nriagu, J.O.; Soon, Y.K. *Geochim. Cosmochim. Acta.* **1984**, 49, 823-834.
10. Fry, B. In. *Stable Isotopes in Ecological Research.* Rundel, P.W.; Ehleringer, J.R.; Nagy, K.A., Eds., Springer-Verlag: New York 1989. pp 445-453.
11. Wieder, R.K.; Lang, G.E.; Granus, V.A. *Limnol. Oceanogr.* **1985**, *30*, 1109-1115.
12. Canfield, D.E.; Raiswell, R.; Westrich, J.T.; Reaves, C.M.; Berner, R.A. *Chem. Geol.* **1986**, *54*, 149-155.
13. Giblin, A.E.; Likens, G.E.; White, D.; Howarth, R.W. *Limnol. Oceanogr.* **1990**, *35*, 852-869.
14. Dornblaser, M.; Gilbin, A.E.; Fry, B.; Peterson, B.J. *Biogeochem.* **1994**, *24*, 129-144.
15. Giblin, A.E.; Likens, G.E.; Howarth, R.W. *Limnol. Oceanogr.* **1991**, *36*, 1265-1271.
16. Howarth, R.W.; Merkel, S. *Limnol. Oceanogr.* **1984**, 29, 598-608.
17. Yanagisawa, F.; Sakal, H. *Anal. Chem.* **1983**, *55*, 985-987.
18. Mariotti, A.; Germon, J.C.; Hubert, P.; Kaiser, P; Letolle, R.; Tradieux, A.; Tardieux, P. *Plant Soil.* **1981**, *62*, 413-430.

19. Kling, G.W.; Giblin, A.E.; Fry, B.; Peterson, B.J. *Limnol. Oceanogr.* **1991**, *36*, 106-122.
20. Sokul, R.R.; Rohl, F.J. 1981. *Biometry.* W.H. Freeman and Company.
21. Nriagu, J.O.; Coker, R.D. *Nature* **1983**, *303*, 692-694.
22. Thode, H.G.; Dickman M.D.; Rao, S.S. *Arch. Hydrobiol. Suppl.* **1987**, *74*, 397-422.
23. Mitchell, M.J.; Owen, J.S.; Schindler, S.C. J. Paleolimnol. **1990**, *4*, 1-22.
24. Cook, R.B.; Kelly, C.A. In. *Sulfur cycling in terrestrial systems and wetlands.* Howarth, R.W.; Stewart, J.B., Eds. Wiley: 1992.
25. Kelly, C.A.; Rudd, J.W.M. *Epilimnetic Biogeochem.* **1984**, *1*, 63-77.
26. Lovely, D.R.; Klug, M.J. *Geochim. Cosmochim. Acta.* **1986**, *50*, 11-18.
27. Rudd, J.W.M.; Kelly, C.A.; Furutani, A. *Limnol. Oceanogr.* **1986**, *31*, 1281-1292.
28. White, J.R.; Gubala, C.P.; Fry, B.; Owen, J.; Mitchell, M.J. *Geochim. Cosmochim. Acta* **1989**, *53*, 2547-2559.
29. Jorgensen, B.B. *Limonol. Oceanogr.* **1977**, *22*, 814-832.
30. Fry, B.; Ruf, W.; Gest, H.; Hayes, J.M. *Chem. Geol. (Isotope Geosci. Sec.)* **1988**, 205-210.
31. Toran, L.R.; Harris, R.R. *Geochim. Cosmochim. Acta* **1989**, *53*, 2341-2348.
32. Westrich, J.T.; Berner, R.A. *Limnol. Oceanogr.* **1984**, *29*, 236-249.
33. Rees, C.E. *Geochim. Cosmochim. Acta* **1973**, 37, 1141-1162.
34. Schidlowski, M.; Hayes, J.M.; Kaplan, I.R. In: *Earth's Earliest Biosphere*; Schopf, J.W., Ed. Princeton University Press, 1983, pp 149-186.
35. Ohmoto, H.; Felder, R.P. *Nature* **1987**, *328*, 244-246.
36. Ohmoto, H.; Kakegawa, T.; Lowe, D.R. *Science.* **1993**, *262*, 555-557.

RECEIVED April 17, 1995

THERMOCHEMICAL SULFATE REDUCTION

Chapter 23

Kinetic Controls on Thermochemical Sulfate Reduction as a Source of Sedimentary H_2S

Martin B. Goldhaber[1] and Wilson L. Orr[2]

[1]U.S. Geological Survey, Mail Stop 973, Denver Federal Center, Denver, CO 80225
[2]Earth and Energy Science Advisors, P.O. Box 3729, Dallas, TX 75208

Laboratory experiments with aqueous ammonium sulfate in the presence of H_2S and toluene are reported which show measurable SO_4^{2-} reduction in 4 to 30 days at 175-250°C. Reduction rates increase with both increasing temperature and H_2S pressure but reduction was not measurable on our experimental timescale without H_2S initially present. An activation energy of 96 (±16) kJ/mole was estimated from the data. These results (and other published studies) indicate that thermochemical sulfate reduction (TSR) is difficult to document below 200°C on a laboratory time scale unless ΣS (i.e. SO_4^{2-} + H_2S) is initially very high (>.5M) and pH is low (<2). Even though these conditions may fall outside the range found in nature, extrapolation of kinetic data to more typical natural values predicts that the formation of high-H_2S natural gas and sulfide mineral deposits of the Mississippi Valley Type at 100-200°C can readily occur at geologically reasonable rates.

Reduction of SO_4^{2-} to form H_2S is important in many sedimentary settings. At temperatures of 0-50°C, well characterized bacterial metabolic processes catalyze the simultaneous biogenic reduction of SO_4^{2-} and oxidation of organic matter (*1*). Recently, this bacterial reduction pathway has been shown to extend to temperatures approaching or even slightly exceeding 100°C (*2*). At substantially higher temperatures of >250°C the available thermal energy is sufficient to allow rapid chemical and isotopic equilibrium between SO_4^{2-} and H_2S (*3*). The subject of this paper is H_2S generation in the intermediate

0097–6156/95/0612–0412$12.00/0

temperature range of 100-200°C; too warm for bacterial processes and too cool for rapid chemical equilibrium. This intermediate temperature range is of interest because it characterizes the formation of sour (H_2S-bearing) gas fields (*4*) and the genesis of many large sediment-hosted ore deposits (*5*). Because of their geologic similarities, we will consider the origin of H_2S in sour gas fields and the so-called Mississippi Valley Type (MVT) ore deposits; both are hosted in carbonate rocks and formed over the temperature range of interest.

The process generating H_2S in this intermediate temperature range has been termed thermochemical sulfate reduction (TSR; *4*). TSR is analogous to the bacterial reduction pathway in that it is a kinetically controlled reaction involving the coupled oxidation of organic matter and reduction of SO_4^{2-}. Although considerable success has been attained in documenting the occurrence and consequences of TSR from field studies (*e.g. 4,6-8*), understanding the critical geologic and geochemical controls has been hindered by the inability to experimentally reproduce TSR under laboratory conditions comparable to those deduced from field sites. Previous successful experimental studies required temperatures above 220°C (*9*), and many involved pH values and reactant concentrations outside the range for most natural systems. One study (*9*) was designed to mimic natural systems. Temperatures ranged from 90-190 °C, and a variety of reductant concentrations and, several possible catalytic agents, were employed. Yet no evidence for TSR was found despite using a sensitive radiosulfur tracer technique to detect low extent of reaction.

In this paper we present experimental results with conditions selected to allow measurable rates of TSR on a convenient time scale at 175-250°C. These data were presented previously in abstract form (*10*). We consider the significance of the data in comparison with other experimental studies in the literature and the extrapolation of these studies to conditions characteristic of naturally occurring TSR in geologic settings. Specifically, we attempt to reconcile apparent differences in laboratory and field observations.

Experimental

TSR was studied at 175 to 250°C in a 500 cc stainless steel autoclave containing $(NH_4)_2SO_4$, H_2S, H_2O, and Toluene with reaction times of 92 to 620 hours. The vessel initially contained 100 cc of sulfate solution whose concentration ranged from .05 to .48 molar. A large molar excess of toluene (20 cc) relative to sulfate was added. Nitrogen was used to flush air from the system, to vary total charge pressure, and to flush H_2S from the autoclave following the completion of each run. Initial H_2S was added from a cylinder to the nitrogen filled reactor to yield final pressures of 10 to 225 psi. Most runs were made with 200 psi H_2S. The total charge pressure was adjusted with N_2 to 300 to 1000 psi at room temperature. The initial and final quantities of

SO_4^{2-} were measured gravimetrically with an accuracy of $\pm 1\%$. Quantities of H_2S produced were not measured because amounts formed were small relative to starting amounts and losses due to corrosion of the stainless steel vessels prohibited accurate measurements. The reactor was controlled at the desired temperature ($\pm 3^\circ C$) for the desired time. Heating and cooling times were short (1-2) hours with respect to reaction times.

Results

Results of runs conducted at $250^\circ C$ are contained in Table I. Also shown in this table are calculated first order rate constants and the half life of the reaction in hours. The treatment of the results as first order in sulfate is an assumption because only initial and final concentrations were measured. However, previous time series studies (*e.g. 11-13; see also 3*) have deduced

Table I

Experiments with Variable H_2S Pressure at $250\ ^\circ C$						
Time; Hours	Initial H_2S	Initial SO_4^{2-} mmoles	Final SO_4^{2-} mmoles	% SO_4^{2-} reduction	Rate Constant	Half-Life
128	0	48.2	48.8	~0	--	--
120	10	48.1	40.8	15	1.4×10^{-3}	500
120	25	48.1	36.9	23	2.2×10^{-3}	315
120	50	48.2	22.1	54	6.5×10^{-3}	106
120	100	48.1	24.7	49	5.6×10^{-3}	124
141	100	46.9	19.4	59	6.3×10^{-3}	111
240	200	48.2	12.1	75	5.8×10^{-3}	120
240	200	24.9	4.5	82	7.1×10^{-3}	98
92	200	24.4	11.5	53	8.2×10^{-3}	85
120	200	48.2	15.6	68	9.4×10^{-3}	74
168	(~225)?	5.1	(0.5)?	89	--	--

first order kinetics with respect to SO_4^{2-}. With this assumption, the first order rate constant k is given by:

$$k=(\frac{1}{t})\ln(\frac{C_0}{C_t})$$

where C_0 and C_t are the initial and final concentrations respectively. The half life of the reaction is:

$$t_{\frac{1}{2}}=\frac{0.693}{k}$$

Reduction rates were quite rapid at 250°C and dependant on the pressure of H_2S. Without initial H_2S, the reaction was not measurable (<1%) in five days. The first order rate constants increase sharply with increasing amounts of initial H_2S in the range of 10 to 200 psi (Figure 1).

An additional observation is that elemental sulfur does not accumulate with excess toluene present, but is formed in apparently large amounts at 250°C if toluene is absent and H_2S is initially present. This observation implies that the experiments described here took place in the elemental sulfur stability field, but that the elemental sulfur was reduced in the presence of toluene.

Table II summarizes results for all temperatures studied. Reaction was observed to occur at temperatures as low as 175°C, although with substantially increased half-life. This to the best of our knowledge is the lowest temperature at which TSR has been experimentally demonstrated, and the first to document TSR at a temperature similar to those inferred from field studies. The temperature dependance of the rates corresponds to an activation energy of 96 ± 16 kJ/mole (23 ± 4 kcal/mole).

Table II

Variation of SO_4^{2-} Half-Life with Temperature at Constant H_2S Pressure[a]		
Temperature °C	Number of Runs	Half-Life; Hours
250	4	94±20
225	1	313
200	2	1150±360
175	1	1740
[a] Initial Load Pressure of 200 PSI		

Discussion

Overall Pathway for Reduction

Before further considering the experimental results, it is useful to consider the key reactions involved in TSR. In an earlier study (6), it was proposed that TSR occurs by way of two successive generalized reactions, each of which must involve a series of more primitive steps. Reactive organic matter is generalized as (CH_2) in these equations.

$$SO_4^{2-} + 3\ H_2S \leftrightarrow 4\ S^0 + 2\ H_2O + 2\ OH^- \tag{1}$$
$$4\ S^0 + 1.33\ (CH_2) + 2.66\ H_2O \rightarrow 4\ H_2S + 1.33\ CO_2 \tag{2}$$
$$\text{---------\ --}$$
$$SO_4^{2-} + 1.33\ (CH_2) + 0.66\ H_2O \rightarrow H_2S + 1.33\ CO_2 + 2\ OH^- \tag{3}$$

This pathway was in part deduced from earlier work by Toland (14-15), who verified equations 1 and 2 experimentally. More recent work has reconfirmed these results (13). The SO_4^{2-}/H_2S reaction (equation 1) was previously postulated to be the slow or rate determining step in TSR (6) when adequate amounts of suitable organic matter were present. The reactive S^0 intermediate may actually be any one of a number of intermediate oxidation state sulfur species such as polysulfides/hydropolysulfides (S_x^{2-}/HS_x^-), or thiosulfate ($HS_2O_3^-/S_2O_3^{2-}$) in addition to elemental sulfur. The reaction between organic matter and "S^o" does not directly produce CO_2; it leads initially to organic acids (14), which then may undergo subsequent decarboxylation reactions (7), to yield the observed association in sour gas fields between H_2S and CO_2 (8).

In a detailed mechanistic analysis of isotope exchange between SO_4^{2-} and H_2S (which is related to the reduction pathway), Ohmoto and Lasaga (3) concluded that the overall rate of reduction was proportional to the product of $\sum[SO_4^{2-}] \cdot \sum[H_2S]$ and the exchange rate was also proportional to the sum of the concentrations of $[SO_4^{2-}] + [H_2S]$, (i.e., $\sum[S]$). The exchange was also a complex function of pH, with substantially increased reaction rates at low pH. The role of total $\sum[S]$ in solution and pH controls on the overall reaction was also discussed by Trudinger et al. (9) who pointed out that if equation 1 is rate controlling, the stability of S^o (or related polysulfide compounds) would be critical to achieving rapid reduction rates. They emphasized that the elemental sulfur stability field expands with increasing temperature, increasing $\sum[S]$, and decreasing pH.

Evaluation of the Experimental Results

Based on the foregoing, three of the major variables controlling the rate of reduction are related to the initial total sulfur concentration in solution ($\sum[S]$), temperature, and pH. In this section, we compare experimental data on TSR on the basis of these variables. We make no claim that $\sum[S]$ is itself a kinetic parameter, but rather, that it is a convenient variable closely related to rate-controlling parameters allowing comparison of a diverse group of experiments. Figure 2 shows schematically the available experimental data including the results from this study on a plot of $\sum[S]$ versus the temperature of the runs. Experiments which successfully produced H_2S are shown in a cross hatched pattern and unsuccessful experiments are shown as a solid field. For plotting purposes, the added solid $S°$ in (*13*) was recalculated to its equivalent solution concentration. The justification for this is given below.

This figure shows that the successful experiments represent either higher total initial aqueous sulfur or higher temperatures or both compared to those that failed to yield observable reduction. It is evident that the 175°C run of this study (Table II) which resulted in measurable SO_4^{2-} reduction, occurred at substantially higher $\sum[S]$ than the unsuccessful experiments of (*9*) at similar temperatures. It is also noteworthy that the unsuccessful runs of Drean (*16*), although conducted at elevated temperature, employed low initial $\sum[S]$ conditions.

An important distinction not evident from Figure 2 is whether H_2S, a required reactant in equation 1, was present at the beginning of the experiments in the various studies. Most successful experiments, this study, and (*15*), started with initial sulfide and showed the necessity of reduced sulfur species to obtain relatively rapid reaction. Likewise, Kiyosu and Krouse (*13*) initially added $S°$ which produced H_2S quickly at temperature by reaction with acetic acid (equation 2); the rate of H_2S formation exceeded that of sulfate reduction and the conversion of $S°$ to H_2S was nearly quantitative. Most unsuccessful experiments in Figure 2 (11-12, 16) contained only SO_4^{2-} with various types of organic matter. The unsuccessful experiments by Trudinger et al. (*9*) had either no initial sulfide or very low concentrations as well as a low $\sum[S]$. The limited success of Kaiser (*17*) with high $\sum[S]$ and no initial sulfide suggests slow H_2S production directly from dextrose and sulfate but, in general, the direct reduction of sulfate by organic matter is known to be orders of magnitude slower than observed TSR rates below 300 °C (*4,6, 14-17, 19, 21*).

Also shown on Figure 2 is a field for the conditions of genesis of Mississippi Valley Type (MVT) sedimentary ore deposits. This field (crosshatched rectangle) has been generalized from data contained in a number of fluid inclusion studies of these ores (*18*). The sulfide component of a subset of these ores has been inferred to have formed via TSR based on sulfur isotope systematics (*5*); although non-TSR sulfide sources are implicated in many cases this distinction is not reflected in the MVT field in Figure 2. The negative experimental results of Trudinger (*9*) directly overlap the field for MVT ores. A possible resolution of this seeming paradox is discussed below.

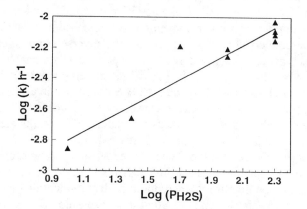

Figure 1. Effect of initial H_2S pressure on first-order rate constants for SO_4^{2-} removal during thermochemical sulfate reduction. All data at 250 C°.

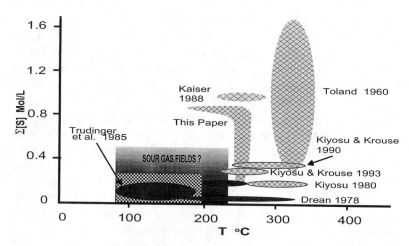

Figure 2. Summary of experimental data on thermochemical sulfate reduction expressed as fields on a plot of total S (H_2S +SO_4^{2-}) in solution vs. temperature of the experiments. Successful experiments are shown in a cross-hatched pattern, unsuccessful ones in a solid darker grey. Likely fields for the formation of MVT type ore deposits and sour gas deposits are shown respectively by a checkerboard rectangle and a gradationally colored gray rectangle. The H_2S concentration for the experiments of this study was calculated using the data of (27).

A corresponding field for the condition of origin of sour gas fields analogous to the MVT field is also shown as gradational shades of gray. Though the temperatures involved can be constrained by fluid inclusion and burial history data (*4,19*) and are in fact similar to those shown on Figure 2 for MVT ore formation, the corresponding $\sum[S]$ is poorly known, and that is the reason for the shading. It is likely that $\sum[S]$ may be somewhat higher than those shown for MVT ores because most sour gas fields are associated with sulfate-bearing evaporites (*19*), and would, after only a moderate extent of reaction, also contain significant H_2S concentrations. There may, thus, exist a difference in a major parameter, $\sum[S]$ controlling the conditions of H_2S generation in these two environments.

A second way to look at the experimental TSR data is in terms of overall rates, in this case expressed as moles of sulfate reduced/liter/year (Figure 3).   In order to facilitate this comparison, the data have been recalculated to a common set of conditions; 0.025M SO_4^{2-} and pH 4.   These conditions were chosen to be representative of ore fluids implicated in the formation of MVT ore deposits (*20*), and the pH may also be appropriate for oil field brines such as those of sour gas fields (*e.g. 7*). The assumptions involved in the recalculation of the data deserve elaboration. The recalculation of rates to the stated SO_4^{2-} concentration is straightforward, and is based on the assumption of first order kinetics (i.e. the rate is directly proportional to initial concentration of SO_4^{2-}). The pH correction is much more problematical. We have assumed that the pH dependence is similar to that estimated for the isotope exchange reaction (*3*), which decreases by one order of magnitude per pH unit between pH 1 and 3 and is constant above 4. We have simply assumed a one order of magnitude decrease in rate per pH unit below 4, although this will introduce some error between pH 3 and 4. The assumed pH dependence was applied to initial run pH's as calculated by computer modeling of the run conditions from calculations in (*3*) and (*7*). In the absence of other information, the pH values for (*13*) were assumed to be 2 based on 25°C measurements. Although this pH correction adds considerable uncertainty to the plotted rates, it is our feeling that it, nonetheless, gives a more reasonable picture of <u>in situ</u> rates than the very low pH values employed in many of the experimental procedures. The effects of variable concentrations of H_2S and nature of organic matter were not explicitly accounted for, although the role of these variables is addressed qualitatively below.

Four separate studies are represented in Figure 3; Kiyosu (*11*) and later Kiyosu and Krouse (*12*) completed two series of experiments at various temperatures using dextrose and acetic acid respectively as the organic reductants and no initial added H_2S . A later set of runs by Kiyosu and Krouse (*13*) with acetic acid employed similar conditions to (*12*) but included initial elemental sulfur in molar amounts similar to the initial SO_4^{2-} . The fourth set of experimental data (this study), used toluene as the organic reductant. The

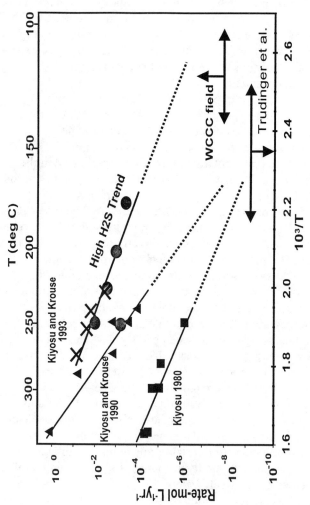

Figure 3. Plot of rate of sulfate reduction expressed as moles reduced per liter of water per year vs. $10^3/T$. The experiments have all been normalized to a sulfate concentration of 0.025M and pH 4 as discussed in the text. The rates for the Whitney Canyon Carter Creek field are taken from (7). The results from this study are shown as filled circles; the results Kiyosu and Krouse as X's.

results from this study are presented as 5 points. The two points plotted at 250°C correspond to the <u>extremes</u> in initial H$_2$S pressures employed (Table I), with the highest initial H$_2$S pressure resulting in the higher rate. This is equivalent to the overall spread in rates as a function of H$_2$S pressure shown in Figure 1. Thus, although initial H$_2$S concentration is not explicitly taken into account in our overall rate, Figure 1 and Figure 3 together suggest the range of rates which can be expected when P$_{H2S}$ is varied with other parameters held constant. Additional points are plotted for our results over the temperature range 175-250°C and 200 psi H$_2$S , which corresponds to the higher range of H$_2$S pressures we studied.

Also plotted as a generalized region are the results of Trudinger et al., (*9*). Even though these were negative runs, it is possible to calculate a detection limit for H$_2$S for their experimental design and therefore a <u>maximum</u> possible rate for their experiments. The arrow towards lower rates indicates that these are maximum values. Although not specifically corrected to pH 4 and 0.025M SO$_4^{2-}$, many of the experiments did, in fact, fall in this general range.

The final data plotted is from a sour gas field; the Whitney Canyon Carter Creek (WCCC) field in the overthrust belt of Wyoming, USA. This rate was derived from geologic constraints and the temperatures from fluid inclusion studies on phases tied paragenetically to TSR. Details are given in (*7*). The range of values plotted for WCCC in Figure 3, to our knowledge, are the only estimates of TSR rates in a sour gas field. They are minimum values because they do not reflect probable H$_2$S losses by leakage from the field or to form FeS$_2$.

Several interesting points emerge from Figure 3. The rates calculated from Kiyosu's first two data sets (*11-12*), although substantially different at elevated temperatures (>200°C), both extrapolate to near the maximum permitted values for the Trudinger et al.'s experiments (*9*) at lower temperatures in the 150°C range. The data for runs with the lowest P$_{H2S}$ from this study are comparable to results of Kiyosu and Krouse (*12*) at 250°C, and thus approach values derived from H$_2$S -free conditions. When extrapolated to 150°C, the rates for the two data sets lacking initial H$_2$S (*11-12*) are on the order of 10^{-9} moles/liter/year, and <10^{-10} moles/liter/year at 100°C (i.e. .001 to .0001 micromoles/year). These rates are slow even for processes on geologic time scales.

A separate trend of higher rates is made up of the recalculated results for high H$_2$S runs from this paper and the later study of Kiyosu and Krouse containing added S° (*13*). These points extrapolate to near the geologically estimated rates for the WCCC field. At 150°C, rates of reduction for this upper trend extrapolate to approximately two to five orders of magnitude greater than for the two H$_2$S-free Kiyosu data sets, with the relative difference increasing with decreasing temperature. The generally similar slopes of the rates estimated from this study and that of (*13*), are reflected in similar calculated activation energies; 96 kJ/mol (23 kcal/mol) for this study, and 121 kJ/mol (28.8 kcal/mol) for (*13*). In contrast the activation energies for the earlier Kiyosu studies are larger; 140 kJ/mol (33.6 kcal/mol) for (*11*) and 208 kJ/mol (49.7 kcal/mol) for (*12*).

We believe that the elevated rates found in this study compared to the initially H_2S-free studies of Kiyosu and Kiyosu and Krouse (*11-12*) are related in large part to the initially high H_2S concentrations employed here. It is evident from the dependance of the rate constant on P_{H2S}, (Figure 1), and the absence of observable reaction in the absence of H_2S as a reactant (Table I), that elevated P_{H2S} is important for increasing SO_4^{2-} reduction rates. The similarly high recalculated rates from (*13*) are equivalent to those presented here; it is evident from the results of (*13*) that the initial charge of free sulfur was very rapidly reduced to H_2S by organic matter (equation 2) and thus produced an initial high H_2S concentration. In keeping with the trend in Figure 1, the SO_4^{2-} reduction rate increased with the initial amount of S^o reflecting early conversion to H_2S added (*13*). Accordingly, we have labeled the upper line on Figure 3 the 'high H_2S trend'.

Implications for TSR in Sedimentary Environments
The information contained in Figures 2 and 3 may be combined to postulate geologic controls on the origin of H_2S in sour gas fields and MVT ores. We will consider sour gas fields first.

In many ways sour gas fields are optimum sites for TSR. This is true for several reasons, some of which are related to the characteristics of their carbonate host-rocks. Carbonate rocks are very low in iron (*19*). Thus H_2S is not removed as pyrite (FeS_2) as it is in clastic rich rocks, but instead, tends to accumulate in solution. During early diagenesis this buildup of aqueous H_2S may lead to sulfur rich kerogens (*21*). With increasing depth of burial, thermal degradation of some of this sulfur-rich organic matter may, at temperatures of >70° C, release relatively small amounts of H_2S (*19*). Thus H_2S may initially be available as a reactant (e.g. equation 1) even prior to the onset of TSR (*28*). The other two reactants for TSR, SO_4^{2-} and organic matter are also frequently abundant in sour gas fields. Sulfate is present as anhydrite as noted above, and organic matter is also present irrespective of indigenous kerogen by virtue of the fact that sour gas fields are also hydrocarbon reservoirs. As Figure 2 shows, high $\sum[S]$ is favorable for TSR. And as illustrated dramatically in Figures 1 and 3, systems with initial H_2S seem to follow a trend of substantially increased reduction rates. The actual reduction rates in sour gas fields may actually be slightly higher than portrayed in Figure 3 because this Figure was predicated upon a SO_4^{2-} concentration of 0.025M whereas at 125° C and a NaCl concentration of 2 moles/kg H_2O, the concentration of SO_4^{2-} in equilibrium with anhydrite is nearly twice that high (*22*).

The rates predicted by the high H_2S trend for temperatures between 120-175°C in Figure 3 are more than adequate to generate large sour gas fields in geologically reasonable time periods (7) provided that a substantial portion of the produced H_2S accumulates in the reservoir and is not lost by leakage or FeS_2 formation. Note, however, that based on the high H_2S trend of rates in Figure 3, there is potentially a two order of magnitude difference between rates for TSR at 120 and 150°C (i.e. ~10^{-8} and ~10^{-6} moles/liter/year respectively), and an additional order of magnitude drop-off in rate at ~100°C. The corresponding times required for generation of comparably sized sour gas fields likewise differ by two orders of magnitude at 120 and 150°C. Thus we predict that formation of geologically "young" sour gas fields may require temperatures at the high end of this range.

The situation for H_2S generation by TSR in MVT ores is less favorable for several reasons. For most MVT ores, there commonly is no evidence for local sulfate evaporites, nor is there evidence that they were initially petroleum reservoirs (although there are exceptions to these generalizations). Furthermore, interpretation is complicated by a long-standing debate as to whether the sulfide sulfur in MVT deposits was formed locally- at the immediate site of ore deposition, or alternatively carried in from a distance with migrating ore-fluids.

We will first consider a local H_2S source. Local generation of H_2S by TSR may be further subdivided into two sub-cases. The first is that H_2S accumulated at the site of ore deposition over a long period of time. This is essentially equivalent to the formation of a sour gas field in close proximity to the ore. This mechanism has been previously advocated (23). The constraints on this process are the same as described above for formation of sour gas fields. The advantage for sulfide mineral generation is that the required H_2S can accumulate over a geologically long period of time even given the slow rates involved.

A second sub-case which has been widely invoked (*e.g.* 24), is that an initially SO_4^{2-} and metal bearing (but H_2S-free) fluid undergoes TSR utilizing indigenous organic matter during its transit through the ore deposit. The results of our analysis (Figure 3) coupled with probable flow rates of fluids through the ore deposits tend to rule out this possibility. Recent flow modeling calculations (25-26) have derived flow rates for MVT ore fluids on the order of 1-10 meter/year. For a 100 meter thick ore-body the residence time of this fluid in the deposit would be 10-100 years. The reduction rates for initially H_2S-free fluid (lower trends in Figure 3) predict rates in the range of 10^{-8} to 10^{-10} moles/liter/year for these conditions. The resulting maximum concentrations of H_2S would only be 10^{-6} to 10^{-9} moles/liter H_2S (i.e. 1 to .001 micromolar); concentrations which are far too low to form ore deposits of the observed size in geologically reasonable times.

The second possibility is that the H_2S was formed far from the ore deposit and carried in with the migrating ore fluid. Leaving aside persistent questions of the possibility of co-transport of metals and H_2S in such a fluid which are outside the scope of this paper, it is still possible to address the efficacy of a distal source. Once again, we recognize two possibilities. The first is that the migrating H_2S was originally generated in a distant "sour gas field" over a long time period, and at some point this accumulated sulfide was transported to the site of ore deposition via a fluid migration episode. This is a viable possibility, but is constrained by potential removal of H_2S along the transport path from the distant sour gas field.

The second possibility is that H_2S was generated within an aquifer along the ore fluid flow path. Given the rates appropriate for an initially low H_2S fluid at 150°C (i.e. ~10^{-8} moles/liter/year), and fluid flow rates of 1 meter/year (*25-26*) long flow paths on the order of 50 - 200 km are required to generate H_2S concentrations in the range of 10^{-3} M necessary for ore formation. Long fluid flow paths on this scale are now believed to characterize MVT or fluid transport (*25-26*), making this also a viable alternative.

Referring back to Figure 2, we can now address the lack of agreement between geologic conditions of MVT ore genesis, and the experimental conditions under which TSR has been verified. We believe that TSR may be too slow to occur on the short time scales of a single pass of ore fluid through an individual ore body (i.e. hundreds of years or less). However, even for initially low H_2S fluids, times on the order of tens or hundreds of thousands of years such as those involved in transport of a packet of ore fluid in a regional aquifer, or even longer residence times such as those associated with formation of sour gas fields, are sufficient to generate adequate H_2S for ore genesis.

Acknowledgments
We wish to acknowledge Michele Tuttle and Andrew Nicholson and two unidentified revewiers who provided useful comments on earlier versions of the manuscript.

Literature cited
1.   Goldhaber, M. B.; Kaplan, I. R. In *The Sea*; Goldberg, E., Ed. Wiley Interscience: New York, **1974,** Vol 5, pp 569-655.
2.   Jørgensen B. B.; Isaksen, M.F.; Jannasch, H. W. *Science*, **1992,** *258,* 1756-1757.
3.   Ohmoto, H.; Lasaga. A. C. *Geochim. Cosmochim. Acta,* **1982,** *46,* 1727-1745.
4.   Machel, H. G. In *Diagenesis of Sedimentary Sequences*, Marshall, J. D., Ed. Geological Society Special Publication No. 36; Geological Society London, **1987,** 15-28.

5.   Ohmoto, H.; Kaiser C. J.; Geer, K. A. In *International Conference on Stable Isotopes and Fluid Processes in Mineralization*; Herbert H. K.; Ho S. E. Eds. Geological Society of Australia Special Publication 23, **1990,** 70-120.

6.   Orr, W. L. Am. Assoc. Petrol. Geol. Bull.; **1974,** *58,* 2295-2318.

7.   Nicholson, A. D. Unpublished PhD Dissertation, Colorado School of Mines; **1993,** 241 pp.

8.   Krouse, H. R.; Viau, C. A.; Eliuk, L. S.; Ueda, A.; Halas, S. *Nature,* **1988,** *333,* 415-419.

9.   Trudinger, P.A.; Chambers, L.A.; Smith, J. W. *Canadian J. Earth Science,* **1985,** *22,* 1910-1918.

10.   Orr, W. L. *Abstracts with Program*; 95th GSA Meeting, New Orleans, **1982,** *14,* 580.

11.   Kiyosu, Y. *Chemical Geology,* **1980,** *30,* 47-56.

12.   Kiyosu, Y.; Krouse, H.R. *Geochemical J.,* **1990,** *24,* 21-27.

13.   Kiyosu, Y.; Krouse, H.R. *Geochemical J.* **1993,** *27,* 49-57.

14.   Toland, W.G.; Hagmann, J.B.; Wilkes, J.B.; Brutschy, F.J. *J. Am. Chem Soc.,* **1958,** *80,* 5423-5427.

15.   Toland, W. G. *J. Am Chem Soc.,* **1960,** *82,* 1911-1916.

16.   Drean, T. A., Unpublished MS Thesis, Pennsylvania State University **1978,** 90p.

17.   Kaiser C. J. Unpublished PhD thesis, Pennsylvania State University; **1988**

18.   Heyl, A. F. In *International Conference on Mississippi Valley Type Lead-Zinc Deposits*; Kisvarsanyi,G,; Grant, S. K.; Pratt, W. P.; Koenig, J. W. Eds. University of Missouri-Rolla Press: Rolla MO., **1983,** 27-60.

19.   Orr, W. L. In *Advances in Organic Geochemistry;* Campos, R.; Goni J. Eds. Enadisma, Madrid, Spain; **1977,** 571-597.

20.   Leach, D. L.; Plumlee, G. S.; Hofstra, A. H.; Landis, G. P.; Rowan, E. L.; Viets, J.B. *Geology,* **1991,** *19,* 348-351.

21.   Orr, W. L; Sinninghe Damsté, J. S. In *Geochemistry of Sulfur in Fossil Fuels*; Orr, W. L.; White, C. M. Eds. American Chemical Society Symposium Series 429, American Chemical Society: Washington D.C., **1990,** *2-29.*

22.   Holland, H. D.; Malinin, S. D. In *The Geochemistry of Hydrothermal Ore Deposits*, Barnes, H. L. Ed. Wiley-Interscience: New York, **1979,** 461-508.

23.   Anderson, G. M. *Economic Geology* **1991,** *86,* 909-926.

24.   Leventhal, J. S. *Economic Geology* **1990,** *85,* 622-631.

25.   Bethke, C. M. *Economic Geology,* **1986,** *81,* 233-249.

26.   Garven, G.; Ge, S.; Person, M. A.; and Sverjensky, D. A. *Am. J. Science,* **1993** *293,* 497-568.

27.   Kozintseva, T. N. *Geochemistry International,* **1964,** *1,* 750-756.

28.   Rospondek, M. J.; de Leeuw, J. W.; Bass, M.; van Bergen, P. F.; Leereveld, H. *Org. Geochem.* **1994,** *21,* 1181-1191.

RECEIVED June 14, 1995

Chapter 24

Controls on the Origin and Distribution of Elemental Sulfur, H_2S, and CO_2 in Paleozoic Hydrocarbon Reservoirs in Western Canada

Ian Hutcheon[1], H. Roy Krouse[2], and Hugh J. Abercrombie[3]

[1]Department of Geology and Geophysics and [2]Department of Physics and Astronomy, University of Calgary, Calgary, Alberta T2N 1N4, Canada
[3]The Institute of Sedimentary and Petroleum Geology, 3303 33rd Street, Northwest, Calgary, Alberta T2L 2A7, Canada

H_2S and CO_2 are observed in natural gas in the Western Canada Sedimentary Basin (WCSB) and concentrations increase with depth and temperature. Paleozoic waters are saline (70-250 g/L TDS) and approach equilibrium with anhydrite, the probable sulfur source for H_2S. The ^{13}C composition of hydrocarbon gases associated with H_2S is consistent with hydrocarbon oxidation. Molar volumes of anhydrite and H_2S are similar at reservoir temperatures and pressures. Calculations show that 0.35 X H_2S could be formed by 1 volume % anhydrite. These observations are consistent with a closed-system origin for H_2S.
   Modeling of thermochemical sulfate reduction (TSR) shows that 75% of the porosity created by dissolution of anhydrite in a closed system is lost to calcite precipitation. Moving H_2S away from the site at which TSR takes place creates more porosity, however the isotopic data do not support an open system model.

Oil and gas reservoir rocks in the uplifted Western Canada Sedimentary Basin (WCSB) range in age from Cambrian to Tertiary. The basin can be divided broadly into a deeper Paleozoic section comprised of limestone and dolomite and a shallower, predominantly clastic, Mesozoic section. The Paleozoic section of the basin hosts numerous oil and gas pools and the stratigraphy, sedimentology and petroleum geology have been extensively studied. High H_2S concentrations occur in hydrocarbon gas pools in this part of the basin, generally in deeper reservoirs. The origin of H_2S, CO_2 and native sulfur is the subject of this paper.
   We examined the H_2S and CO_2 content of natural gases and chemical and isotopic data published for formation waters and gases in the Alberta portion of the WCSB in order to recognize possible controls on the origin, distribution and amount of H_2S. H_2S has economic importance because it decreases the value of natural gas and may play a role in both the formation of dissolution porosity and destruction of porosity by cementation during diagenesis. H_2S may originate by microbial or inorganic processes (1) and, as a reactive gas, the concentrations may be affected by reactions with enclosing rocks and associated waters.

0097–6156/95/0612–0426$12.00/0

There are numerous compositional analyses of natural gas in the Western Canada Sedimentary Basin (WCSB) publicly available through the Alberta Energy and Utilities Board (AEUB, formerly the Alberta Energy Resources Conservation Board) pool average gas data base (2) and the CO_2, H_2S, N_2 and He content is routinely reported. Isotopic analyses of hydrocarbon and non-hydrocarbon gas components in reservoirs are reported from various geological settings (3) and it has been suggested (4) that high H_2S concentrations in Paleozoic reservoirs originate from thermal (abiological) sulfate reduction. Analyses of waters are abundant in the AEUB data, but most tend to be incomplete with Ca, Mg, Cl, SO_4 and alkalinity measured directly and Na+K concentrations reported as the difference in charge balance between cations and anions. A few published studies (5-9) have reported high quality chemical and isotopic analyses of waters from the WCSB. These data are used in this paper to help define the possible origins of H_2S in deeply buried carbonate reservoirs of Paleozoic age in the WCSB.

Water and Gas Compositions in Western Canada

General Trends in Gas Chemistry. The partial pressure of H_2S and CO_2 in both Paleozoic and Mesozoic reservoir rocks tends to increase with temperature and, therefore, depth. Figure 1 shows pH_2S for Devonian and Mississippian age reservoirs and at temperatures from 50 to 150°C, pH_2S tends to be higher in Devonian reservoirs. Devonian reservoirs tend to have lower pCO_2 than Mississippian reservoirs (Figure 2). The trend of increasing CO_2 and H_2S with temperature is consistent with the origin of both gases being related to devolatilizaton of minerals, such as calcite and anhydrite, or devolatilizaton of organic matter, which may contain up to 11wt. % sulfur (10, 11). The concentrations of both CO_2 and H_2S could be affected by reaction of carbonate, sulfate and sulfide minerals.

High CO_2 contents occur in quartz-rich sandstone reservoirs in southeastern Alberta in the Mesozoic part of the section that overlies Paleozoic carbonates. The hydrodynamic regime and regional variations in water and gas chemical and isotopic compositions suggest that the CO_2 in this area is the result of hydrocarbon oxidation that accompanies bacterial sulfate reduction (9, 12).

Distribution of H_2S and CO_2. General trends in the distribution of H_2S and CO_2 were considered by preparing maps that divide the Paleozoic part of the stratigraphic section into Devonian and Mississippian units, similar to the broad divisions used by Hitchon (13-15) in his classic study of natural gas in western Canada.

The H_2S concentrations in Devonian reservoirs are highest near the disturbed belt along the foothills of the Rocky Mountains, ranging from 0.2 mole fraction (X) and 0.6 X with rarely observed occurrences up to 0.85 X, such as in the Bearberry pool. These extreme values occur in the deepest part of the section at present depths below surface that range from 2000-4000 m, reservoir temperatures in the range of 80-120°C, and reservoir pressures between 30 and 60 MPa. Due to uplift and erosion following the Laramide orogeny, maximum temperatures were higher because burial depths were probably more than 1900 m greater in the past (16). The distribution of CO_2 parallels that of H_2S, although concentrations are typically lower, with the highest values ranging from 0.05 to 0.10 X.

The patterns of distribution of H_2S and CO_2 for reservoirs of Mississippian age are similar to those observed in Devonian reservoirs, with high H_2S and CO_2 near the more deeply buried part of the section in the foothills. H_2S concentrations are lower in Mississippian reservoirs than in Devonian reservoirs, although locally variations may be large.

General Trends in Water Chemistry. The processes that form H_2S and CO_2 almost certainly affect the formation waters and it is important to understand the distribution of variations in formation water chemistry. There are published studies of formation water chemical and isotopic compositions in the WCSB (5-9) in different areas in

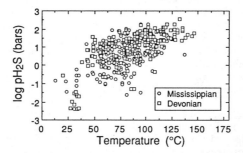

Figure 1. Correlation of partial pressure of H_2S and temperature in reservoirs of Mississippian and Devonian age. pH_2S tends to be higher in Devonian reservoirs as compared to Mississippian reservoirs.

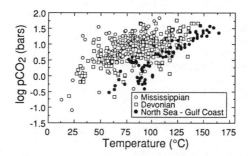

Figure 2. Correlation of partial pressure of CO_2 and temperature in reservoirs of Mississippian and Devonian age. At temperatures between 40 and 80°C in the WCSB, pCO_2 shows an increase of at least two orders of magnitude compared to the regular trend displayed by reservoirs in the North Sea and Gulf Coast (22).

units ranging in age from Cambrian to Tertiary. Waters in the WCSB tend to become more saline with age and depth, with waters in rocks that are Mesozoic and younger being less saline than waters in Paleozoic rocks and with a relatively continuous increase in salinity with increasing age. This difference in chemistry has persisted since uplift of the basin prior to the Eocene, suggesting that fluid flow that results from differences in topography, while recognized at depths up to 1 or 2 km (9, 12, 17), has not completely flushed the basin and erased compositional gradients.

Paleozoic rocks in the WCSB are dominantly carbonates associated with evaporites, and the Mesozoic rocks are dominated by clastic rocks. The chemical composition of formation waters in Mesozoic and Paleozoic rocks is somewhat different with waters in Tertiary and Cretaceous rocks having relatively low salinity $Na-Cl/HCO_3$ waters in contrast to waters in Devonian and Mississippian age rocks that are relatively higher salinity $Na/Ca-Cl$ waters. Published data (6) show that waters in Paleozoic rocks have sulfate concentrations that range from 0 to about 1300 mg/L.

Origin of Sulfur, H_2S and CO_2

Volume Considerations.   The source of sulfur in reservoirs subjected to thermochemical sulfate reduction is interpreted to be anhydrite (4, 18), which is abundant in Paleozoic rocks in the WCSB. To determine if anhydrite is a feasible source of sulfur for thermochemical sulfate reduction, the solubility of anhydrite was calculated at temperatures from 25 to 125°C from published data (19). Figure 3 shows the product of the activities of the aqueous ions Ca and SO_4 determined from water analyses from (6), compared to the solubility of anhydrite. Formation waters at temperatures greater than 80°C approach saturation with anhydrite, consistent with an anhydrite sulfate source.

Accepting anhydrite as the sulfur source, an estimate of the volume of anhydrite required to form the observed H_2S concentrations will show if H_2S can be accounted for by a local reservoir scale source (closed system) or if transport (open system) is required. Calculations of H_2S molar volumes were completed for temperatures from 60-120°C and pressures from 20 to 60 MPa (200-600 bars). This encompasses present day conditions in H_2S bearing reservoirs in Alberta, although burial depths were at least 1900m deeper (17) and temperatures and pressures were higher in the past. The volume of H_2S was calculated using $PV = ZRT$ and the critical temperature (373.5 °K), critical pressure (8.937 MPa, or 89.37 bars) and compressibility (20). The effect of mixing H_2S with methane and other gases on the volume of H_2S has been ignored, but the non-ideal gas volume terms, estimated from critical properties are included. For the range of temperatures and pressures, volumes of H_2S range from approximately 40 to 60 cm^3/mol (Figure 4). The molar volume of anhydrite is 45.94 cm^3/mol. One mole of anhydrite is required to form one mole of H_2S and the similarity of molar volumes suggests a given volume of anhydrite will produce approximately an equivalent volume of H_2S.

Assuming a starting porosity of 5% before thermochemical sulfate reduction begins, the apparent volume fraction of H_2S in the gas can be calculated from the molar volumes of CO_2, H_2S, CH_4, C_2H_6 and C_3H_8 at the temperature and pressure of interest and the mole fractions of each gas. For 100 cm^3 rock the porosity contains 0.05 x 0.2295 cm^3 = 1.14759 equivalent cm^3 of H_2S, which, divided by the volume per mole (89.3 cm^3/mol), gives 0.01285 moles of H_2S. One mole of anhydrite (45.94 cm^3/mol) is required to produce one mole of H_2S, thus the volume of anhydrite required to produce 0.35 X H_2S is approximately 0.59 cm^3, or 0.59 volume percent anhydrite in the rock. Because the volumes of all the gases change as a function of temperature, pressure and the gas composition, the relative volumes of anhydrite and H_2S will vary slightly, but it is safe to assume that between 0.5 and 1.0 % anhydrite in a rock with 5% porosity is sufficient to produce up to 0.35 X H_2S.

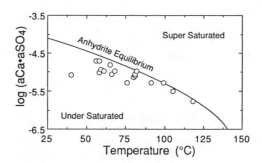

Figure 3. Formation waters from Paleozoic rocks approach saturation with anhydrite at temperatures in the range of 80°C. The equilibrium constant for anhydrite is shown by the curved line, the open circles are ion activity products for water compositions from (6).

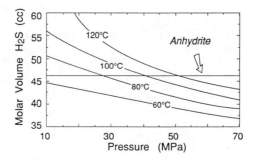

Figure 4. The volume of H_2S as a function of temperature and pressure. Volume of anhydrite shown for comparison.

This simple calculation suggests that H_2S can be derived from local sources of anhydrite and, although long distances of migration of H_2S or sulfate are not precluded, they are not necessary. In the West Hackberry Dome of the Louisiana Gulf Coast transport of fluids over an area of about 1.5 by 1.5 km has been suggested (*21*), however this study examined sandstones with no internal source of anhydrite. Anhydrite is present in WCSB Paleozoic carbonate rocks as nodules and pore fillings and dissolution of anhydrite is observed in some places (*4, 18*). The importance of local derivation, as opposed to long distance migration, depends on the abundance of anhydrite before sulfate reduction; a question that can only be resolved by a systematic, detailed, petrographic study of H_2S bearing intervals and careful comparison to non-H_2S bearing intervals, which is beyond the scope of this paper.

H_2S and CO_2. There is a linear trend of log pCO_2 with temperature in the U.S. Gulf Coast and North Sea (*22*), but pCO_2 concentrations observed in WCSB reservoirs are higher (Figure 2). It has been suggested (*23*) that pCO_2 is controlled by clay-carbonate buffering during diagenesis of lithoclastic rocks and the high values imply there is no pH buffering assemblage in Mississippian and Devonian carbonates of the WCSB. High CO_2 values (*9*) in the WCSB occur in quartz sandstones with no buffer assemblage. As can be seen from reaction (1), sulfate reduction can be expected to result in increased CO_2. The ratio of CO_2 to H_2S during sulfate reduction is dependent on the composition of the oxidized hydrocarbon and, as reaction (2) shows, the ratio increases as the C:H ratio of the hydrocarbon increases.

$$CH_4 + 2H^+ + SO_4^{2-} \Leftrightarrow H_2S + CO_2 + 2H_2O \tag{1}$$

$$\tfrac{4}{7}C_2H_6 + 2H^+ + SO_4^{2-} \Leftrightarrow H_2S + \tfrac{8}{7}CO_2 + \tfrac{12}{7}H_2O \tag{2}$$

C-H bonds for alkanes become weaker as the C:H ratio increases, and it is expected that longer chain alkanes would be more easily oxidized than short chain alkanes. Detailed examination of the chemical and isotopic composition of natural gas in the WCSB associated with high H_2S concentrations (*4*) supports the oxidation of higher carbon number alkanes in preference to methane. It has been noted (*24*) that TSR results in increased H_2S and CO_2 in the gas and depletion in the amount of saturated hydrocarbons in condensate liquids. If hydrocarbons are oxidized during TSR, it would be expected that during sulfate reduction, CO_2 concentration would tend to increase as the amount of H_2S increases. Present day gas compositions may not always show this relationship, nor do they necessarily reflect the composition of the oxidized hydrocarbon, for several reasons. The oxidized hydrocarbon is unlikely to be a single compound with a fixed C:H ratio. Further, both CO_2 and H_2S will participate in water-rock interactions, such as precipitation and dissolution of sulfur, carbonate, sulfide, and sulfate minerals, and this will affect the concentrations. Finally, both CO_2 and H_2S can form, or be removed, by other processes unrelated to reduction of sulfate and oxidation of hydrocarbons.

Sulfur. The volumetrically most significant sulfur compounds observed in hydrocarbon reservoirs in which TSR is interpreted to have occurred are H_2S gas, anhydrite ($CaSO_4$), solid or liquid sulfur, and dissolved sulfate. Sulfur is observed to be associated with H_2S (*4, 18, 25*). Sulfur has an intermediate oxidation state between H_2S and sulfate, but whether native sulfur is formed by oxidation of H_2S after TSR, or during the TSR reaction is not known. Thermodynamic data for liquid sulfur are not available, so the stability of sulfur relative to the other minerals noted above was calculated in reaction (3) using solid sulfur. Because most gases associated with H_2S-bearing reservoirs contain CO_2, and anhydrite and calcite are commonly associated with TSR, it is possible to construct equation (4), which shows the oxidation of H_2S to solid sulfur in the presence of calcite and anhydrite.

$$H_2O + \frac{1}{2}S_{2\,(s)} \Leftrightarrow H_2S + \frac{1}{2}O_{2\,(g)} \tag{3}$$

$$CaCO_3 + 3H_2O + 2S_{2\,(s)} \Leftrightarrow CaSO_4 + CO_2 + 3H_2S \tag{4}$$

Assuming unit activity of water, and that anhydrite and calcite are pure minerals, the stability of pure solid sulfur can be calculated from the reservoir gas composition and the equilibrium constant for reaction (4) which can be expressed as $(fCO_2){\cdot}(fH_2S)^3$. $K_{(4)}$ was calculated using SUPCRT92 (19) and thermodynamic data for sulfur (26). The gas compositions for H_2S-bearing reservoirs were obtained from the AEUB and the fugacity coefficients of CO_2 and H_2S were assumed to be unity. Figure 5 shows the results for Mississippian and Devonian rocks. Values of $(fCO_2){\cdot}(fH_2S)^3$ are mostly greater than the value for equilibrium between calcite+sulfur and anhydrite $(K_{(4)})$ and lie within the calcite-sulfur stability field.

Isotopic Data

H and C isotopic compositions of hydrocarbon gases from natural gases associated with H_2S in Devonian and Mississippian age reservoirs shows a systematic change in $\delta^{13}C$ values of light hydrocarbon gas as a function of an extent of reaction parameter (4), with C_2 and higher compounds oxidized in preference to C_1. Calcite has $\delta^{13}C$ values as low as -30‰, due to incorporation of carbon from hydrocarbon oxidation.

The separation of isotopic values $(\Delta^{13}C)$ of light hydrocarbon gases may be an indicator of the degree of organic maturity of the source matter from which gas is generated (27). At low levels of organic maturity the maximum separation between C_1 and C_2 is given as about 16‰ $\Delta^{13}C$, however hydrocarbon gases associated with high H_2S (4) show separations ranging up to 18‰ $\delta^{13}C$, greater than predicted, especially since the gases were probably generated at high levels of organic maturity. The maximum $\Delta^{13}C$ would be expected to be at low levels of maturation, and the difference would be expected to decrease as maturation increased, thus increasing pH_2S should be accompanied by decreasing $\Delta^{13}C$. Figures 6 and 7 show $\Delta^{13}C$ for various gas pairs plotted versus pH_2S for Mississippian and Devonian reservoirs. There are no trends in the data and $\Delta^{13}C$ for most gas samples is far greater than would be predicted, suggesting that some other process, possibly oxidation, has influenced the carbon isotopic composition of the gas components.

We are not aware of any method by which the relative effect on $\Delta^{13}C$ from maturation of the source material can be quantitatively separated from the effect of oxidation. Since C-C bonds would be expected to be weaker for gas components of higher carbon number, oxidation would preferentially attack the higher carbon number gases and would break ^{12}C bonds preferentially, leading to a relatively higher rate of increase in ^{13}C for the unreacted higher carbon number gases. Maturation and oxidation change $\Delta^{13}C$ in the same way, and it is not obvious how the relative influence of such effects on $\Delta^{13}C$ can be discerned.

Figure 8 shows the variation of $\delta^{18}O$ with sulfate concentration for WCSB waters in rocks of various ages. Waters with low sulfate and negative δ ^{18}O are found in Cretaceous rocks, consistent with the suggestion that an influx of meteoric water is responsible for promoting bacterial sulfate reduction (9, 12). Waters that have low sulfate (and high pH_2S, not shown), but have positive $\delta^{18}O$ are found in Devonian or Mississippian age rocks and are interpreted to have experienced water-rock interaction during thermochemical sulfate reduction.

Reaction Path Modeling

The volume comparison (Figure 4) suggests that anhydrite and H_2S can "trade places" and that only small volumes of anhydrite are required to produce relatively high H_2S

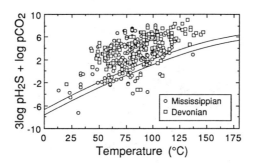

Figure 5. The stability of calcite+sulfur versus anhydrite. Gases in Mississippian and Devonian reservoir tend to be in chemical equilibrium with calcite+sulfur, rather than anhydrite.

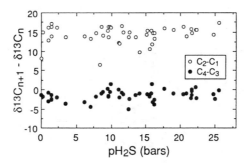

Figure 6. The difference in $\delta^{13}C$ ($\Delta^{13}C$) between light hydrocarbon gases in reservoirs in Mississippian rocks versus the partial pressure of H_2S. There is no obvious dependence between $\Delta^{13}C$ and pH_2S.

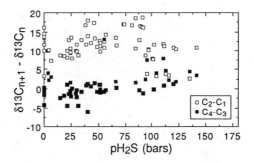

Figure 7. The difference in $\delta^{13}C$ ($\Delta^{13}C$) between light hydrocarbon gases in reservoirs in Devonian rocks versus the partial pressure of H_2S. At higher values of pH_2S, $\Delta^{13}C$ for C_2-C_1 decreases as pH_2S increases.

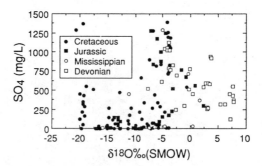

Figure 8. $\delta^{18}O$ (SMOW) versus sulfate for WCSB waters. Low sulfate waters enriched in ^{18}O compared to SMOW and interpreted to reflect water-rock interactions are observed in Devonian rocks. Low sulfate waters depleted in ^{18}O compared to SMOW are observed in Cretaceous rocks.

in gases at reservoir conditions. Anhydrite, in addition to providing a volumetrically appealing source for H_2S, has the potential for generation of porosity by dissolution. However, CO_2 is released in the proposed TSR reactions, Ca must be released if anhydrite dissolves as a sulfur source (reaction 4), and the reduction of sulfate to H_2S is likely to be a strong pH buffer, implying that TSR will be accompanied by calcite and dolomite precipitation. Saddle dolomite has been suggested to form during TSR (28). Isotopic data show that petrographically late calcite contains CO_2 from hydrocarbon oxidation (4). A reaction path model including anhydrite and dolomite was used (29) to examine the reduction of sulfate in a treatment of Mississippi Valley type (MVT) ore deposits, but did not include calcite as a potential product. It was concluded that the precipitation of sparry dolomite via a reaction similar to (5) is a possible result of sulfate reduction in the presence of anhydrite, organic matter and dolomite. Similar model calculations focused on gas compositions have been performed (Nicholson and Goldhaber, *personal communication*).

$$2\,CH_4 + CaSO_4 + Mg^{2+} \Leftrightarrow CaMg(CO_3)_2 + Ca^{2+} + 2\,H_2O + 2\,H_2S \qquad (5)$$

Reaction (2) is a strong pH buffer and calcite should be precipitated as anhydrite is dissolved during the reduction of sulfate. Some of the porosity gained by anhydrite dissolution potentially is thus lost to calcite precipitation. The relative changes in volume of calcite, anhydrite and porosity during the progress of reaction (6) was calculated at 100°C (Figure 9) using EQ3/6 (30, 31):

$$CH_4 + CaSO_4 \Leftrightarrow CaCO_3 + H_2O + H_2S \qquad (6)$$

As a rough number, for every four volumes of anhydrite converted to H_2S, three volumes of calcite are precipitated, leaving only one volume as porosity. The ratio of volume gained by anhydrite dissolution to volume lost by calcite precipitation requires that a large volume of anhydrite, approximately five times the amount required to generate the observed H_2S, be dissolved in order to produce more than 1% porosity increase by TSR.

If H_2S gas rises into a zone in which it can be re-oxidized, the H^+ stored in H_2S during sulfate reduction is released (29) and can be an effective agent of dissolution. Examining reaction (2) we see that the introduction of H_2S which is then allowed to re-oxidize to sulfate, away from the site at which it was generated, releases H^+. The amount of dolomite dissolution that could result from migration of H_2S can be considered by a calculation in which H_2S is allowed to react with dolomite. The results of such a calculation are shown in Figure 10 which simulates the release, but not the oxidation, of H_2S. Porosity gain can be 1-2% by volume since no calcite forms. The amount of H_2S involved in the chemical model is more than four times the amount of CH_4 allowed to react in the simulation used to construct Figure 9. In other words, the demand for anhydrite and organic matter to produce significant dissolution is very high and some mechanism of focusing the H_2S is required.

It is possible that H_2S produced by TSR may be transported into reservoirs of the WCSB by fluid flow, moving H_2S away from the site of generation and allowing porosity to be produced without the loss of porosity due to calcite precipitation. However all the evidence suggests that hydrocarbons in the reservoir are oxidized during TSR and that these reservoirs approach closed system behavior. It is possible that a complex cycling of fluids on a reservoir scale is involved during TSR and that some zones of a reservoir develop higher porosity than others.

Summary and Conclusions

The concentrations of H_2S and CO_2 in natural gas accumulations in Paleozoic carbonate rocks in the WCSB increase with increasing depth and temperature. H_2S

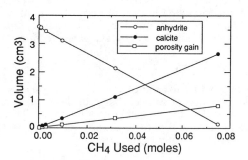

Figure 9. The changes in volume of anhydrite, calcite and porosity during TSR, calculated using EQ3/6 (*30, 31*).

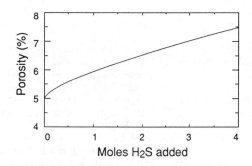

Figure 10. The dissolution of dolomite by H_2S modeled using EQ3/6.

concentrations in reservoirs in Devonian age rocks tend to be higher than in reservoirs in Mississippian age rocks and in both groups, H_2S concentrations are highest in the deeper westernmost parts of the basin. CO_2 concentrations parallel H_2S concentrations, except in the more shallow Mesozoic rocks in the eastern and southern parts of the WCSB where CO_2 results from bacterial processes.

Waters in Paleozoic rocks are more saline than waters in Mesozoic rocks and have higher sulfate concentrations. The waters in Paleozoic age rocks tend to be in chemical equilibrium with anhydrite above 80°C and anhydrite is the probable sulfur source for H_2S in Paleozoic reservoirs. The volumes of anhydrite and H_2S are similar in the temperature range from 80-120°C and pressures from 30-60 MPa, suggesting that the production of H_2S in the WCSB could be a relatively closed system process, although an open system process is not precluded. To produce concentrations up to 35% H_2S in reservoir rocks with initially 5% porosity would only require an initial volume of anhydrite between 0.5 and 1%. Reservoir gas compositions suggest that, within the limits of the thermodynamic data, the gases are in equilibrium with calcite and sulfur and that anhydrite should not be stable in the gas zone.

The C isotopic composition of hydrocarbon gases associated with H_2S is consistent with the oxidation of these gases during the thermochemical reduction of sulfate (4). The compositions of waters in Paleozoic rocks show that high H_2S and low sulfate concentrations are correlated with relatively ^{18}O-enriched waters, suggesting extensive water-rock interaction during thermochemical sulfate reduction. The reduction of sulfate from anhydrite during the oxidation of hydrocarbons should result in precipitation of calcite as anhydrite dissolves. Closed system modeling of such a reaction shows that porosity increases created by the dissolution of anhydrite are compensated for by volume decreases caused by calcite precipitation. If more open system conditions exist, and H_2S can migrate into rocks away from the site at which TSR takes place, more porosity can be created, however the isotopic data for calcite cements and hydrocarbon gases do not strongly support an open system model. At the reservoir scale, it is possible that some cycling occurs and that zones of high porosity are developed as TSR progresses.

Acknowledgments

The maps of the distribution of H_2S and CO_2 in this paper were produced by John Cody and we thank him for his assistance. The isotopic data for hydrocarbon gases were obtained with the support of Shell Canada Resources, particularly C. A. Viau, and their assistance and permission to use the data are much appreciated. Hutcheon and Krouse received financial assistance from the Natural Sciences and Engineering Research Council (NSERC) of Canada and Hutcheon received support from the Petroleum Research Fund. Acknowledgment is made to the donors of the Petroleum Research Fund, administered by the American Chemical Society, for partial support of this research. Some of the work reported in this paper was completed while the first author was on an NSERC Senior Industrial Fellowship sponsored by Shell Canada, The University of Calgary, and NSERC. Hans Machel and an anonymous reviewer provided comments that substantially improved this manuscript.

Literature Cited

1. Orr, W.L. *Am. Assoc. Petrol. Geol. Bull.* **1974**, *58*, 2295-2318.
2. Alberta Energy and Utilities Board (AEUB) *Alberta Resources of Gas;* , (5 volumes), Calgary, AB, Energy Resources Conservation Board Report ST87-35. 1987.
3. Krouse, H. R. *Stable Isotope Research in Support of More Effective Utilization of Gas Fields in Alberta,* Alberta - Canada Energy Resource Research Fund, Agreement U-30, 1983, December.

4. Krouse, H.R.; Viau, C.A.; Eliuk, L.S.; Ueda, A.; Halas, S. *Nature* **1988**, *333*, 415-419.
5. Hitchon, B.; Friedman, I. *Geochim. Cosmochim. Acta* **1969**, *33*, 1321-1349.
6. Hitchon, B.; Billings, G.K.; Klovan, J.E. *Geochim. Cosmochim. Acta,* **1971**, *35*, 567-598.
7. Connolly, C.A.; Walter, L.M.; Baadsgaard, H.; Longstaffe, F.J. *App. Geochem.* **1990**, *5*, 375-396.
8. Connolly, C.A.; Walter, L.M.; Baadsgaard, H.; Longstaffe, F.J. *App. Geochem.* **1990**, *5*, 397-414.
9. Cody, J. *Geochemistry of formation fluids in the Mannville Group (Lwr. Cret.), Southern Alberta: Sources, controls and water-rock interaction of carbon dioxide rich fluids.* Unpublished MSc thesis, Department of Geology and Geophysics, University of Calgary, Calgary, AB, 1993, 129 p.
10. Orr, W. L. *Org. Geochem.* **1986**, *10*, 499-516.
11. Orr, W. L. and Damste, J. S., In *Geochemistry of sulfur in petroleum systems;* Orr, W. L. and White, C. M. Eds.; 1990, ACS Symposium Series 429, p. 2-19.
12. Cody, J.C.; Hutcheon, I. *Bull. Can. Petrol. Geol.* **1994**, *42*, 15 pp.
13. Hitchon, B. *Jour. Can. Petrol. Tech.* **1963**, *2*, 60-76.
14. Hitchon, B. *Jour. Can. Petrol. Tech.* **1963**, *2*, 100-116.
15. Hitchon, B. *Jour. Can. Petrol. Tech.* **1963**, *2*, 165-174.
16. Nurkowski, J.R. *Am. Assoc. Petrol. Geol. Bull.* **1984**, *68*, 285-295.
17. Toth, J. and Corbet, T., In *Fluid Flow in Sedimentary Basins and Aquifers;* Goff, J.C. and Williams B.J.P. Ed.; Geological Society Special Publication, 1987, Vol. 34, p. 45-56.
18. Machel, H.G., In *Diagenesis of Sedimentary Sequences;* Marshall, J.D. Ed.; Geological Society Special Publication, 1987, Vol. 36; p. 15-38.
19. Johnson, J.W.; Oelkers, E.H.; Helgeson, H.C. *Comp. Geosci.* **1992**. *18*, 899-947.
20. Weber, H.C. and Meissner, H.P., *Thermodynamics for Chemical Engineers;* John Wiley and Sons, New York, NY, 1959, 507 p.
21. McManus, K.M.; Hanor, J.S. *Geology* **1993**, *21*, 727-730.
22. Smith, J.T.; Ehrenberg, S.N. *Mar. Petrol. Geol.* **1989**, *6*, 129-135.
23. Hutcheon, I. Shevalier, M.; Abercrombie, H.J. *Geochim. Cosmochim. Acta* **1993**, *57*, 1017-1027.
24. Claypool, G.E.; Mancini, E.A. *Am. Assoc. Petrol. Geol. Bull.* **1989**, *73*, 904-924.
25. Heydari, E.; Moore, C.H. *Geology* **1989**, *17*, 1080-1084.
26. Robie, R.A.; Hemmingway, B.S.; Fischer, J.R. *Thermodynamic properties of minerals and related substances at 298.15 K and 1 bar pressure and at high temperature.* United States Geological Survey Bulletin 1452, 1978, 456 pp.
27. James, A.T. *Am. Assoc. Petrol. Geol. Bull.* **1983**, *67*, 1176-1191.
28. Machel, H.G. *Geology* **1987**, *15*, 936-940.
29. Anderson, G.M.; Garven, G. *Econ. Geol.* **1987**, *82*, 482-488.
30. Wolery, T.J. *EQ3NR, A computer program for geochemical aqueous speciation-solubility calculations: Theoretical manual, user's guide, and related documentation (version 7.0).* Lawrence Livermore National Laboratory, UCRL-MA-110662 PTIII. 1992, 246 p.
31. Wolery, T.J.; Davelar, S.A. *EQ6, A computer program for reaction path modeling of aqueous systems: Theoretical manual, User's guide, and related documentation (Version 7.0).* UCRL-MA-110662 PTIV. 1992, 338 p.

RECEIVED June 20, 1995

Chapter 25

Devonian Nisku Sour Gas Play, Canada: A Unique Natural Laboratory for Study of Thermochemical Sulfate Reduction

H. G. Machel[1], H. Roy Krouse[2], L. R. Riciputi[3], and D. R. Cole[3]

[1]Department of Geology, University of Alberta, Edmonton,
Alberta T6G 2E3, Canada
[2]Department of Physics and Astronomy, University of Calgary,
Calgary, Alberta T2N 1N4, Canada
[3]Chemical and Analytical Science Division, Oak Ridge National
Laboratory, Oak Ridge, TN 37831

The Upper Devonian Nisku Formation in the subsurface of south-central Alberta, Canada, produces oil, sweet and sour gas condensate. Each pool constitutes a small natural laboratory with respect to deep burial diagenesis. Petrographic features, stable isotopes of several solids (calcite, dolomite, anhydrite, elemental sulfur, pyrite, marcasite), isotope and bulk composition of gas condensates, and brine compositions of several sour gas pools indicate that thermochemical sulfate reduction was responsible for the formation of the sour gas that is presently pooled in the Nisku. The data also show that thermochemical sulfate reduction took place in the gas-water transition zone(s), which can be as narrow as only a few meters.

The Upper Devonian Nisku Formation in the subsurface of south-central Alberta, Canada, is an oil, sweet and sour gas condensate play (*1-3*). A unique feature of this play is that hydrocarbons and sour gas are contained in numerous closely spaced pools that have been essentially isolated hydrodynamically from one another since hydrocarbon entrapment, as shown by variable initial reservoir pressures and gas compositions. Hence, each pool constitutes a small natural laboratory with respect to deep burial diagenesis. Study of these pools provides several new clues regarding the process of thermochemical sulfate reduction (TSR), as well as criteria to distinguish thermochemical from bacterial sulfate reduction (BSR).

This paper reports the spatial distribution and some compositional data of various late-diagenetic solids (calcite, dolomite, anhydrite, elemental sulfur, Fe-sulfides) and fluids (oil, hydrocarbon gases, carbon dioxide, hydrogen

0097–6156/95/0612–0439$12.00/0

sulfide, formation waters) that were involved in sulfate-hydrocarbon redox-reactions. Preliminary results of this study have been documented in (*1-7*). More detailed work on the inorganic and particularly the organic geochemistry of the sulfur-bearing phases of the Nisku is in progress and will be reported in forthcoming papers.

Geologic Setting and Hydrocarbon Distribution

The Nisku play is located in south-central Alberta, Canada (*1-3*; Figure 1). The Nisku is the shallowest of the many sour gas plays in Devonian carbonates of the Western Canada Sedimentary Basin, all of which are located in the deeper part of the basin (Figures 1,2: other examples are Obed, Strachan, Ricinus, Bearberry, Harmattan, Swan Hills). Due to the regional structural SW dip, the subsurface depths of the Devonian carbonates range from zero in the northeasternmost part of the basin to more than 6000 m in the foothills region of the Rocky Mountains, which runs roughly along the southwestern border of Alberta just inside the province (Figures 1,2).

In the main study area, subsurface depths of the Nisku range from about 3000 to 4500 m, and the Nisku contains oil in its updip [NE] and sour gas in its downdip [SW] part (Figure 3). Reefs that form the reservoir facies occur as an elongate shelf margin complex and as a series of small reefs, with diameters of 1 - 2 km and thicknesses of 50 to 100 m, on the slope toward the basin in the NW. These reefs are surrounded by relatively impermeable calcareous shales and appear to be hydrologically 'isolated' from one another, as shown by their virgin pressures (original reservoir pressures prior to production) that vary drastically, even for pools that are located very close to one another (Figure 3). The shallowest pools have pressures close to hydrostatic, whereas the deepest pools have pressures far in excess of hydrostatic (compare the virgin pressures to pressures expected based on present day depths at the normal hydrostatic gradient of about 10 MPa/km). Also, there is no pressure drawdown from pool to pool during production.

Virgin H_2S contents, i.e., H_2S contents at discovery and prior to production of the pools, also vary drastically. In the isolated pools, virgin H_2S contents generally increase downdip, whereas the virgin H_2S contents increase upwards in the upper of the two closures (sub-pools) in the elongate shelf margin reef complex (Figure 3). On the slope toward the basin, the boundary between oil pools updip and gas pools downdip is sharp (line **a** in Figure 3), but significant (>1%) concentrations of H_2S occur only downdip of line **b** (Figure 3). The elongate shelf margin reef complex straddles lines **a** and **b** and contains H_2S both updip and downdip of these lines.

General Burial and Diagenetic History

Nisku reefs underwent a complicated diagenetic history that forms a

Figure 1. Location map of some sour gas plays in the subsurface of Alberta, Canada. A-B marks cross section line of Figure 2.

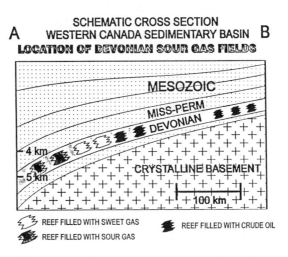

Figure 2. Schematic dip section through the Western Canada Sedimentary Basin (see Figure 1 for location). Devonian reefs constitute the principal reservoir facies in the Paleozoic section [there also are many clastic plays in the Mesozoic section]. In the shallow parts of the basin reefs are oil pools that, with increasing depth, give way to sweet and then sour gas pools. At present, the shallowest sour gas pools are at about 3.5 to 4 km.

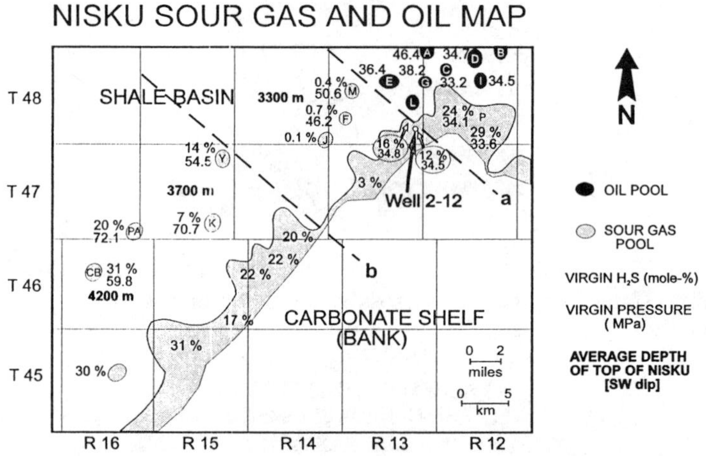

Figure 3.　　Nisku study area, showing distribution of oil versus gas, virgin pressures, H₂S concentrations (mole-%), and average depths of the top of the pools. Line **a** separates oil from gas pools, line **b** is the 'sour gas line' or 'TSR line'. See text for further explanation.

paragenetic sequence consisting of at least 19 distinctive phases (*1-2*). During shallow burial to about 300 m in the Late Devonian, the carbonate sediments were lithified to limestones by different types of calcite cements. During burial to about 300 to 1500 m in the Late Devonian and Mississippian, most Nisku reefs were pervasively dolomitized to so-called grey matrix dolomite, and up to 20% per reef of replacive anhydrite was emplaced, mostly in the form of cm-sized, milky-white nodular masses. After a long period of minor deposition, non-deposition and/or erosion, the Nisku reefs were buried fairly rapidly from about 2000 m to maximum depths of 4500 to 6000 m (depending on their location within the study area) during the Late Cretaceous and Early Tertiary. During this time and burial interval, oil migrated into the system and sweet gas and sour gas condensate formed, as well as late diagenetic saddle dolomite, calcite, sulfur, and some Fe-sulfide (*1-4*). These phases are identified as late-diagenetic on the basis of paragenetic relationships and their spatial distribution relative to the oil-water and gas-water contacts (see below). After maximum burial, the study area was uplifted and some of the overburden eroded, resulting in the present burial depths that are about 1500 to 2000 m less than the maximum burial depths.

Petrography and Spatial Distribution of Late Diagenetic Phases

Most wells that penetrated the Nisku contain 3 to 8 of the 19 diagenetic phases recognized in the Nisku. In some locations, however, almost all phases are developed. Well 2-12 (see Figure 3 for location) exemplifies the spatial distribution of the late diagenetic phases in the sour gas part of the Nisku play (Figure 4). This well is cored from 3179 m to 3264 m, and contains sour gas condensate in its upper half and water in its lower half, with a gas-water transition zone from about 3210 to 3230 m (Figure 4). Most of the host rock consists of grey matrix dolomite. Near the top of the core, replacive white anhydrite nodules (Figure 4: sample 1, enlargement 1a) are partially infiltrated and replaced by bitumen (enlargement 1b) and by white calcite (enlargement 1c; calcite is stained pink [grey in 1c] with Alizarin-red, in order to enable easy differentiation from anhydrite and dolomite). Anhydrite nodules are associated with bitumen-coated vugs (sample 2) to a depth of about 30 m below the top of the core. White saddle dolomite occurs from 3211 m downward, first scattered in molds and vugs (sample 3), then occluding such secondary pores (sample 5). Between 3221-3227 m, elemental sulfur (sample 4: S°) occurs intergrown with saddle dolomite. Near the bottom of this core, large calcite crystals occur as cements and as partial replacements of saddle dolomite (sample 6; calcite is black because of crossed polars extinction position). Fe-sulfides occur in two distinctive forms: as relatively small crystals and clusters of pyrite, mainly in the lower parts of well 2-12 and most other Nisku wells, and (b) as relatively large splays of pyrite and/or marcasite crystals that appear to replace some of the replacive anhydrites.

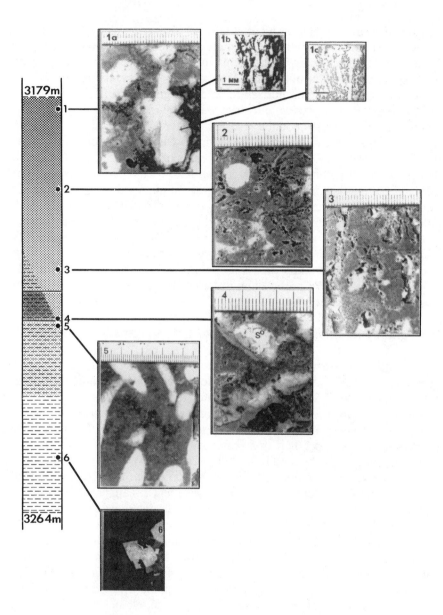

Figure 4.   Simplified litholog of well 2-12 (see Figure 3 for location), displaying the vertical distribution of the most important features formed during, or related, to thermochemical sulfate reduction. Fine stipple is gas saturation, horizontal dashes mark water saturation, with overlapping interval between about 3210-3230 m. The petrographic features are explained in the text.

Geochemistry

Methods   Minerals for geochemical analyses were selected by microscopic investigation of thin sections stained with Alizarin-red. Powder samples of calcite, dolomite, anhydrite, elemental sulfur and a 2 cm large pyrite nodule were extracted with a dental drill for XRD, stable and radiogenic isotope analyses.

Oxygen and carbon isotope analyses of carbonates were conducted using procedures modified from (8). Sulfur isotope ratios of powder samples of anhydrite, pyrite, and elemental sulfur were measured using the procedures outlined in (9). Crystals of pyrite, marcasite, and anhydrite ranging from a few to several hundred micrometers in size were analyzed *in situ* for sulfur isotopes using a Cameca 4f ion microprobe. Details of the analytical procedures for the ion microprobe are documented in (6,7) and Riciputi et al. (*Geochim. Cosmochim. Acta*, in press).

Standard fluid inclusion heating and freezing runs were performed on calcite and saddle dolomite samples. The analytical procedures are documented in (1) and (10).

Gases and formation waters were obtained from three sour gas wells. The bulk composition of the gases was determined by gas chromatography, and carbon isotope compositions of CO_2, H_2S, CH_4, C_2H_6, and C_3H_8 by the methods outlined in (9). Formation waters were analyzed partially in the field (pH, temperature, conductivity) and partially in the lab (major and minor anions and cations, D and O isotopes) by the methods described in (11). [Note: Gases and oils have also been retrieved from about 10 other Nisku pools. These samples form the basis of an extensive investigation of the organic geochemistry of the Nisku play (B.K. Manzano, M.Sc. thesis, University of Alberta, in prep.)].

All analytical results can be obtained in tabulated form from the senior author upon request. In this paper, only some of our data are presented in graphical form (see below). Analytical precisions are always better than the widths of the symbols used in these graphs. Analytical accuracies have been calibrated against NBS or internal standards and are better than 0.1‰ for the stable isotopes of all powder samples. Average errors for the ion probe sulfur isotope analyses of Fe-sulfides vary between 0.4 and 0.8‰ (1σ), whereas errors for the sulfate analyses are ≈2‰.

Results   The stable isotope values of saddle dolomite in well 2-12 are lowest near the top of the water-saturated zone and increase downwards, whereby $\delta^{18}O$ ranges from about -8 to -6 ‰PDB and $\delta^{13}C$ ranges from about -12 to +2 ‰PDB (Figure 5). In contrast, the $\delta^{18}O$ and $\delta^{13}C$ values of the grey

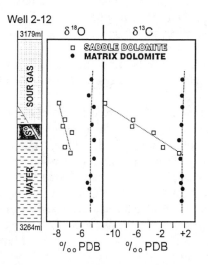

Figure 5.   Stable isotope composition of saddle dolomite and grey matrix dolomite from well 2-12. Column on left corresponds to that in Figure 4. See text for further explanation.

matrix dolomite samples are invariant with depth and are about -5 and +2 ‰PDB, respectively (Figure 5). Late sparry calcite also has depleted isotope values: $\delta^{18}O \approx$ -6 to -8 ‰PDB, $\delta^{13}C \approx$ -12 to -18 ‰PDB (not shown).

The homogenization temperatures of primary fluid inclusions of saddle dolomite range from 110 to 180 °C, with two median values of about 125 °C and 145 °C; freezing point depressions of these inclusions scatter between about -5 to -25 °C (Figure 6). The freezing point depressions of saddle dolomite from well 2-12 correspond to salinities of about five times the salinity of seawater (Figure 6).

Ion microprobe analyses of anhydrites yielded $\delta^{34}S$ values of about +24 to +28 ‰CDT (Figure 7). Powder samples of about 100 more anhydrite samples from various Nisku reefs yielded $\delta^{34}S$ values in the same range (*1*).

Ion microprobe $\delta^{34}S$ data of disseminated pyrite crystals from the syndepositionally well oxygenated reef facies in well 2-12 display two populations (Figure 7, left: the isotopically heavy population is from the sulfur-bearing interval). The $\delta^{34}S$ value of one powder sample from a large pyrite nodule from much lower in this well, retrieved from the bank facies below the reef, is -11 ‰CDT. Powder samples of two crystals of elemental sulfur from the S°-interval in this well yielded $\delta^{34}S$ of about +24 ‰CDT (Figure 7). $\delta^{34}S$ data were also furnished for well 2-19, which is located near well 2-12. Ion microprobe data from disseminated pyrite and marcasite crystals from the syndepositionally less well oxygenated facies near the bottom of the reef facies display an upward decreasing trend (Figure 7, right). The lowest values in both cores are about 45 to 50 ‰ lighter than the anhydrites.

Stable isotope data of the sour gas components are also distinctive: $\delta^{34}S$ values of H_2S are around +20 ‰CDT and approach those of Nisku anhydrites; $\delta^{13}C$ values of CH_4, C_2H_6, and C_3H_8 are about -41, -29 and -27 ‰PDB, respectively; $\delta^{13}C$ values of CO_2 are about -7 ‰PDB (Figure 8, top). Nisku formation waters tend to be brines with relatively heavy $\delta^{18}O$ and δD values (Figure 8, bottom). Water retrieved from one sour gas well has an extraordinary composition compared to the other Nisku pools: it is nearly potable and has an unusually positive D isotope ratio (Figure 8: P).

Discussion

The late diagenetic solid phases in well 2-12 have a distinctive spatial distribution relative to that of the fluid phases in this well (Figure 4): saddle dolomite occurs only in the water-bearing part of the pool, suggesting that it formed after the gas-water contact, which really is an approximately 20 m thick transition zone, arrived at its present position [dolomite can only form from aqueous solution]; elemental sulfur occurs only in the gas-water zone and intergrown with saddle dolomite, suggesting that it also formed after the gas-

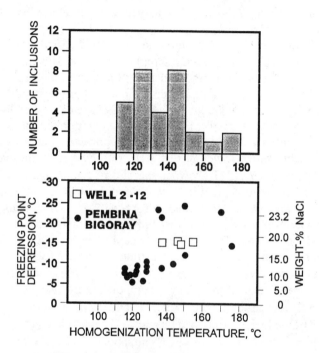

Figure 6.   Plot of fluid inclusion data of saddle dolomites from well 2-12 and several other wells (named Pembina and Bigoray, located about 20 to 50 km updip from the study area shown in Figure 3; data are from own analyses and from *10*). Homogenization temperatures greater than about 150°C are suspect [150°C approximated the maximum burial temperature] and probably are from leaked inclusions.

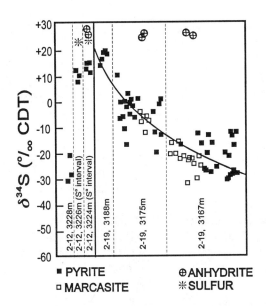

Figure 7. δ³⁴S values of ion microprobe analyses of Nisku pyrites, marcasites, anhydrites, and elemental sulfur from wells 2-12 and 2-19 (located close to well 2-12). Samples from well 2-12 show two populations, samples from well 2-19 decrease upwards. See text for further explanation.

SOUR GAS ISOTOPE DATA

GAS POOL		CO_2	H_2S	CH_4	C_2H_6	C_3H_8
F	0.7 % 46.2	- 6.9	+17.5	- 41.5	- 29.6	- 27.5
M	0.4 % 50.6	- 6.9	+20.7	- 41.7	- 29.5	- 25.8
Well 2-12	12 % 34.5	- 7.1	+22.1	- 38.7	- 29.1	- 27.5
		$\delta^{13}C$ ‰PDB	$\delta^{34}S$ ‰CDT	$\delta^{13}C$ ‰PDB	$\delta^{13}C$ ‰PDB	$\delta^{13}C$ ‰PDB

FORMATION WATER DATA

GAS POOL		Na^+	Cl^-	pH	δD	$\delta^{18}O$
F	0.7 % 46.2	44.8	70.0	6.9	- 56.2	+4.4
M	0.4 % 50.6	27.7	50.0	7.0	- 34.1	+2.4
P	29 % 33.6	4.6	0.1	5.4	+84.7	+4.3
		g/l	g/l		‰ SMOW	‰ SMOW

Figure 8. Sour gas isotope and formation water data. The pools are identified by their letters, H_2S (mole-%) contents, and initial pressures (MPa), which match those in Figure 3. See text for further explanation.

water zone arrived at its present position [elemental sulfur is known to form during the waning stages of TSR as a by-product (5,12,13)]; solid bitumen is common only in the upper 20 - 30 m of the pool, suggesting former pooling of oil in the uppermost part of the reef; late diagenetic replacive calcite is present throughout the pool but occurs as a replacement of anhydrite only in the uppermost, gas-bearing upper part. These phenomena suggest the following scenario:

(1) oil migration into the pool;

(2) *in situ* thermal maturation and cracking to gas condensate and solid bitumen, with concurrent hydrocarbon expansion and pressure increase, such that the gas-water contact/transition zone was progressively displaced downward;

(3) TSR was instigated during advanced stages of gas condensate formation, generating H_2S and further pushing the gas-water zone downward; dissolution of anhydrite and/or partial replacement by calcite in the upper part of the pool;

(4) cessation of TSR, forming elemental sulfur where the gas-water zone reached its present position;

(5) saddle dolomite formation largely after the gas-water zone reached its present position.

These processes took place while the study area underwent burial from about 2000 to about 5500 m during the Late Cretaceous to Early Tertiary, which is known to be the period during which the basin underwent regional tilting and rapid subsidence (2,14). The study area was then partially uplifted and some 1500 to 2000 m of overburden were removed (15,16), resulting in the present subsurface depths of the Nisku pools. During and after uplift the pressures in these pools partially reequilibrated to hydrostatic values commensurate with the present burial depths, as indicated by the lowest pressured pools having hydrostatic pressures (Figure 3). Those pools with overpressures are apparently so well sealed that they maintain some or all of the overpressures they had near maximum burial.

The geochemical data presented above add a further level of resolution to this scenario. Saddle dolomite formed by two interacting processes, i.e., pressure solution, liberating Ca^{2+}, Mg^{2+}, and CO_3^{2-} ions from matrix dolomite, and TSR, forming CO_3^{2-} ions with depleted $\delta^{13}C$ values (4). The stable isotope data of the saddle and matrix dolomites (Figure 5) suggest that TSR was still active while the uppermost saddle dolomite crystals formed. The depleted $\delta^{13}C$ values of saddle dolomite against the background of normal marine $\delta^{13}C$ values of the matrix dolomite demonstrate the incorporation of oxidized organic carbon, and the upward trend towards lower $\delta^{13}C$ values shows that the influence of TSR increased upwards. The upward decreasing $\delta^{18}O$ values may have one of two causes: an upward increase in temperature by about 10 °C over the 20 m depth interval in question, or, more likely, an upward increase in the percentage of oxygen in the dolomite that was derived from reduction of sulfate ions [the source sulfate has $\delta^{18}O$ values of about -10 ‰PDB (*1*)].

The fluid inclusion data show that the saddle dolomite crystals probably formed at temperatures around 125 - 145 °C, which is the range of the two maxima in the data set (Figure 6, top). Values higher than about 145°C are attributed to stretching and/or leaking of fluid inclusions, considering that such values are relatively rare and that 145°C is close to the maximum burial temperature of the sequence, as estimated from thermal maturation and isotope data (*15,16*). The salinities of most fluid inclusions, as determined from freezing experiments (Figure 6, bottom) are comparable to those determined from most present formation waters (Figure 8: Pools F: 0.7%/46.2 and M: 0.4%/50.6, respectively; see also analyses of other formation waters retrieved from the Devonian (*17,18*)). This finding further indicates a late-diagenetic origin of the saddle dolomites from fluids that were very similar in composition to the present Nisku formation waters. [The one sample of near potable formation water retrieved from sour gas Pool P (Figures 3,8) does not represent true formation water but steam that condensed out of the sour gas at the surface.]

The $\delta^{34}S$ distribution of the ion microprobe data is consistent with BSR as the dominant process of Fe-sulfide formation in the Nisku Formation. The lowest $\delta^{34}S$ values are about 50 ‰ lighter than the source sulfates (Figure 7) which is expected for BSR (*5,9*). The upward trend toward isotopically lighter values in the lower part of well 2-19 is best explained by closed-system Rayleigh fractionation (*9*). Apparently, downward diffusion could not keep up with BSR and the system became increasingly closed with respect to BSR and SO_4^{2-} replenishment [this probably took place nearly syndepositionally at burial depths of only a few cm to m, at a time when the bulk of the reef above had not yet been deposited]. The bimodal distribution of the data from well 2-12, however, suggests that at least in this well TSR has been a locally important process in Fe-sulfide formation (see also Riciputi et al., *Geochim. Cosmochim. Acta*, in press). The 15 to 18 ‰ difference between $\delta^{34}S$ of the isotopically heavy pyrite population and anhydrite in well 2-12 is consistent with the experimentally determined TSR sulfate-sulfide fractionation at T ≈ 120 - 150 °C (*19*). The fact that most Fe-sulfide appears to have a biogenic origin suggests that most iron was fixed very early in the diagenetic history by BSR. When TSR did affect these pools, most Fe was fixed, precluding the formation of significant amounts of Fe-sulfide even though excess sulfide was and is available.

The relatively high $\delta^{34}S$ values of H_2S in Pools F, M, and V (= well 2-12) (Figure 8, top) and of elemental sulfur in well 2-12 further suggest that these phases were formed by TSR. However, the H_2S contents in Pools F and M (0.7% and 0.4%, respectively) and the adjacent Pool J (0.1% H_2S) are so low that they do not require TSR, suggesting that TSR did not occur updip of line **b** (Figure 3). The H_2S in these pools probably originated from BSR and/or thermal cracking of NSO compounds, and the high $\delta^{34}S$ values of their H_2S may be the result of Rayleigh fractionation. Considering that much of the H_2S formed via BSR during early diagenesis apparently was bound as Fe-sulfide, the H_2S presently contained in Pools F and M probably is derived from thermal cracking.

In support of these interpretations, the $\delta^{13}C$ values of CH_4, C_2H_6, and C_3H_8 from Pools F and M (Figure 8) do not show a signal indicative of TSR but are quite typical for sweet gas condensates (20). The $\delta^{13}C$ values of CO_2 in these pools are much lower than those of the host rock dolomite (compare to Figure 5), but they are within about 2-3‰ of calculated equilibrium values with associated dissolved bicarbonate and calcite and/or dolomite. Hence, there appears to be only a slight admixture of ^{13}C-depleted CO_2 in Pools F and M, consistent with CO_2-generation via decarboxylation during thermal maturation.

The $\delta^{13}C$ values of CH_4, C_2H_6, C_3H_8, and CO_2 in the TSR pool of well 2-12 are similar to those of Pools F and M, except for the somewhat higher value for methane (Figure 8, top). This phenomenon is best explained by updip migration, through the elongated shelf margin reef complex, of gases derived from much deeper in the section, presumably adding ^{13}C-enriched CO_2 and ^{13}C-depleted hydrocarbon gases to the system, thereby reversing the $\delta^{13}C$ signal expected for TSR pools (20). Various organic geochemical data (B.K. Manzano, M.Sc. thesis, University of Alberta, in prep.) support the interpretation of updip migration and admixture of gases from deeper in the section to the shelf margin reef complex around well 2-12. Manzano's data of the oils and gas condensates from the hydrologically 'isolated' pools basinward of the shelf edge further demonstrate an increase in thermal maturation from the shallower to the deeper pools, and that TSR took place in all gas condensate pools downdip of line **b** (Figure 3).

Conclusions

The data presented in this paper indicate that the Nisku Formation experienced early-diagenetic bacterial sulfate reduction. Supporting evidence includes: (i) the extremely light sulfur isotope ratios of most disseminated pyrite/marcasite, modified by concomitant closed-system Rayleigh fractionation, presumably during nearly syndepositional downward diffusion of dissolved sulfate; (ii) $\delta^{34}S$ ratios of replacive anhydrites that are heavier than Late Devonian seawater sulfate, and that formed from residual dissolved sulfate and partial reoxidation of bacteriogenic pyrite (1,21). The present reservoir characteristics, however, are due to thermal maturation and thermochemical sulfate reduction which generated H_2S, as well as saddle dolomite, calcite spar, some pyrite, and S°, as by-products in the water-saturated parts of the pools. Similar mineral parageneses are known from other sour gas fields that formed via thermochemical sulfate reduction (*5,12,20,22-23*), but the distinctive spatial distribution of all of these minerals relative to the gas-water contact in one well is not commonly found.

Stable isotope data document oxidation of hydrocarbons to CO_2 and carbonate; fluid inclusion data suggest that saddle dolomite formed from brines at T ≈ 125 to 145 °C; gas condensate compositions suggest a lack of TSR in pools with less than 1% H_2S, the occurrence of TSR in pools with >1% H_2S, and mixing of gases in the shelf margin reef complex. Most of the gas condensate in the shelf margin reef complex updip from line **a** (Figure 3) did not form there but has migrated updip from below the line. TSR took place downdip from line **b** (Figure 3), which could be called the "sour gas line" or "TSR line". Small, apparently isolated pools located between lines **a** and **b** (Figure 3) did not experience TSR. Their gas compositions are due to thermal maturation commensurate with wet gas maturity, and the small amounts of H_2S contained in these pools probably resulted from thermal cracking of NSO-compounds. The elongated shelf margin reef complex is a special case where TSR probably began downdip from line **b** and then propagated updip, such that well 2-12, located between lines **b** and **a**, exhibits features typical for TSR.

Acknowledgments

We are grateful to E.A. Burton for analytical assistance (field and laboratory analyses of the 3 Nisku formation water samples). Constructive discussions with and/or reviews by J. Kaufman, P.A. Cavell, B.K. Manzano, and one anonymous reviewer are much appreciated. This project has been funded by AMOCO Canada (Machel), the Natural Science and Engineering Research Council of Canada (Machel and Krouse), and by the Geoscience Research Program, Office of Basic Energy Research, U.S. Department of Energy under contract DE-AC05-84OR21400 with Martin Marietta Energy Systems, Inc. (Riciputi and Cole).

Literature Cited

1.   Machel, H.G. **1985**, Unpublished Ph.D. Thesis, McGill University, Montreal.

2.   Anderson, J.H.; Machel, H.G. In *Reefs, Canada and adjacent area;* Geldsetzer, H.H.J.; James, N.P.; Tebbutt, G.E., Eds., Canadian Society Petroleum Geologists Mem. No. 13; Calgary, AB, **1988/9**; pp 391-398.

3.   Machel, H.G. *Nisku oil and sour gas play, subsurface of central Alberta*; Canadian Society Petroleum Geologists Short Course Notes, Calgary, AB, **1990**, pp 8-1 - 8-6.

4.   Machel, H.G. *Geology*, **1987**, *Vol 15*, pp 936-940.
5.   Machel, H.G. *Carbonates and Evaporites*, **1989**, Vol 4, pp 137-151.
6.   Riciputi, L.R.; Cole, D.R.; Machel, H.G.; Christie, W.H.; Rosseel, T.M. In *Water Rock Interaction*; Kharaka, Y.K.; Maest, A.S. Eds.; A.A. Balkema, Rotterdam, **1992**, pp 1197-1200.
7.   Riciputi, L.R.; Machel, H.G.; Cole, D.R. *Journal of Sedimentary Research*, **1994**, Vol A64, pp 115-127.
8.   Epstein, S.; Graf, D.L.; Degens, E.T. In *Isotopic and Cosmic Chemistry*; Craigh, H.; Miller, S.L., Wasserburg, G.J. Eds., North Holland Publishing, **1964**, pp 169-180.
9.   Krouse, H.R. *Canada Energy Resource Research Fund Agreement U-30*, **1983**, pp 1-100.
10.   Anderson, J.H. *Unpublished Ph.D. Thesis*, University of Texas at Austin, **1985**, 392 pp.
11.   Machel, H.G.; Burton, E.A. *Journal Sedimentary Research*, **1994**, Vol A64, pp 741-751.
12.   Machel, H.G. In *Diagenesis of sedimentary sequences;* Marshall, J.D. Ed., Geological Society Special Publication, No. 36, London, **1987**, p. 15-28.
13.   Machel, H.G, In *Native Sulfur - Developments in Geology and Exploration;* Wessel, G.R.; Wimberly, B.H. Eds., Society for Mining, Metallurgy and Exploration, **1992**, p. 3-22.
14.   Alberta Research Council *Geological Atlas of the Western Canada Sedimentary Basin*, **1994**, 510 pp.
15.   Longstaffe, F.J., *Journal of Sedimentary Petrology*, **1986**, Vol 56, pp 78-88.
16.   Hacquebard, P.A. In *The origin and migration of petroleum in the Western Canada Sedimentary Basin*; Deroo G.; Powell, T.G.; Tissot, B.M.; McCrossan, R.G.; Hacquebard, P.A. Ed., Geological Survey of Canada Bulletin, **1977**, Vol 262, pp 11-22.
17.   Connolly, C.A.; Walter, L.M.; Baadsgaard, H.; Longstaffe, F.J. *Applied Geochemistry*, **1990**, Vol 5, pp 375-395.
18.   Connolly, C.A.; Walter, L.M.; Baadsgaard, H.; Longstaffe, F.J. *Applied Geochemistry*, **1990**, Vol 5, pp 397-413.
19.   Kiyosu, Y.; Krouse, H.R. *Geochemical Journal*, **1990**, Vol 24, pp 21-27.
20.   Krouse, H.R.; Viau, C.A.; Eliuk, L.S.; Ueda, A.; Halas, S. *Nature*, **1988**, Vol 333, pp 415-419.
21.   Machel, H.G.; Burton, E.A. *Journal Sedimentary Petrology*, **1991**, Vol 61, pp 394-405.
22.   Heydari, E.; Moore, C.H. *Geology*, **1989**, Vol 12, pp 1080-1084.
23.   Kaufman, J.; Meyers, W.J.; Hanson, G.N. *Journal Sedimentary Petrology*, **1990**, Vol 60, pp 918-939.

RECEIVED April 17, 1995

Author Index

Affiliation Index

Oak Ridge National Laboratory, 439
Pennsylvania State University, 194
State University of New York—Stony
    Brook, 168,332
Texas A & M University, 206
U.S. Geological Survey, 412
Universität Hamburg, 93
University of Alberta, 439
University of Bristol, 243
University of Calgary, 426,439
University of Connecticut, 365

University of Delaware, 168
University of Illinois, 378
University of Maryland Center
    for Environmental and Estuarine
    Studies, 224
University of Miami, 260
University of Wales, 168
Western Research Institute, 138
Woods Hole Oceanographic Institution,
    80,138

Subject Index

A

Acid volatile sulfide procedure,
    to determine proportion of types of
    reduced sulfur compounds in
    sediments, 245
Aliphatic hydrocarbon fraction,
    identification in sulfur-rich surface
    sediments of Meromictic Lake Cadagno,
    64,67–71
Alkanoic subunits in sulfur-rich
    geomacromolecules, 93–108
    analytical techniques, 95–96
    chain-length effects, 100,101f
    desulfurization efficacy and
        specificity, 96–99f
    diagenetic history, 100,102,103f
    experimental procedure, 94–95,97f
    precursor-related sulfurization, 102–104
    subunits released by ruthenium tetroxide
        oxidation, 105–109
    sulfur-bound alkyl subunits, 98
    sulfur-bound carboxylic acids, 104–106f
    sulfur content, 98,100,101f
    sulfur incorporation modes, 98,100
Alkyl subunits, sulfur bound,
    identification in sulfur-rich
    geomacromolecules, 93–107
Alkylated thiophenes and thiolanes,
    distribution in sediments from
    Nördlinger Ries, 315,319,320f

Alkylbenzenesulfonate-containing
    detergents, serious foaming problems
    in treatment plants, 367
Alkylthiophene, formation, 24–26
Amorphous ferrous sulfide
    conversion to mackinawite, 178
    precipitated, pyrite formation reactions,
        194–202
Ancient sediments, digestion procedures
    for determination of reduced sulfur
    species, 243–255
Anhydrite
    role in thermochemical sulfate
        reduction, 439–452
    sulfur source for H_2S, 426–436
Anoxic bottom waters, stable sulfur
    isotopic compositions of chromium-
    reducible sulfur, 400,401f,403f,404
Anoxic Chesapeake Bay sediment
    porewater, thiol–inorganic sulfur com-
    pound temporal relationship, 294–308
Anoxic sediments, pyrite formation,
    206–221
Aqueous sulfide oxidation, for
    characterization of transient 2+ sulfur
    oxidation-state intermediate, 280–291
Aragonites, sulfate incorporation, 336–342
Aromatic hydrocarbon fraction,
    identification in sulfur-rich surface
    sediments of Meromictic Lake
    Cadagno, 70

Aromatic sulfur heterocycles, thermal pathways for transformations in organic macromolecules, 110–135

Arthrobacter, use of sulfonate sulfur for growth, 368

Asphaltenes, desulfurization, 74–75

Assimilation, sulfonate sulfur, 368–373

Atmospheric sulfur cycle, concerns, 2

Authigenic iron sulfide, importance of formation in sedimentary paleomagnetic studies, 195

Authigenic pyrite
advantages as geochemical indicator, 194–195
occurrence in modern and ancient marine sediments, 194

Authigenic pyrite formation, reaction mechanisms, 195–196

B

Bacteria, breakdown of sulfonates, 367–368

Bacterial cultures, digestion procedures for determination of reduced sulfur species, 243–255

Bacterial sulfate reduction
H_2S formation, 6,412
role in formation of sulfide minerals, 243–255

Bay of Concepcion, sulfur transformations in early diagenetic sediments, 38–56

Benzothiophene, formation, 26–29*f*

Biogeochemical sulfur cycle,
complexity, 280
probed using ^3S radiolabeling, 348–362

Bitumen, influence of sulfur cross-linking on molecular size distribution of sulfur-rich macromolecules, 80–91

C

C_{35} bacteriohopane tetrol, formation, 130,132

Ca^{2+}, role in hydrogen sulfide oxidation, 263,265*f*,266

Calcite
role in thermochemical sulfate reduction, 439–452
sulfate incorporation, 336–338

Carbon cycle, relationship with sedimentary sulfur cycle, 3

Carbon dioxide
paleozoic hydrocarbon reservoirs in Western Canada
CO_2 distribution, 427
gas chemistry, 427,428*f*
isotopic compositions, 432,433–434*f*
origin of S, H_2S, and CO_2, 429–432
reaction path modeling, 432,435,436*f*
water chemistry, 427,429
role in thermochemical sulfate reduction, 439–452

Carbon supply, role in stable sulfur isotopic compositions of chromium-reducible sulfur in lake sediments, 404–408

Carbonate minerals, sulfate removal, 5

Carbonate rocks, occurrence in sedimentary rock volume of continental crust, 322

Carboxylic acids, sulfur bound, identification in sulfur-rich geomacromolecules, 104–106*f*

Chemical degradation for release of alkanoic subunits in sulfur-rich geomacromolecules, 93–107

Chemical reduction of sulfate during catagenesis, H_2S formation, 9–10

Chesapeake Bay, environmental controls on iron sulfide mineral formation, 224–240

Chesapeake Bay sediment porewater, anoxic, thiol–inorganic sulfur compound temporal relationship, 294–308

Chesapeake Bay water column, benthic oxygen demand, 294

Chromium-reducible sulfur in lake sediments, stable sulfur isotopic compositions, 397–409

Chromous chloride digestions, hot, determination of reduced sulfur species in bacterial cultures and ancient and recent sediments, 243–255